Metal Bonding and Interactions in High Temperature Systems

With Emphasis on Alkali Metals

James L. Gole, EDITOR
Georgia Institute of Technology

William C. Stwalley, EDITOR
University of Iowa

Based on a symposium
jointly sponsored by the
Divisions of Physical Chemistry
and Inorganic Chemistry at the
181st Meeting of the
American Chemical Society,
Atlanta, Georgia,
March 31–April 3, 1981.

ACS SYMPOSIUM SERIES 179

AMERICAN CHEMICAL SOCIETY
WASHINGTON, D. C. 1982

Library of Congress CIP Data

Metal bonding and interactions in high temperature systems with emphasis on alkali metals.
(ACS symposium series, ISSN 0097-6156; 179)

"Based on a symposium jointly sponsored by the Divisions of Physical Chemistry and Inorganic Chemistry at the 181st Meeting of the American Chemical Society, Atlanta, Georgia, March 31–April 3, 1981."
"Symposium on High Temperature Chemistry"—Pref.
Includes bibliographies and index.

1. Alkali metals—Congresses. 2. Metals at high temperatures—Congresses. 3. Metal–metal bonds—Congresses.
I. Gole, James L., 1945- . II. Stwalley, William C., 1942- . III. Symposium on High Temperature Chemistry (1981: Atlanta, Ga.). IV. American Chemical Society. Division of Physical Chemistry. V. American Chemical Society. Division of Inorganic Chemistry. VI. Series.

QD172.A4M2 546'.38 81–20555
ISBN 0–8412–0689–9 AACR2
ACSMC 8 179 1–629 1982

PRINTED IN THE UNITED STATES OF AMERICA

ACS Symposium Series

M. Joan Comstock, *Series Editor*

FOREWORD

The ACS Symposium Series was founded in 1974 to provide a medium for publishing symposia quickly in book form. The format of the Series parallels that of the continuing Advances in Chemistry Series except that in order to save time the papers are not typeset but are reproduced as they are submitted by the authors in camera-ready form. Papers are reviewed under the supervision of the Editors with the assistance of the Series Advisory Board and are selected to maintain the integrity of the symposia; however, verbatim reproductions of previously published papers are not accepted. Both reviews and reports of research are acceptable since symposia may embrace both types of presentation.

CONTENTS

PREFACE

For at least the next decade, high temperature chemistry and technology are destined to play an important role in solving challenges now before the scientific community. Among these are the need to develop alternate energy sources and to institute more efficient materials conversion processes. The symposium upon which this book is based was intended to discuss these needs while providing a good mix of basic and applied high temperature science. Because high temperature chemistry is a broad discipline, a focus on a few key areas was chosen to encourage meaningful participation of the various scientific communities, applied as well as basic, industrial and governmental as well as academic. These areas include interactions of alkali metal atoms and dimer bonding, bonding in small metallic clusters, bonding and interactions of the alkali hydrides and halides and metal atom-water interactions, and applications of alkali metal and alkali compound vapors. These groupings are particularly timely, and their combination offers an intertwining of basic and applied research efforts. All of the invited speakers in the various groupings have completed manuscripts included herein. In addition, about half of the contributed papers to the symposium were sufficiently close in content to the invited papers that these manuscripts are also included here.

The topics considered in this symposium are of considerable current interest. Recent attention has been given to small metal clusters as models for surface imperfections and concomitantly as sites for catalytic activity. In addition, the characterization of small metal clusters may well add insight into the nature of metal–metal bonding in polynuclear organometallic compounds. These species are attracting considerable attention as homogeneous catalysts. The establishment of structural properties and the molecular electronic makeup of small clusters is now within the grasp of the physical chemist. This work promises to aid parameterizations on much larger species and hence offers the important prospect of interpolating between properties of the free atom and dimer and the metallic state. The elucidation of this middle ground between the micromolecular structure of small inorganic compounds and the macromolecular properties of the bulk metallic phase promises new insight that will undoubtedly be pertinent to the development of more efficient materials conversion processes.

Since the alkali metals are theoretically simple ("visible hydrogen"), readily ionized, and strongly interacting with laser or solar light, they represent ideal systems for quantitatively understanding microscopic interconversion mechanisms between photon, chemical, electrical, and thermal energies. Thus the alkali metals and their compounds are playing an ever increasing role in practical applications (for example, sodium lamps, thermionic convertors, magnetohydrodynamics, lasers, and fusion reactors), which is reflected in the chapters on applications of alkali metals. It is hoped that these contributions will help bridge the gap between applied and basic research.

JAMES L. GOLE
Georgia Institute of Technology
Atlanta, GA 30332

WILLIAM C. STWALLEY
University of Iowa
Iowa City, IA 52242

August 1981

BONDING AND INTERACTIONS OF ALKALI METAL ATOMS AND DIMERS

1

Electronic Structure and Spectra of Light Alkali Diatomic Molecules and Their Molecular Cations

D. D. KONOWALOW and M. E. ROSENKRANTZ

State University of New York, Binghamton, Department of Chemistry, Binghamton, NY 13901

All-electron ab-initio calculations of the potential curves and wavefunctions for the eight lowest-lying electronic states of Li_2 and of Na_2 and the six lowest-lying electronic states of Li_2^+ and Na_2^+ are used to predict the spectral features of a variety of transitions. We scale these results to predict qualitatively several aspects of the structure and spectra of K_2, K_2^+, Rb_2, Rb_2^+, Cs_2 and Cs_2^+.

Recently we have reported ab initio all-electron quantum-mechanical investigations of eight low-lying states of each of Li_2 and Na_2 (1-6) and of six low-lying states of Li_2^+ (7,8). The computation of the potential energy curves for the low-lying states of Na_2 at large separations ($15 \leq R \leq 30$ bohr) will be reported shortly (9).

Here, we report, for the first time, ab initio computations of the six lowest-lying electronic states of Na_2^+. These computations utilize the basis set developed to describe the low-lying states of the neutral Na_2 molecules (6) and utilize integrals which have been computed previously (6,9). The molecular energies computed at the single-configuration self-consistent-field (SC-SCF) level are listed in Table I. These SC-SCF computations should provide relatively reliable potential curves for what are effectively one-electron systems. We do not attempt to describe the electron correlation associated with the core electron motions nor that associated with the polarization of the core electrons by the single valence electron. Thus, while dispersion effects are not well described, the first order ion-induced dipole interaction and the major electrostatic interactions of the valence electron are probably reasonably well described at the SC-SCF level. Note in Table II, where we list molecular constants for Na_2^+, that the 1 $^2\Sigma_u^+$ state is bound. Its 1 $^3\Sigma_u^+$ counterpart in the neutral molecule is predicted to be strictly repulsive at the SC-SCF level. The $-C_4R^{-4}$ ion-induced dipole interaction accounts for the difference.

0097-6156/82/0179-0003$05.00/0

TABLE I Potential energy curves[a] for Na_2^+ obtained from all-electron, ab initio computations

R bohr	$1\ ^2\Sigma_g^+$	$1\ ^2\Sigma_u^+$	$^2\Pi_u$	$^2\Pi_g$	$2\ ^2\Sigma_g^+$	$2\ ^2\Sigma_u^+$
4.00	.506383	--	--	--	.327497	--
4.25	.520333	--	--	--	.344265	--
4.50	.531860	.421169	.428153	--	.358824	--
4.75	.541350	--	--	--	.371545	--
5.00	.549070	.439966	.442172	.356177	.382675	--
5.50	.560082	.455180	.451944	.372738	.400960	.321218
6.00	.566501	.467831	.458789	.386396	.415139	.337683
6.20	.568090	--	--	--	.419921	--
6.50	.569656	.478515	.463534	.397796	.426332	.352273
6.80	.570418	--	--	--	--	--
7.00	.570568	.487618	.466749	.407381	.435330	.365310
7.20	.570477	--	--	--	--	--
7.50	.569974	.495409	.468842	.415482	.442688	.376990
8.00	.568396	.502080	.470112	.422352	.448781	.387441
8.50	.566201	.507782	.470787	.428196	.453869	.396754
9.00	.563637	.512625	.471025	.433164	.458115	.404994
10.00	.558144	.520194	.470718	.441043	.464645	.418660
11.00	.552856	.525482	.469832	.446786	.469099	.429165
12.00	.548282	.529089	.468730	.450986	.471975	.437208
13.00	.544730	.531489	.467600	.454058	.473527	.443377
15.00	.539781	.534025	.465611	.457954	.474092	.451794
18.00	.536632	.535150	.463675	.460670	.471013	.458482
21.00	.535612	.535265	.462796	.461714	.467378	.461434
24.00	.535264	.535187	.462480	.462113	.465051	.462587
27.00	.535122	.535106	.462391	.462272	.463872	.462934
30.00	.535053	.535049	.462380	.462343	.463304	.462966

[a]Energies listed in hartree atomic units $e^2/a_o = 219474.6\ cm^{-1}$. All energies have a prefix of -323. Thus, the energy of the $1\ ^2\Sigma_g^+$ state at R = 4 bohr is $-323.506383\ e^2/a_o$.

TABLE II Molecular constants for several electronic states of Na_2^+

State	Data Source	R_e (bohr)	D_e (cm^{-1})	ω_e cm^{-1}	$\omega_e x_e$ cm^{-1}
$1\ ^2\Sigma_g^+$	a	7.02	7796	116.0	0.43
	b	6.69	8230	126	--
	c	6.24	7905	--	--
$1\ ^2\Sigma_u^+$	a	20.3	48.5	9.68	0.45
	c,d	25	43.9	--	--
$^2\Pi_u$	a	9.14	1900	49.9	0.33
	b,d	(9.0)	(2170)	--	--
	c	9.8	1860	--	--
$2\ ^1\Sigma_g^+$	a	14.37	2399	39.1	0.16
	b,d	(12)	(2850)	--	--
	c,d	(16)	(3030)	--	--

[a] Present work.
[b] Reference (10).
[c] Reference (11).
[d] Values in parenthesis have been estimated crudely by us from tabulated data in the reference cited.

At present, the low-lying states of Na_2^+ are better characterized computationally than experimentally, although multiphoton ionization experiments may change that picture eventually. We find reasonably close agreement between the results of our all-electron computations, pseudopotential (10), and model potential (11) computations. The latter two kinds of computations may give more accurate results than our ab initio computations since they may account for at least certain core polarization effects.

The adiabatic potential energy curves for these electronic states calculated in the Born-Oppenheimer approximation, are given in Figure 1. Since we have discussed the choice of basis functions and the choice of configurations for these multiconfiguration self-consistent field (MCSCF) computations (12) previously (1-9), we shall not explore these questions in any detail here. Suffice it to say that the basis set for Li describes the lowest ^2S and ^2P states of the Li atom at essentially the Hartree-Fock level of accuracy, and includes a set of crudely optimized d functions to accommodate molecular polarization effects. The basis set we employed for calculations involving Na is somewhat less well optimized than is the Li basis; in particular, sσ molecular orbitals are not as well described for Na_2 (relatively speaking) as they are for Li_2.

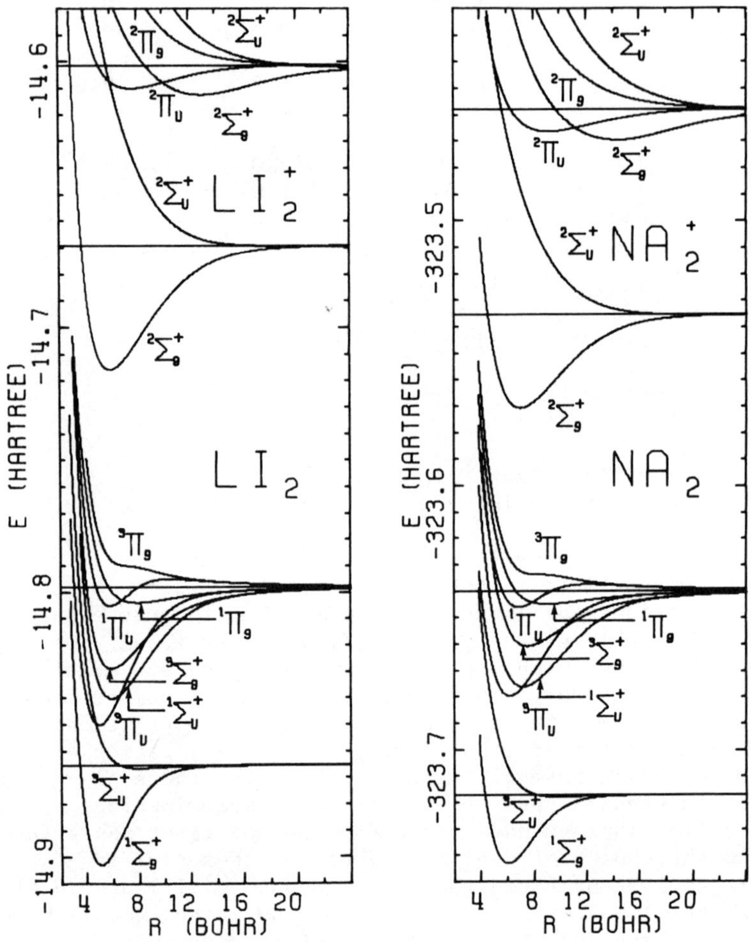

Figure 1. Potential curves for low-lying states of Li_2, Li_2^+, Na_2, and Na_2^+ obtained from the all-electron calculations reported in Refs. 1–9.

Since the research which produced the potential curves in Figure 1 has been (or will be) reported over a span of some six years, it seems appropriate now to make some very general observations about them. First we note that these potentials are expected to be highly accurate at large internuclear separations say, for R>15 bohr. Because of limitations on the kinds of electronic configurations which we could treat, (see Ref. (1) for a discussion), the Π state potential curves, for the neutral molecules are somewhat less accurate than the Σ state curves, especially at internuclear separations of $R < 15$ bohr. Our best results for Σ states of Li_2 and all states of $Li_2{}^+$ are probably in error by 2% or less. The corresponding errors for Na_2 and $Na_2{}^+$ are probably about three times larger. We obtained these error estimates by scaling our potential curves so that they reproduced in the least squares sense the experimental energy level spacing of the $X^1\Sigma^+_g$ and $A^1\Sigma^+_u$ states.

Because they are isovalent molecules, the similarity of the potential curves for comparable states within the Li and the Na systems is, of course, no surprise. As expected, the potential curves for the Na system are shifted to the right of the corresponding curves in the Li system and each of the Na curves is "fatter" and shallower than its Li counterpart. Consequently, the bound states in the Na system are expected to have more bound vibrational levels which are more closely spaced than their Li counterparts. These expectations are borne out for states such as $X^1\Sigma^+_g$, $A^1\Sigma^+_u$ and $B^1\Pi_u$ which are well characterized experimentally and for these and other states which are so far best characterized by computations such as our own.

Note that the lowest Σ_u curve crosses the lowest Π_u curve for both $Li_2{}^+$ and Li_2 but not for $Na_2{}^+$ and Na_2. This can lead to strong predissociation of the Π_u curves in the Li system but not in the Na system. The difference in behavior apparently has three main causes: the first, the 3s-3p level separation is some 2050 cm^{-1} larger in Na than the 2s-2p separation in Li. If we were to translate rigidly upwards the Li 2s + 2p asymptote together with all its curves, by some 2050 cm^{-1}, we would find the $^3\Sigma^+_u$ - $^3\Pi_u$ curve crossing to occur at about 4.1 bohr instead of the actual 4.6 bohr, and the lowest several vibrational levels of the $^3\Pi_u$ state would be relatively free of predissociation. Secondly, the $^3\Pi_u$ state is about 30% deeper in Li_2 than in Na_2. Thus, the $^3\Pi_u$ curve comes much closer to the lowest ns+ns asymptote (and consequently closer to the $^3\Sigma^+_u$ curve) in Li_2 than in Na_2. Thirdly, since Na and Na^+ are substantially more polarizable than Li and Li^+, the repulsive Σ_u curves in the Na system are "softer" (more attractive) than in the Li system. While these qualitative arguments have concentrated on the presence or absence of the $^3\Sigma^+_u$ - $^3\Pi_u$ curve crossing in the neutral systems, the arguments extend in an obvious manner to the $^2\Sigma^+_u$ - $^2\Pi_u$ predissociation (or not) in $Li_2{}^+$ (and $Na_2{}^+$).

Scaled Potential Curves for K_2, Rb_2, Cs_2 and their Molecular Cations

Note that all the alkali atoms have ns ^2S and np ^2P as their two lowest-lying atomic terms, (clearly n is the principal quantum number). This suggests that we should be able to scale our potential curves for the light alkalis in order to estimate, however crudely, the positions and shapes of the corresponding states of K_2, Rb_2, Cs_2 and their molecular cations.

Let us outline a very simple scaling scheme. We define scaled internuclear separation (R_s) and energy (E_s) variables:

$$R_s = \rho R^c = (R_e^x / R_e^c)\, R^c,$$

$$E_s(R_s) = \varepsilon^c(R^c) = (D_e^x / D_e^c)\, E^c(R^c),$$

where ρ and ε are scaling factors for internuclear separation and the dissociation energy, R_e^c and D_e^c are the corresponding values of the calculated binding energy curve $E^c(R^c)$ which is to be scaled and $E_s(R_s)$ is the resulting scaled binding energy curve. It would be typical to take R_e^x and D_e^x to be experimental values if this data were available. This is the sort of scaling that is carried out in applying the Law of Corresponding States (13).

We tested this scheme first by scaling our Li_2 and Li_2^+ potentials so that we estimate the corresponding potentials for Na and Na_2^+. Instead of experimental values of R_e^x and D_e^x we used our values calculated for the $X^1\Sigma_g^+$ state of Na_2 to obtain ρ and ε. Thus we use only the Li_2 and Li_2^+ curves and the R_e and D_e values for the X state of Na_2 to estimate the entire potential curve for each of eight electronic states of Na_2 and six electronic states of Na_2^+. Except for the $^3\Sigma_g^+$ and $^3\Pi_u$ states the scaling worked well as is evident from Figure 2.

Let us now turn to the estimation of the potential curves for low-lying neutral and cationic diatomic molecules for the heavy alkalis. For each molecule we take R_e^x and D_e^x to be the experimental values (14) for the corresponding $X^1\Sigma_g^+$ state. We also adjust the separation of the asymptotes to correspond to the appropriate experimental resonance transition (ns-np) and ns ^2S ionization energies. (We have ignored the spin-orbit splitting (15) of the P state of the heavy alkalis as we had for Li and Na. The single ^2S + ^2P asymptote was made to correspond to the degeneracy-weighted energy of the ^2P state. Clearly this approximation becomes progressively more serious in K_2, Rb_2 and Cs_2 where the comparable splittings are about 58 cm^{-1}, 237 cm^{-1} and 554 cm^{-1}, respectively.) The scaled potential curves are shown in Figures 2, 3 and 4. It is clear that the scaled curves which are based on our Li system --- and those based on our Na system --- agree reasonably well with each other except for the curves for the $^3\Sigma_g^+$ and $^3\Pi_u$ states. (This discrepancy was expected in view of our comparison of the Li and Na systems.) In view of our earlier remarks, it appears that the predictions based on Li may be the

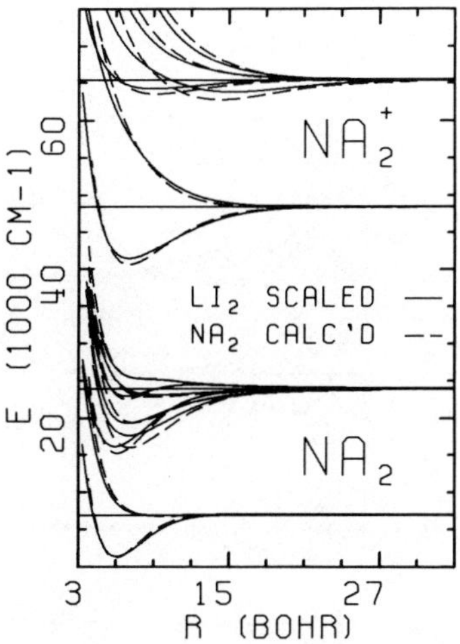

Figure 2. A comparison of the potential curves for Na_2 and Na_2^+ computed ab initio with those obtained by scaling the corresponding curves of Li_2 and Li_2^+. Identity of curves is shown in Fig. 1.

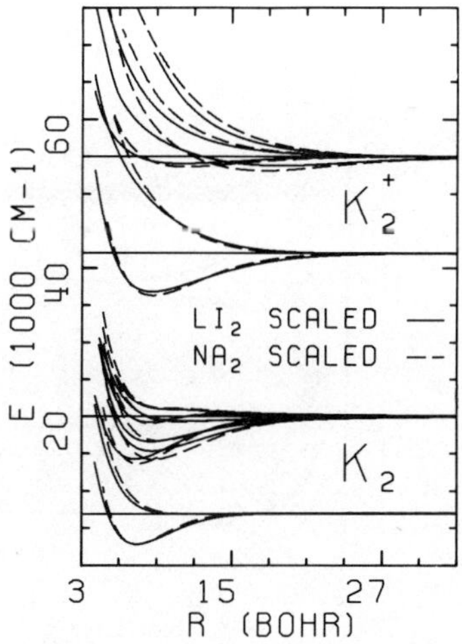

Figure 3. Estimated potential energy curves for K_2 and K_2^+ obtained by scaling computed potentials for Li_2 and Na_2. Identity of curves is shown in Fig. 1.

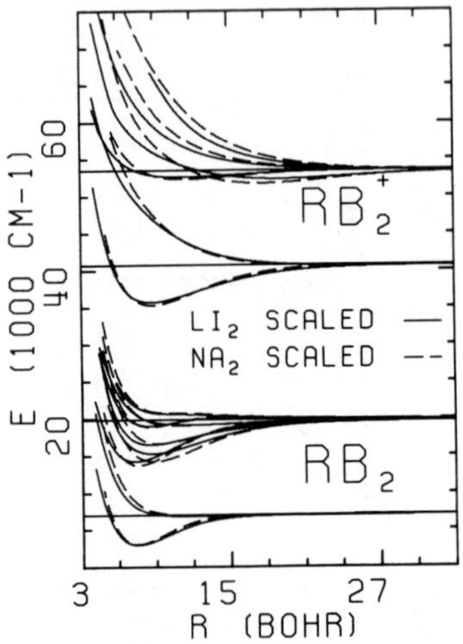

Figure 4. Estimated potential energy curves for Rb_2 and Rb_2^+ obtained by scaling computed potential for Li_2 and Na_2. Identity of curves is shown in Fig. 1.

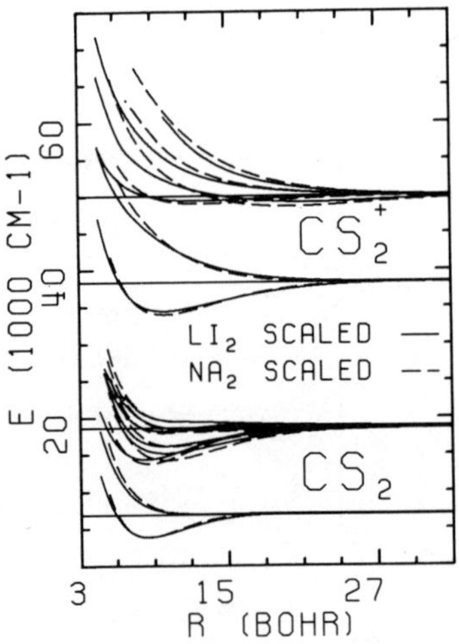

Figure 5. Estimated potential energy curves for Cs_2 and Cs_2^+ obtained by scaling computed potentials for Li_2 and Na_2. Identity of curves is shown in Fig. 1.

more reliable. There is no question that these estimates for the potentials for K_2 (and Rb_2 and Cs_2) are easy to obtain. The real question is to assess the reliability of predictions based on these curves.

First note that the $^3\Sigma_u^+$ state of K_2 is predicted not to predissociate the $^3\Pi_u$ state at all strongly since the curves don't cross at low energies. This question is important, for example, in terms of understanding possible A→X and B→X laser emission in K_2, and in terms of understanding the use of K_2 as a working fluid in a solar powered engine (16). Curiously enough, these triplet curves are predicted not to cross no matter whether we base our prediction on Li_2 (where these curves do cross) or on Na_2 (where they don't cross). Evidently the shrinkage of the two curves toward their respective asymptotes in the course of scaling the binding energy curves is sufficient to overcome the shift of the ns + ns - ns + np asymptotes closer together in the systems for the heavy alkalis than they lie in either the Li or Na systems.

Recently, Bhaskar and coworkers (17) have attributed an IR absorption band between 1.1 and 1.6 μ to the $^3\Sigma_g^+ \leftarrow {}^3\Sigma_u^+$ transition in K_2. Clearly, our scaled curves suggest that such a band system could exist with reasonable Franck-Condon factors in that wavelength region. (It would, in fact, be possible to obtain scaled transition dipole moment functions, (see below), calculate Franck-Condon factors based on them and our scaled potential curves, and make more detailed quantitative estimates of the $^3\Sigma_g^+ \leftarrow {}^3\Sigma_u^+$ band system in K_2.)

Before turning from our crude considerations on the heavy alkalis, let us make a final prediction. That is that there shall be observed in K_2 vapor an absorption feature which corresponds to the $^3\Pi_g \leftarrow {}^3\Sigma_u^+$ absorption which lies somewhat to the blue of the atomic resonance line $\nu_{sp} \cong 13000\ cm^{-1}$; we predict its peak to occur at 710nm. [We have learned at this symposium (18) that a feature in the K_2 absorption spectrum at 720 nm has been attributed to this transition.] This feature corresponds to the absorption identified by Koch, Stwalley, and Collins (19) in Li_2, and by Woerdman and deGroot (18) in Na_2. The feature in Li was found at 588 nm, while our calculations predicted it to be at around 595 nm. We estimate the $^3\Pi_g \leftarrow {}^3\Sigma_u^+$ absorption peak to lie at 546 nm in Na_2 while it is observed (18) at 551.5 nm. In Rb_2 such a feature will occur at about 727 nm and in Cs_2 the corresponding feature will occur at about 800 nm, according to rough estimates we've obtained from our scaled potential curves.

Spectroscopic Considerations

Recall that a spectrum corresponding to a one-photon transition is essentially a plot of intensity (of emission or absorption) as a function of the difference in energy (usually expressed as wavelength or frequency) between the two states involved in the transition. The difference potential (the difference between two potential energy curves) provides rough estimates of transition energies. For diatomic molecules, it is straightforward to solve the Schrödinger equation for the vibrational-rotational energy eigenvalues and corresponding nuclear motion wavefunctions once the electronic potential curve is known and thus obtain refined information about transition energies. (Clearly, we invoke the Born-Oppenheimer approximation here.) It is also straightforward to calculate the electronic transition dipole moment function

$$D(R) = \int \psi^*_{e,i}\ \mu\ \psi_{e,j}$$

in terms of the electronic wavefunctions for the states i and j involved in the transition, and the dipole moment operator μ. Intensities of a spectral transition are proportional to the square of the integral (20)

$$R_{nk} = \int \Phi_n^* [D(R)] \Phi_k$$

where Φ_n denotes the vibrational-rotational wavefunction for the state n. Frequently, investigators approximate the function D(R) by its asymptotic value D(∞) which corresponds to the transition dipole moment of the relevant atomic transition. Clearly, this approximation, which we term the atomic approximation, is worthless for molecular transitions which are allowed but which correspond asymptotically to forbidden atomic transitions. It is even incorrect to assume that such molecular transitions will be weak. We have found transitions among excited states of Li_2 (21) which are forbidden asymptotically yet which have transition dipole values as large as 24 atomic units (=61 Debeye)! Even where the transition is allowed asymptotically, the atomic approximation can lead to substantial error in the predicted intensities. It is not unusual for D(R) to deviate by 20-30% from its asymptotic value at internuclear separations where the overlap of the vibrational wavefunctions Φ_n and Φ_k is appreciable. Thus, intensities which are calculated by using the atomic approximation for such transitions may be in error by 40-70%. A paper is in preparation (22) which will report the D(R) values for all dipole allowed transitions among most of the states which are depicted in Figure 1.

In the following paragraphs we give selected examples of the use of our wavefunctions and potential curves to predict or confirm various spectroscopic features of the alkalis. We know of plans to observe Li_2^+ spectra in at least two laboratories (23, 24) so some predictions regarding the spectra appear to be in order. Julienne (25) has used our wavefunctions for Li_2^+ to calculate the electronic transition dipole moment function corresponding to the $A^2\Pi_u - X^2\Sigma_g^+$ transition and to calculate the matrix element $\langle {}^2\Sigma_u^+ \mid L_x - iL_y \mid {}^2\Pi_u \rangle$ needed to determine the rate of predissociation of the ${}^2\Pi_u$ state by the ${}^2\Sigma_u^+$ state. Since the spin-orbit splitting between the $^2P_{1/2}$ and $^2P_{3/2}$ states of Li is only 0.34 cm^{-1} (15), the $^2\Pi_u$ state is characterized by Hund's case b coupling. Thus the $^2\Sigma_u^+$ state only predissociates the rotational levels which give rise to P and R branches; the Q branch is not predissociated.

Julienne has used the calculated transition dipole moment function to obtain the radiative lifetimes for the $^2\Pi_u \rightarrow {}^2\Sigma_g^+$ transition. He finds, for example, that the radiative lifetime of the v'=0 Q branch emission is about 12 ns and increases fairly rapidly with vibrational level. He also finds that radiation and predissociation rates for the P and R transitions are comparable for certain rotational-vibrational levels.

Our curves for the molecular ions $Na_2^+ \ldots Cs_2^+$ show that the $^2\Sigma_u^+$ and $^2\Pi_u$ potentials do not cross at an energy where the $^2\Pi_u$ state is bound. Thus, the interesting predissociation predicted to be present in Li_2^+ will not be present in these heavier alkalis. The origin of the difference in predissociation behavior can be understood by a simple extension of the arguments we've put forward in discussing the $^3\Pi_u$ - $^3\Sigma_u^+$ interaction in the neutral molecules.

The spectroscopy of the singlet manifold in the light alkalis is relatively well-developed in part since the molecular ground state is a singlet. We have obtained transition dipole moment functions for transitions in the singlet manifold which are in reasonable agreement with the limited experimental data which is available (26,27). We have not yet undertaken a line-by-line comparison of calculated and experimental Franck-Condon factors (which are more readily available than are the D(R)). Although it is a straightforward task, it is a particularly time consuming one.

It has been noted by Wellegehausen (28,29) that one can obtain good laser emission in the $A^1\Sigma_u^+$-$X^1\Sigma_g^+$ and $B^1\Pi_u$-$X^1\Sigma_g^+$ systems of Li_2 and Na_2, only weak laser action in these systems of K_2 but, so far, no laser emission has been observed in Rb_2 and Cs_2 . The results presented here, together with unpublished calculations on excited states of Li_2 (21) suggest that self absorption of any laser emission should become a more probable event the heavier the alkali dimer. Thus, it appears that Wellegehausen's observations might be explained by such self absorptions.

By contrast to the singlet manifolds of the alkalis, low-lying members of the triplet manifolds have probably been much better characterized computationally. This is understandable since the ground state of the triplet manifold, $^3\Sigma_u^+$, is very weakly bound (D_e = 300 cm^{-1} = 420 K for Li_2, and D_e = 180 cm^{-1} = 250 K for Na_2, for example). Thus at the temperatures ordinarily used to obtain alkali vapor, one expects to find a very small fraction of bound $^3\Sigma_u^+$ molecules. (Thus, it is not uncommon for spectroscopists to term the $^3\Sigma_u^+$ state "the repulsive state" and question only whether it effectively predissociates the $^3\Pi_u$ or else immediately dismiss it from further consideration. They have missed the opportunity to investigate some fascinating spectroscopy.) There is apparently a sufficient amount of $^3\Sigma_u^+$ available in Li, Na, K, and Cs vapors for at least two kinds of transitions involving it to have been observed (17,18,19,30). Bhaskar and coworkers (17) have observed in potassium vapor an IR absorption band between 1.1 and 1.6μ which they attribute to the $^3\Sigma_g^+ \leftarrow {}^3\Sigma_u^+$ transition. Zouboulis and coworkers (30) identify a 1.25 - 2.5μ band with this transition in Cs_2. Our scaled potential curves for K_2, for example, show that absorption in this region could indeed be attributable to that triplet transition. One would expect to observe a band with some structure as expected for a transition that is dominated by absorption from the $^3\Sigma_u^+$ continuum to bound levels of the $^3\Sigma_g^+$ state. Bound-bound transitions corresponding to transitions from very low v" to rather high v' > 40 should be detectable at wavelengths $\lambda < 0.9\mu$. We expect the emission characteristics of the $^3\Sigma_g^+ - {}^3\Sigma_u^+$ transition in K_2 to be similar to those we have predicted (31) for Li_2 and Na_2. Here, we would expect the peak continuous emission intensity to occur near 1.1μ. If the $^3\Sigma_g^+$ state of K_2 could be populated sufficiently rapidly, the $^3\Sigma_g^+ \rightarrow {}^3\Sigma_u^+$ transition would comprise a tunable near infrared excimer laser with its peak about midway between that of the corresponding excimer lasers based on Li_2 $\lambda \sim 1.3\mu$ and on Na_2 $\lambda \sim 0.83\mu$. These estimates for K_2 are necessarily very rough since they are based on our scaled potential curves. It does seem unlikely that we are in error by substantially more than 0.1μ, however. A similar analysis of Rb_2 and Cs_2 could be made in terms of our scaled potentials.

The $^3\Pi_g \leftarrow {}^3\Sigma_u^+$ transition in the triplet manifold has apparently been observed in Li_2 (19) and in Na_2 (17,24). Koch and coworkers (19) observed a bound-free-bound transition which they attributed in part to $^3\Pi_g \leftarrow {}^3\Sigma_u^+$ absorption in Li_2. They obtained a peak absorption at around 588 nm with substantial structure to the red. In Figure 6, we show the photodissociation cross section (32) as a function of the frequency (and wavelength) of radiation exciting the bound-free $^3\Pi_g \leftarrow {}^3\Sigma_u^+$ transition in Li_2. We predict a peak slightly to the blue of 595 nm which agrees nicely with the observed peak at 588 nm.

Veza and coworkers (17) claim the first direct experimental confirmation of the $^3\Pi_g$ state in Na_2. They attribute the satellite band in the B $^1\Pi_u$ - X$^1\Sigma_g^+$ band system at 551.5 nm to the maximum in the $^3\Pi_g - {}^3\Sigma_u^+$ difference potential and thus to that transition. We estimate, extremely crudely, that the satellite should lie at around 546 nm.

It is clear that *ab initio* computations, especially of the potential curves and wavefunctions for triplet states has proven to be useful in the interpretation of a variety of spectral features in Li_2 and Na_2. It appears that our semi-empirical treatment of the heavier alkalis may prove useful until we are able to treat these systems *ab initio*. We have launched a more detailed

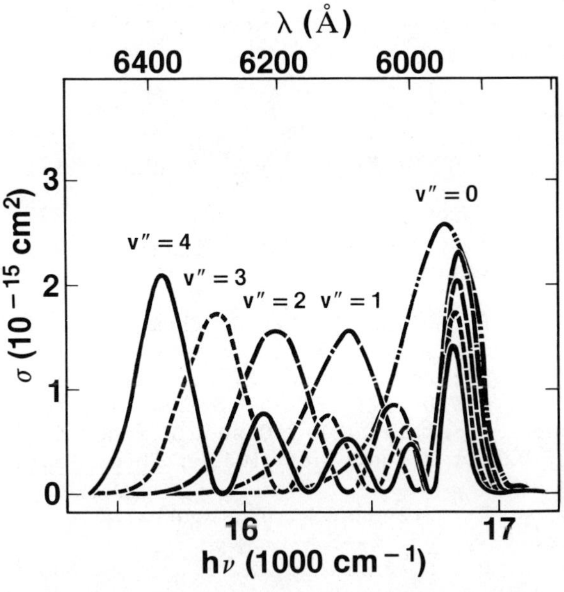

Figure 6. The calculated photodissociation cross section for the $^3\Pi_g \leftarrow {}^3\Sigma_u^+$ *transition in* Li_2. *The peak agrees with that reported in Ref.* 19.

study of the alkalis which aims to treat the heavier alkalis, heteronuclear molecules, more highly excited electronic states, and molecular cations and anions.

Acknowledgments

We are grateful to A. C. Wahl, B. Liu, W. J. Stevens, D. Neumann, P. S. Julienne, and R. J. LeRoy for providing us computer codes and useful advice which have proven invaluable in the course of our research. Thanks to our experimental colleagues P. Kusch, M. M. Hessel, W. C. Stwalley, W. Demtröder, G. Pichler, R. A. Bernheim, and T. Lucatorto for vivid demonstrations and/or patient explanations of some of the mysteries of spectroscopy of the alkalis. D. S. Hochhauser has calculated some of the potential energy curves for Na_2^+ and G. R. Radlauer has obtained the scaled potential curves for the heavy alkalis and prepared Figures 1-5.

Literature Cited

1. Olson, M. L.; Konowalow, D. D., Chem. Phys. Lett. 1976, 44, 281-4.
2. Olson, M. L.; Konowalow, D. D., Chem. Phys. 1977, 21, 393-9.
3. Olson, M. L.; Konowalow, D. D., Chem. Phys. 1977, 22, 29-39.
4. Konowalow, D. D.; Olson, M. L., J. Chem. Phys. 1977, 67, 590-2.
5. Konowalow, D. D.; Olson, M. L., J. Chem. Phys. 1979, 71, 450-7.
6. Konowalow, D. D.; Rosenkrantz, M. E.; Olson, M. L., J. Chem. Phys. 1980, 72, 2612-5. Note: The entry in Table III for the $B^1\Pi_u$ state at R=9.0 should read -0.0010319.
7. Konowalow, D. D.; Rosenkrantz, M. E., Chem. Phys. Lett. 1979, (61), 489-94.
8. Konowalow, D. D.; Stevens, W. J.; Rosenkrantz, M. E., Chem. Phys. Lett. 1979, 66, 24-8.
9. Konowalow, D. D.; Rosenkrantz, M. E., to be published.
10. Bardsley, J. N.; Junker, B. R.; Norcross, D. W., Chem. Phys. Lett. 1976, 37, 502-6.
11. Cerjan, C. J.; Docken, K. K.; Dalgarno, A., Chem. Phys. Lett. 1976, 38, 401-4.
12. Das, G.; Wahl, A. C., 1972, Bison-MC: A FORTRAN Computing System for MCSCF Calculations on Atoms, Diatoms and Polyatoms, Argonne National Laboratory Report ANL-7955.
13. Hirschfelder, J. O.; Curtiss, C. F.; Bird, R. B.; "Molecular Theory of Gases and Liquids", Wiley, New York, 1954.
14. Huber, K. P.; Herzberg, G., "Molecular Spectra and Molecular Structure IV. Constants of Diatomic Molecules" Van Nostrand Reinhold, 1979.
15. Moore, C. E., Nat. Stand. Ref. Data Ser., Nat. Bur. Stand. (U.S.) 35 Vols. I and II, 1971.

16. Mattick, A. T.; Hertzberg, A.; Decher, R.; Lau, C. V., J. Energy 1979, 3, 30-9.
17. Bhaskar, N. D.; Zouboulis, E.; McClelland, T.; Happer, W., Phys. Rev. Lett. 1979, 62, 640-2.
18. Woerdman, W.; deGroot, J. J., American Chemical Society Division of Physical Chemistry, Abstracts of the 181st National Meeting 1981, paper 144.
19. Koch, M.; Stwalley, W. C.; Collins, T., Phys. Rev. Lett. 1979, 42, 1052-4.
20. Herzberg, G., "Molecular Spectra and Molecular Structure I. Spectra of Diatomic Molecules" Van Nostrand, Princeton, 1950.
21. Konowalow, D. D.; Fish, J. L., to be published.
22. Konowalow, D. D.; Rosenkrantz, M. E.; Hochauser, D. S., to be published.
23. Demtröder, W.; Eisel, D., 1980 private communication.
24. Weiner, J., 1981 private communication.
25. Julienne, P. S.; Konowalow, D. D.; Rosenkrantz, M. E., Abstracts of the 34th Molecular Spectroscopy Conference, The Ohio State University, June 1979 paper TF11.
26. Hessel, M. M.; Smith, E. W.; Drullinger, R. E., Phys. Rev. Lett. 1974, 33, 1251-4.
27. Demtröder, W.; Stetzenbach, W.; Stock, M.; Witt, J., J. Mol. Spectrosc. 1976, 62, 382-9.
28. Wellegehausen, B., IEEE J. Quantum Elect. 1979, QE-15, 1108-30.
29. Wellegehausen, B., private communication.
30. Zouboulis, E.; Bhaskar, N. D.; Vasilakis, A.; Happer, W. J. Chem. Phys. 1980, 72, 2356-63.
31. Konowalow, D. D.; Julienne, P. S., J. Chem. Phys. 1980, 72, 5815-8.
32. Julienne, P.S.; Konowalow, D. D., to be published.

RECEIVED August 26, 1981.

Selective Photolysis and Photoionization of Alkali Metal Dimers

C. B. COLLINS, F. W. LEE, and P. A. VICHARELLI
University of Texas at Dallas, Box 688, Richardson, Texas 75080
D. POPESCU and I. POPESCU
Central Institute of Physics, Bucharest/Magurele, Romania

Progress with a multiphoton technique is reported for the measurement of relative cross sections for the photolysis of simple molecules into particular product channels. In this work two independently tunable dye lasers were used to sequentially dissociate molecules of Cs_2 and CsKr for wavelengths in the range 420 to 660 nm, and then to excite the resulting products to determine the relative cross sections for the photolysis of the Cs_2 and CsKr molecules into each of the lowest four of the energetically possible product states.

The dramatic growth occurring over the past few years in laser chemistry and laser isotope separation has refocused interests upon dissociative processes in molecules. Collectively, these interests are traceable to the pragmatic goals of producing appreciable populations of selected atomic or molecular states having useful reactive properties or isotopic content. From this perspective, it is natural that photodissociation of some parent molecule would appear to be the ideal means for obtaining a desired product.

In the past, this type of selective photolysis has been accomplished by exciting a rather complex precursive molecule with infrared laser radiation at a frequency which could select some initial vibrational mode of absorption that would ultimately lead to the desired product. Many of the impressive results that have been achieved with this approach have been summarized in a recent review (1). Unfortunately, two general difficulties have persisted in the conventional approaches. The first has been the tendency to obtain pyrolysis of the parent molecule rather than photolysis. The second is that if viewed as reactive species, laser photons are extremely expensive "substances," their cost being elevated further by the degree of spectral purity and stability that must be maintained. Both

0097-6156/82/0179-0019$05.00/0

factors provide motivation for the consideration of alternative methods for the preparation of populations of selected states that could reach practical magnitudes at reasonable cost.

One alternate technique which is beginning to present some appeal is the excitation of electronic states of simple parent molecules that can preferentially dissociate into the desired product states. The first example of state-selective photolysis to be induced by a laser was first reported by the Saclay group in 1973 (2). They observed the dissociation of H_2 through the laser excitation of a dissociative intermediate state occurring as a component step in a multiphoton ionization process. In 1974 we reported (3) the spectroscopy of such a process occurring in Cs_2 that had been examined with the output from a tunable dye laser. As in the H_2 work, it was found that multiphoton absorption could occur with high probability through a resonant intermediate state that was electronically excited and unstable against dissociation on the time scale of the multiphoton process. Because of these early experiments, it seemed reasonable to hope that such electronic transitions might occur over sufficiently broad bands of wavelengths so that relatively inexpensive photons, from filtered incoherent radiation, for example, might be able to produce appreciable populations of discrete product states. However, a more general application of such techniques to other systems would require detailed knowledge of the dissociating state to be excited, of the states of the products with which it correlates, and of the dynamics of the dissociation process which would determine the extent to which the product channels might become mixed. Unfortunately, practically nothing is known about the dissociative electronic states of most molecules.

The purpose of this paper is to present a progress report describing a rather general technique for the study of such electronically excited, dissociative states of simple molecules. We have recently applied this technique to a study of the photolysis of Cs_2 dimers and CsKr excimers in order to resolve two general questions: (1) whether relatively broad dissociation bands exist in the spectra that might populate selected product states, and (2) whether such photolytic spectra could lead to useful estimates of the potential curves for the dissociative states involved.

This technique has been recently used in a study of the photolysis of Cs_2 with extremely encouraging results (4-7). A relatively broad band was found which selectively inverted the product populations in the 5D(3/2) and 6P(1/2) states to such an extent that even continuous laser action should be sustained by those products with adequate optical pumping at wavelengths within that band. In addition, the detail that was found was sufficient to locate three new repulsive states of Cs_2 and to correlate them with their atomic limits (6). Those results are reviewed here for comparison to new data obtained for the photolysis of CsKr excimers at visible wavelengths. The sensitivity and resolution achieved in this work attest to the promise held by this new multiphoton technique for the detailed characterization of the repulsive states of simple molecules that are becoming of critical interest in a variety of applications.

Method

Ideally, the photolysis of a simple molecule can be accomplished by inducing a transition to an electronically excited state which is unstable for one of several possible reasons. A typical situation is shown in Figure 1. This figure has been drawn to illustrate the photolysis of Cs_2, but the situation for CsKr would be essentially analogous if one of the component Cs atoms were replaced by a Kr atom everywhere in the figure. The principal difference would be that the dissociation energy of the initial molecular state would be greatly reduced, therefore, giving a much shallower minimum in the ground state than is shown in this figure. Photolysis could occur through the excitation of a transition, such as (1a), to a bound state which can subsequently predissociate to produce Cs^*. The principal restriction in this case is that there must be an intersection, at a suitable energy, of the potential curve for that bound state with a curve for a repulsive state of the proper symmetry. Another possibility for the production of Cs^* by photolysis of Cs_2 would be the excitation of a transition, such as (1b), directly to a repulsive state correlated with the same dissociation limit.

In the Born-Oppenheimer approximation, the relative importance of channels (1a) and (1b), together with their dependence on wavelength would depend upon the matrix elements for the transition between the electronic states, the Franck-Condon factors, the Hönl-London factors, and upon the probabilities for spontaneous dissociation of the excited state formed. In principle, except for the last one, these are well known quantities whose product is the transition probability for that particular absorption band of Cs_2. When multiplied by the last quantity, and with an adjustment of numerical constants it becomes the cross section for the photolysis of Cs_2 into Cs^* + Cs. It is the measurement of this cross section that lies at the focus of this work.

The products of dissociation are detected through the absorption of a second photon in a transition such as (2) in Figure 1. However, instead of attempting to detect the slight attenuation of an illuminating beam of wavelength corresponding to transition (2), the strategy of using the second transition to populate a relatively long-lived Rydberg state, Cs^{**}, is employed. Then, as had been demonstrated in much earlier work, if that Cs^{**} state were sufficiently close in energy to the

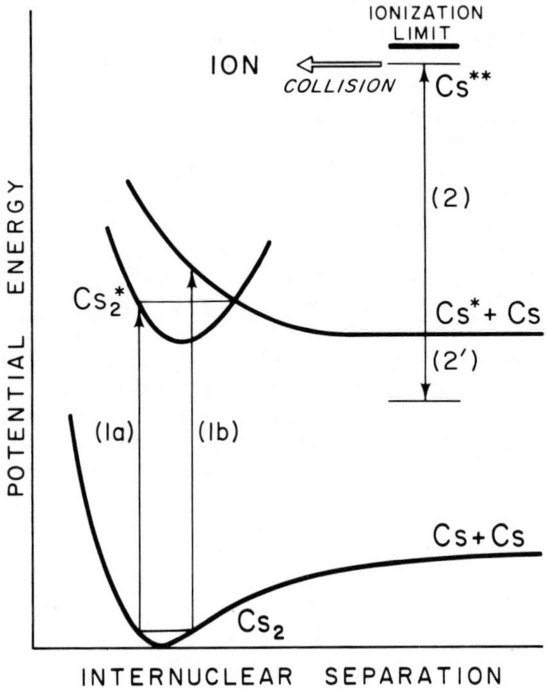

Journal of Chemical Physics

Figure 1. Schematic representation of the multiphoton process for studying the selective photolysis of Cs_2. Typical molecular potentials are plotted as functions of internuclear separation with radiative transitions shown by arrows. The first transition causing photolysis is designated 1a for predissociation or 1b for direct dissociation. The second transition used to excite the dissociation products, Cs, to the readily ionized state, Cs** is 2 and 2′ shows a possible transition for fluorescent detection (5).*

ionization limit, it would have essentially unit probability for collisional ionization by simple inelastic collisions with the majority species in the medium. The resulting ionization is then detected in comparison to a background ionization which tended to be intrinsically low.

Experimental

The apparatus used in this work is shown schematically in Figure 2. Two independently tunable, pulsed dye lasers were used to provide, first, the photolysis pulse and, subsequently, the pulse of illumination needed to excite the dissociation products to readily ionized Rydberg states. The dye lasers were pumped by a single nitrogen laser operating at a power level of 300 kW in a 10 nsec pulse. The output from the pump laser was geometrically split into two portions that were separately focused onto the two dye lasers. The output beams from the dye lasers were aligned to be spatially collinear, but were temporally dispersed so that the one used to dissociate the population of parent molecules arrived first at the absorption cell by about the duration of the pulse. Provisions were available for the adjustment of the temporal separation of the pulses but for this work the interpulse delay was set to the minimum pracaticable value of 3 nsec to reduce the possibility of collisionally mixing the product populations. As expected, with a reversed temporal sequence the effect was completely absent. The linewidth of the output from the dye lasers was of the order of 0.01 nm and corresponded approximately to the resolution with which the wavelength could be set.

The actual detection of the ionization produced in the photolysis channel was accomplished by including an in situ thermionic ion detector in the vapor cell. The particular arrangement is shown schematically in Figure 3. This method has been shown to be very effective for the products of the photoionization of metal vapors and has been used in all of the previous studies in which hybrid multiphoton resonances have been reported. Essentially, the technique uses the ionization in the vapor as a distributed grid to control a space-charge limited current flowing between a cathode and an anode arranged to contain the absorbing vapor between them. Sensitivities to the production of one ion per second have been achieved in this manner. In the experiment reported here, the detecting diode was operated in a cesium vapor at a pressure of 0.13 Torr for the studies of photolysis of CsKr and the partial pressure of cesium was kept the same but a density of 0.132 Amagat of Kr was added to the vapor cell.

In the experiment, the wavelength of the photolysis pulse from the laser was set to successive values which were uniformly spaced across the spectral region studied. Then, for each value of λ_1 the wavelength of the delayed pulse was tuned

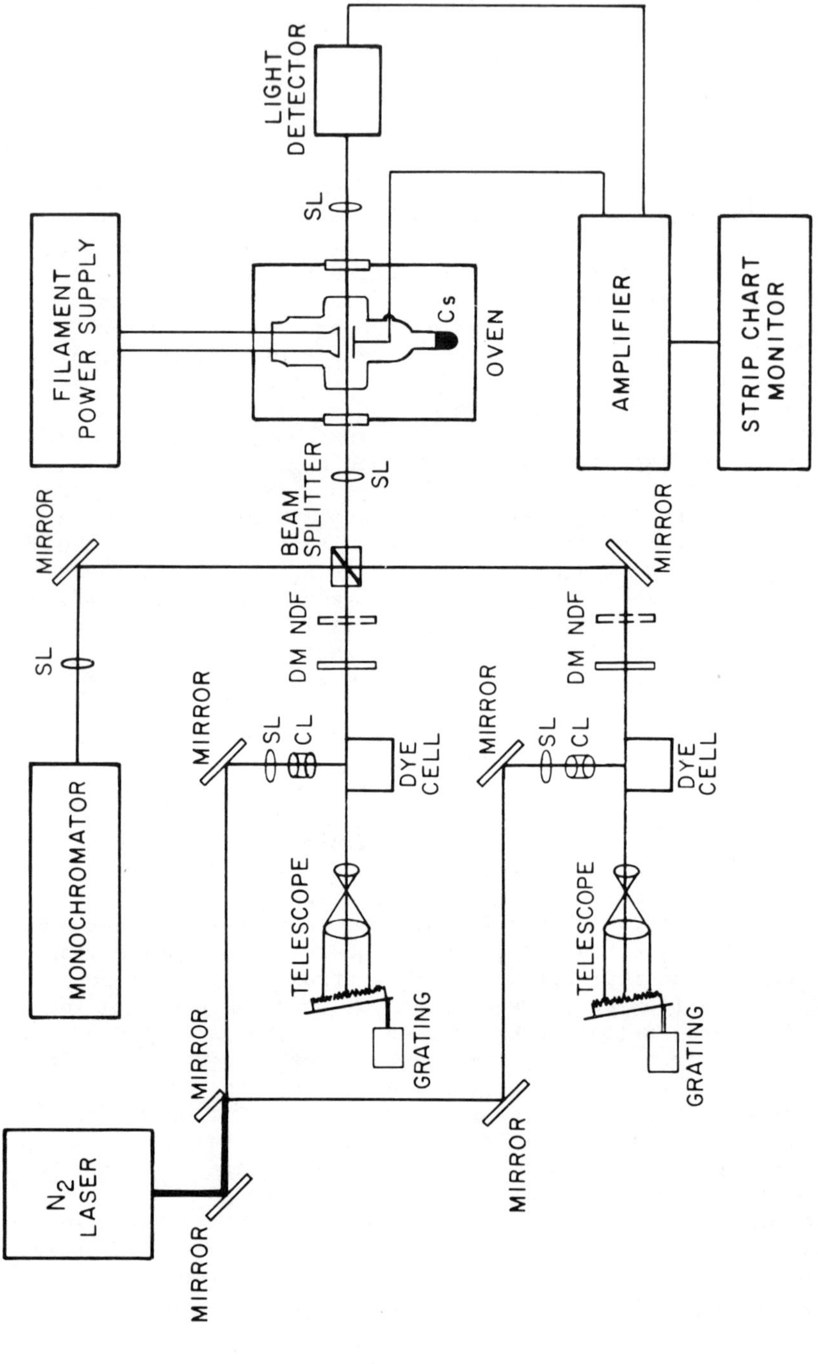

Journal of Chemical Physics

Figure 2. Schematic representation of the experimental apparatus. Key: CL, cylindrical lens; DM, dielectric mirror; SL, spherical lens; and DF, neutral density filter.

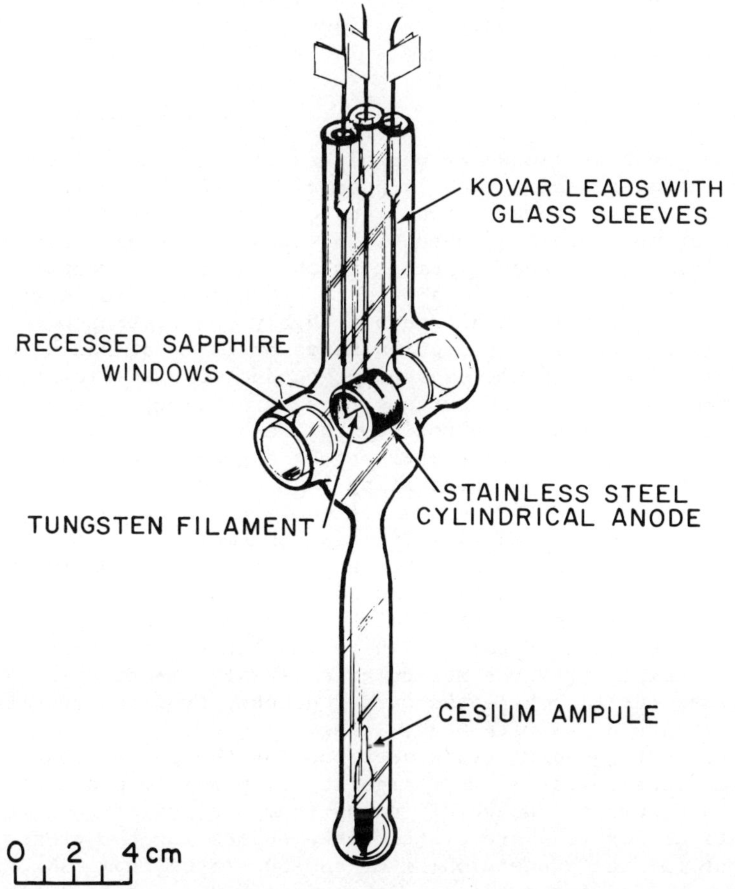

Figure 3. Adsorption cell showing details of the thermionic diode employed to detect photoionization (5).

across a small interval containing the atomic transition used to detect a particular dissociation product. Immediately before and after each scan of λ_2 across the atomic line, the pulse energies of the lasers were determined by blocking each in order to allow only one at a time to illuminate the detector. Then, in order to obtain a background, the same wavelength range for the delayed pulse was scanned again with the photolysis beam blocked.

Discussion

The cross sections for photolysis of Cs_2 into all product states that were energetically possible to reach from the ground state are shown as functions of the transition energy of the parent molecule in Figure 4. Several remarkable features are immediately apparent. First is that with one exception all of the photolysis channels appear to be concentrated in wavelength into three spectral bands, roughly correlating with the red, yellow, and blue absorption bands of Cs_2 that had been reported (8) as a result of studies conducted with classical techniques. Second, and perhaps more surprising, is that the occurrence of strongly selective, but relatively broad bands for the photolysis of Cs_2, seems to be a general phenomenon. Finally, the total indetectability of any photolysis bands for the production of Cs(6P) atoms for energies below 19,500 cm^{-1} must be considered remarkable, as is the small production of such populations above that threshold. These complex features have been analyzed in detail and have led to the identification of three new repulsive potentials of Cs_2 (6). These potential curves are shown in Figure 5, where the light lines represent curves obtained from the molecular constants reported in the literature (9-14), while the curves deduced from the results of this work are drawn with heavy lines.

The corresponding cross sections for the photolysis of CsKr into the lowest four, energetically possible products are shown in Figure 6. Though differing from the scales to which the data of Figure 4 are plotted, the scales for the cross sections for the production of 6P and 5D states from CsKr are the same to facilitate direct comparison between them. No production of 6P(1/2) was observed at any visible wavelength so the cross section for that channel of photolysis can be concluded to be less than 1.0 in the units of Figure 6. As in the case of the photolysis of Cs_2, the photolysis channels appear to be concentrated into three spectral bands correlating with the red, yellow and blue absorption bands of CsKr that had been reported (15) as a result of classical studies conducted at orders of magnitude higher inert gas density. In fact, except for the red band for the production of 5D atoms at wavelengths near 15,800 cm^{-1}, the similarity between the photolysis bands of Figures 4 and 6 is remarkable to an extent raising questions

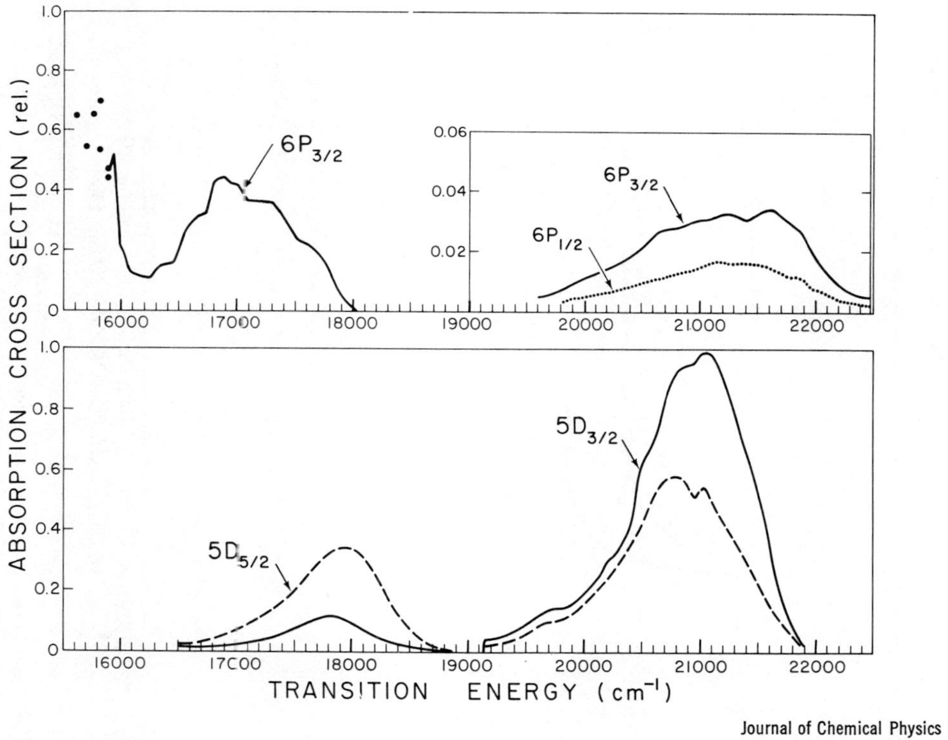

Figure 4. Relative cross sections as function of transition energies for the selective photolysis of Cs into each of the energetically possible products. Product states have $J = 5/2$ (– – –), $J = 3/2$ (——), and $J = 1/2$ (· · ·). Data points plotted for the production of 6P(3/2) atoms at small transition energies record values obtained in a wavelength region containing structure on a scale small compared to the size of the 2.5-nm sampling interval over which measurements were made (6).

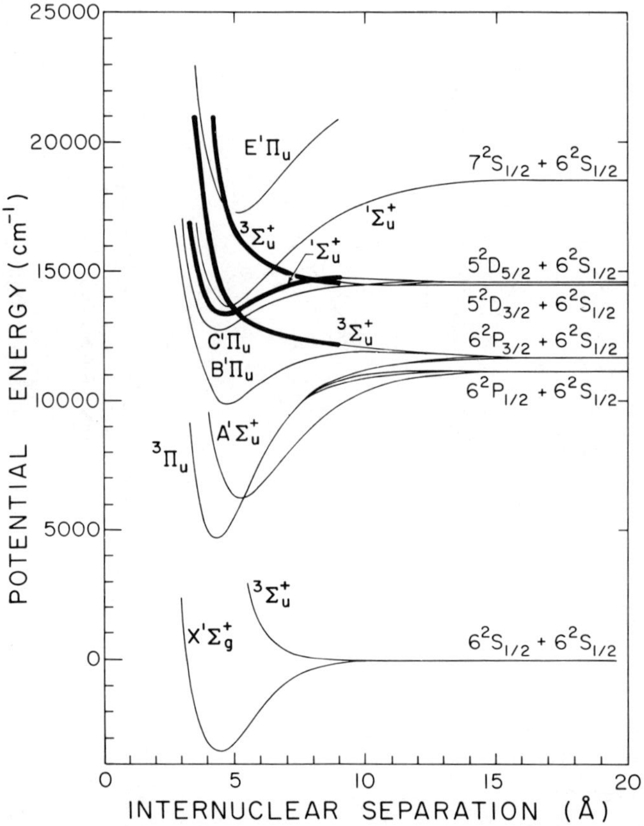

Journal of Chemical Physics

Figure 5. Potential curves for the excited states of Cs_2. Light lines show potentials that were constructed from molecular constants obtained from the literature. Heavy lines plot potentials approximated from the results of this work (6).

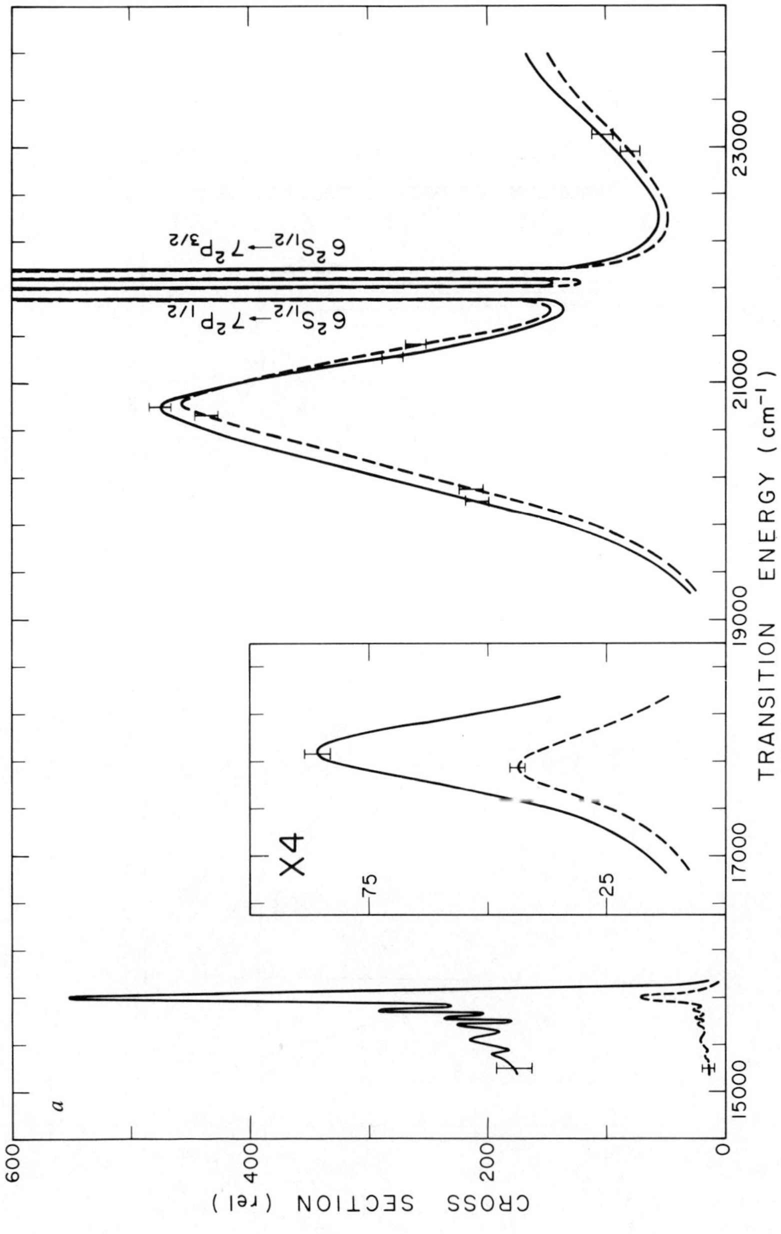

Figure 6. Relative cross sections as functions of transition energies for the selective photolysis of CsKr into each of the lowest four, energetically possible product states. Error bars record representative scatter obtained during successive remeasurement. Key: (6a) ——, $Cs(5^2D_{5/2})$; and – – –, $Cs(5^2D_{3/2})$; (6b) ——, $Cs(6^2P_{3/2})$.

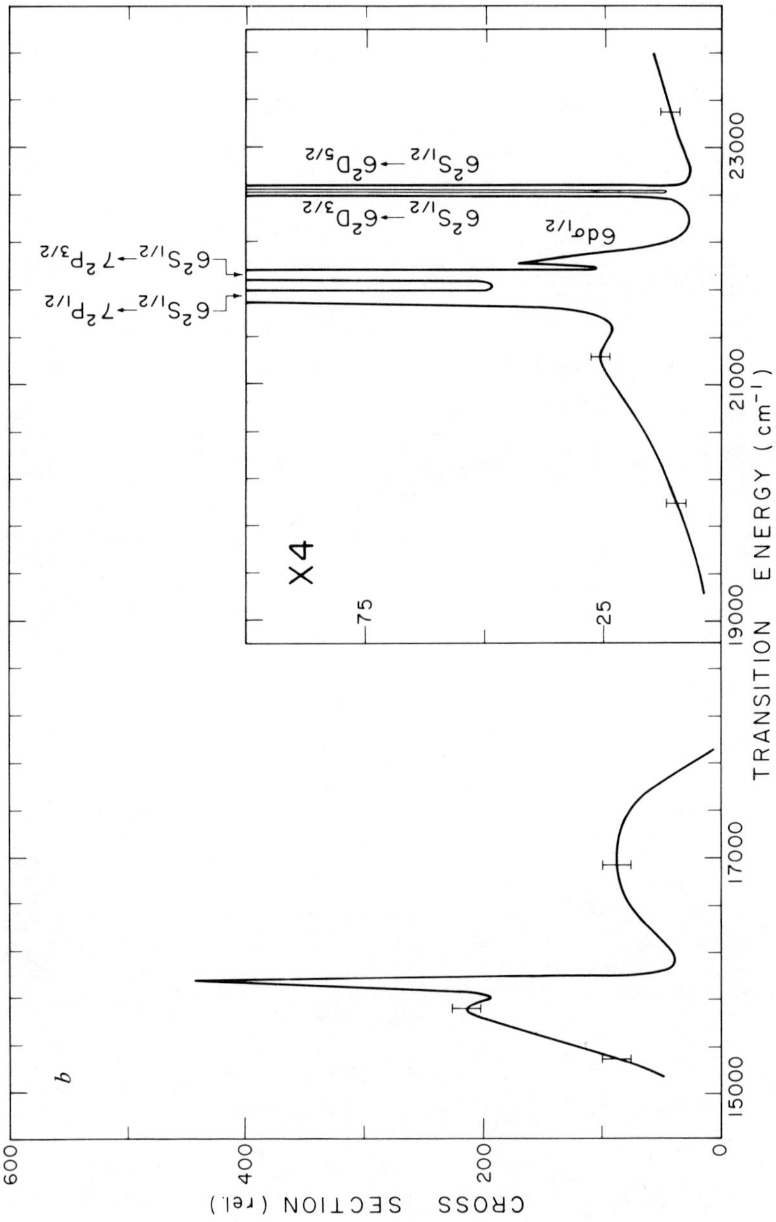

Figure 6. Continued.

about the identify of the actual parent molecules. It seems that most of the features of Figure 6 could be the result of the photolysis of Cs_2Kr, principally along the Cs-Cs band. In such a system it would be reasonable to expect that the spectrum would resemble that of Cs_2, red-shifted by the increasing potential found in the polarization energy of the more excited states. This would explain all of the features that we observe, with the exception of the red band shown at the left of the figure. This red band can be confidently attributed to photolysis of a parent CsKr molecule.

The larger photolysis band in the red region of Figure 6 giving Cs 5D(5/2) agrees in general appearance with the absorption band recently attributed (16) to CsKr (5D) except that higher contrast is seen in the oscillations shown in this figure. This is a reasonable difference and demonstrates the advantages of the multiphoton technique for studying photolysis. In the previous work (16) the occurrence of photolysis had been detected by the observation of fluorescence from 6P(3/2) resulting from the cw excitation of the vapor. It had been assumed that only the collisional and radiative cascading from the 5D(5/2) product state contributed to the 6P(3/2) population and so the fluorescence signal was expected to have been proportional to the single photolysis channel producing 5D(3/2). Unfortunately, in that work fluorescence was also detected from the direct channel shown in Figure 6 that produced 6P(3/2) atoms as a result of the photolysis of Cs_2Kr. Since they could not distinguish between the direct and indirect channel they should have obtained an absorption band equivalent to the sum of the selective cross sections shown in Figure 6. That summation would have reduced the contrast shown by the detailed variations seen in this work.

Concluding Remarks

The photolysis spectra presented in this work clearly support the statement made in the introduction that relatively broad band irradiation of Cs_2 and CsKr could lead to the selective population of even a single fine structure component of the possible dissociation products. In addition, the band of wavelengths over which this selectivity is maintained is relatively broad, being about 25 nm in half width. Such results illustrate the type of pragmatic benefit which could not have been anticipated from the potential curves for Cs_2 that were available prior to this work. Clearly, the implementation of this new two-photon technique has provided a powerful insight into the identity and location of several dissociative states of Cs_2. So, it seems that this technique has proven itself to be a very sensitive tool for the study of selective photolysis of simple molecules through electronically excited states.

Literature Cited

1. Cantrell, C. D.; Freund, S. M.; Lyman, J. L. "Laser Handbook," Vol III(b), edited by M. L. Stitch, North Holland Co., New York, 1979.
2. LuVan, M.; Mainfray, G.; Manus, C.; I. Tugov, Phys. Rev. 1973, A7, 91.
3. Collins, C. B.; Johnson, B. W.; Mirza, M. Y.; Popescu, D.; Popescu, I; Phys. Rev. 1974, A10, 813.
4. Collins, C. B.; Anderson, J. A.; Lee, F. W.; Vicharelli, P. A.; Popescu, D.; Popescu, I.; Phys. Rev. Lett. 1980, 44 139.
5. Collins, C. B.; Anderson, J. A.; Popescu, D.; Popescu, I.; J. Chem. Phys. 1981, 74, 1053.
6. Collins, C. B.; Lee, F. W.; Anderson, J. A.; Vicharelli, P. A.; Popescu, D.; Popescu, I.; J. Chem. Phys. 1981, 74 1067.
7. Collins, C. B.; Anderson, J. A.; Lee, F. W.; Vicharelli, P. A.; Popescu, D.; Popescu, I.; "Proceedings of the International Conference on Lasers '79" edited by V. J. Corcoran, STS Press, McLean, VA, 1980.
8. Gupta, R.; Happer, W.; Wagner, J.; Wennmyr, E.; J. Chem. Phys 1978, 68, 799.
9. Kato, H.; Yoshihara, K.; J. Chem. Phys 1979, 71, 1585.
10. Kusch, P.; Hessel, M. M.; J. Mol. Spectrosc. 1969, 32, 181.
11. Kusch, P.; Hessel, M. M.; J. Mol. Spectrosc. 1968, 25, 205.
12. Benedict, R. P.; Drummond, D. L.; Schlie, L. A.; J. Chem. Phys. 1977, 66, 4600.
13. Dickinson, H. O.; Rudge, M. R.; J. Phys. 1970, B3, 1448.
14. Höning, G.; Czajkowski, M.; Stock, M.; Demtröder, W.; J. Chem. Phys. 1979, 71, 2138.
15. Moe, G.; Tam, A. C.; Happer, W.; Phys. Rev. 1976, A14, 349.
16. Sayer, B.; Ferray, M.; Visticot, J. P.; Lozingot, J.; J. Phys. 1979, B12, 227.

RECEIVED August 26, 1981.

3

Emission and Absorption Spectroscopy of High Pressure Sodium Discharges

J. P. WOERDMAN
Philips Research Laboratories, 5600 MD Eindhoven, The Netherlands

J. J. DE GROOT
Philips Lighting Division, 5600 MD Eindhoven, The Netherlands

The far quasistatically broadened wings of the Na D transition give important contributions to the emission and absorption spectrum of a high-pressure sodium discharge. We report experimental and theoretical spectra and discuss in particular the Na_2 triplet satellite at 551.5 nm and the Na_2 singlet satellite at 805 nm. Extension of this type of spectroscopy to other alkali metal vapors is discussed.

A good understanding of the emission and absorption spectra of high-density alkali vapors is important for designing efficient high-pressure alkali metal vapor lamps (1,2). Other interests in such spectra stem from the work on alkali dimer lasers (3) and from studies into resonant interaction between excited state and and ground state alkali atoms (4). We will concentrate here on high-pressure sodium (HPS) discharges in local thermal equilibrium (LTE). For such discharges the emission coefficient ϵ, as a function of the frequency ν, is determined by the sodium vapor pressure p_{Na}, the radial temperature profile T(r) and the absorption line profile $k(\nu)$ of the transitions involved (5). The spectral radiance of the discharge can be calculated from $\epsilon(\nu)$ and the one-dimensional radiative transfer equation. For diagnostic purposes the reverse route is followed: p_{Na} and T(r) are derived from the emission spectrum (6,7). Such LTE calculations of the spectral radiance have been performed previously (5), using a dispersive, resonantly broadened profile for the Na D transition. In the far wings of the Na D line, however, the calculated spectral radiance disagreed with experiment for $p_{Na} \gtrsim 100$ Torr (see Figure 1). Strong bands occur at 551.5 nm and 805 nm. Figure 2 illustrates the pressure dependence of these emission bands; the 551.5 nm band is so strong that self-reversal occurs at $p_{Na} = 1000$ Torr (Figure 2c). Both bands have been observed (5,8,9,10) and interpreted (11,12,13) previously. The 551.5 nm band is due to a maximum in the Na_2 potential difference $^3\Pi_g - {}^3\Sigma_u^+$ and the 805 nm band is due to a minimum in the Na_2 potential difference $A^1\Sigma_u^+ - X^1\Sigma_g^+$. Clearly, in a model for calculating the spectral radiance, $k(\nu)$ should incorporate the short-range NaNa interactions for $p_{Na} \gtrsim 100$ Torr.

0097-6156/82/0179-0033$05.00/0

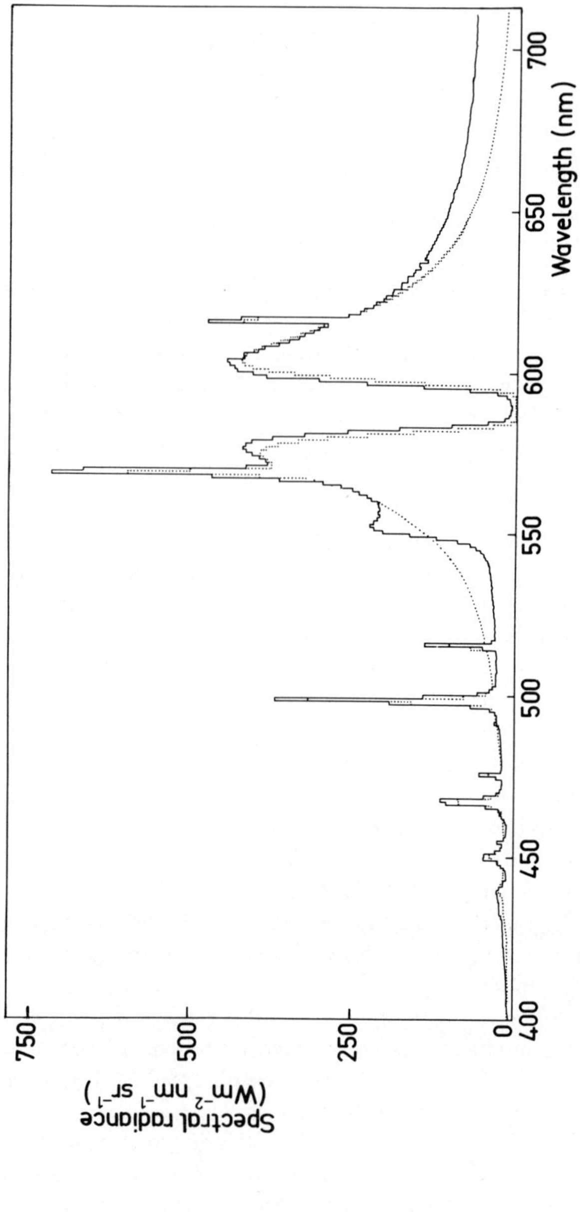

Figure 1. Comparison between the measured (——) and calculated (· · ·) spectral radiance of a high-pressure Na discharge; the Na vapor pressure is 300 Torr, the tube diameter 7.6 mm, the axis temperature 3700 K, and the wall temperature 1600 K. The calculation (5) is based on LTE; resonance broadening is assumed for the NaD transition and Stark broadening of the upper level for the nonresonance Na transitions. Both the measured and the calculated spectra are based on 1-nm integration intervals.

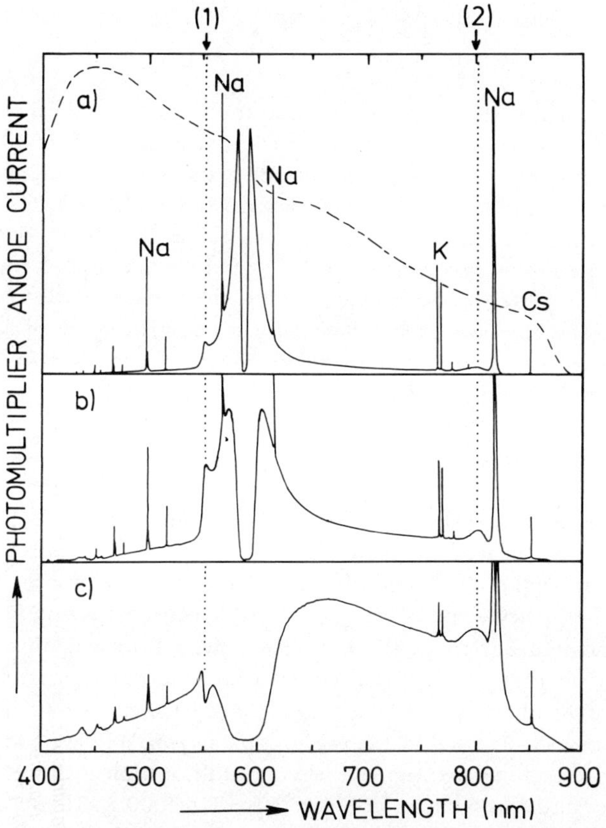

Figure 2. Experimental emission spectra of a high-pressure Na discharge. The tube diameter is 7.6 mm, the axis temperature 3000 K, the wall temperature 1400–1800 K, and the sodium vapor pressure 100 (a), 300 (b) and 1000 (c) torr. The spectra have not been corrected for the instrument spectral sensitivity indicated by ---. Arrows 1 and 2 indicate the classical satellites. A number of Na nonresonance lines and the K and Cs resonance lines (due to trace impurities) are indicated.

Quasistatic calculation of the absorption spectrum

Recent ab initio calculations of the relevant Na_2 potentials by Konowalow, Rosenkrantz and Olson (14), henceforth abbreviated as KRO, permit such an approach. In the quasistatic or classical Franck-Condon approximation the binary contribution to k(ν,T) is given by (15,16)

$$k(\nu,T) = \frac{\pi^2 e^2}{2mc} f_{12} g_1 [Na]^2 R^2 \left|\frac{dR}{d\nu}\right| \exp-\left(\frac{V(R)}{k_B T}\right) \qquad (1)$$

where f_{12} = oscillator strength of Na_2 transition, g_1 = statistical weight of Na_2 lower state, [Na] = Na density, R = internuclear separation of Na_2, V(R) = ground state potential energy of Na_2, k_B = Boltzmann constant and e, m, and c have their usual meaning. Satellite bands occur for extrema in $\nu = \nu(R)$, i.e. for $|dR/d\nu| \rightarrow \infty$. For V(R) and $|dR/d\nu|$ we use values based on Table III of KRO. For the oscillator strength f_{12} we use the atomic values (12,17) and we take $g_1 = 1$ for singlet states and $g_1 = 3$ for triplet states. In order to "smooth" the classical satellites ($|dR/d\nu| \rightarrow \infty$) the calculated spectra have been folded with a Gaussian curve of width Δ. Since in the experimental spectra the width of the 551.5 nm band is ~ 7 nm (FWHM) we choose Δ = 7 nm (FWHM). More sophisticated quasistatic approaches (18,19), requiring knowledge of $d^2\nu/dR^2$, may yield the shape of the satellite bands (see e.g. (20)).

The spectrum k(ν,T) thus calculated should be corrected, since, as pointed out by KRO, the asymptotic value ν(∞) from their calculations should be increased by 1050 cm^{-1} (i.e. by a factor of 1.066) in order to obtain the correct experimental Na D transition frequency. Previously (12) our correction procedure has been to multiply the frequency scale of the calculated spectrum by this factor of 1.066. It has been pointed out to us (D.D. Konowalow, private communication) that a more appropriate correction procedure is to increase the separation of the KRO potential curves involved by 1050 cm^{-1}, i.e. to add 1050 cm^{-1} to the frequency scale of the spectrum as calculated in the previous paragraph. We have adopted this latter procedure and a resulting spectrum of the reduced absorption coefficient $k(\nu,T)/[Na]^2$ is given by the fully drawn curves in Figure 3 for T = 2000 K. The spectrum contains four partly overlapping contributions due to the four optically allowed Na_2 transitions in the visible and near-infrared part of the spectrum, namely $A^1\Sigma_u^+ \leftarrow X^1\Sigma_g^+$, $B^1\Pi_u \leftarrow X^1\Sigma_g^+$, $^3\Sigma_g^+ \leftarrow {}^3\Sigma_u^+$ and $^3\Pi_g \leftarrow {}^3\Sigma_u^+$. When comparing with our previous (12) calculations the effect of the new correction procedure is a slight "compression" of the spectrum towards the center of the Na D line.

Experiments

We have measured emission and absorption spectra of an HPS sodium discharge within a sapphire tube. The tube has an inside diameter of 7.6 mm, is terminated

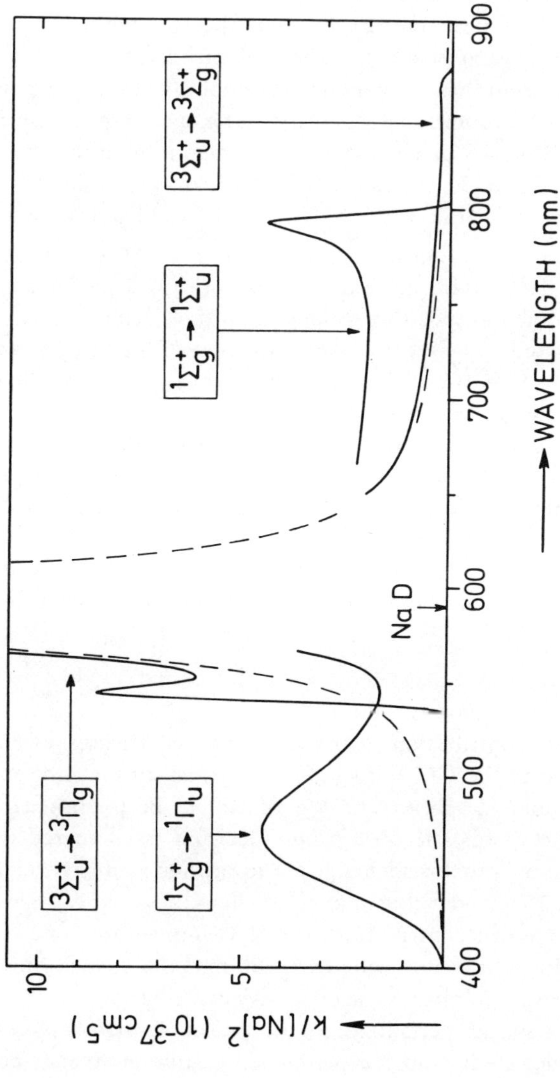

Figure 3. The drawn curves represent the reduced absorption coefficient k/[Na]^2 as a function of wavelength, calculated for 2000 K from Table III of Ref. 14, using Eq. 1, and adding 1050 cm^{-1} to all transition frequencies. The spectrum has been folded with a Gaussian curve 7 nm wide (FWHM) to "smooth" the satellites. The four contributions from the four optically allowed Na$_2$ transitions are shown separately; they should be added to obtain the complete spectrum. – – – represents k/[Na]^2 corresponding to a dispersive, resonantly broadened NaD line profile, extrapolated from the data of Ref. 24.

with two tungsten electrodes and is mounted inside an evacuated quartz bulb. The discharge is AC operated (50 Hz) at currents between 2A and 5A (rms). The wall temperature is typically 1500 K (21) and the axis temperature varies between 3000 K (at current reversal) and 4000 K (at current maximum); the radial temperature profile is roughly parabolic (5,21). All spectra have been taken at current reversal in order to minimize the relative contribution of the non-resonance lines. Further experimental details have been given in (12). Emission spectra have already been discussed (Figures 1,2). Transmission spectra have been obtained for $100 \leq p_{Na} \leq 1000$ Torr. From these spectra we first established that the absorption coefficient $k(\nu)$ is roughly proportional to $[Na]^2$. The spectrum of the reduced absorption coefficient $k/[Na]^2$ is given in Figure 4. There is good agreement with the calculated spectrum (Figure 3), even slightly better than when using our original correction procedure (12). The peak of the $B^1\Pi_u \leftarrow X^1\Sigma_g^+$ band and the satellites of the $^3\Pi_g \leftarrow {}^3\Sigma_u^+$ and $A^1\Sigma_u^+ \leftarrow X^1\Sigma_g^+$ transitions occur at 470 nm, 545.2 nm and 793 nm, respectively, in the calculated spectrum and at 490 nm, 551.5 nm and 805 nm, respectively, in the experimental spectrum. In particular the calculated peak absorption of the $^3\Pi_g \leftarrow {}^3\Sigma_u^+$ satellite agrees very well with that of the 551.5 nm band in the experimental spectrum ($k = 10 \times 10^{-37} [Na]^2$ cm^{-1}), thus confirming the validity of the $^3\Pi_g$ and $^3\Sigma_u^+$ potential curves of KRO.

Unlike the triplet contributions to the absorption spectrum, the singlet contributions have a strongly temperature-dependent absorption coefficent k. A fit between the experimental spectrum of Figure 4 and calculated spectra as in Figure 3 yields T = 1900 (± 200) K for the $B^1\Pi_u \leftarrow X^1\Sigma_g^+$ maximum and T = 2700 (± 200) K for the $A^1\Sigma_u^+ \leftarrow X^1\Sigma_g^+$ satellite. The difference can be explained by the different temperature dependences of k in conjunction with the radial temperature profile of the discharge.

Discussion and conclusions

When the quasistatic contribution to the absorption coefficient, as described by eq. (1), is incorporated in the LTE model (5) a more realistic emission spectrum of the HPS discharge results (de Groot and Woerdman, to be published). This is already evident when the quasistatic absorption spectrum is compared with the extrapolated dispersive, resonantly broadened Na D absorption profile (see Figure 3). In the far red wing ($\lambda > 650$ nm) the contribution of the $A^1\Sigma_u^+ \leftarrow X^1\Sigma_g^+$ transition is much larger than the dispersive contribution; in a LTE model the same holds for the emission spectrum. In the far blue wing ($\lambda < 560$ nm) the contribution of the $^3\Pi_g \leftarrow {}^3\Sigma_u^+$ transition likewise dominates the dispersive contribution.

A few additional remarks pertaining to the triplet spectroscopy of alkali metal vapors can be made. Under most experimental conditions (vapor cell, heat pipe) the population of the repulsive triplet ground state $^3\Sigma_u^+$ is negligible as compared to that of the bound singlet ground state $X^1\Sigma_g^+$. As a consequence, the triplet bands are usually partly or completely overlapped by the much stronger singlet bands, thus impeding quantitative measurements of triplet absorption coefficients.

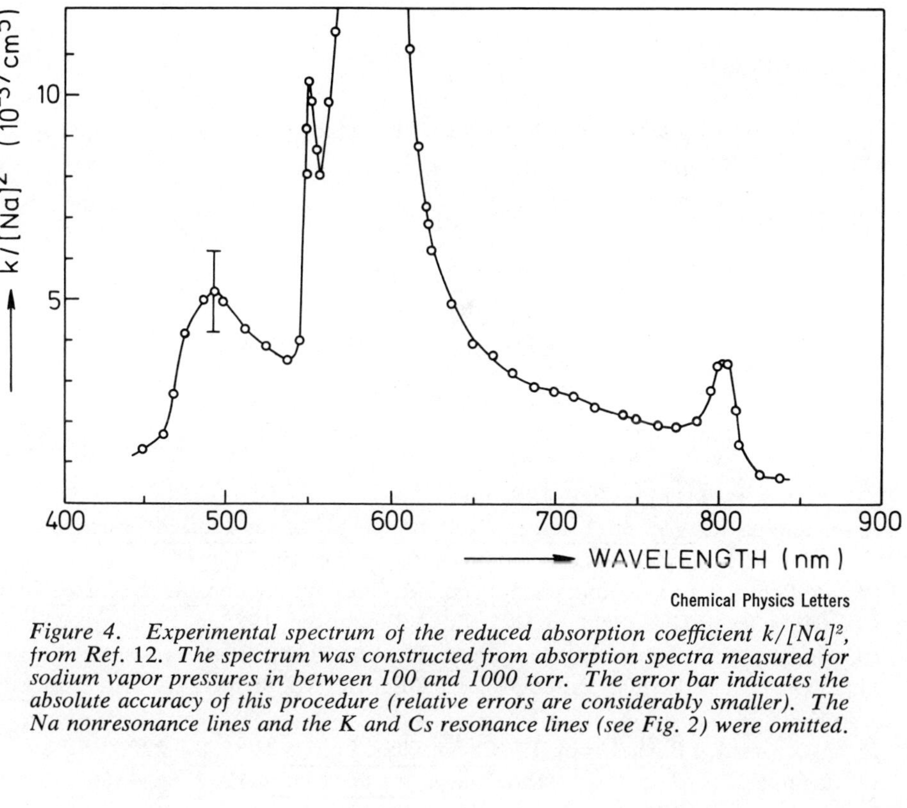

Chemical Physics Letters

Figure 4. Experimental spectrum of the reduced absorption coefficient k/[Na]^2, from Ref. 12. *The spectrum was constructed from absorption spectra measured for sodium vapor pressures in between 100 and 1000 torr. The error bar indicates the absolute accuracy of this procedure (relative errors are considerably smaller). The Na nonresonance lines and the K and Cs resonance lines (see Fig. 2) were omitted.*

In a high-pressure discharge, however, the high temperatures suppress the singlet spectrum and thus facilitate observation of the triplet spectrum. Using a sapphire discharge tube quantitative measurements of the triplet absorption coefficient are possible, as demonstrated in this work. We are presently trying to produce a high--pressure lithium lamp in order to observe the $^3\Pi_g \leftarrow {}^3\Sigma_u^+$ transition of Li_2 and to compare with the KRO calculations. This Li_2 transition cannot be observed in a heat pipe because of singlet overlap (11); it has been observed indirectly in two-photon Li_2 absorption (23). For K_2 an emission band near 720 nm in the blue wing of the resonance transition is rather similar to the 551.5 nm band of sodium, thus suggesting the transition $^3\Pi_g \rightarrow {}^3\Sigma_u^+$ as its origin (8); absorption spectra are lacking in this case.

Acknowledgment

We thank Prof. D.D. Konowalow for his valuable comment on the appropriate correction procedure.

Literature Cited

1. Wharmby, D.O. Proc. IEE 1980, 127, 165.
2. Zollweg, R.J. ACS Symposium Series 1981 (present volume).
3. Wellegehausen, B. ACS Symposium Series 1981 (present volume).
4. Movre, M.; Pichler, G. J. Phys. B 1980, 13, 697.
5. de Groot, J.J.; van Vliet, J.A.J.M. J. Phys. D 1975, 8, 651.
6. Teh-Sen Jen; Hoyaux, M.F.; Frost, L.S. J. Quant. Spectrosc. Radiat. Transfer 1969, 9, 487.
7. Bartels, H. Z. Phys. 1950, 128, 546.
8. Schmidt, K. Proc. 6th Internat. Conf. on Phenomena in Ionized Gases (Hubert, P.; Crémieu-Alcan, E., Eds.) Vol. 3, Paris, 1963, p.323.
9. de Groot, J.J.; van Rooyen, J. Proc. Int. Conf. on Phenomena in Ionized Gases (Hölscher, D.G.A.; Schram, D.C., Eds.), Eindhoven, 1975, p. 135.
10. Sorokin, P.P.; Lankard, J.R. J. Chem. Phys. 1971, 55, 3810.
11. Veža, D.; Rukavina, J.; Movre, M.; Vujnović, V.; Pichler, G. Opt. Commun. 1980, 34, 77.
12. Woerdman, J.P.; de Groot, J.J. Chem. Phys. Letters 1981, 80, 220.
13. Lam, L.K.; Gallagher, A.; Hessel, M.M. J. Chem. Phys. 1977, 66, 3550.
14. Konowalow, D.D.; Rosenkrantz, M.E.; Olson, M.L. J. Chem. Phys. 1980, 72, 2612.
15. Vasilakis, A.; Bhaskar, N.D.; Happer, W. J. Chem. Phys. 1980, 73, 1490.
16. Gallagher, A. in "Topics in Applied Physics", Vol. 30 (Rhodes C.K., Ed.) Springer, Berlin 1979, p. 135.
17. Woerdman, J.P. J. Chem. Phys. (submitted).
18. Sando, K.M.; Wormhoudt, J.C. Phys. Rev. A 1973, 7, 1889.
19. Baylis, W.E.; Szudy, J. J. Quant. Spectrosc. Radiat. Transfer 1975, 15, 641.

20. Veža, D.; Movre, M.; Pichler, G. J. Phys. B 1980, 13, 3605.
21. de Groot, J.J. Proc. 2nd IEE Conf. on Gas Discharges, London, 1972, p. 124.
22. York, G.; Scheps, R.; Gallagher, A. J. Chem. Phys. 1975, 63, 1052.
23. Koch, M.E.; Stwalley, W.C.; Collins, C.B. Phys. Rev. Letters 1979, 42, 1052.
24. Niemax, K.; Pichler, G. J. Phys. B 1975, 8, 179.

RECEIVED August 26, 1981.

Photoabsorption Measurements of Atomic and Molecular Potassium

C. Y. R. WU

Department of Physics and Earth and Space Sciences Institute, University of Southern California, Los Angeles, CA 90007

Photoabsorption measurements of atomic potassium in the 50-70 nm region and molecular potassium in the 210-400 nm region have been carried out by using a conventional heat-pipe oven. An experimental procedure which has been proven to be effective in reducing the experimental uncertainty is described. The trend of absorption cross sections, i.e., increasing as the photon energy increases in the 50-70 nm region, supports theoretical results which indicate that the correlation effects between the 4s and 3p electrons are important though the correlation effects may not be strong enough to significantly alter the total oscillator strength of every individual orbital electron. No vibrational structure associated with the photoionization of molecular potassium in the UV region is observed. Further experiments utilizing different technique are in progress.

Recently in our laboratory we have initiated a program to study the photoabsorption processes of metal vapors throughout the UV and EUV region. Our research interests are (1) to obtain the absolute cross section measurement of atomic and molecular metal vapors, (2) to study the photoionization processes of molecular metal species, and (3) to study the photodissociation processes of molecular metal ions. Several experimental methods such as the heat-pipe absorption spectroscopy, photoionization mass spectroscopy, and electron-ion coincidence technique, will be used in the study. This report summarizes our first experiment using heat-pipe absorption spectroscopy.

Experimental

The photoabsorption of atomic and molecular potassium were measured in the extreme ultraviolet (EUV, 50-70 nm) and ultraviolet (UV, 210-400 nm) region. In the first experiment the

0097-6156/82/0179-0043$05.00/0

atomic line emissions of Ar, O, and N ions were produced in a repetitive spark discharge and were used as the light source. An Aℓ metal thin-film was used (1) to separate the heat-pipe absorption cell from the vacuum monochromator and (2) as an optical window with transmission in the 17-76 nm region. In the second experiment, using a 1KW Hg arc lamp as the light source, the absorption in the 210-400 nm region was measured. A quartz window was used to separate the heat-pipe cell and the monochromator.

A heat-pipe tube (1), 4 feet long and 3-1/3 inches in diameter, is made of 304 stainless steel and has a wall thickness of 65 mils. The temperature profile of the heat-pipe oven is measured by 21 thermocouples attached to the outside surface of the heat-pipe and by 8 movable thermocouples inside the heat-pipe. The heat-pipe was typically operated at a pressure of 0.1 Torr with a 50 cm long isothermal zone. Figure 1 shows a measured temperature profile. Since the directly measured temperature at the center region is high, caused by direct exposure of the thermocouples to the heater, the temperature in this region (dotted line) is interpolated from the temperature readings at both ends. A constant temperature for the metal at the zone is assumed. This is confirmed by nearly constant temperature readings obtained with the movable thermocouples inside the heat-pipe.

When the heat-pipe is in a steady state the pressure of the metal vapor is equal to that of the buffer gas. The vapor pressure in the isothermal zone is thus determined by measuring the buffer gas pressure with an MKS Baratron capacitance manometer.

The photoabsorption cross section is measured by the split-beam technique, i.e., by simultaneously monitoring the light intensities both before and after the photon beam passes through the heat-pipe oven. The spectrometer bandwidth was set at 0.08 nm in the present measurement.

Data Reduction

The main difficulties in the absolute photoabsorption cross section measurements are the determination of the number densities of the metal vapors, the presence of metal clusters, and the outgassing problem. The number density can now be fairly accurately determined by using the heat-pipe technology. The presence of a small percentage of metal dimer or other outgassing products could lead to a noticeable uncertainty in the absolute atomic cross section determination since the molecular cross section can be as high as 10-100 times larger than that of atom (2). To minimize such uncertainty, we find that it is necessary to operate the heat-pipe oven at the lowest temperature possible $\sim$500°K (0.1 Torr He buffer gas pressure), as we have done in the present work. To further reduce the absorption due to the residual outgassing products a specific experimental procedure has been carried out in our measurements. Namely, after recording the

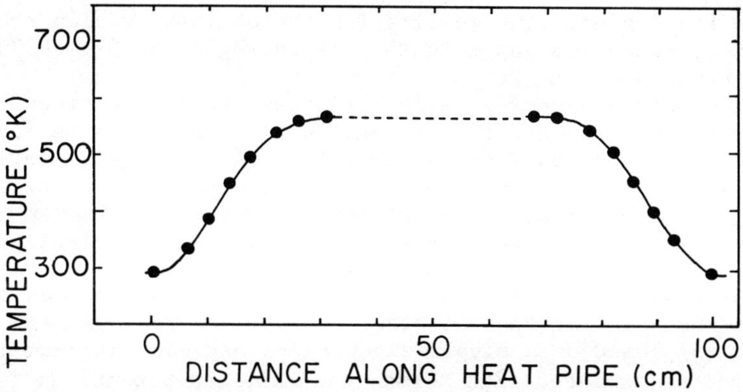

Figure 1. Temperature profile of the heat-pipe oven.

current ratio, (i/i_o), of the incident and the transmitted light intensities and the temperature and pressure readings of the heat-pipe, we rapidly cool the heat-pipe to near room temperature (typically in ∿ 20 minutes). By this time, the metal vapor pressure is reduced to practically zero. The current ratio, $(i/i_o)_f$, is again measured and assumed to contain all the outgassing products. The photoabsorption cross section is thus given by

$$\sigma = [\ln(i/i_o)_f - \ln(i/i_o)]/[n_o \int_o^L (P(x)/T(x))dx]$$

where $n_o = 9.6586$ is a constant, $P(x)$ is the pressure in Torr, $T(x)$ is the temperature in °K, and L is the absorption path length of metal vapors.

Results and Discussion

The EUV Region. The results for the photoabsorption cross sections of potassium vapor in the energy region 50.8 nm - 76.0 nm are shown in Figure 2.

The electronic configuration of potassium in its ground state is KK $3s^2 3p^6 4s$ and the thresholds for ionizing a 4s and a 3p electron are at 285.7 nm (4.339 eV) and 48.1 nm (25.74 eV), respectively. From the sum rule and the independent subshell assumption the oscillator strength from 3p will be a factor of 6 over that of 4s. In other words, the integrated photoionization cross section for the 3p electron will be a factor of six larger than that of 4s if there is no electron correlation between 3p and 4s electrons, or electrons of other subshells. The calculated cross sections (3) based on a single-electron theory and using the atomic wavefunctions computed with Herman and Skillman potentials (4) are also indicated in Figure 2. Hudson and Carter (5) and Marr and Creek (6) measured the photoionization cross section from the threshold of the 4s ionization potential to 68.0 nm and to 124.0 nm, respectively. Driver (3) measured the cross section from 50.0 to 35.0 nm, i.e., the 3p continuum. The cross sections of some of the autoionizing lines between 65.0 - 50.0 nm were also measured (4,5).

Our data mainly cover the gap between that of Hudson and Carter (5) and Driver (3). In the region where our data overlap with those of Hudson and Carter, the results are in reasonable agreement, although our data are consistently higher than those of Hudson and Carter. The reason may be due to the different experimental techniques employed.

Our measured data show that the photoabsorption cross section of potassium increases monotonically as photon energy increases. This is expected from the 4s - 3p electronic correlation effect as discussed by Chang (7). Without such a correlation effect the cross section should decrease toward shorter wavelengths as predicted by the single-electron theory.

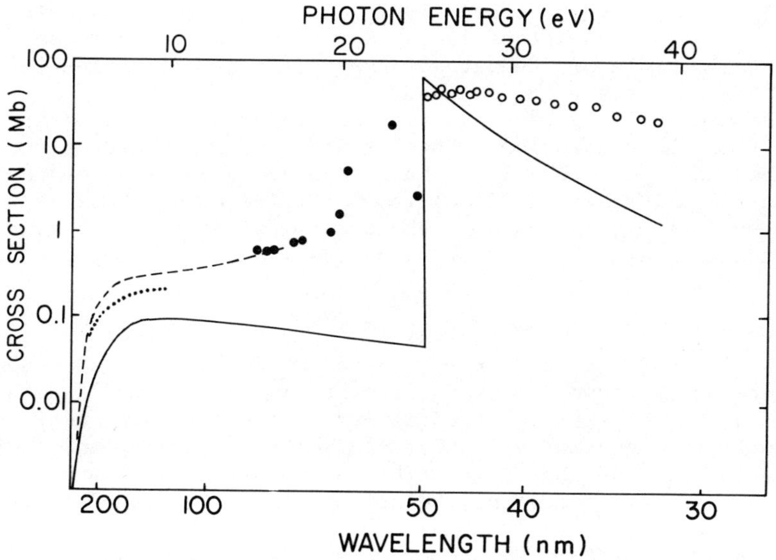

Figure 2. Summary of experimental cross sections for atomic K. Key: ●, present work; ○, Ref. 3; – – –, Ref. 5; · · ·, Ref. 6. The calculated result (——) based on a single-electron model is also shown for comparison (3).

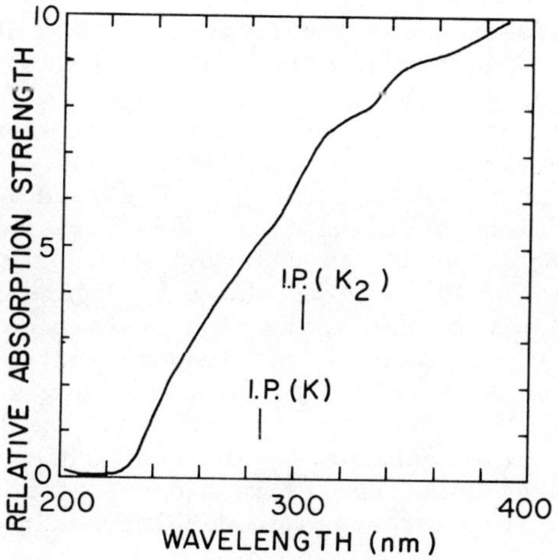

Figure 3. The cross sections of K molecule in the 210–400 nm region.

The data point at 50.8 nm measured from the present experiment seems to suggest that the cross section of the 4s continuum will join the 3p continuum with a sharp step increase. This may support the argument made by Driver that the correlation effect between 4s and 3p electrons is not strong enough to increase the sum of the oscillator strength of the 4s electron signifantly beyond the expected value of one.

The unusually large cross section measured at 55.5 nm may result from the absorption by one of the higher members of the autoionizing lines. The line shapes and cross sections of these autoionizing lines are difficult to determine using the line source such as the one used in the present experiment. A synchrotron radiation source may provide an ideal continuum light source for such detailed studies.

The UV Region. The results for the relative cross sections of potassium vapor in the energy region of 210-400 nm are shown in Figure 3. Because of strong absorptions in this wavelength region the vapor pressure of potassium in the heat-pipe could not be reduced to the point such that the heat-pipe operation condition could still be sustained. The isothermal zone is thus not established. Therefore, only relative data are measured in the present work.

Hudson and Carter (8) and Creek and Marr (2) measured the cross section of the potassium molecule up to 300 and 350 nm, respectively. The shapes of their absorption curves do not agree with our results. A constant absorption in the ∿ 230-290 nm region was observed by the previous investigators (2,8) while a sharp increase in absorption starting from ∿ 230 nm can be seen in Figure 3. The origin of this discrepancy is not known.

Recent experimental and theoretical results show the dissociation energies for the ground states of Li_2^+ (9,10), Na_2^+ (11,12, 13), and K_2^+ (12,13) are 10324, 7920, and 6404 cm^{-1}, respectively. Thus, one would expect to see some vibrational structures associated with the photoionization transition. In fact, several autoionizing Rydberg states of K_2 have been observed in a sequential two-photon ionization technique (14). However, in the total absorption plot no structure is observed on the high energy side of the ionization threshold of K_2 as shown in Figure 3. A possible explanation is that the photoionization cross sections are smaller than those of other processes, e.g., Rydberg and photodissociative states of the neutral K_2 molecule, such that any structure associated with the photoionization transition may be buried in the total absorption cross sections. A study of this problem using the photoionization mass spectrometry in a one-photon excitation process is currently in progress in our laboratory.

Acknowledgements

The author wishes to express his gratitude to Drs. T. N. Chang, D. L. Judge, C. C. Kim, and N. Shen for helpful discussions and assistance in various stages of this work.

Literature Cited

1. Vidal, C. R.; Cooper, J. J. Appl. Phys. 1969, 40, 3370-4.
2. Creek, D. M.; Marr, G. V. J. Quant. Spectrosc. Radiat. Transfer. 1968, 8, 1431-6.
3. Driver, R. D. J. Phys. B. 1976, 9, 817-27.
4. Herman, F.; Skillman, S. "Atomic Structure Calculations"; Englewood Cliffs, N.J., Prentice-Hall, 1963.
5. Hudson, R. D.; Carter, V. L. J. Opt. Soc. Am. 1967, 57, 1471-4.
6. Marr, G. V.; Creek, D. M. Proc. Roy. Soc. A. 1968, 304, 233-44.
7. Chang, T. N. J. Phys. B. 1975, 8, 743-50.
8. Hudson, R. D.; Carter, V. L. Phys. Rev. 1965, 139, A1426-8.
9. Mathur, B. P.; Rothe, E. W.; Reck, G. P.; Lightman, A. J. Chem. Phys. Letters, 1978, 56, 336-8.
10. Konowalow, D. D.; Rosenkrantz, M. E. Chem. Phys. Letters. 1979, 61, 489-94.
11. Bardsley, J. N.; Junker, B. R.; Norcross, D. W. Chem. Phys. Letters, 1976, 37, 502-6.
12. Leutwyler, S.; Hofmann, M.; Harri, H-P.; Schumacher, E. Chem. Phys. Letters, 1981, 77, 257-60.
13. Habitz, P.; Schwartz, W. H. E. Chem. Phys. Letters, 1975, 34, 248-52.
14. Leutwyler, S.; Herrmann, A.; Woste, L.; Schumacher, E. Chem. Phys. 1980, 48, 253-67.

RECEIVED August 26, 1981.

5

Interaction Between Cs*(7S, $5D_{5/2}$) and Rare Gas Atoms

B. SAYER, M. FERRAY, J. P. VISTICOT, and J. LOZINGOT

Centre d'Etudes Nucléaires de Saclay, Service de Physique Atomique, 91191 Gif-sur-Yvette, Cedex, France

Information concerning the interaction potentials of Cs*(7S, 5D)-rare gas pairs is obtained by interpreting the temperature dependence of the 6S-7S,5D far wing and satellite profiles. A sensitive laser fluorescence technique is used to obtain the absorption coefficient of the mixture. The collision induced oscillator strength, a rapidly varying function of the interatomic distance in the case of such forbidden transitions, is also deduced. Experimental potentials and oscillator strengths are compared with available calculated values.

In the quasistatic theory, the profile of the far wings of an atomic line is closely related to the interatomic potentials at relatively short interatomic distances (6-12 a.u.). According to the Franck-Condon principle, the radiative transition occurs at a wavelength λ, when the distance R between the two atoms is such that the difference between upper and lower potentials $V_u - V_l$ is equal to hc/λ. The absorbed (or emitted) light intensity per unit wavelength is thus proportional to $4\pi R^2$ f(R) exp(-V/kT) dR/dλ where V is the potential of the initial level and f(R) the oscillator strength of the transition. An Arrhenius plot (logarithm of the intensity as a function of 1/T) normally yields a straight line, the slope of which gives V(λ). The potential of the other level is then obtained by adding (or subtracting) hc/λ. If the dependence of one of the potentials versus R is known, the dependence of the other potential can be derived. Moreover the absolute measurement of the intensity of the far wing allows the determination of the oscillator strength f for each value of λ and of R if λ(R) is monotonic. This is the extension from well-known molecular spectroscopic methods of determining potentials to the transient situation of colliding atoms.

0097-6156/82/0179-0051$05.00/0

Choice of Transitions

We choose to study the forbidden transitions of the cesium atom between the ground state 6S and the excited states 7S and $5D_{5/2}$ in the presence of rare gases (RG) (1, 2, 3). The reasons for this choice are the following:

1. Alkali-rare gas interaction potentials are easier to calculate. For the pairs Cs(7S)-RG and Cs(5D)-RG, the Pascale and Vandeplanque curves (4) and those of Czuchaj and Sienkiewicz (5) are available.
2. The interaction with the perturber leads to a mixing of the wavefunctions which makes the transition slightly allowed. Under these conditions the interpretation of experimental results is a little more complicated, but information on the value of the oscillator strength of the collision induced transition can be obtained and compared with the predictions of Pascale (6) and Czuchaj (7).
3. The shapes of the calculated Cs-rare gas potential curves show an avoided crossing of the curves associated with the 7S and $5D_{5/2}, m=\frac{1}{2}$ (5d Σ) states (Figure 1). Such a situation characterizes a strong coupling between the two states and the shape of the calculated potential curves is then very sensitive to the calculation method. The comparison between theoretical predictions and experimental results is expected to be of particular interest in this range of interatomic distances.
4. In its interpretation a line wing profile is obviously assumed not to overlap with other wings. This applies to those associated with 6S-7S and $6S-5D_{5/2}, m=\frac{1}{2}$ except in a narrow spectral region for the case of helium perturbers (8).
5. Another condition, related to the possibility of simply interpreting the temperature dependence of the wing profile, is that one point in the spectrum should be related to one single point of the potential curve. In the present case this is always verified except in the neighborhood of extrema of the difference ΔV between the potentials of the upper and lower states of the transition. However this region, which corresponds to the so-called satellite, is not amenable to interpretation by the quasistatic theory and will be considered later on.

Experimental Method

To determine the temperature dependence of the far wing profile, we have either to work at high atomic density or to use a sensitive method of measurement. In the present situation the concentration of both constituents must not be too high because of the risk of strong absorption originating from either alkali dimers or alkali atoms interacting simultaneously with two perturbers. To avoid these effects we choose the following experimental conditions: $N_{Cs} \leq 10^{14}$ cm^{-3}, $N_{RG} \leq 10^{19}$ cm^{-3}. Thus it is necessary to use a very sensitive measurement method. For this purpose the absorption spectrum has been determined by a laser fluorescence technique, described in detail elsewhere (8). Briefly it is as follows: the absorption of a photon photoexcites a cesium atom which is perturbed by a rare gas atom from its ground state to the 7S or $5D_{5/2}, m=\frac{1}{2}$ state. The main de-excitation path of this atom is the radiative cascade through the only allowed transitions: 7S or 5D → 6P → 6S. To detect the absorption we observe the fluorescence at 852.1 nm ($6P_{3/2} \rightarrow 6S_{1/2}$). The sensitivity of this method is such that it is possible to detect an absorption of about 10^{-7} per cm. The experimental device is simple: the beam of a CW tunable dye laser passes through a Pyrex cell containing the cesium-rare gas mixture and the fluorescence light is detected at a right angle. The far wing profile $k(\lambda)$ is obtained when the wavelength of the laser is swept.

Results

The results obtained are shown in Figure 2 as well as Figure 1. The V_{6S} potential is deduced from the slope of the straight line representing $\log\ k(\lambda) = \phi(1/T)$ for various values of λ. The upper state potential $V_u = V_{6S} + hc/\lambda$ is plotted as a function of V_{6S}. This rather unusual representation is used here in order to take the best advantage of experimental results, the only uncertainties which appear being due to experimental error and to the validity of quasistatic theory. Experimental points are plotted in Figure 2 with inclined error bars because the uncertainties in V_u and V_{6S} are obviously the same. The points of the upper part correspond to the 6S-7S transition. It is to be noted that the depth of the potential well decreases from Xe to Ne (1000 cm^{-1} for Xe, 700 cm^{-1} for Kr, 400 cm^{-1} for Ar). If a well exists for Ne its depth is less than 50 cm^{-1}. Considering the points corresponding to $6S\text{-}5D_{5/2}, m=\frac{1}{2}$ (lower part of Figure 2), we observe that this transition occurs in a potential region of the ground state and of the excited state which is more repulsive when the rare gas is lighter. It is interesting to note that, as we hoped, the potentials are well determined in the region of the avoided crossing, except perhaps in the case of xenon for which the 6S-7S transition seems to occur a little farther than the crossing. For the three other rare gases the smallest difference between the potentials is almost independent of the nature of the perturber, about 1700 cm^{-1}. This difference seems slightly less for xenon.

In Figures 1 and 2 the calculated potentials are also

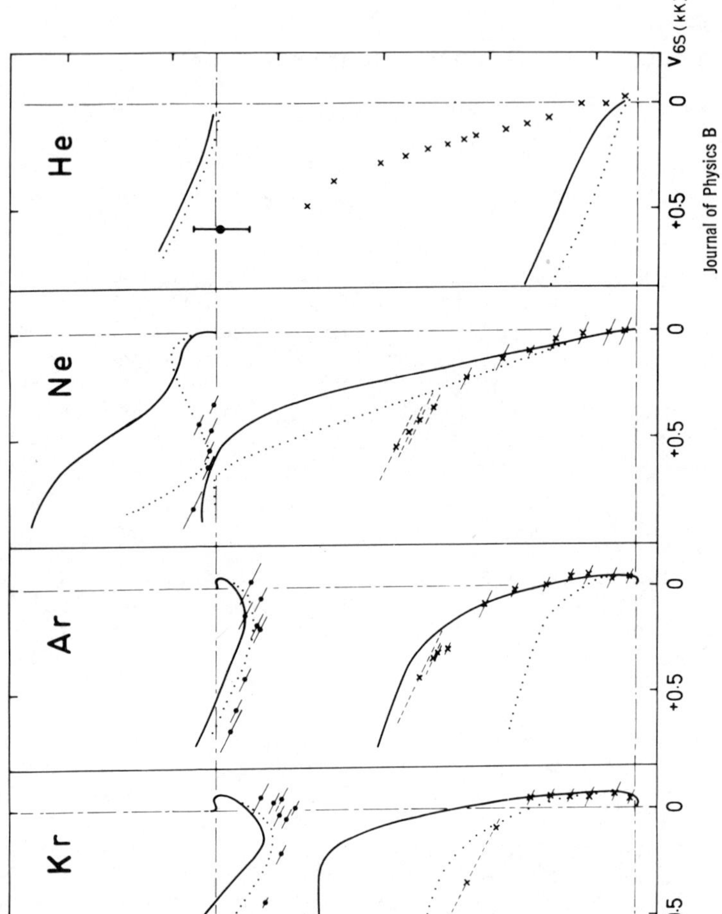

Figure 1. Potentials of the 7S (•) and $5D_{5/2,m=1/2}$ (×) states for Cs rare gas pairs as a function of the potential of the ground state (2, 3); comparison with the results in Ref. 4 (——) and in Ref. 5 (· · ·).

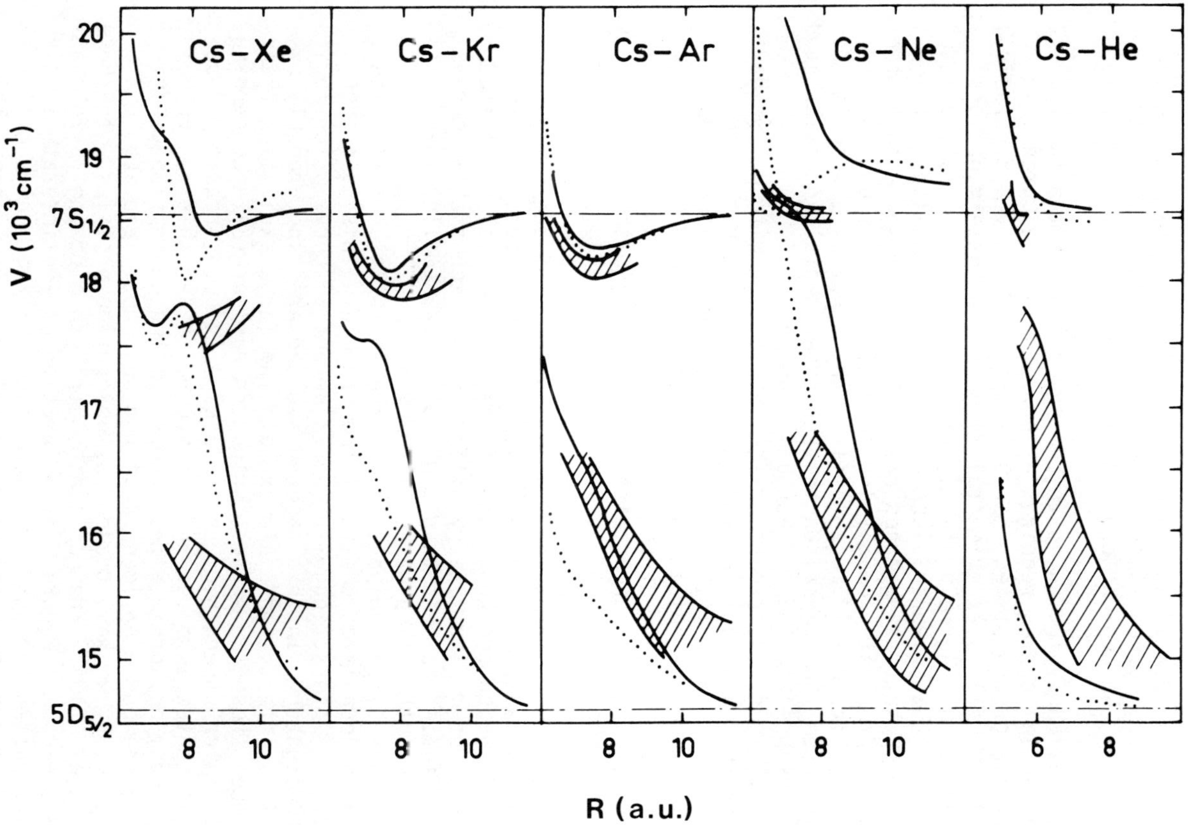

Journal of Physics B

Figure 2. Potential curves of the 7S and $5D_{5/2, m=1/2}$ states of Cs–rare gas pairs (2); comparison between experimental results (▨) and curves calculated in Ref. 4 (——) and in Ref. 5 (· · ·).

plotted. For those corresponding to $5D_{5/2,m=\frac{1}{2}}$, various situations are observed: good agreement with the Pascale-Vandeplanque potentials for Ar and with those of Czuchaj and Sienkiewicz for Kr, but all the calculated potentials appear too repulsive for Xe and Ne. In these last two cases the minimum energy between the two theoretical curves is much smaller than that determined experimentally. No avoided crossing is observed on the experimental curves.

As previously indicated the determination of the excited state potential curves requires a knowledge of the potential curve of the ground state. For this potential curve, theoretical (4, 5, 9) as well as experimental (10, 11, 12) determinations exist, but do not agree perfectly among themselves. Taking into account both these imprecisions and the experimental uncertainties, we propose cross-hatched areas in Figure 1 where the potential curves are most probably located. The plot of potential versus R in Figure 1 is more familiar than that as a function of V_{6S}, but makes the comparison between experimental and calculated curves less precise than is actually the case.

For the particular case of the interaction with He, we have to be careful when applying the quasistatic theory to such a light perturber as helium. Nevertheless the tentative use of this theory leads to the results given in Figures 1 and 2. The potentials of the $5D_{5/2,m=\frac{1}{2}}$ state differ widely from those calculated and we think that the uncertainty due to the limitations of the quasistatic theory cannot explain such a discrepancy. Very little information has been obtained for the 7S state because unlike the situation with other rare gases, the wing of the 6S-7S transition probably arises from a very narrow interatomic distance region. Although the knowledge of the 7S potential is limited to one point, the minimum energy difference between the two curves can be estimated as less than 1000 cm^{-1}, i.e. smaller than that obtained for other rare gases (1700 cm^{-1}).

In Figure 1 the 5D potential curve exhibits an inflection point in the quasistatic part of the profile. This is evidence for a strong interaction with the 7S potential curve which is its closest neighbor. In this case an avoided crossing probably exists. This interaction could explain the large excitation transfer between these two states, suspected earlier (8). Concerning the oscillator strength of the collision induced transitions between the states involved and the ground state, this can be determined from an absolute measurement of the absorption coefficient. Since our experimental method does not yield absolute values of $k(\lambda)$, we have scaled our absorption profiles with those of Moe et al. (13). The oscillator strength f of the 6S-7S transition is given in Figure 3 as a function of the interatomic distance. f passes through a maximum value which is in general well predicted by the calculations of Pascale and of Czuchaj. This value, of the order of magnitude of a few times 10^{-2}, indicates that the transition is almost allowed in a limited region of R.

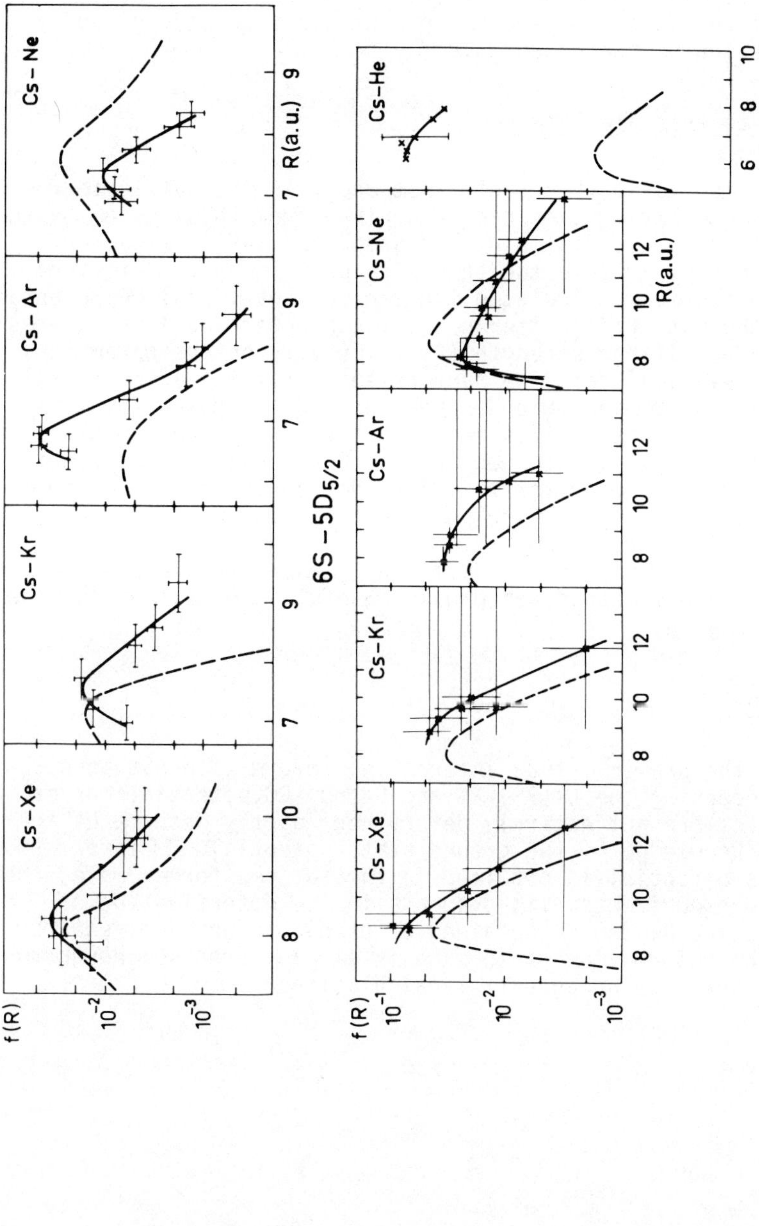

Journal of Physics B

Figure 3. Oscillator strength of the Cs(6S-7S) and Cs(6S-5D$_{5/2}$) transitions (1–3) induced by collision with various rare gases, comparison with values calculated in Ref. 6 (——) and in Ref. 7 (· · ·). The horizontal error bars are very broad when the absorption occurs in a flat part of the 6S potential curve.

The agreement with calculations is particularly good for Xe, for which the calculated potentials are relatively far from our determined values. We may note that the oscillator strength is very sensitive to the mixing of the 7S and 6P wavefunctions responsible for the transition moment, while the potential curves are mainly sensitive to the mixing of the wavefunctions of the 7S and 5D states in the region involved. Information obtained from the oscillator strength appears complementary to that given by the comparison of the potentials.

Extension to the Satellite Region

The quasistatic theory does not apply in the satellite region, but some information can be derived from this region by the use of the semiclassical method of Szudy and Baylis (14).

In a recent article the satellite profile and its temperature dependence are calculated for various analytical representations of the potential difference between upper and lower states and of the oscillator strength (15). By varying the parameters of these analytical forms we have tried to fit the experimental profile and its temperature dependence. For example, the left hand curve of Figure 4 gives the best fit of the experimental profile (dotted line) obtained for the Cs-Ar pair. The potential of the upper state used for this fit (dotted line) is plotted in Figure 4b and is seen to extrapolate to the one deduced from the quasistatic interpretation (full line). Such agreement is unobtainable (15) if other quite different parameters are used, and consequently this method can provide useful information from the satellite region.

Conclusion

From the present study interesting information has been obtained concerning the Cs(7S)-RG and Cs(5D)-RG potential curves. These curves are not entirely determined, partly because of the imprecise knowledge of the ground state potential. If information about this potential is obtained in the future, for example, from scattering experiments, improvements in the determination of $V_{7S}(R)$ and $V_{5D}(R)$ can be obtained by using the present results. However the information given here is already precise enough to guide new interaction potential calculations.

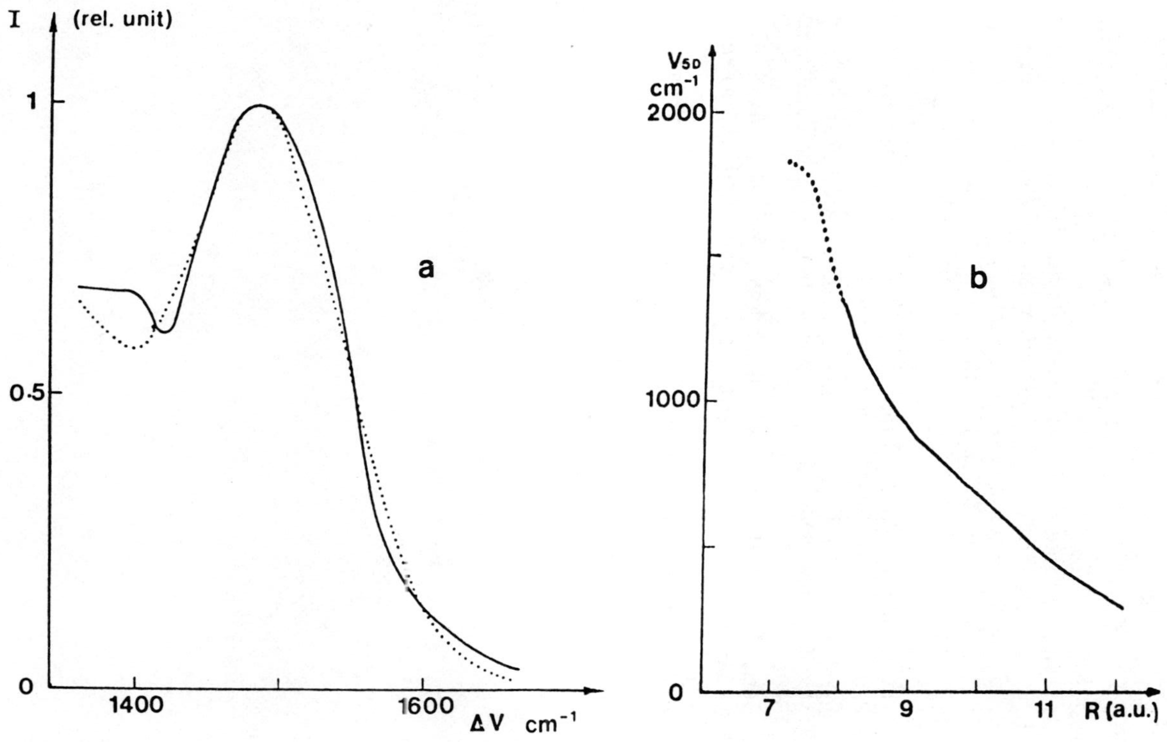

Journal of Physics B

Figure 4. a: Comparison between experimental (· · ·) and calculated (——) satellite profiles for the Cs ($6S_{1/2}$–$5D_{5/2}$)–Ar transition (15). b: Potential of the excited state corresponding to the calculation.

Literature Cited

1. Sayer, B.; Ferray, M.; Lozingot, J. J. Phys. B: Atom. Molec. Phys. 1979, 12, 227.
2. Sayer, B.; Ferray, M.; Visticot, J. P.; Lozingot, J. J. Phys. B: Atom. Molec. Phys. 1980, 13, 177.
3. Ferray, M.; Visticot, J. P.; Lozingot, J.; Sayer, B. J. Phys. B: Atom. Molec. Phys. 1980, 13, 2571.
4. Pascale, J.; Vandeplanque, J. J. Chem. Phys. 1974, 60, 2278.
5. Czuchaj, E.; Sienkiewicz, J. Z. Naturforsch. 1979, 34a, 694.
6. Pascale, J. J. Chem. Phys. 1977, 67, 204.
7. Czuchaj, E. Z. Physik 1979, A292, 109.
8. Sayer, B.; Ferray, M.; Visticot, J. P.; Lozingot, J. J. Chem. Phys. 1978, 68, 3618.
9. Baylis, W. E. J. Chem. Phys. 1969, 51, 2665.
10. Buck, U.; Pauly, H. Z. Physik 1968, 208, 390.
11. Malerich, C. J.; Cross, R. J. J. Chem. Phys. 1970, 52, 386.
12. Hedges, R. E. M.; Drummond, D. L.; Gallagher, A. Phys. Rev. A 1972, 6, 1519.
13. Moe, G.; Tam, A. C.; Happer, W. Phys. Rev. A 1976, 14, 439.
14. Szudy, J.; Baylis, W. E. J. Quant. Spectrosc. Radiat. Transfer 1975, 15, 641.
15. Visticot, J. P.; Szudy, J.; Sayer, B. J. Phys. B (in press).

RECEIVED August 26, 1981.

6

Numerical Model for the Effect of a Spatial Temperature Gradient on Chemical Reactions in a Knudsen Gas

J. P. WOERDMAN and S. S. ESKILDSEN[1]
Philips Research Laboratories, Eindhoven, The Netherlands
W. J. J. REY
Philips Research Laboratory, Brussels, Belgium

We derive a numerical model for the effect of a spatial temperature gradient on the local equilibrium of a chemical reaction in a low--density gas (Knudsen regime). The gas consists of two constituents and the chemical reaction is assumed to take place at the walls of the container. The numerical results are compared with experimental results on the equilibrium 2Na - Na_2. From the comparison it follows that the chemical accommodation coefficient for a Na_2 wall collision is essentially equal to 1.

The work described here was motivated by a number of experiments (1,2) by the first author, showing that for sodium vapor in glass cells the chemical reaction 2Na - Na_2 occurs at the walls of the cell. From these experiments information on surface catalysis can be deduced, which is not easily obtained otherwise. Therefore the need was felt for a numerical model for a quantitative interpretation of the experimental data.

We will first briefly describe the experiments. A glass cell contains liquid sodium metal in an appendix at temperature T_0 (see inset of Figure 1). The main body of the cell has a cylindrical shape (length L, diameter d) and is at a temperature $T > T_0$. The sodium vapor contained in the cylinder consists of Na atoms and Na_2 molecules. Typically, in the temperature range of interest (T_0 about 500 K and T = 530 K - 620 K) the sodium vapor consists almost exclusively of Na atoms; the fraction of Na_2 molecules is 5×10^{-3} - 5×10^{-4}. In the experiments the sodium vapor density is low enough to ensure that the mean free path for two-body collisions of the types Na - Na and Na - Na_2 is larger than the dimensions of the cylinder. We are therefore in the "Knudsen regime": the Na and Na_2 particles move through the cylinder like "billiard balls", only colliding with the walls. The relative density of Na_2 molecules is measured locally in the cylinder using Na_2 fluorescence excited

[1] Current address: Institute of Physics, University of Aarhus, Aarhus, Denmark.

0097-6156/82/0179-0061$05.00/0

by an argon laser beam. An axial temperature difference $T_2 - T_1$ across the cylinder gives rise to a density gradient of the Na_2 molecules. The value of this density gradient depends on the efficiency of the surface to catalyze the reaction. We will now introduce a numerical model to deduce this efficiency from the measured density gradients.

Adsorption, scattering and desorption

We assume that Na atoms and Na_2 molecules hitting the surface are either directly scattered back into the volume or are adsorbed, chemically equilibrated ($2Na \longleftrightarrow Na_2$) at the surface and subsequently desorbed. We introduce the chemical accommodation coefficient p (3); p is the probability that a wall collision proceeds through the "reactive" adsorption-desorption channel and $1-p$ the probability that it proceeds through the "non-reactive" scattering channel. p_1 and p_2, referring to incident Na atoms and Na_2 molecules, respectively, are the free parameters of the model. p_1 and p_2 are assumed to be independent of surface temperature T, angle of incidence θ and particle velocity v_{th}. The Na_2 fraction of the particle flux ($cm^{-2}\ s^{-1}$) desorbing from a surface element dS_j at temperature T_j is q_j, the Na fraction $1-q_j$. In our model

$$q_j = QT_j^{1/2} \exp(E_{diss}/kT_j) \tag{1}$$

where $E_{diss}/k = 8664K$ (4) and the prefactor Q is a constant, independent of T_j. For a cell at uniform temperature T, eq. (1) leads to a Na_2 volume density given by $q/v_{th} \propto \exp(E_{diss}/kT)$.

We introduce Φ_{1j} and Φ'_{1j} as the Na fluxes ($cm^{-2}\ s^{-1}$) incident on, respectively leaving from a surface element dS_j, integrated over the solid angle 2π; likewise, Φ_{2j} and Φ'_{2j} are introduced as Na_2 fluxes (see Figure 2a). In a stationary situation we have

$$\Phi'_{1j} = (1-p_1)\Phi_{1j} + (1-q_j)p_1\Phi_{1j} + 2(1-q_j)p_2\Phi_{2j} \tag{2}$$

$$\Phi'_{2j} = (1-p_2)\Phi_{2j} + q_jp_2\Phi_{2j} + \tfrac{1}{2}q_jp_1\Phi_{1j} \tag{3}$$

Assuming that p_1 and p_2 are of the same order of magnitude and using that $\Phi_{1j} \gg \Phi_{2j}$, $q_j \ll 1$, leads to decoupling of the Na and Na_2 fluxes

$$\Phi'_{1j} = \Phi_{1j} \tag{4}$$

$$\Phi'_{2j} = (1-p_2)\Phi_{2j} + Aq_j \tag{5}$$

From particle number conservation it follows that Φ_{1j} and therefore $A(\equiv \tfrac{1}{2}p_1\Phi_{1j})$ is constant, i.e. independent of T_j. Note further that for physical reasons $p_1 = 0$, and therefore $A = 0$, if we assume $p_2 = 0$.

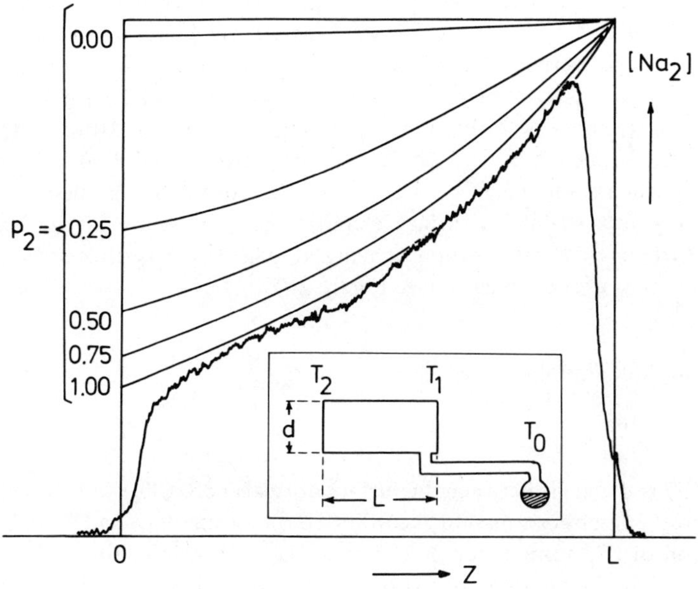

Figure 1. Experimental and calculated Na_2 gradients along the axis of a cell with $L/d = 1.91$, $T_0 = 503$ K, $T_1 = 532$ K, and $T_2 = 609$ K. [Na_2] refers to the fraction of molecules in the level $v = 4$, $J = 39$. The parameter p_2 is the chemical accommodation coefficient of a Na_2 surface collision.

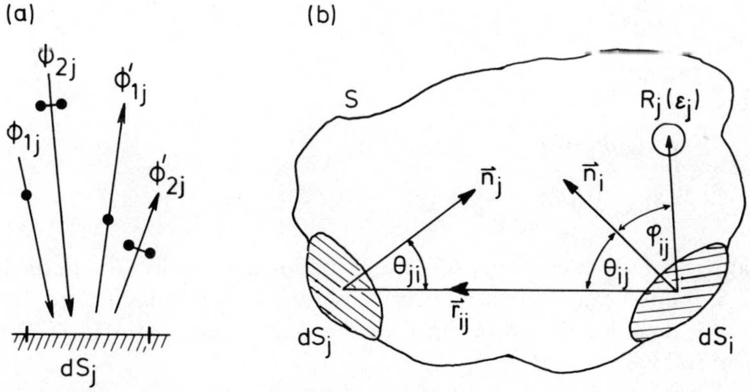

Figure 2. a: Surface element dS_j with incident particle fluxes ϕ_{1j} (Na) and ϕ_{2j} (Na_2) and leaving particle fluxes ϕ'_{1j} (Na) and ϕ'_{2j} (Na_2). b: Arbitrary surface S. Surface normals $\vec{n}_i$ and $\vec{n}_j$ are erected at the surface elements dS_i and dS_j. The point R_j is situated in the volume; it is surrounded by a sphere of radius ϵ_j.

For the angular distribution of the Na_2 molecules leaving the surface we will assume the Knudsen Cosine Law to be valid, that is the emitted flux varies as $\cos\theta$, where θ is the angle between the surface normal and the flux direction. This cosine dependence is a necessary consequence of the Second Law of Thermodynamics for the case $p_2 = 1$ (i.e. only desorption, no scattering), see e.g. (5). Although for $p_2 < 1$ the Cosine Law is not necessarily satisfied by the scattered and desorbed Na_2 fluxes separately we will assume this to be the case (see (2) for further discussion).

The Na_2 flux striking a surface element dS_j is made up of the Na_2 fluxes leaving all surface elements dS_i and arriving at dS_j. The Na_2 flux received at dS_j coming from dS_i is proportional to (see Figure 2b)

$$\Phi'_{2i} \frac{\cos\theta_{ji}\cos\theta_{ij} dS_i}{(r_{ij})^2} \tag{6}$$

where θ_{ij} (θ_{ji}) is the angle between the surface normal at dS_i (dS_j) and the connecting vector r_{ij} ($\cos\theta_{ij}$ describes emission according to the Cosine Law and $\cos\theta_{ji}$ describes the projection of dS_j normal to r_{ij}). The total Na_2 flux striking the surface element dS_j is obtained by integrating eq. (6) over the whole cell surface. If this surface is divided into a number of finite elements ΔS_i we can write

$$\Phi_{2j} = \frac{1}{S_j} \sum_i \int_{\Delta S_i} \frac{\Phi'_{2i}\cos\theta_{ji}\cos\theta_{ij}}{(r_{ij})^2} dS \tag{7}$$

where S_j is a normalization constant. We define a matrix S with elements s_{ij} by writing

$$\Phi_{2j} = \sum_i s_{ij}\Phi'_{2i} \tag{8}$$

$$s_{ij} = \frac{1}{S_j} \int_{\Delta S_i} \frac{\cos\theta_{ji}\cos\theta_{ij}}{(r_{ij})^2} dS \tag{9}$$

(note that $s_{ii} = 0$ if the elements ΔS_i are infinitesimally small). The normalization constants S_j follow from particle number conservation: if Φ'_{2i} is equal to some constant for all i, then Φ_{2j} should be equal to the same constant (for all j). From this requirement follows

$$\sum_i s_{ij} = 1 \tag{10}$$

(If the elements ΔS_i are infinitesimally small we find $S_j = \pi$). Rewriting eqs (5) and (8) in matrix notation gives an implicit equation for Φ_2

$$[I + (p_2 - 1)S]\vec{\Phi}_2 = SA\vec{q} \quad (11)$$

where I is the identity matrix. The (square) matrix S depends only on the geometry of the cell and the vector $\vec{q}$ is determined by the chemical equilibrium constant at the local wall temperature according to eq. (1). Thus the Na_2 flux striking each surface element ΔS_i can be calculated from eq. (11).

The Na_2 volume density

We now proceed to find the expression for the Na_2 density (cm^{-3}) in a volume point R_j (see Figure 2b). This density is denoted d_{2j}. We consider a small sphere of radius ϵ_j centered at R_j. The number of Na_2 molecules passing per second through the sphere around R_j and coming from the surface element dS_i is proportional to

$$\Phi'_{2i} \frac{\cos\varphi_{ij} dS_i}{(r_{ij})^2} \quad (12)$$

In comparison with eq. (6) the factor $\cos\varphi_{ji}$ is missing in eq. (12) since the projected cross-sectional area of the sphere is constant (namely $\pi\epsilon_j^2$), independent of the direction of r_{ij}. We assume that Na_2 molecules leaving a surface element dS_i (scattering + desorption) have $v_{th} \propto T_i^{1/2}$. The Na_2 density at R_j is then given by

$$d_{2j} = \frac{1}{V_j} \sum_i T_i^{-1/2} \int_{\Delta S_i} \frac{\Phi'_{2i} \cos\varphi_{ij}}{(r_{ij})^2} dS \quad (13)$$

where V_j is a normalization constant. We define a matrix V with elements v_{ij} by writing

$$d_{2j} = \sum_i T_i^{-1/2} v_{ij} \Phi'_{2i} = \sum_i T_i^{-1/2} v_{ij} (Aq_i + (1 - p_2)\Phi_{2i}) \quad (14)$$

$$v_{ij} = \frac{1}{V_j} \int_{\Delta S_i} \frac{\cos\varphi_{ij}}{(r_{ij})^2} dS \quad (15)$$

The constants V_j follow from a consideration similar to that leading to eq. (10)

$$\sum_i v_{ij} = 1 \quad (16)$$

(if the elements ΔS_i are infinitesimally small we find $V_j = 4\pi$).

The procedure to calculate the Na_2 volume densities d_{2j} is now as follows: Φ_{2i} is calculated from eq. (11), using eqs (1, 9, 10), and is then substituted in eq. (14) to calculate d_{2j}, using eqs (1, 15, 16). Note that AQ acts as a scaling factor, the value of which is irrelevant, since we are interested in relative densities d_{2j} only.

The relative Na volume densities d_{1j} can be calculated using the same formalism. Assuming that a Na atom leaving a surface element dS_i has $v_{th} \propto T_i^{1/2}$ we find

$$d_{1j} = \sum_i T_i^{-1/2} v_{ij} \quad (17)$$

Results

For the numerical calculation we usually divided the cell surface (cylinder + end windows) into 14 elements of roughly equal area. Finer surface division yielded essentially the same results. The Na_2 density was calculated for a grid of volume points R_j in cylindrical cells with L/d = 0.48, 0.90, 1.91, 3.14 and 18.7. The temperature distribution T_j across the cylinder surface was taken from experiment (typically $\Delta T = 70$ - 80K between the end windows). The calculations were done for $p_2 = 1.00$, 0.75, 0.50, 0.25 and 0.00. In the experiment we measure the density of Na_2 in the level v = 4, J = 39 (fluorescence excited by the 465.8 nm argon laser line) instead of the total Na_2 density (2). In order to calculate $[Na_2]_{4,39}$ we have multiplied q_j, taken from eq. (1), by the correction factor (6)

$$\frac{[Na_2]_{v,J}}{[Na_2]_{total}} = \frac{(1-\exp(-E/kT_j))}{\exp(vE/kT_j)} \frac{B}{kT_j}(2J+1)\exp(-BJ(J+1)/kT_j) \tag{18}$$

with v = 4, J = 39, E = 159 cm^{-1}, B = 0.15 cm^{-1}, k = 0.6952 and T_j in K. The corrected value of q_j was then substituted in eqs (11, 14). The fraction of Na_2 in v = 4, J = 39 varies between 1.06×10^{-3} to 1.12×10^{-3} in the relevant temperature range (530 - 620 K).

Figure 1 shows $[Na_2]_{4,39}$ calculated along the axis of a cell with L/d = 1.91, for $\Delta T = 77$ K; the Na_2 gradient is independent of p_1, as is evident from eqs (4, 5). For $p_2 = 0.00$ a very small Na_2 gradient results; it is due to non-reactive wall collisions, which have been assumed to be thermalizing. The calculated Na_2 gradient is maximum for $p_2 = 1$; this gives the best fit with experiment. More computational results of the model are given in (2), together with experimental results for comparison. All experimental results can be fitted with $p_2 = 1$. See also (2) for a critical discussion of the various assumptions made in the derivation of the present model.

Acknowledgment

We thank K. Weiss for helpful discussions.

Literature Cited

1. Woerdman, J.P. J. Mol. Structure 1980, 61, 221.
2. Woerdman, J.P. Chem. Phys. (submitted).
3. Langmuir, I. J. Am. Chem. Soc. 1916, 38, 1145.
4. Verma, K.K.; Vu, T.H.; Stwalley, W.C. J. Mol. Spectroscopy 1981, 85, 131.
5. Clausing, P. Ann. Physik 1930, 4, 533.
6. Herzberg, G. "Molecular Spectra and Molecular Structure"; van Nostrand Reinhold, New York, 1950, 2nd ed. p. 121.

RECEIVED August 26, 1981.

METALLIC BONDING EXEMPLIFIED BY ALKALI AND OTHER SMALL METAL CLUSTERS

ESR of Stationary and Pseudorotating Alkali Trimer Molecules

D. M. LINDSAY, DENISE GARLAND, FRANCES TISCHLER, and G. A. THOMPSON

City University of New York, Department of Chemistry, City College, New York, NY 10031

The relaxation theory of ESR linewidths is specialized to the case of pseudorotating K_3. Trimer molecules are assumed to fluctuate randomly between the three stable configurations which correspond to static Jahn-Teller distortions from D_{3h} symmetry. A simulated ESR spectrum is shown to be in quite good agreement with the linewidth alternation observed for K_3 in argon at low temperatures. Measured linewidths give approximate correlation and transverse relaxation times, 10^{-8} sec $\lesssim \tau_c \lesssim 10^{-10}$ sec and 10^{-8} sec $\lesssim T_2 \lesssim 10^{-6}$ sec, for the temperature range 4.2 K $\lesssim T \lesssim 20$ K. The temperature dependence of τ_c is well represented by either exponential or power law behavior. Possible pseudorotation mechanisms, including a temperature dependent tunneling, are outlined.

Small metal clusters have received considerable attention because of their possible involvement as "active sites" in a variety of catalyzed reactions. Although not particularly noted for their catalytic activity, alkali clusters have a simple chemical composition and may, therefore, model the more complicated systems in a manner analogous to the role played by the hydrogen atom in atomic structure. Less emphasized is the fundamental nature of alkali clusters per se. Since the ground state of H_3 is not chemically bound, alkali trimers are the most elementary species which can exhibit a Jahn-Teller interaction.

In previous electron spin resonance (ESR) studies of matrix isolated Na_3 (1, 2) and K_3 (3, 4), alkali trimers have been shown to be chemically bound and well described both by simple bonding ideas (1, 3) and by the more sophisticated calculations recently employed for Li_3 (5), Na_3 (6) and K_3 (7). For the potassium trimer in argon, two distinct ESR spectra are observed (3). An obtuse angled isomer corresponds to one of three static Jahn-Teller distortions from D_{3h} symmetry, and is surprisingly similar to the

0097-6156/82/0179-0069$05.00/0

structure reported for Na_3 (1). A "pseudorotating" isomer is both qualitatively and quantitatively consistent with fast exchange between the three "stationary" trimer geometries. In a preliminary report on the trimer lineshape, the pseudorotating spectrum was shown to exhibit an "alternating linewidth" effect, as would be expected for an exchanging radical (4). In this paper, trimer linewidths are reported and discussed from the viewpoint of a purely phenomenological model involving a stochastic modulation of the trimer hyperfine splitting. Linewidth expressions, derived for the particular case of pseudorotating K_3, lead to a simulated ESR spectrum in quite good agreement with that observed. Approximate values for the trimer correlation and transverse relaxation times are computed. Possible exchange mechanisms are outlined in the last section.

Experimental

The experimental details have been reported elsewhere (1, 3). Briefly, matrices are formed by codeposition of excess argon with atomic potassium on a sapphire plate mounted inside an ESR cavity which is itself attached to a variable temperature liquid helium dewar. Cluster formation occurs during deposition and is accomplished by warming the sapphire surface above a nominal deposition temperature of 4.2 K. For spectra shown here, temperature measurements were made with a calibrated carbon resistor and are judged accurate to within ± 5%.

Trimer Spectra

Alkali trimer molecules have a relatively flat potential surface, whose principal features are a conical peak surrounded by a trough containing three potential wells (3, 5, 6). Figure 1 shows the trimer potential surface plotted as a function of Q_x and Q_y, the doubly degenerate pair of normal mode coordinates for D_{3h} symmetry (3). The central peak corresponds to a degenerate, $^2E'$ ground state of equilateral triangular geometry. The three potential minima have obtuse angled isosceles geometry ($^2B_2$ symmetry) and arise from three possible Jahn-Teller distortions from D_{3h} symmetry. Adjacent minima are separated by relatively low energy saddle points of acute isosceles geometry ($^2A_1$ symmetry). As shown by the broken line in Figure 1 this situation permits the intraconversion of obtuse triangular forms without passing through the energetically remote equilateral triangle ("pseudorotation").

In Reference 3, two distinct ESR spectra were identified for potassium clusters in argon matrices. Both spectra have doublet ground states ($S = \frac{1}{2}$) and a well resolved hyperfine (hf) structure arising from the Fermi contact interaction of the unpaired electron spin with three $I = \frac{3}{2}$ nuclei. Seven groups of four transitions each were assigned to a potassium trimer of $^2B_2$ symmetry whose apical and two equivalent terminal atoms have hf splitting

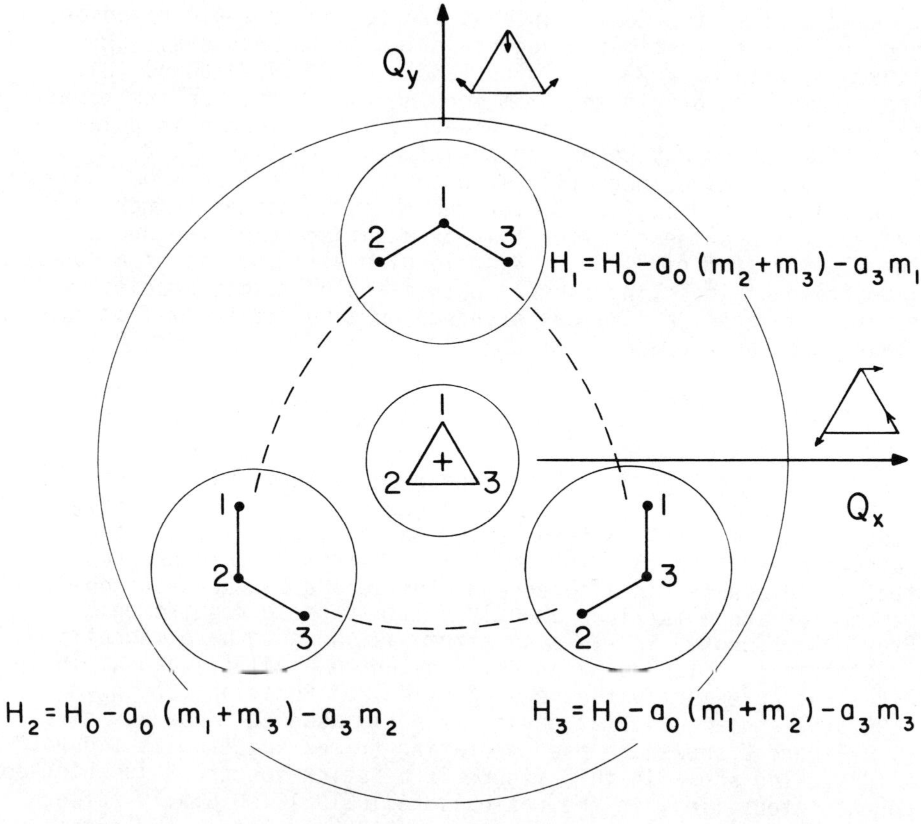

Figure 1. Normal coordinate and real space description of pseudorotating K_3. Circles represent energy contours for the trimer potential surface (3).

constants a_3 and a_0, respectively. Powder spectra were fit to

$$H_i = H_0 - a_0 M - a_3 m_i \qquad (1)$$

in which $m_i = m_1$, m_2 or m_3 is the spin projection quantum number for the apical nucleus and M is the sum of the corresponding quantities for the remaining two atoms. The term $H_0 = g_e/g_0 \, H_e$, where H_e and g_e denote the resonance field and g value for a free electron, describes the Zeeman interaction for an isotopic g tensor, $\tilde{g} = g_0 \tilde{1}$. For K_3 (estimated uncertainties in parentheses), g_0 = 1.9985(5) with a_0 = 39.0(3) gauss and a_3 = ± 4.7(2) gauss (3). The a_0 constant originates from bonding 4s orbitals on the terminal atoms and is positive; a_3, however, arises from spin polarization effects and may be either positive or negative (1, 2, 3).

Figure 1 gives magnetic field positions, derived from Equation (1) with i = 1, 2 or 3, for the three distorted trimer geometries. Since these differ only by a numbering of the nuclei, the ESR spectrum of "frozen" K_3 molecules will consist of a superimposition of indistinguishable obtuse angled trimer species. If, however, adjacent minima can exchange on a sufficiently fast time scale, then an average spectrum,

$$\langle H \rangle = \tfrac{1}{3} \sum_{i=1}^{3} H_i = H_0 - \tfrac{1}{3}(2a_0 + a_3)M' \qquad (2)$$

for which all three ^{39}K nuclei appear equivalent, will be observed. Since $M' = m_1 + m_2 + m_3$ ranges from $+\tfrac{9}{2}$ to $-\tfrac{9}{2}$ in integral steps, a ten line hf pattern having $\langle a \rangle = \tfrac{1}{3}(2a_0 + a_3)$ and a g value identical to that for the "frozen" trimer would be predicted. As discussed in Reference 3, the second ESR spectrum observed for argon matrices exactly fulfills these requirements. Previously denoted K_n, this spectrum was shown to have magnetic parameters g_0 = 1.9990(5) and hf constant a = 24.5(7) gauss, in excellent agreement with the predictions of Equation (2), g_0 = 1.9985(5) and a = 24.3(3) gauss, if a_3 is assumed to be negative.

Figure 2 shows the pseudorotating trimer spectrum at T = 34.2 K (4). Also shown in this figure is a "stick spectrum" in which component intensities, in the ratio 1:3:6:10:12:12:10:6:3:1, reflect the M' degeneracy of three equivalent $I = \tfrac{3}{2}$ nuclei in the high field limit (3, 4). Magnetic field positions for the simulated spectrum were computed using Equation (2) and the experimental parameters of Reference 3. A broadening of the low field transitions is observed at all temperatures and most likely arises from the trapping of radicals in more than one matrix site. Aside from this asymmetry in the linewidths, the agreement between computed and observed intensity ratios is quite good. At low temperatures, however, component intensities differ markedly from the expected 1:3:6:10:12:12:10:6:3:1 pattern. As shown for T = 4.2 K in Figure 3a the $M' = \pm \tfrac{7}{2}$, $\pm \tfrac{5}{2}$ and $\pm \tfrac{1}{2}$ components show a pronounced line broadening whereas $M' = \pm \tfrac{9}{2}$ and $\pm \tfrac{3}{2}$ remain narrow.

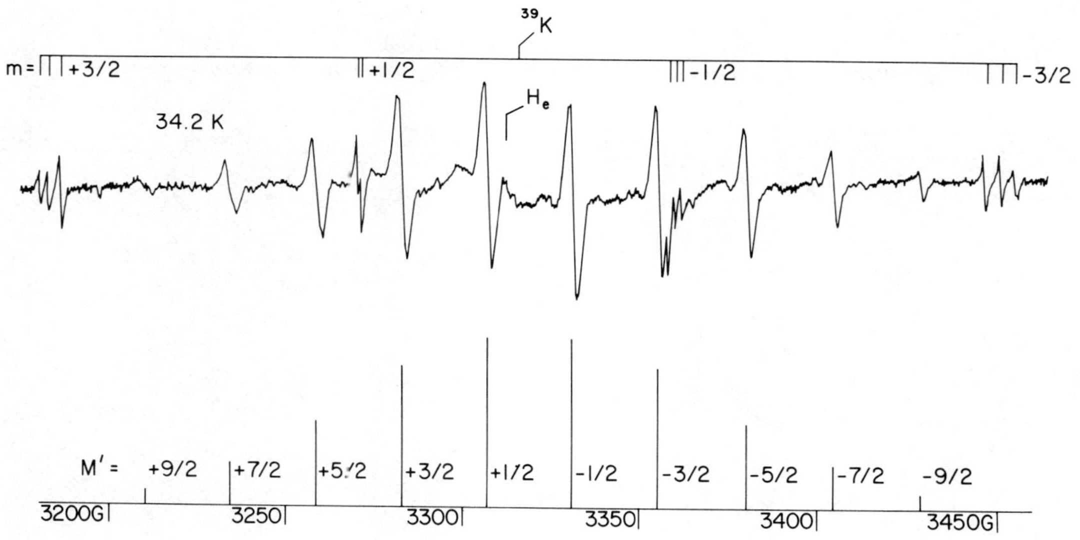

Figure 2. ESR spectrum of pseudorotating K_3 in an Ar matrix at 34.2 K. The "stick spectrum" shows the predicted intensity distribution for three equivalent ^{39}K nuclei. Also indicated are K atom resonances for several matrix sites. $H_e = 3312.3$ G is the resonance field of a free electron for a cavity frequency of 9.2824(2) GHz.

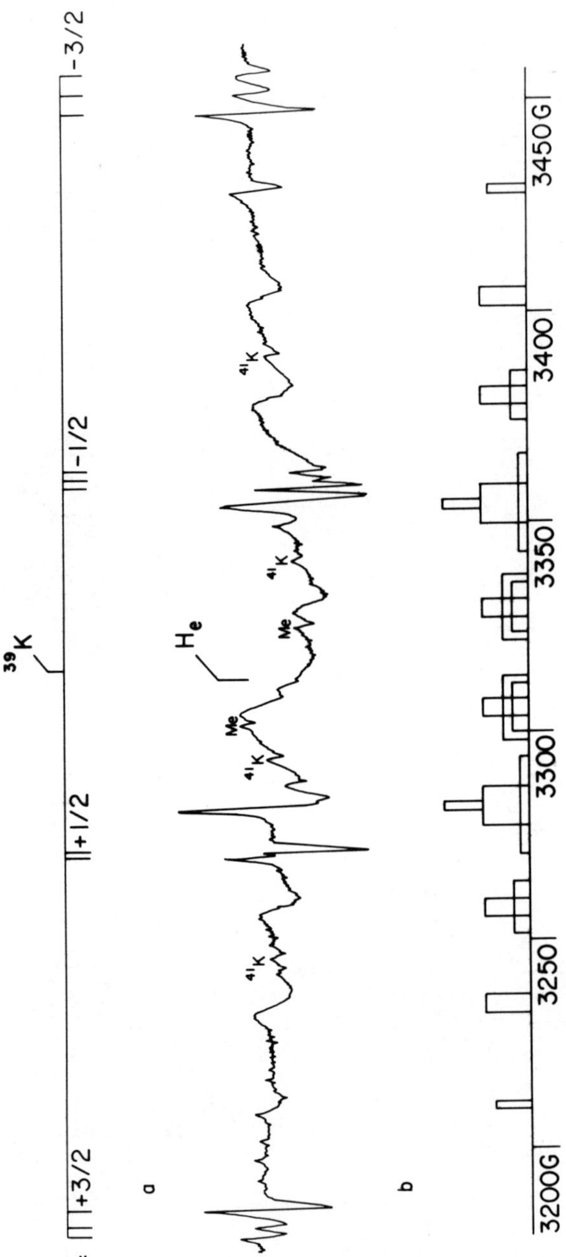

Figure 3. Comparison of observed (a) and simulated (b) spectra for pseudorotating K_3 in Ar matrices at 4.2 K. See also Fig. 2.

In Reference 4, this "alternating linewidth" effect (8) was shown to be consistent with a temperature dependent pseudorotation in K_3. A correlation diagram, which relates transition fields for the three obtuse angled trimer geometries, qualitatively predicts line broadening for $M' = \pm \frac{7}{2}, \pm \frac{5}{2}$ and $\pm \frac{1}{2}$ when pseudorotation is relatively slow. At high temperatures fast exchange leads to the motionally averaged spectrum shown in Figure 2.

Trimer Linewidths

The linewidth analysis for pseudorotating K_3 closely follows the more familiar treatment of relaxation mechanisms in both the NMR and ESR of chemically exchanging systems (8 - 12). Physically, the electron spin interacts with both the static, laboratory magnetic field and a random, fluctuating magnetic field arising from an isotropic hf coupling with the nuclear spins of the trimer. Hyperfine coupling constants are not explicitly time dependent. Rather they assume different values according as to possible occupancies of the three $^2B_2$ minima shown in Figure 1. Thus the hf interaction is time dependent only if the trimer is assumed to "jump" randomly from one well to another.

Two terms may contribute to the trimer linewidth. Diagonal elements of the hf interaction produce a secular modulation of the magnetic levels of both electron spin states without causing transitions. Off diagonal elements, which cause (lifetime broadening) transitions between Zeeman levels, give rise to a non-secular contribution. For either effect to be significant, the hf interaction must have "sufficient" magnitude. Non-secular terms face the additional requirement that the Fourier spectrum of the interaction be rich at the Larmor frequency, $\omega_0/2\pi = 9.3$ GHz. For K_3 in argon, the exchange rate for jumping between adjacent $^2B_2$ minima is never sufficiently fast for non-secular terms to make an appreciable contribution.

Several formalisms have been applied to relaxation in exchanging radicals. Principal among these are modifications of the classical Bloch equations (8, 12) and the more rigorous quantum mechanical theory of Redfield *et al.* (8 - 11). When applied in their simplest form, as in the present case for K_3, both approaches lead to the same result. Since the theory has been elegantly described by many authors (8 - 12), only those details which pertain to the particular example of K_3 will be presented here. Secular terms contribute to the ESR linewidth (Γ) and transverse relaxation time (T_2) by an amount

$$\Gamma = T_2^{-1} = \int_0^\infty G(\tau)\, d\tau \qquad (3)$$

where the correlation function $G(\tau) = \langle h_i h_j \rangle$ is the appropriate ensemble average for exchanging trimer species, and,

$$h_i = H_i - \langle H \rangle \qquad (4)$$

is defined, through Equations (1) and (2), in order to make $\langle h_i \rangle$ vanish. Although Equation (3) is strictly applicable to absorption spectra having Lorentzian lineshapes, its use in connection with a powder spectrum will have only a minor effect on the conclusions of this paper. The several approximations applied to deriving Equation (3) are well described in Reference 10. Most significant is the neglect of perturbation terms past second order. As a consequence, expressions derived in this section will be valid over most, but not all, of the temperature range employed in the ESR experiments.

Figure 4 shows the "three jump model" used in deriving $G(\tau)$ for pseudorotating K_3. Jumping between adjacent wells (i = 1, 2 or 3) is assumed to be a stationary, Markov process which may be characterized by a single, phenomenological rate constant λ. The three potential minima are assumed to be identical so that the probability of occupying any one well at a particular time τ is given by $P_i(\tau) = \frac{1}{3}$. Under these circumstances

$$G(\tau) = \sum_{i,j=1}^{3} h_i h_j P_i(o) P_{ij}(\tau) \tag{5}$$

where the $P_{ij}(\tau)$ are conditional probabilities relating the occupancies of wells i and j at the times $\tau = o$ and $\tau = \tau$, respectively. The $P_{ij}(\tau)$ may be formally derived by application of the Kolmogorov equations of probability theory (13). For the situation represented in Figure 4,

$$\begin{aligned} dP_{ii}/dt &= - 2\lambda P_{ii} + 2\lambda P_{ij} \\ dP_{ij}/dt &= - \lambda P_{ij} + \lambda P_{ii} \end{aligned} \tag{6}$$

analogous to the more familiar expressions often encountered in the elementary theory of chemical reactions. For initial conditions $P_{ii}(o) = 1$ and $P_{ij}(o) = 0$, the simultaneous Equations (6) have a solution

$$\begin{aligned} P_{ii}(\tau) &= \tfrac{1}{3}[1 + 2e^{-\tau/\tau_c}] \\ P_{ij}(\tau) &= \tfrac{1}{3}[1 = e^{-\tau/\tau_c}] \end{aligned} \tag{7}$$

where $\tau_c = (3\lambda)^{-1}$ is a characteristic, correlation time for trimer exchange. Substitution of 4 and 7 in the expression 5 gives

$$G(\tau) = \tfrac{2}{9}(a_o = a_3)^2 f(M', k) e^{-\tau/\tau_c} \tag{8}$$

where

$$f(M', k) = M'^2 = e(m_1m_2 + m_1m_3 + m_2m_3) \tag{9}$$

with $k = m_1m_2m_3$. The function $f(M', k)$ is listed in Table I.

Table I. Parameters appearing in linewidth variation of pseudorotating K_3, Equation (10) of text.

M'	k	f(M', k)	D_k
±9/2	27/8	0	1
±7/2	9/8	1	3
±5/2	3/8	1	3
	-9/8	4	3
±3/2	1/8	0	1
	-3/8	3	6
	-27/8	9	3
±1/2	-1/8	1	3
	3/8	4	3
	-9/8	7	6

Also shown in this table are the number (D_k) of transitions (m_1, m_2, m_3) which contribute to a given k. Applying Equation (3) to individual hf components M' = $+\frac{9}{2}$,--,$-\frac{9}{2}$ gives

$$(M', k) = \tfrac{2}{9}(a_0 - a_3)^2 \cdot f(M', k) \cdot \tau_c \qquad (10)$$

where the last expression has been notated to show that exchange contributions to the trimer linewidth will, in general, be a sum of Lorentzian functions k. Equation (10) also shows that ESR linewidths are temperature dependent only through a corresponding variation in the correlation time τ_c.

Results for K_3

Table II shows the temperature variation in the peak-to-peak

Table II. Measured linewidth variation for pseudorotating K_3.

	Γ(gauss)			Γ(gauss)	
T(K)	M' = $-\frac{7}{2}$	M' = $-\frac{5}{2}$	T(K)	M' = $-\frac{7}{2}$	M' = $-\frac{5}{2}$
4.2	4.55	4.85	15.2	2.67	2.67
6.7	4.36	4.36	15.3	2.87	2.87
7.6	4.06	3.56	16.2	2.97	2.77
9.5	3.17	3.17	17.1	2.48	2.87
11.7	2.97	3.07	19.1	2.77	2.77
			20-35	2.18	2.57

linewidth for the M' = $-\frac{7}{2}$ and M' = $-\frac{5}{2}$ components for pseudorotating K_3. After an initial annealing of the matrix, measured linewidths (Γ) are independent of the sample history and, within experimental error, depend only upon temperature. Above T ~ 20 K, ESR linewidths remain approximately constant and are represented by average values, Γ_0, as shown in the last entry of Table II. Γ_0 is taken to be a measure of temperature independent contributions to the ESR linewidth. Although powder spectra are most often inhomogeneously broadened, the exchange contribution to the trimer linewidth is assumed to be given by $\Gamma - \Gamma_0$. Owing to the 10-20% uncertainty in Γ, there is considerable scatter in these differences, particularly at higher temperatures. However, the linewidth data do give approximate values for τ_c and T_2 and allow an experimental test of the "three jump model" introduced in the preceding section.

Figures 3a and 3b show, respectively, observed and simulated spectra for the pseudorotating trimer at T = 4.2 K. Absorption linewidths for the "bar spectrum" were derived from Equation (10), with f(M', k) from Table I and $\frac{2}{9}(a_0 - a_3)^2 = 3.4 \times 10^{15}$ sec^{-2} from Reference 3. The correlation time, $\tau_c = 3.8 \times 10^{-9}$ sec, was chosen to give the measured linewidth for $M' = -\frac{7}{2}$, $\Gamma = \Gamma_0 + \Gamma(M', k) = 4.6$ G from Table II. Relative areas in Figure 3b reflect the D_k degeneracy given in Table I. For ESR transitions, $(m_1, m_2, m_3) = (\pm\frac{3}{2}, \pm\frac{3}{2}, \pm\frac{3}{2})$ and $(\pm\frac{1}{2}, \pm\frac{1}{2}, \pm\frac{1}{2})$, f(M', k) = 0 and trimer exchange is not predicted to contribute to the linewidth. Accordingly, bar spectra for these transitions correspond to $\Gamma = \Gamma_0 = 2.2$ G with ESR intensities equal to that measured for $M' = -\frac{9}{2}$. Aside from the matrix site effects noted earlier, the general features of the observed spectrum are quite well represented in Figure 3b.

Trimer linewidths were compared with both an exponential, $\tau_c = \tau_0 e^{\delta/T}$, and a power law, $\tau_c = \tau_0 T^{-n}$, temperature dependence in the correlation time τ_c. Figure 5 shows the least squares fit for $M' = -\frac{7}{2}$. Both plots represent the experimental linewidth equally well (correlation coefficient, r = 0.90) and the present data are not sufficiently accurate to distinguish between the two interpretations. For reasons given below, the data point at T = 4.2 K was not included in this fit. Similar results (but with r = 0.98) pertain to $M' = -\frac{5}{2}$. Although, as shown in Figure 3b, two k components contribute to this transition, it is most likely that the broad, f(M', k) = 4 and narrow, f(M', k) = 1, features are sufficiently well separated for the latter to dominate the observed $M' = -\frac{5}{2}$ linewidths.

Table III gives the computed parameters for both $M' = -\frac{7}{2}$ and

Table III. Least squares parameters (defined in text) for trimer correlation times (τ_c). Estimated errors are given in parentheses.

	$\tau_c = \tau_0\ e^{\delta/T}$		$\tau_c = \tau_c\ T^{-n}$	
M'	$\tau_0 \times 10^{11}$(sec)	δ(K)	$\tau_0 \times 10^{8}$(sec K^n)	n
$-\frac{7}{2}$	17(5)	16(3)	3.4(21)	1.5(3)
$-\frac{5}{2}$	5.8(11)	21(2)	18(11)	1.9(2)

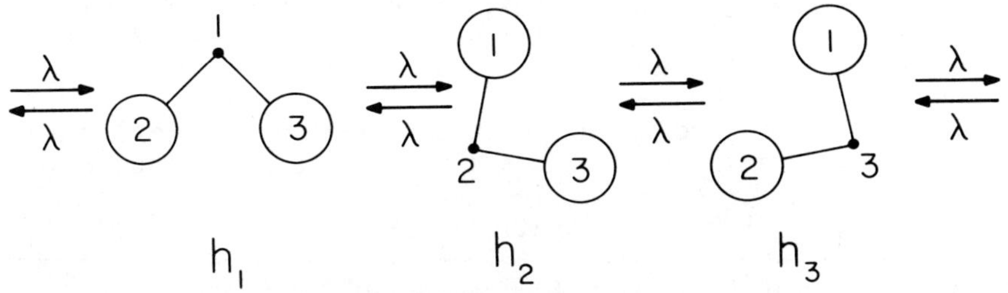

Figure 4. Three-jump model for pseudorotating K_3. See Eq. 4 for definition of h_i.

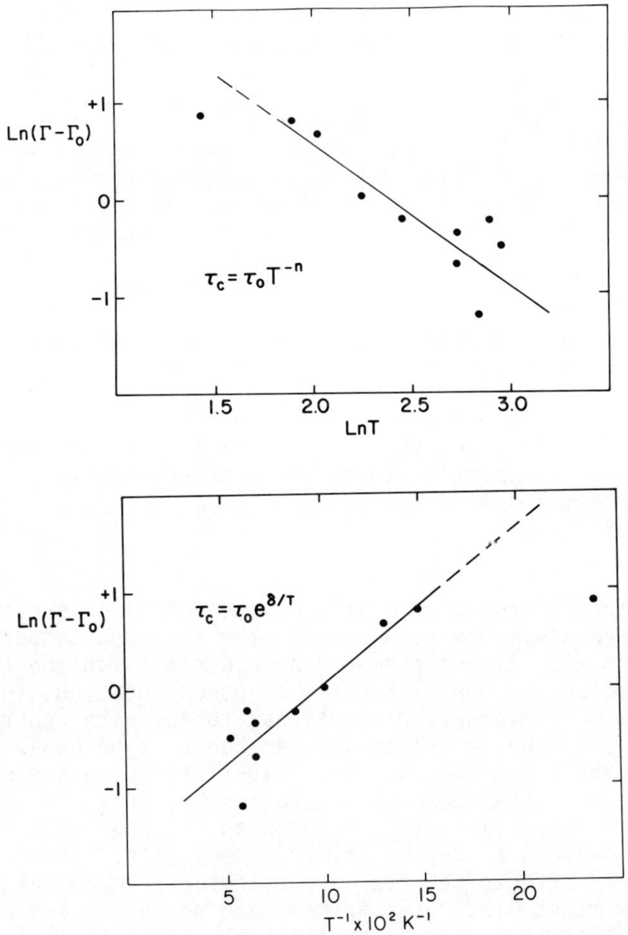

Journal of Chemical Physics

Figure 5. Comparison of measured ESR linewidths (Γ) with assumed temperature dependence in trimer correlation times (τ_c) a, $\tau_c = \tau_o T^{-n}$ b, $\tau_c = \tau_o e^{\delta/T}$.

$M' = -\frac{5}{2}$. Estimated uncertainties, in parentheses, represent one standard deviation in the least squares fit to the experimental data. Table IV summarizes several parameters of interest for

Table IV. Temperature variation in correlation time (τ_c), secular linewidth, $\Gamma(M', k)$, and transverse relaxation time (T_2) for $M' = -\frac{7}{2}$. The Larmor frequency $\omega_0/2\pi = 9.3$ GHz. For an explanation, see text.

T (K)	$\tau_c \times 10^{10}$ (sec)	$\Gamma(M', k)$ (gauss)	$T_2 \times 10^8$ (sec)	$(\tau_c/T_2)^{\frac{1}{2}}$	$\omega_0\tau_c$
4	93	11	3.2	0.54	540
6	24	2.9	12	0.14	140
8	13	1.6	23	0.08	76
10	8.4	1.0	35	0.05	49
15	4.9	0.6	60	0.03	29
20	3.8	0.5	77	0.02	22

pseudorotating K_3. These data were calculated from Equation (10), with $f(M', k) = 1$ and $\frac{2}{9}(a_0 - a_3)^2 = 3.4 \times 10^{15}$ sec^{-2} from Reference 3, together with the least squares τ_0 and δ from Table III. These example results permit a check on several of the approximations alluded to earlier. A necessary condition for second order perturbation theory to be valid is $(\tau_c/T_2)^{\frac{1}{2}} << 1$ (10). As shown by the data in column 5 of Table IV, this approximation breaks down at low temperatures and predicted linewidths become much larger than those actually observed. The ratio of the non-secular to the secular contribution to the trimer linewidth is of order $(1 + \omega_0^2\tau_c^2)^{-1}$ (8 - 12). As shown by the last entry in Table IV line broadening by non-secular terms is quite negligible over the temperature range T = 4-20 K.

Discussion

Contingent upon the details of the potential surface for K_3, several mechanisms may give rise to the observed temperature dependence in τ_c. If the ground vibronic state for the trimer is of order kT below the energy barrier to pseudorotation, then the correlation time (inversely proportional to the rate constant for exchange) would be expected to show Arrhenius type behavior. Alternatively, the trimer ground state might lie above the pseudorotation barrier, as has been suggested for Li_3 (5). In this situation a fast, but temperature independent, exchange would be predicted. However, since the trimer potential surface is markedly flat, matrix effects may play a particularly important role. Thus the matrix might "lock" the trimer into an obtuse angled geometry at some temperatures but be sufficiently fluid to allow pseudorotation at others.

For Na_3, which is expected to be more closely analogous to the potassium trimer than Li_3, Martin and Davidson (6) estimate a zero point vibrational energy $\omega \sim 85$ cm^{-1} with a barrier to pseudorotation, $2\beta \sim 200$ cm^{-1}. Particularly critical for Jahn-Teller systems (14), is the distortion parameter ρ_0 (Δ in the notation of (6)). Referring to Figure 1, ρ_0 is the radial distance from the conical peak to the average energy minimum in the trough region of the trimer potential surface. For Na_3, $\rho_0 \sim 0.5$ Å (6). If vibrational frequencies, potential parameters, and bond distances scale according to, respectively, the force constant, bond energy, and internuclear distance of the corresponding alkali dimers, then, very approximately, $\omega \sim 65$ cm^{-1} with $2\beta \sim 150$ cm^{-1} and $\rho_0 \sim 0.6$ Å for K_3. Consequently the ground vibronic level of K_3 is some 85 cm^{-1} below the pseudorotation barrier and, for $4 \lesssim T \lesssim 35$ K, exchange can only occur by quantum mechanical tunneling. This rate may be approximately calculated using a WKB method similar to that applied to the more familiar case of ammonia inversion (15). Using the previously estimated parameters for K_3 (reduced mass, $\mu = \frac{2}{3}m = 26$ amu) and numerical integration of an assumed cosine potential we find a tunneling frequency of order 1 GHz.

Although temperature dependent tunneling is not, of course, to be expected for gas-phase molecules, its occurence for transition metal complexes in the solid phase has been well documented (14, 16, 17). Even in carefully prepared crystals, random strains are of sufficient magnitude to make the previously assumed equivalence of potential minima no longer valid. If, for K_3 in argon, matrix effects contribute a similar blurring of the trimer energy levels, then tunneling between adjacent wells no longer conserves energy. Transitions from the vibronic ground state of one well to the corresponding level of its inequivalent neighbor must now be stimulated by phonon absorption or emission. As given elsewhere (14, 16, 17), the several possible mechanisms correspond to the direct Raman or Orbach processes more familiarily applied to spin lattice relaxation. In all three cases this situation leads to an attenuation of the gas-phase tunneling frequency and complex, generally hyperbolic, temperature behavior for τ_C. In certain limiting situations either an exponential or power law temperature dependence is predicted. Consequently our results for K_3 in argon are also consistent with a trimer zero point energy well below the pseudorotation barrier.

Acknowledgments

This work was supported by the City University of New York PSC-BHE Research Award Program and by the National Science Foundation under grant number CHE 79-13260. Acknowledgment is made to the Donors of The Petroleum Research Fund, administered by the American Chemical Society, for partial support of this research.

Literature Cited

1. Lindsay, D. M.; Herschbach, D. R.; Kwiram, A. L. Mol. Phys. 1976, 32, 1199.
2. Lindsay, D. M.; Herschbach, D. R.; Kwiram, A. L. Mol. Phys. 1980, 39, 529.
3. Thompson, G. A.; Lindsay, D. M. J. Chem. Phys. 1981, 74, 959.
4. Thompson, G. A.; Tischler, F.; Garland, D.; Lindsay, D. M. Surf. Sci. 1981, in press.
5. Gerber, W. H.; Schumacher, E. J. Chem. Phys. 1978, 69, 1692.
6. Martin, R. L.; Davidson, E. R. Mol. Phys. 1978, 36, 1713.
7. Dietz, E. R. Phys. Rev. 1981, A23, 751.
8. Atherton, N. M. "Electron Spin Resonance"; Wiley: New York, 1973.
9. Freed, J. H.; Fraenkel, G. K. J. Chem. Phys. 1963, 39, 326.
10. Abragam, A. "The Principles of Nuclear Magnetism"; Oxford: London, 1961.
11. Slichter, C. P. "Principles of Magnetic Resonance"; Harper and Row: New York, 1963.
12. Carrington, A.; McLachlan, A. D. "Introduction to Magnetic Resonance"; Harper and Row: New York, 1967.
13. Feller, W. "An Introduction to Probability Theory and Its Applications", Wiley: New York, 1967; Volume I.
14. Englman, R. "The Jahn-Teller Effects in Molecules and Crystals"; Wiley: New York, 1972.
15. Townes, C. H.; Schawlow, A. I. "Microwave Spectroscopy"; McGraw-Hill: New York, 1955.
16. Ham, F. S. in "Electron Paramagnetic Resonance", edited by Geschwind, S.; Plenum: New York, 1972.
17. Williams, F. I. B.; Krupka, D. C.; Breen, D. P. Phys. Rev. 1969, 179, 255.

RECEIVED September 18, 1981.

8

Preparation, Electronic Spectra, and Ionization of Metal Clusters

E. SCHUMACHER, W. H. GERBER, H.-P. HÄRRI, M. HOFMANN, and E. SCHOLL

University of Bern, Institute for Inorganic and Physical Chemistry, CH-3000 Bern 9, Switzerland

Metal clusters of Li(<7), Na(<17), K(<13), Na_xK_y(x+y<9), Mg(<5), Zn(<5), Cd(<5) and Hg(<13) have been detected and in part spectroscopically investigated in molecular beams and in inert gas matrices. Supersonic nozzle expansion leads to population maxima at 2,3,4,5 in Hg beams by adjusting nozzle- and oven-temperatures and nozzle shape.

New Li_xH (x<6) clusters are formed by thermal decomposition of LiOH on liquid Li at 1500 K. Li_3H seems to be more stable than its neighbors.

New results by two photon ionization (TPI) spectroscopy of Na_2^+, NaK^+, Li_2, Li_3 and Na_3 are discussed.

The theoretical treatment of Li_3 in the ground and in five electronically excited states allows a complete interpretation of the Li_3 matrix spectrum as well as a more detailed analysis of the Na_3 TPI spectrum.

Na, Na_2 and Na_3 separated from a Na_x beam by a Stern-Gerlach magnet have been collected in a krypton matrix in high purity. The electronic spectra are discussed.

Among the volatile metals investigated it proves most difficult to prepare higher Li clusters. A suitable oven material at 1500 K is TZ-Molybdenum (Ti, Zr, C alloyed Mo, plasma welded) the heating elements being either tungsten or graphite. All alkali-clusters are predominantly destroyed by electron impact but are easily ionized with photons. The IIa,b metals probably form only weakly bound van der Waals-nmers M_n for n<10. Ionization removes an antibonding electron leading to ions stabilized against shedding a neutral fragment. They are therefore observable by elec-

0097-6156/82/0179-0083$06.25/0

tron impact ionization. A cylindrically shaped extension of the nozzle (.5 cm long, .8 cm dia.) enhances the collision probability in the transition region from viscous to molecular flow. This combined with an appropriate temperature difference between oven and nozzle leads to maximum populations in Hg_x beams at $x = 2,3,4$ or 5, Figure 1 (1). Table I shows normal Na_x abundances.

The main trend in the photoionization potentials (PIP) of alkali clusters as a function of size can be interpreted by the excess work function of a conducting curved surface as compared to an infinite flat one. The additive term $.5e^2/R$ (R = radius of sphere) is due to the fact, that the positive image charge only recedes to the center of the sphere instead of to minus infinity as it does with a flat surface when an electron is moved to plus infinity (2), Figure 2. The influence of structure and especially of the discrete electron states of small clusters on the PIP can be treated approximately either by a pseudopotential calculation (3) or the SCF-Xα (4) method. EHT and CNDO/2 have been less successful (5). Using the transition state formalism of the Xα scheme ($\alpha = .73135$) the IP of Na_x ($x<10$) were computed for several high symmetry geometries in either a low- or high-spin configuration. Variation of the assumed structures and of the Na-Na distance have only a small influence on the IP's which closely match the experimental values, Figure 2. This suggests, that the Na_x are floppy molecules for which not only the Born-Oppenheimer but even the adiabatic approximations break down. The successful gross representation of the PIP by a spherical drop model or a jellium calculation (6) lends support to this hypothesis. The M_3 particles have in all cases studied (Li_3, Na_3, K_3, Na_2K, NaK_2) a PIP of nearly 1 eV smaller than that of the dimers. This is due to the antibonding or non-bonding character of the uppermost orbital in either D_{3h} or C_{2v} symmetry.

Heteroclusters of Li/H have been observed (Li_2H, LiH_2 and Li_2H_2 see ref. (7)). At 1000 K and higher the reaction

$$2Li(\ell) + LiOH(\ell) = Li_2O(s) + LiH(\ell) \qquad \Delta G° = -33.2 \text{ kcal/mol}$$

with excess Li produces Li_xH clusters which can be detected by photoionization with a Xe-arc lamp without filters (in air) (8), Figure 3. Observation of the deuterides confirms the identification. At 1533 K (450 torr Li, .2 mm nozzle) Li_3H is about as abundant as Li_3, Figure 4. This class of heteroatom clusters is interesting for the study of the ionic-metallic bond transition, and for understanding the series H_3 (9), LiH_2 (10), Li_2H (11) and Li_3 (12). C.H.Wu and H.R.Ihle (10) find a very stable LiH_2 molecule ($Li + H_2 = LiH_2$ $\Delta H°=-23$ kcal/mol). M/e=8 or 9 for $^6LiH_2$ or $^7LiH_2$ are not detectable in our experiments by photoionization, in accord with the high IP = 6.3 eV (10). Calculations and spectroscopy of these particles are in progress at Bern (13).

TABLE I

Cluster populations in a Na supersonic nozzle beam with photoionization mass spectrometry, nozzle d = 0.25 mm, abundances in amps.

T_n (K)	T_0 (K)	Na $\cdot 10^{-9}$	Na_2 $\cdot 10^{-9}$	Na_3 $\cdot 10^{-10}$	Na_4 $\cdot 10^{-11}$	Na_5 $\cdot 10^{-11}$
973	893	1.13	0.16	0.03	0.07	----
1003	933	1.66	0.37	0.12	0.44	0.03
1063	968	2.62	0.74	0.33	1.69	0.14
1103	998	2.2	1.20	0.58	3.50	0.40
1143	1033	2.88	1.74	0.78	5.90	0.70
1163	1073	3.22	2.50	1.10	8.20	1.10
1183	1103	3.45	3.30	1.10	9.7	1.5

T_n nozzle temperature, T_0 oven temperature. The Na/Na_2 ratio has been calibrated with a Langmuir-Taylor detector and known equilibrium mixtures. For the higher clusters the MS signals are directly given without corrections (1). Neglecting fragmentation, the MS signals are estimated to be proportional within a factor of 2 to the true abundances.

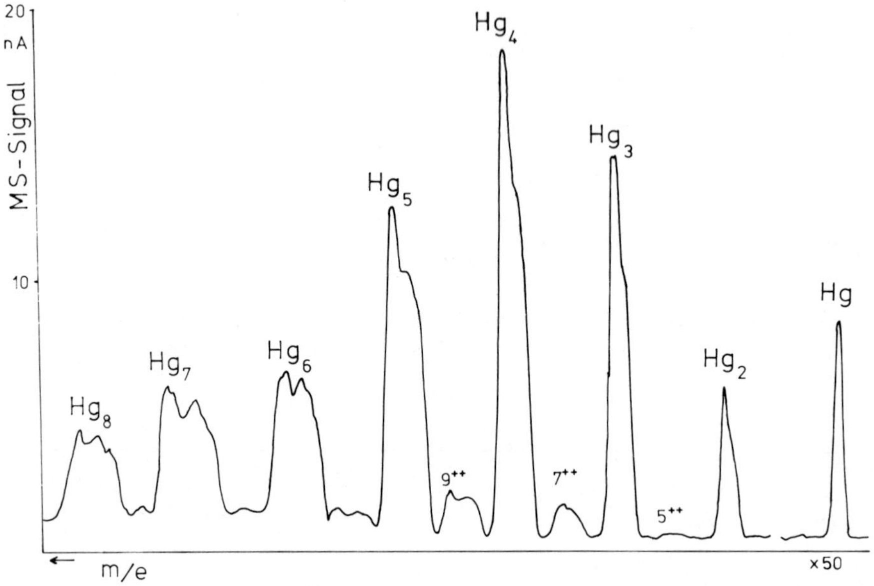

Figure 1. Mass spectrum by electron impact ionization of Hg_x in a supersonic nozzle beam with oven pressure of 7 bars, nozzle temperature 1143 K (1).

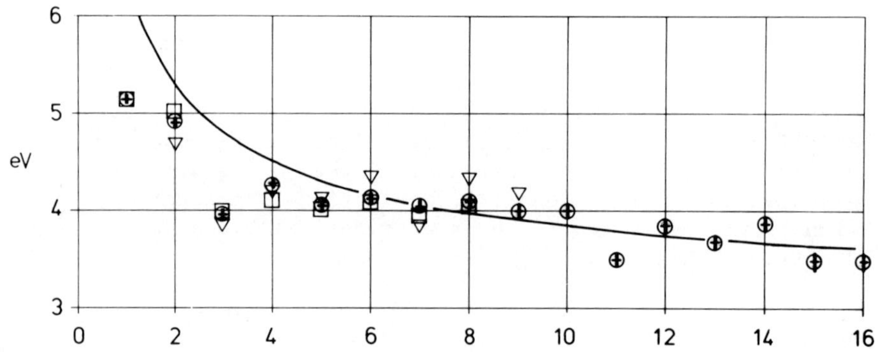

Figure 2. Single-photon photoionization potentials for Na_x (2, 33). The curve is the work function of a conducting spherical drop (2). Key: ▽, SCF-Xα-SW (4); □, pseudo-potential ionization potentials (3); and ⊕, experimental values.

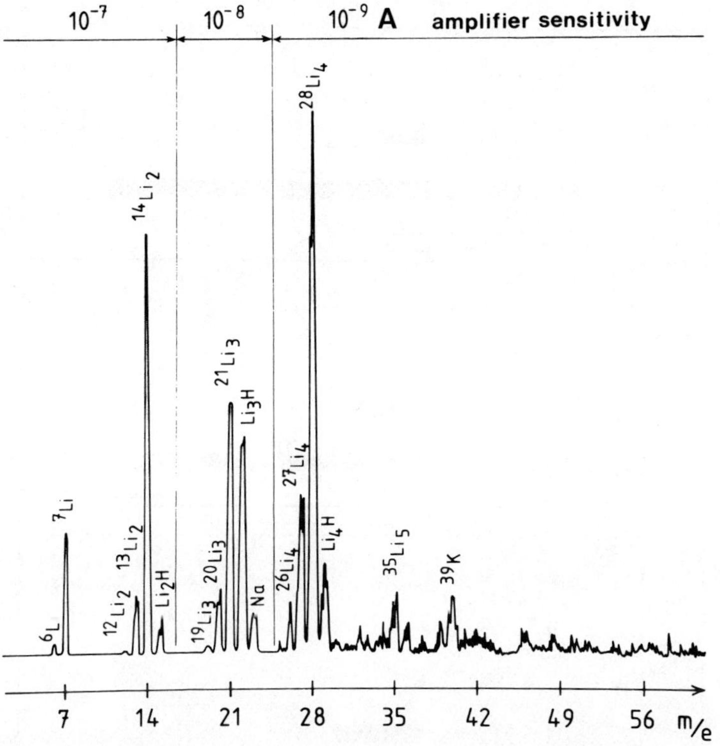

Figure 3. Mass spectrum by photoionization of Li_x and Li_xH clusters in a supersonic beam at 1533 K (430 torr, 0.2 mm nozzle) (8).

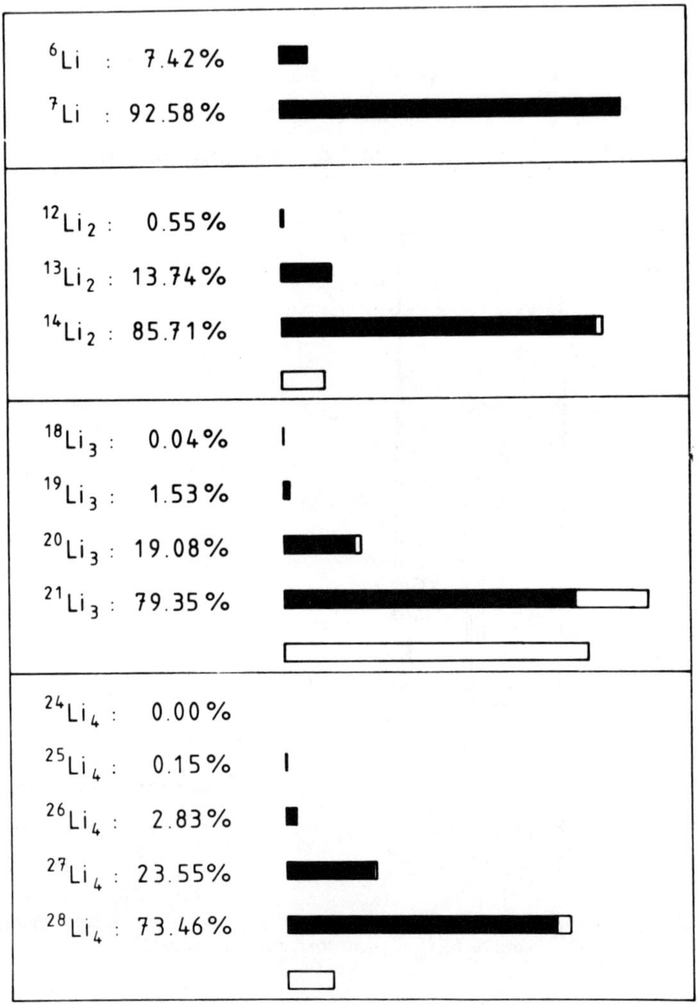

Figure 4. Abundances of Li_x and monohydride clusters in the MS of Fig. 3. Key: ▭, *Li_xH.*

Li clusters up to Li_6 are detected under the same conditions as mentioned above. There is no indication of anomalously low stabilities for Li_3 or Li_5 which are missing in the gas clustering series, Li_2 to Li_{15}, of Kimoto and Nishida (14). Table II summarizes data on Li clusters.

Particle-specific absorption spectroscopy in molecular beams can be realized by putting an optical signature on the ion-current of a single m/e channel in the mass-spectrometer (15,16, 17). The spectra of neutral excited states are obtained by tuning a dye laser over the excitation region and simultaneously probing the excited state population with an ionizing second laser or a broad band light source within the ion source of a quadrupole mass-spectrometer as shown in Figure 5. Reversing the role of the two lasers information about ionization or ion states is available by excitation to a fixed rovibronic neutral state of a molecule and scanning the ionization region with a second laser (18), Figure 6. If the (stabilized) power of the exciting laser is small enough to prevent saturation effects, TPI-spectroscopy gives correct absorption intensities (16). Since the ionization efficiency of the second laser is the bottleneck of the TPI process (15) high power density (photonfluence) is necessary to harvest a high fraction of the excited state population for the m/e ion current. Done intracavity this second step may be able to offer single particle observability combined with total isotope-specificity, two properties not easily met with laser induced fluorescence alone (19). Na_3 $^2E'' \rightarrow {}^2E'$ fluorescence can be picked out within the 100 times more intense vibronic progression of the Na_2 A → X transitions (1). This yields information about the ground state gyrovibrational modes, whereas TPI spectroscopy gives excited state information. Problems arise with the mode structure of tunable lasers as shown in Figure 7 for the A(v'-5) ← X(v" = 0) rotational band in $^7Li_2$. For resolution on the order of $\pm$ 0.4 cm^{-1}, the output mirror piezo crystal is wobbled longitudinally with an 800 cycle sine-wave in order to broaden the laser bandwidth quasihomogeneously. A second complication is reflected on the neutral state spectra by the highly structured ionization-cross-section, e.g. for Na_2 in Figure 8. A broad-band ionizing light source averages over the autoionization peaks and eliminates spikes (17). Resonance photoionization via very intense molecular autoionization transitions (AIT) can, however, give a 100-fold increase in the detection efficiency of the TPI spectra. It requires the exact lock-in of the ionizing laser into a specific AIT during the scan of the intermediate states, an experiment not yet performed but possible (20). Finally, spectral perturbations are easily detected in TPI spectra. They often lead to an anomalous line intensity in addition to frequency shifts because the perturbed state has a longer lifetime and hence is more efficiently ionized (16). Photon momentum deflection spectroscopy of molecules

TABLE II

Li_x clusters: ground state, structure, ionization energy

	Structure (bohr)	IP, eV
$Li(^2S)$		5.39
$Li_2(^1\Sigma_g^+)$	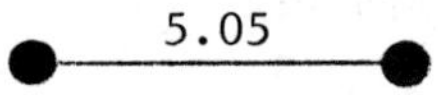	5.1 ± 0.1 4.85 ± 0.1 (12) 5.15* (36) 5.17 (38)
$Li_3(^2E')$* config. $(a_1')^2(e')^1$	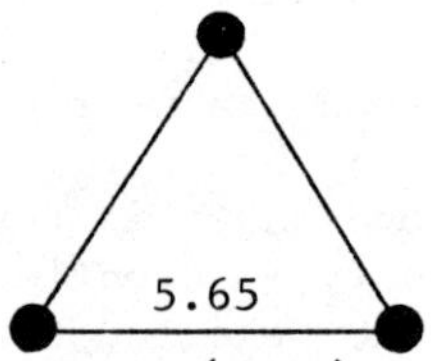 D_{3h} (mean)	4.1 ± 0.1 4.35 ± 0.2 (12) 4.0* (26) 3.95* (37)
$Li_4(^1A_{1g})$ config. $(a_{1g})^2(b_{2u})^2$	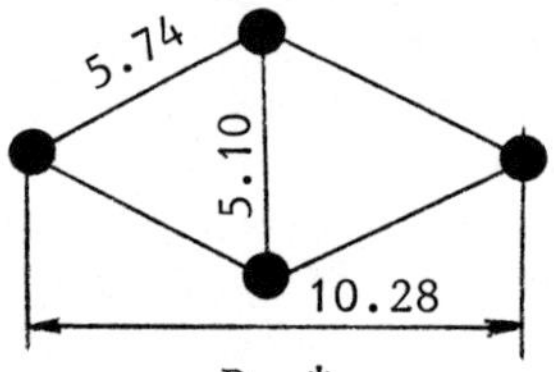 D_{2h}*	4.2 ± 0.2 4.65* (36)
Li_5		<4.5
$Li_6(^1A_{1g})$* config. $(a_{1g})^2(e_u)^4$	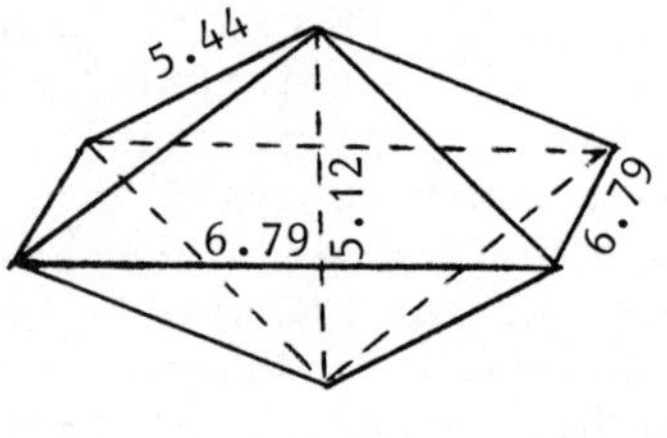	4.60* (36)
Li_∞	bcc 6.63	2.4 ± 0.1 (39)

* calc.

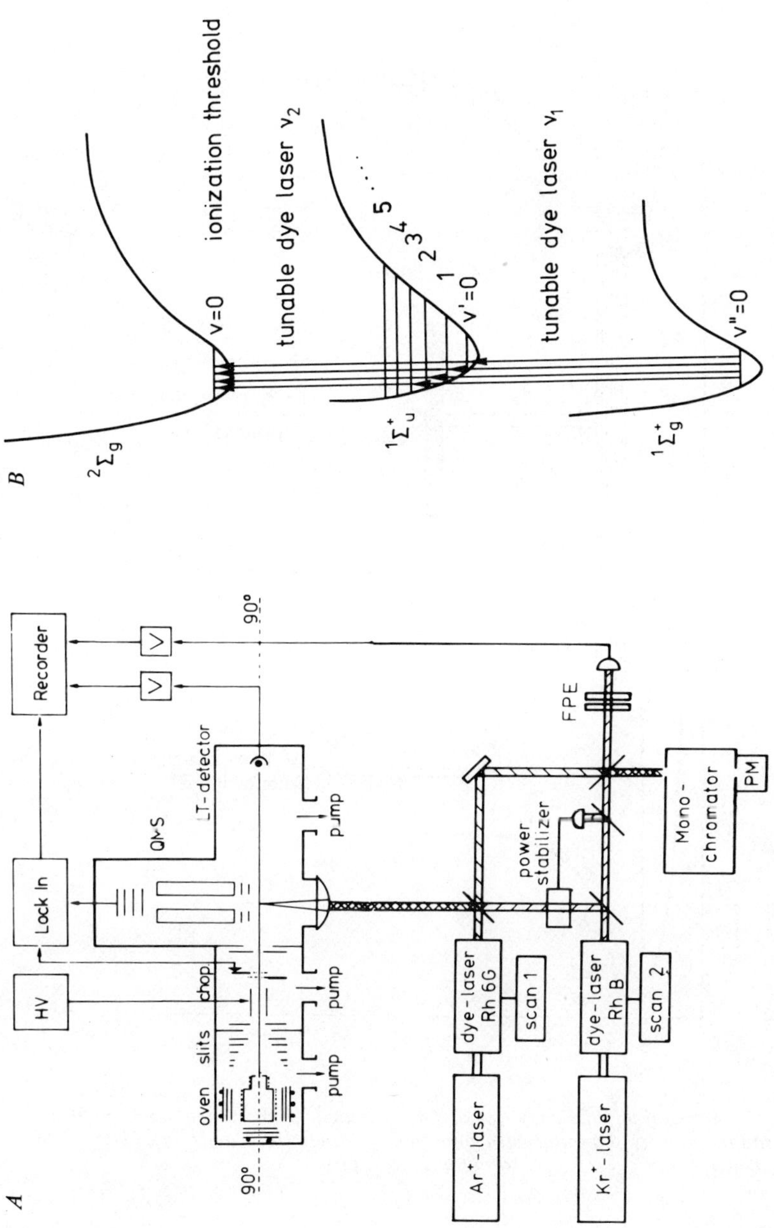

Figure 5. Absorption spectroscopy by two-photon ionization mass spectrometry, apparatus (A) and schematic for a dimer (B). Substates of the intermediate electronic state are spectroscopically investigated.

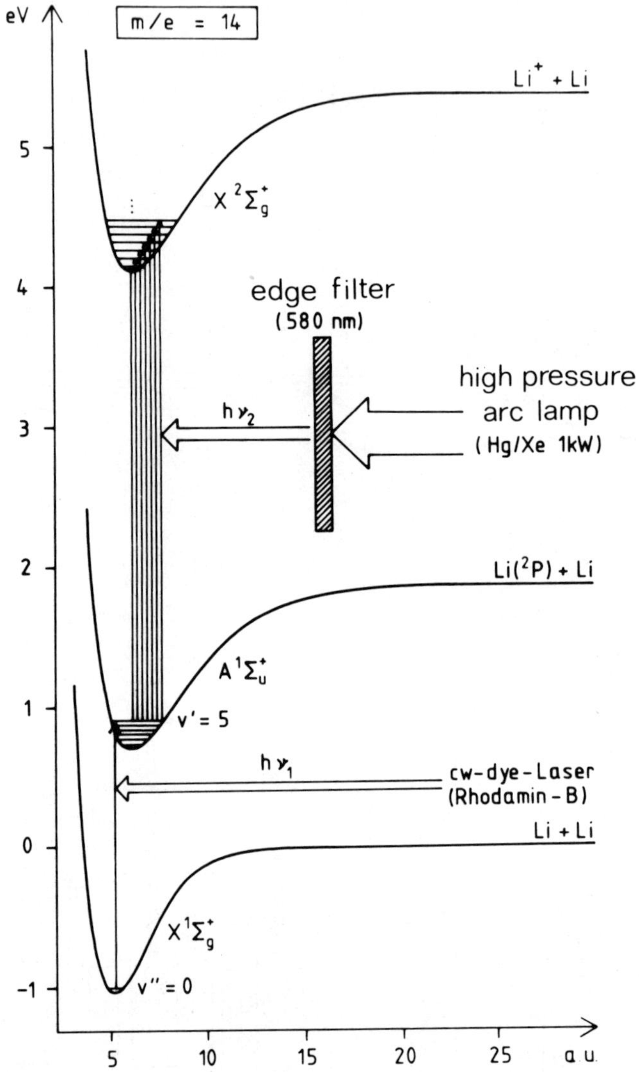

Figure 6. Ionization efficiency spectrum, obtained from a neutral gyrovibronic intermediate state by scanning with photon $h\nu_2$, either from a second laser, monochromator or edge filter.

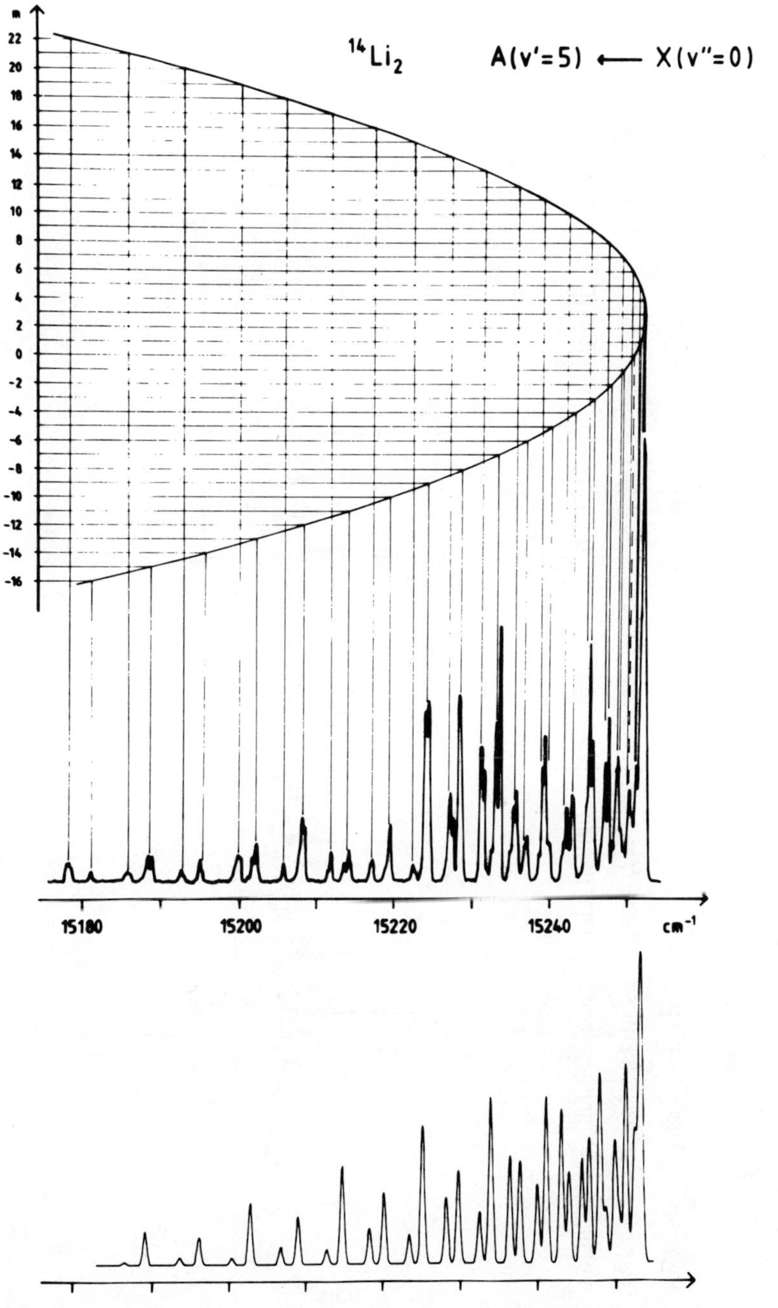

Figure 7. Rotational band of the $^7Li_2$ $A(v' = 5) \leftarrow X(v'' = 0)$ transition. The computer simulation shows clearly that the experiment exhibits several artifacts. Calculated spectrum at 75 K (resolution 0.5 cm^{-1}).

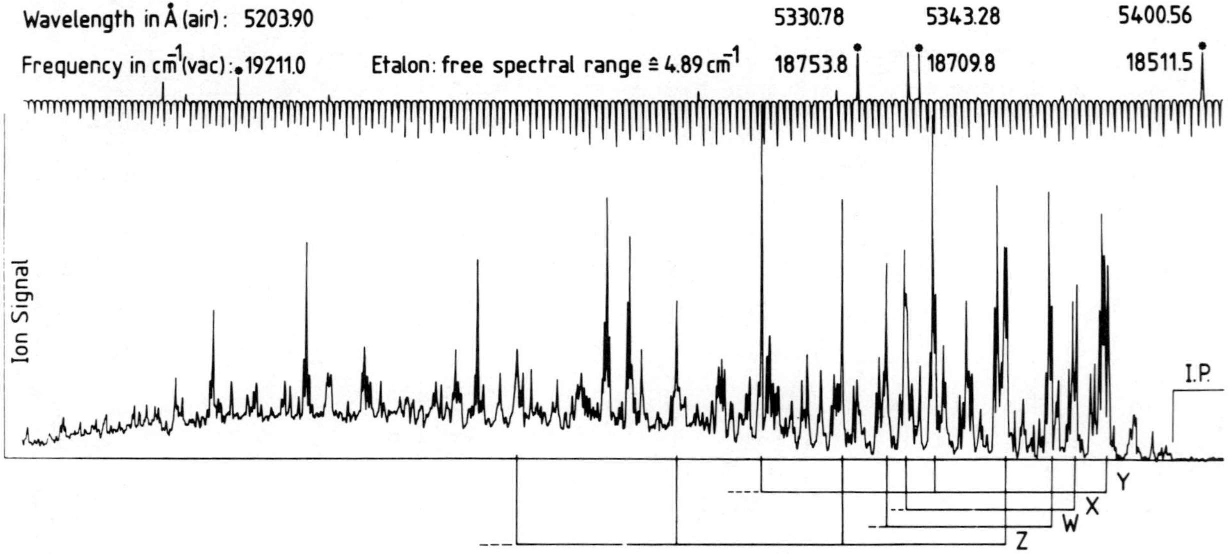

*Figure 8. Ionization efficiency spectrum of Na_2 in the $^1\Pi_u$ ($v' = 5$, $J'' = 0$–6) states (22). I.P. Ionization threshold. Key: *, optogalvanic Ne-lines.*

is a single photon absorption spectroscopy devoid of several of the complications mentioned (21). However, its sensitivity is much lower than that of TPI spectroscopy.

Figure 9 shows five TPI spectra of the $^{39}K_2$ B ← X band with stagnation pressures from 20 to 620 torr correlating with vibrational temperatures of 133 to 57 K. Corresponding rotational temperatures for the Na_2 A(v'=9) ← X(v"=0) band as in Figure 10 vary from 70 to 30 K with p from 25 to 105 torr. The temperatures are obtained by a computer simulation of each spectrum with a Gaussian fit of the Boltzmann factor to the measured spectrum. Each degree of freedom separately shows thermal equilibrium within ± 10 K or less. Rotationally cool beams are essential for the identification of the ionization threshold with the TPI-experiment as shown for the Na_2 B(v'=5,J'=0-6) intermediate state in Figure 8. This leads to the most precise IP's known for Na_2, K_2 and NaK (18,22). Table III summarizes single PI, TPI and electron impact (EI) data for these molecules. It is evident, that PI and probably EI determine vertical IP's, TPI (lower) adiabatic IP's. The structure of the ionization efficiency curve is surprisingly rich in sharp detail especially if the intermediate state has a higher vibrational quantum number. For the isotopic molecules $^{39}K_2$ and $^{39}K^{41}K$ the curves are widely different and ratios of the ionization efficiency of more than 100 have been found for certain frequencies (18). Apart from the continuum intensity, which decreases with v' autoionizing Rydberg states, their gyrovibrational substates are involved. No complete analysis exists yet. Experiments with higher resolution and with a defined single rovibronic intermediate state are under way.

A careful investigation of the predicted lowest electronic transitions in Li_3 has not yet produced useful spectroscopic information due to a still marginal abundance of Li_3 in our experiments. With Na_3 the spectral regions below (to 400 nm) and above (to 850 nm) the known transitions (23) have been scanned with high sensitivity. No new electronic states have been detected by TPI spectroscopy although theory predicts at least two more. The lower one at about 480 nm is predicted to be dissociative leading to $Na_2(^1\Sigma_g^+)$ and $Na(^2P)$. It has probably been identified by J.Gole et al. (24) by monitoring the Na fluorescence, whereas in our TPI experiments neither the m/e=69 nor the m/e=23 channels gave useful spectroscopic information. Below we give evidence of the presence of that transition in a low temperature matrix.

The theoretical treatment of the dynamic Jahn-Teller effect in Li_3 with the PNO-CEPA-CI ab initio method (25) has been refined in the groundstate and extended to the lowest four electronically excited states in the PNO-CI approximation(26,27). Two cross sections for the adiabatic hypersurfaces for the groundstate $^2E'$ and the lowest excited state, $^2E''$, are shown in Figure 11. The flat groundstate surface exhibits a roughly triangular trough

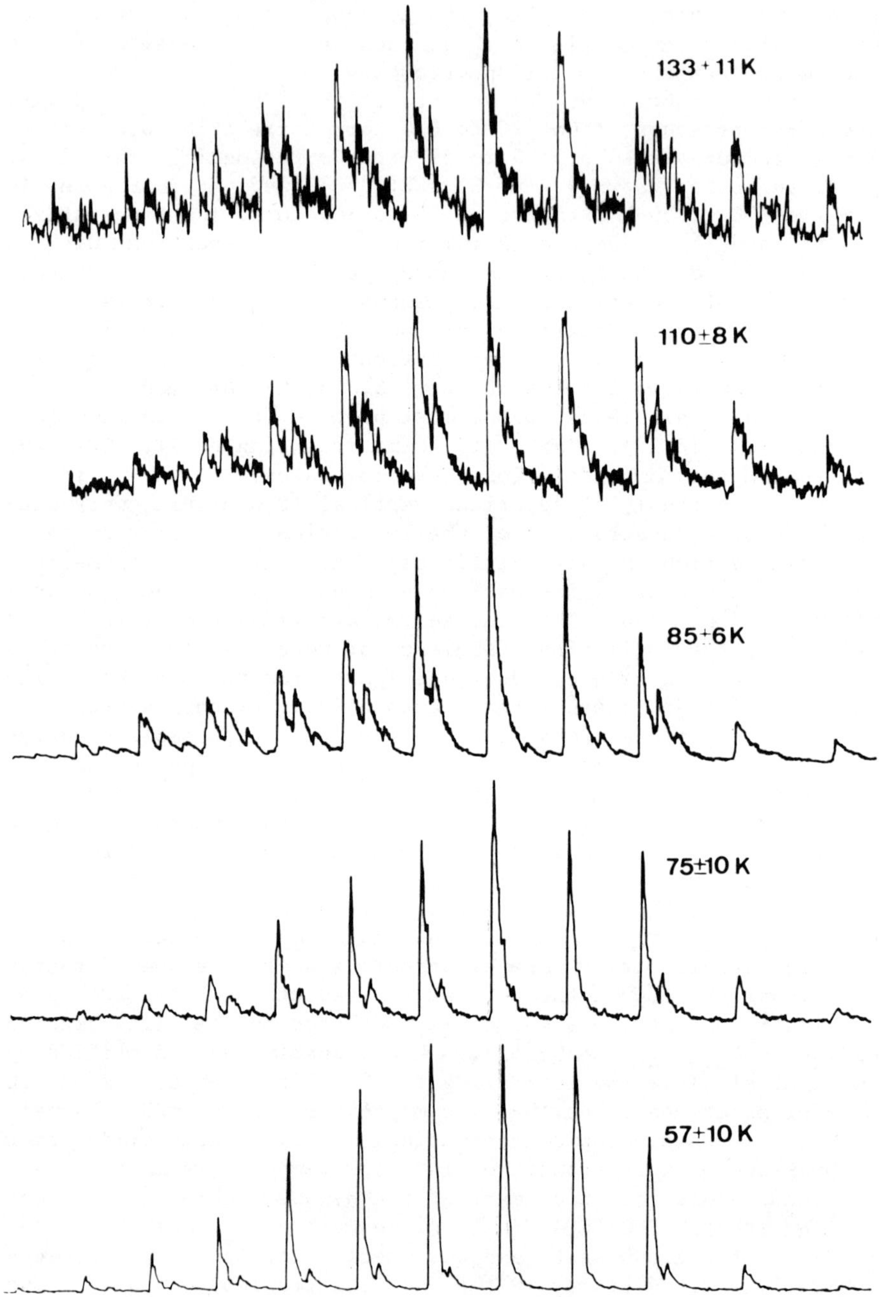

Figure 9. Vibrational progression of $^{39}K_2$ in the $B \leftarrow X$ band as measured by TPI spectroscopy. The parameter is the vibrational temperature chosen by oven pressures of 20, 60, 200, 360, and 620 torr from top to bottom (20).

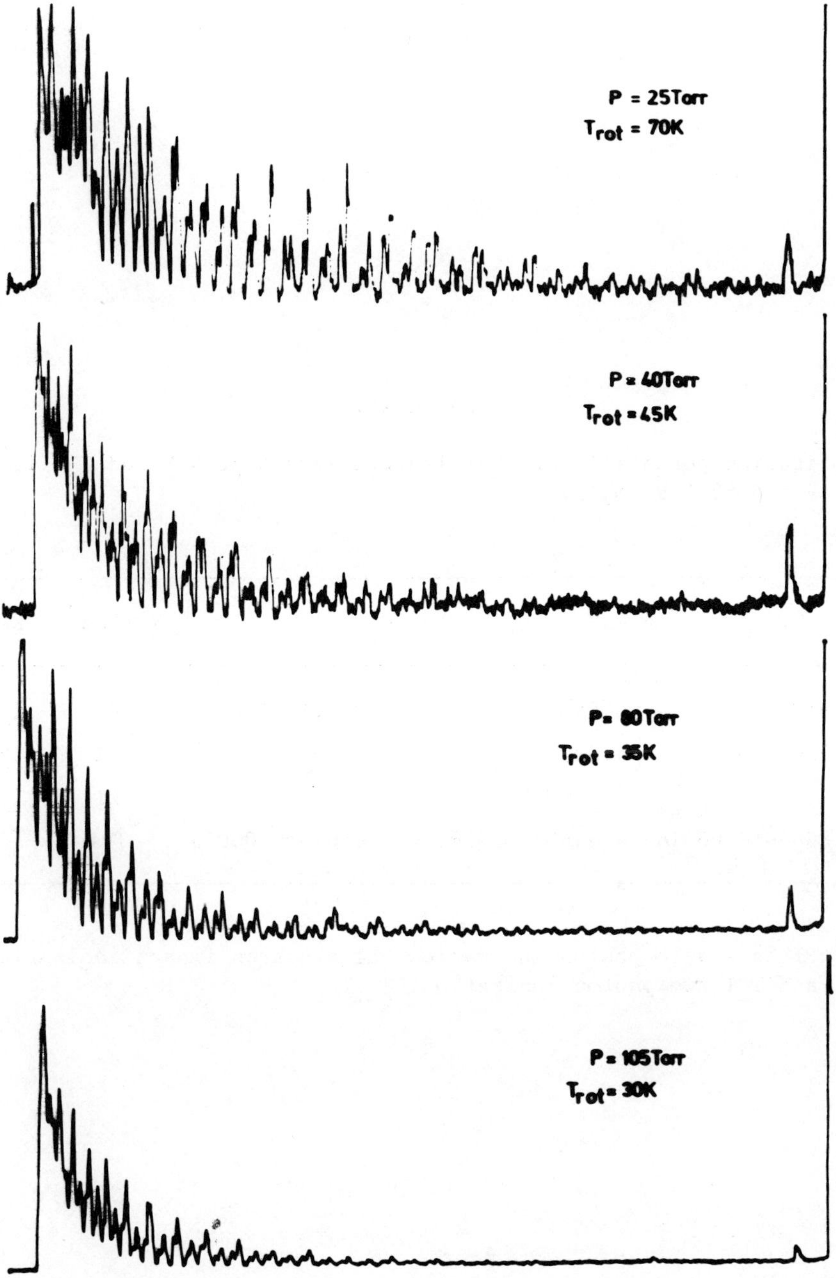

Figure 10. Rotational band of the $Na_2(v' = 9) \leftarrow X(v'' = 0)$ transition by TPI spectroscopy. The parameter is the rotational temperature with respect to oven pressure (20).

TABLE III

Ionization potentials of the alkali dimers Na_2, NaK and K_2 (in eV, 1 eV = 8065.479 cm^{-1}).

Na_2	NaK	K_2	Method[a)]	Ref.
4.9±0.1	4.5±0.1	4.0±0.1	PI	(34)
4.934±0.011	4.52±0.05	4.05±0.05	PI	(2)
	4.57±0.2		EI	(35)
4.88898±0.00016	4.41636±0.00017	4.06075±0.00015	TPI	(22)

a) PI is single photon ionization; EI electron impact ionization; and TPI two photon ionization.

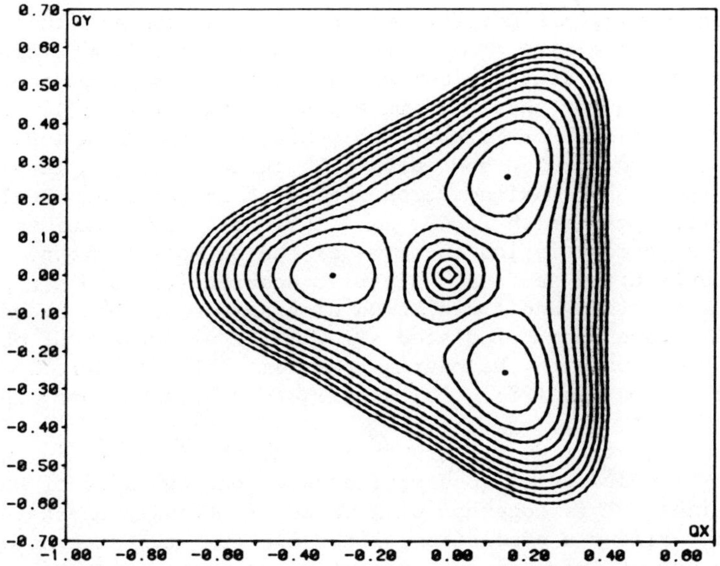

Li$_3$ E$''$ state (contour line spacing 50 cm^{-1})

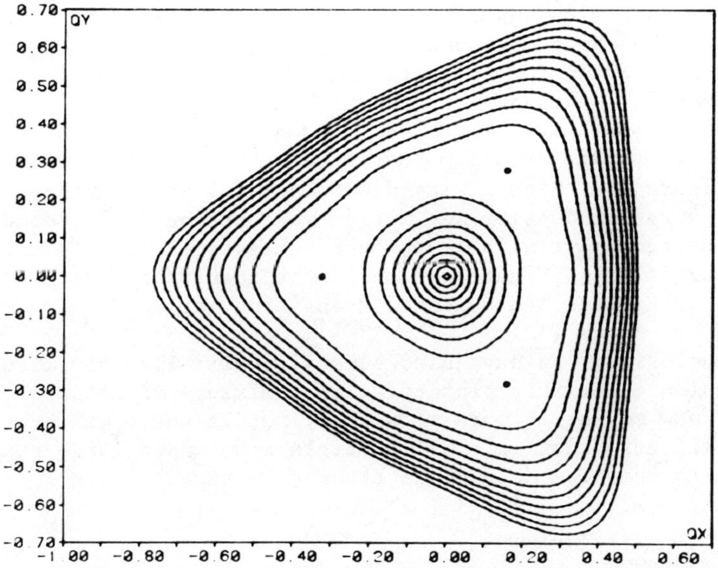

Li$_3$ ground state (contour line spacing 50 cm^{-1})

Figure 11. Cross sections of the adiabatic energy hypersurfaces of the ground state $^2E'$ and the lowest excited state $^2E''$ of Li_3 in the two degenerate normal modes Q_x, Q_y (basis of irreducible E'-representation) as coordinates (27).

with an energy minimum for an isosceles triangle with an apex angle of 71° (points). This is separated by a shallow barrier from an adjacent less deep minimum energy triangle with an apex angle of 54°. Neither of these minima support a vibrational level, i.e. Li_3 is a fluxional or "floppy" molecule, whose vibrations cannot be classified as C_{2v}-species. Only three vibronic levels are below the center maximum (crossing point of the Jahn-Teller E_- and E_+ surfaces) which is 400 cm^{-1} above the lower minimum. This justifies a description in terms of the D_{3h} point group. Figure 12 presents cross sections of five doublet-states and Figure 13 the MO schematic and correlation diagram. From this the data of Table IV have been calculated allowing an assignment of observed electronic bands in the matrix spectrum of Li clusters by T.Welker and T.P.Martin (28). For details of the arguments refer to (23,29).

Applying scaling factors from the comparison of Li_2 and Na_2 to the Li_3 parameters, good estimates of the behvaior of Na_3 can be obtained. This together with an SCF-Xα transition state calculation allows a prediction of the spectral transitions and an assignment of the measured absorption spectrum of Na_3 (23). The excellent groundstate calculation on Na_3 by Martin and Davidson (30) gives also estimations on excited states which, however, do not correlate with observation nor the scaled Li_3 data. In order to locate the above mentioned missing TPI Na_3 band at 480 nm, our experimental setup for collecting, performed spectrally and mass-spectrometrically, analyzed cluster beams in an inert gas matrix at 4 K (31) and was equipped with a Stern-Gerlach magnet. This allowed us to separate the paramagnetic Na and Na_3 from the diamagnetic Na_2, Figure 14. High purity dimer matrices are shown in Figure 15. A very low density Na_3 matrix was obtained by cocondensing the separated partial beam through a mask with Kr onto the sapphire target. In Figure 16 the first result of this endeavour clearly exhibits the expected transition at 440 nm, matrix-shifted by -40 nm (32).

Conclusion: We have used several, partially new methods to study single particle properties in a mixture of metal clusters. We are now moving to more refractory metals where we expect to see rigid structure at small particle size and a large range of chemical properties depending strongly on size.

Acknowledgments

This work has been sponsored by the Swiss National Science Foundation for Scientific Research, Grant No. 2.444-0.79.

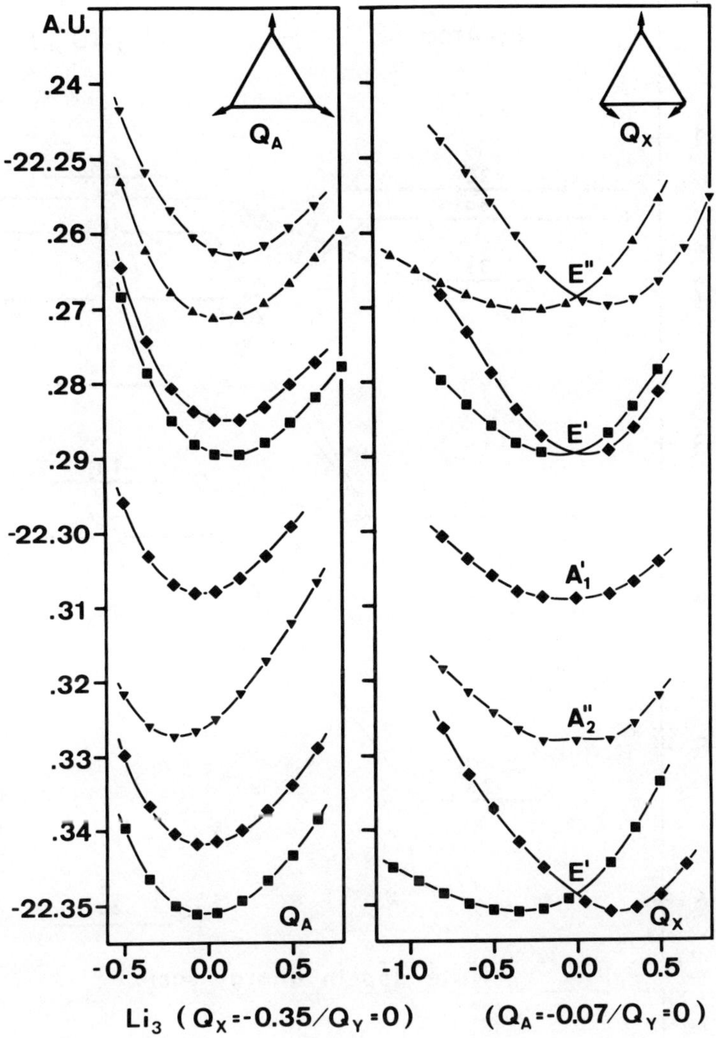

Figure 12. Cross sections of the adiabatic electronic energy hypersurfaces of Li_3. The lowest five doublet states with the electronic configuration $(1a_1')^2(1e')^4(2a_1')^2(x)^1$ are shown (27). Key: ■, B_2; ◆, A_1; ▲, A_2; *and* ▼, B_1.

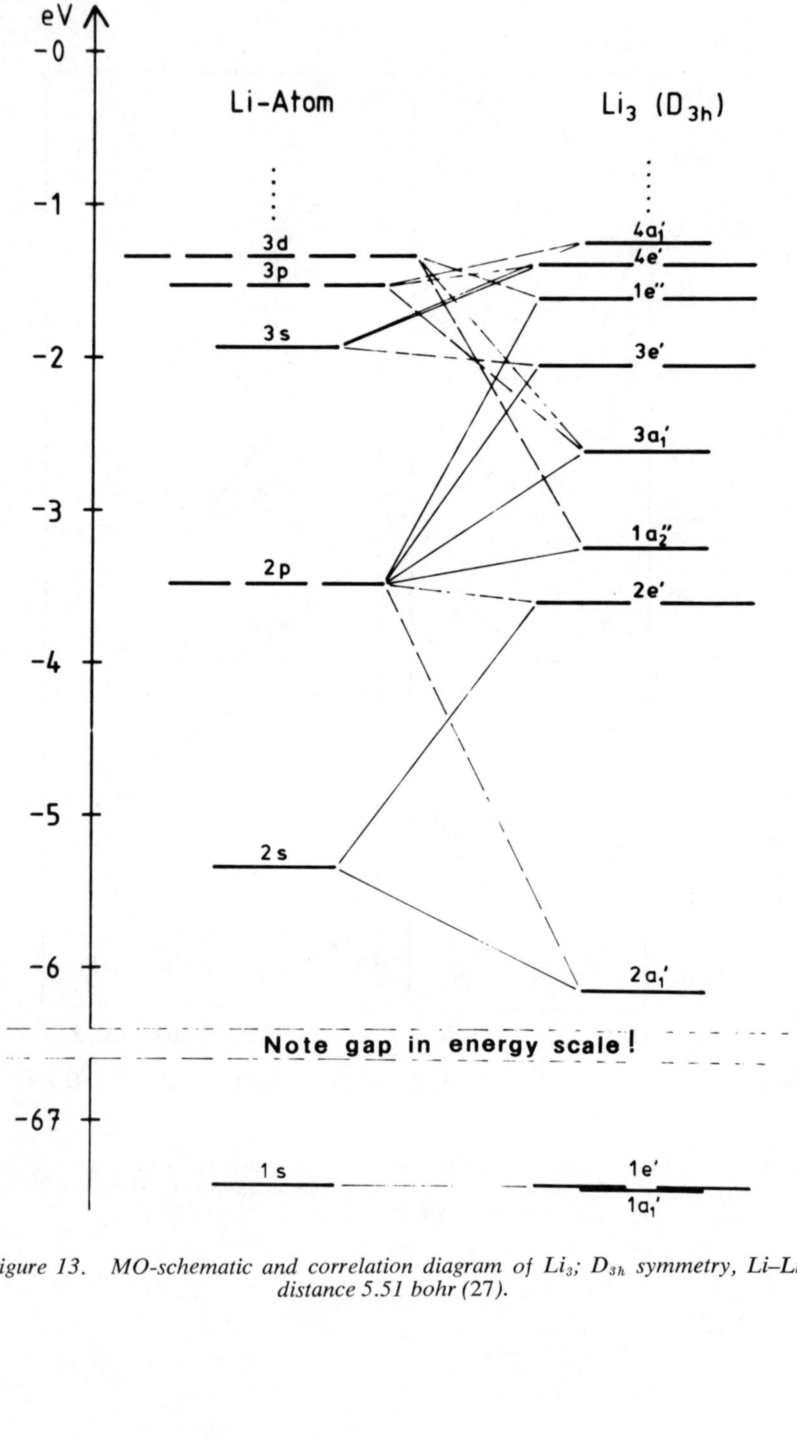

Figure 13. MO-schematic and correlation diagram of Li_3; D_{3h} symmetry, Li–Li distance 5.51 bohr (27).

TABLE IV

Li_3 doublet states

No.	Configuration	State	Energy $[cm^{-1}]$	Li_3-Xe-Matrix Absorption
0	$(2a_1')^2(2e')^1$	$^2E'$	0	(groundstate)
1	$(2a_1')^2(1a_2'')^1$	$^2A_2''$	5'000	
2	$(2a_1')^2(3a_1')^1$	$^2A_1'$	9'260	
3	$(2a_1')^2(3e')^1$	$^2E'$	13'520	13'890
4	$(2a_1')^2(1e'')^1$	$^2E''$	17'550	18'050
5*	$(2a_1')^2(4e')^1$	$^2E'$	≈20'300	21'050
6**	$(2a_1')^1(2e')^1(1a_2'')^1$	$^2E''$	≈23'400	23'310

* : in gas dissociation into $Li(^2S) + Li_2(X)$; bond energy approx. 20 kcal mol^{-1}.

**: dissociation into $Li(^2S) + Li_2(^1\Pi_u)$ or $Li(^2P) + Li_2(^1\Sigma_g^+)$, metastable; bond energy approx. -8 kcal mol^{-1}.

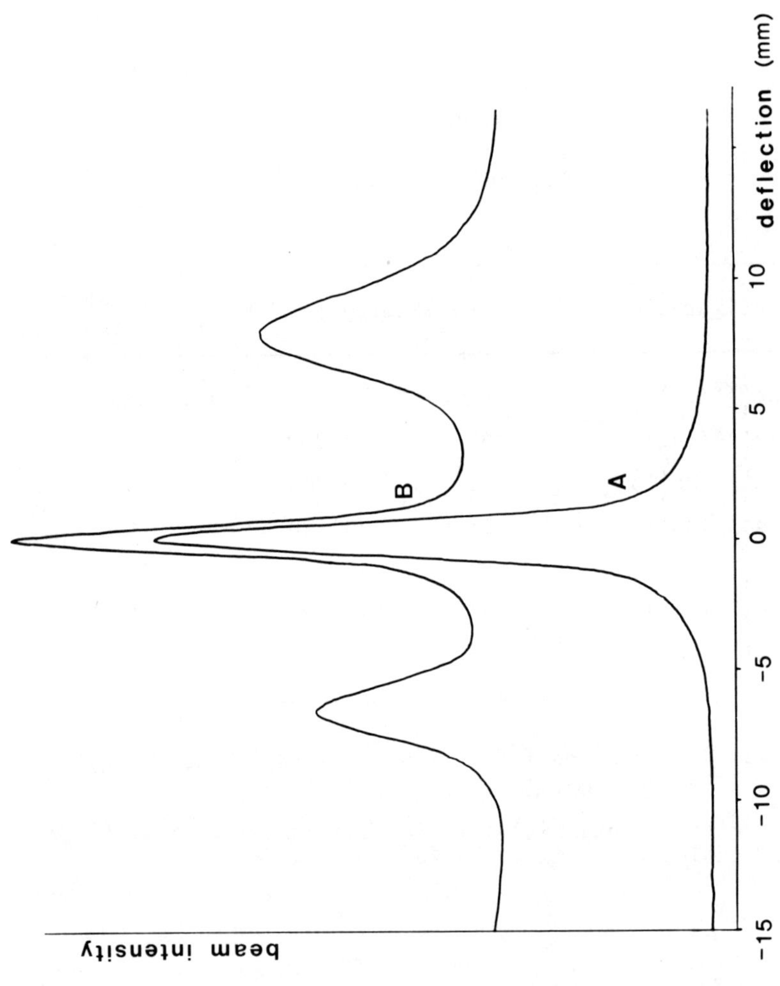

Figure 14. Molecular beam intensity profiles of a Na_x molecular beam, without (A) and with (B) Stern–Gerlach magnetic field on. The undisplaced beam is Na_2 (+ traces of Na_4), the displaced intensities are mainly Na atoms, the profile is the Maxwell–Boltzmann distribution mapped out onto the deflection angle.

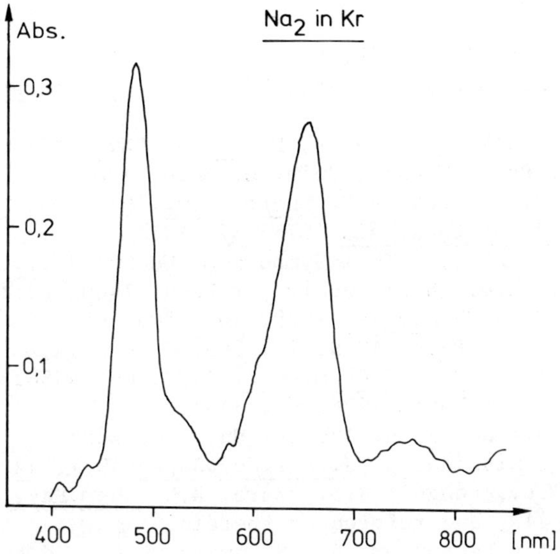

Figure 15. Matrix spectrum of a high purity Na_2 beam. At the lower left flank of the right Na_2 A ← X band wiggles from a small population of atoms are visible. They are formed during the measurement of the spectrum by photolysis (1).

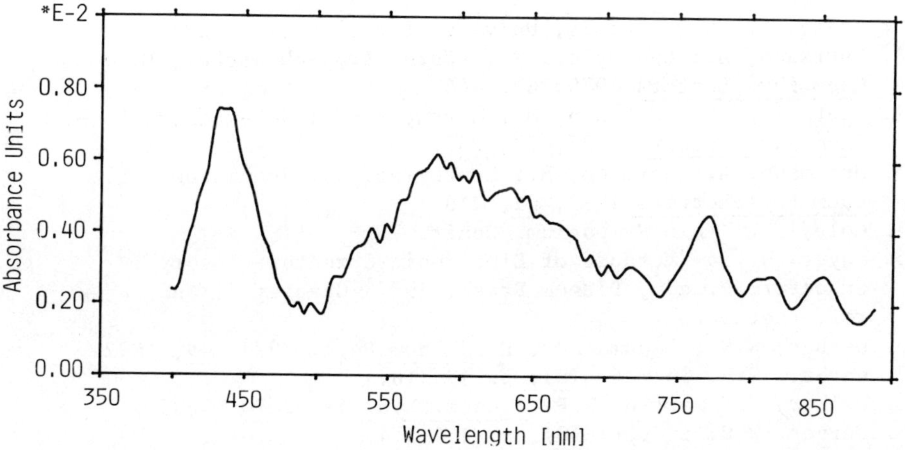

Figure 16. Absorption spectrum of Stern–Gerlach enriched Na_3 in Kr (M/R = 23'000). Notice the small optical density. The peak at 440 nm corresponds to a photodissociative gas phase transition (1).

Literature Cited

1. Hofmann, M. Ph.D. Thesis, Univ. of Bern, 1980.
2. Herrmann, A.; Schumacher, E.; Wöste, L. J.Chem.Phys. 1978, 68, 2327.
3. Flad, J.; Stoll, H.; Preuss, H. J.Chem.Phys. 1979, 71, 3042.
4. Scholl, E. Ph.D. Thesis, Univ. of Bern, 1980.
5. Baetzold, R.C.; Mack, R.E. Inorg.Chem. 1975, 14, 686.
6. Chini, M. J. of Catalysis 1975, 37, 187.
7. Wu, C.H.; Ihle, H.R. This Symposium, Confr. 14.7.
8. Gerber, W. Ph.D. Thesis, Univ. of Bern, 1980; p. 151.
9. Herzberg, G. J.Chem.Phys. 1979, 70, 4806.
10. Wu. C.H. J.Chem.Phys. 1979, 71, 783.
11. Wu, C.H.; Ihle, H.R. J.Chem.Phys. 1977, 66, 4356.
12. Wu, C.H. J.Chem.Phys. 1976, 65, 3181.
13. Huber, D. planned Ph.D. Thesis, Univ. of Bern.
14. Kimoto, K.; Nishida, I. J.Phys.Soc.Japan 1977, 42, 2071.
15. Feldman, D.L.; Lengel, R.K.; Zare, R.N. Chem.Phys.Letters 1977, 52, 413, and references therein.
16. Herrmann, A.; Leutwyler, S.; Schumacher, E.; Wöste, L. Chem.Phys.Letters 1977, 52, 418.
17. Herrmann, A.; Leutwyler, S.; Schumacher, E.; Wöste, L. Helv.Chim.Acta 1978, 61, 453.
18. Leutwyler, S.; Herrmann, A.; Wöste, L.; Schumacher, E. Chem.Phys. 1980, 48, 253.
19. Zare, R.N.; Dagdigian, P.J. Science 1974, 185, 739; Zare, R.N. Far.Disc.Chem.Soc. 1979, 67, 7.
20. Wöste, L. Ph.D. Thesis, Univ. of Bern, 1978; p. 221.
21. Herrmann, A.; Leutwyler, S.; Wöste, L.; Schumacher, E. Chem.Phys.Letters 1979, 62, 444.
22. Leutwyler, S.; Hofmann, M.; Härri, H.-P.; Schumacher, E. Chem.Phys.Letters 1981, 77, 257.
23. Herrmann, A.; Hofmann, M.; Leutwyler, S.; Schumacher, E. Chem.Phys.Letters 1979, 62, 216.
24. Gole, J.L. This Symposium, Confr. 75.
25. Meyer, W. in "Methods of Electronic Structure Theory"; Schaefer, H. ed., Plenum Press: 1977; Chapter 11 and ref. therein.
26. Gerber, W.H.; Schumacher, E. J.Chem.Phys. 1978, 69, 1692.
27. Gerber, W.H. in ref. (8), p. 19-116.
28. Welker, T.; Martin, T.P. J.Chem.Phys. 1979, 70, 5683.
29. Gerber, W.H. in ref. (8), p. 165-174.
30. Martin, R.L.; Davidson, E.R. Mol.Phys. 1978, 35, 1713.
31. Hofmann, M.; Leutwyler, S.; Schulze, W. Chem.Phys. 1979, 40, 145.
32. see ref. (1), and Hofmann, M.; Schumacher, E. Chem.Phys. 1981, in preparation.

33. Herrmann, A. Ph.D. Thesis, Univ. of Bern, 1978; p. 166.
34. Robbins, E.J.; Leckenby, R.E.; Willis, P. Advan.Phys. 1967, 16, 739.
35. Zmbov, K.F.; Wu, C.H.; Ihle, H.R. J.Chem.Phys. 1977, 67, 4603.
36. Beckmann, H.-O.; Koutechý, J.; Botschwina, P.; Meyer, W. Chem.Phys.Letters 1979, 67, 119.
37. Gole, J.L.; Childs, R.H.; Dixon, D.A.; Eades, R.A. J.Chem.Phys. 1980, 72, 6386.
38. Mathur, B.P.; Rothe, E.W.; Reck, G.P.; Lightman, A.J. Chem.Phys.Letters 1978, 56, 336.
39. Handbook of Chemistry and Physics, 53rd ed., CRC Publishing Company, 1972.

RECEIVED August 26, 1981.

9

Energetics of Small Metal Clusters

K. A. GINGERICH

Texas A&M University, Department of Chemistry, College Station, TX 77843

The Knudsen effusion method in conjunction with mass spectrometric analysis has been used to determine the bond energies and appearance potentials of diatomic metals and small metallic clusters. The experimental bond energies are reported and interpreted in terms of various empirical models of bonding, such as the Pauling model of a polar single bond, the empirical valence bond model for certain multiply-bonded diatomics, the atomic cell model, and bond additivity concepts. The stability of positive ions of metal molecules is also discussed. Examples of the influence of recent experimental and theoretical results on the improvement of dissociation energies obtained from equilibrium measurements are given.

The knowledge of the properties of diatomic metals and small metallic clusters is of particular importance due to their current use as models in catalytic activity. The intense current interest in these quite novel species is evidenced by the attention they received at the Faraday Symposium No. 14. (1)

The most commonly measured property for these types of molecules is their dissociation energy or atomization energy. According to a recent review (2) these have been experimentally determined for approximately 50 homonuclear diatomic metal molecules, 15 polyatomic metal molecules (including germanium but excluding silicon and antimony), 110 diatomic intermetallic compounds and more than 20 polyatomic intermetallic molecules. The most reliable values for the bond energies of the majority of these molecules have been obtained by using Knudsen effusion mass spectrometry (2). For some of them, notably the alkali dimers, and Group IIA and Group IIB dimers, as well as for some of their intermetallic diatomic molecules, optical spectroscopic methods have provided the best or only determination of dissociation energies (3,4).

0097-6156/82/0179-0109$05.00/0

Energetics of small metal clusters also include their ionization potentials, electron affinities and the dissociation energies of the corresponding positive and negative ions. Quite little is known about these properties (4), but significant advances have recently been made by Schumacher and associates, who measured precise ionization potentials of more than 20 alkali metal clusters (5). We will, in the following, place emphasis on the bond energies of the neutral diatomic metals and small metal clusters, touching only briefly on appearance potentials and on dissociation energies of molecular ions.

The Knudsen effusion mass spectrometric method will be briefly described and illustrated. The influence of various assumptions about the geometry and electronic structure of the molecules on the third law values of the reaction enthalpies obtained will be discussed. The development and use of empirical models for calculating dissociation and atomization energies will be illustrated and their value as a bridge between experimental values and semi-empirical and theoretical calculations demonstrated.

Knudsen Cell Mass Spectrometric Method

The Knudsen cell mass spectrometric method is well established and has been described in many reviews, as can be seen in references (6-10). It is an important method for equilibrium vapor studies of high temperature systems for temperatures up to approximately 3000 K. There is no other method presently available that permits the measurement of bond energies of minor molecular vapor components at such high temperatures.

Table I. Relative ion currents or partial pressures observed in the saturated vapor over metals for which polyatomic molecules have been observed

Element (M)	T (K)	Electron energy (eV)	I(M)	I(M_2)	I(M_3)	I(M_4)	Ref.
Li	903	7	1	1.37×10^{-3}	2.6×10^{-5}		(11)
Cu	1849	24	1[a)]	1.98×10^{-3}	2.3×10^{-8}		(12)
Ag	1500	24	1[a)]	2.29×10^{-3}	3.4×10^{-8}		(12)
Au	1883	24	1[a)]	2.56×10^{-3}	5.9×10^{-8}		(12)
Sn	1651	20	1	2.26×10^{-3}	1.8×10^{-4}	4.0×10^{-5}	(13)
Pb	1105	17	1[a)]	1.54×10^{-4}	1.5×10^{-6}	4×10^{-8}	(14)
Bi	923	18	1	1.18	1.06×10^{-2}	2.4×10^{-2}	(15)

[a)] Data corresponds to relative partial pressures.

Metal molecules typically are minor components of the equilibrium vapor above condensed metals as Table I (11-15) illustrates. Shown here are only those metals for which polyatomic molecules have been measured under equilibrium conditions. For most of the transition metals, including the lanthanides and actinides, the concentration of the diatom relative to hat of the atomic species is between 10^{-4} and 10^{-7} (10).

These low molecular concentrations in the vapor, especially above transition metals, are the major reason why the investigation of various other properties of metal molecules have proven to be such a challenge. Experimental methods had to be devised, such as controlled synthesis of metal molecules in inert gas matrices at cryostatic temperatures, or the enhancement of relative molecular concentrations through nozzle expansion followed by subsequent optical spectroscopic or other physical-chemical experimental studies. The success of such methods and the resulting physical property measurements will be the subject of several presentations in this symposium.

In the Knudsen effusion mass spectrometric method a molecular beam of the vapor effusate from the Knudsen cell is sampled and ionized in the ion source region, usually by electron bombardment. The ions are then accelerated and their mass analyzed. For the identification of the ions and their correlation with the corresponding molecular precursors, their mass-to-charge ratio, isotopic abundance distribution, appearance potential, ionization efficiency and intensity profile in the molecular beam are usually measured. From the measured ion currents the corresponding partial pressures are obtained using established calibration procedures. Knowing the partial pressures the equilibrium constants for the chemical reactions of interest can be determined. The corresponding reaction enthalpies can be evaluated by the third law method which requires the knowledge of absolute entropies or from the temperature dependence of the equilibrium constant (second law method). Evaluation by the second law method requires measurement over a sufficiently large temperature range. Often this is not possible in the case of metal molecules, due to their small concentration in the equilibrium vapor. Ideally, one applies both second and third law methods and agreement in both reaction enthalpies obtained supports the reliability of the results.

The measurements of the heat of sublimation of gold and the dissociation energy of diatomic gold, as summarized in Table II, are an example of the high accuracy that can be obtained by the Knudsen cell mass spectrometric method under favorable conditions (16). The Knudsen effusion mass spectrometric study by Kordis et al. (16) for the heat of vaporization of gold is in excellent agreement with the results from an interlaboratory study that has been the basis for the establishment of the vapor pressure standard of gold (17). This mass spectrometric study yielded in addition second and third law entropy values which are

Table II. Thermodynamic evaluation of the heat of sublimation of gold and the dissociation energy of Au_2

Reaction	Temp. range (K)	Second law ΔH°_{298} kJ	Second law ΔS°_{298} JK^{-1}	Third law ΔH°_{298} kJ	Third law ΔS°_{298} JK^{-1}
Au(s,l) = Au(g)	1338-1715	336.1±2.1	132.3±1.3	367.3±1.3	132.9
		368.8±2.1[a]		367.0±0.9[a]	
Au_2(g) = 2 Au(g)	1405-1950	223.2±2.1	91.4±1.3	223.7±1.0	91.6

[a] Selected NBS values from interlaboratory study (17).

[b] Selected values (in kJ mol^{-1}): $D^\circ_{298}(Au_2) = 223.4 \pm 2.1$ or $D^\circ_o(Au_2) = 221.3 \pm 2.1$.

in excellent agreement and which are not given in the interlaboratory study. Likewise the measurements yielded second and third law values for the enthalpy and entropy of dissociation of Au_2(g) that are in excellent agreement. Aiding in this study of the gold diatom was (a) the in situ temperature calibration at the melting point of gold under NBS test conditions for the gold melting point standard, (b) the fact that the molecular constants of Au_2 are well-known (4) and (c) the fact that a temperature range of more than 500 K could be covered in the measurement of the dissociation reaction.

Table III. Dissociation energy, D°_o, of Hg_2. Values are in kJ mol^{-1}.

D°_0	Methods		Ref.
2.1 - 14.6	Thermochemical, using		(19)[a]
1.5±2	absorption spectra to		(3)
6.8 - 8.8	follow equilibria		(4)[a]
7.1±1.9	3rd-law,	Knudsen cell mass	(18)
8.0±4.2	2nd-law,	spectrometry[b]	

[a] Winans and Heitz (19) evaluated the same spectroscopic data as did Huber and Herzberg (4) but included corrections for rotation, which according to (4) did not apply.

[b] Using 20 volt electrons.

The recent study by Hilpert (18) of the van der Waals molecule Hg_2, illustrated in Table III (3,4,18,19), is an example of the high sensitivity attainable with the method. The investigation also showed that the equilibria involving a very weakly bonded molecule could be reliably studied although the molecule had been ionized by electrons with an energy of approximately two hundred times that of its dissociation energy. At the same time this study confirmed a recent reevaluation by Huber and Herzberg (4) of thermal spectroscopic literature data.

Selected Atomization Energies of Small Metal Clusters and Diatomic Positive Ions

The experimental bond energies of diatomic metals and small clusters have very recently been reviewed (2). Therefore, only the atomization energies that have since become available have been listed in Table IV. Also included are previously known molecules for which revised values have been reported. The error limits for AuCs and AuRb have been increased by the difference in the value for $D_0^\circ(Au_2)$ in references (20) and (16). For BiIn and Bi_2In the error limits have been estimated as twice the third law standard deviation plus half of the difference between the reported second and third law values (21).

Table IV: Atomization energies, D_0°, of gaseous metal molecules and intermetallic compounds not covered in Reference (2). Values are in kJ mol^{-1}.

Molecule	D_0°	Ref.	Molecule	D_0°	Ref.
Au_3	367±13	(12)	Bi_2Li_2	583.0±15	(24)
AuCs	249±13	(20)	Cu_3	294±13	(12)
AuRb	239±13	(20)	Hg_2	7.5±1[a]	(4,18)
Ag_3	253±13	(12)	LaPt	496±21[a]	(25)
BiIn	147.9±4.3	(21)	LiPb	74.9±8	(23)
Bi_2In	354.2±14	(21)	LiBiPb	346.6±6	(24)
BiLi	150±5	(22,23)	Pb_2	83±1[a]	(14,22)
$BiLi_2$	326±10	(24)	Pt_2	358±15	(26)
Bi_2Li	367.4±7	(24)	PtY	470±12	(26)

[a] Revised.

For Hg_2 the new mass spectrometric value by Hilpert (18) has been taken as the average of the second and third law values, 7.5±2 kJ mol^{-1}, and coincides with the value given in (2). This value was in turn based on the assessment by Huber and Herzberg (4) of previous spectroscopically determined thermochemical data. In view of the excellent agreement of the results obtained from various methods the error limit has been reduced to ±1 kJ mol^{-1}.

As part of the discussion of the energetics of small metal clusters, the ionization potentials, electron affinities and the atomization energies of the corresponding ions are to be included. Of these, the ionization potentials and, when possible, the dissociation energies of the positive ions are briefly considered here. In connection with the mass spectrometric determination of atomization energies of metal molecules no particular emphasis is usually given to the determination of precise appearance potentials. The measurement of precise appearance potentials would require special techniques such as the retarding potential difference method (27), which produces nearly mono-energetic electron beams. However, the application of these methods results in an inherent loss in sensitivity. This is a severe drawback in the case of equilibrium studies of metal molecules since, as a rule, such molecules are minor vapor components and maximum sensitivity is required for their thermodynamic evaluation. However, very precise ionization potentials can be measured using photoionization spectroscopy (5,28). Berkowitz (28) reviewed early work concerning alkali metal dimers. Herrmann et al. (5) have measured the ionization potentials of numerous sodium, potassium and mixed sodium-potassium clusters. For most of these clusters the atomization energies of the neutral molecules are not known. Therefore, the dissociation energies of the corresponding positive ions cannot be calculated.

For selected diatomic metal molecules, the ionization potentials and the dissociation energies of the M_2^+ ions, are given in Table V. The data was taken from Herrmann et al. (5), Huber and Herzberg (4) and Zmbov et al. (29). As can be seen from the table the dissociation energies of the homonuclear ions are always larger than those of the neutral parent molecules. This increase in stability is particularly pronounced for Hg_2^+ and is considered the principal reason that the weakly bonded Hg_2 molecule does not tend to fragment but rather form positive ions when bombarded with 20 eV electrons as inferred from Hilpert's mass spectrometric study (see Table II). Similar stable unipositive ions are expected for the other group IIA and group IIB metals. Thus, there is the possibility of a Knudsen effusion mass spectrometric determination of dissociation energies of the corresponding homonuclear van der Waals dimers. The Tl_2 molecule, recently predicted to be of the van der Waals type (30), apparently also belongs in this category.

Table V: Dissociation Energies of M_2^+ ions. Values are in eV

M_2	$D_0^\circ(M_2)$	I.P. (M)	I.P. (M_2)	$D_0^\circ(M_2^+)$ [a)]	Ref.
Li_2	1.04_6	5.39	5.00	1.44	(4)
LiNa	0.90	5.138	4.94	0.99_4, 1.09_8	(4,29)
Na_2	0.720	5.138	4.90	0.95_8	(5)
NaK	0.62	4.339	4.52	0.44	(5)
K_2	0.514	4.339	4.06	0.79_3	(5)
KLi	0.81	4.339	4.69	0.46	(29)
Rb_2	0.49	4.176	$3.44 \leq$ I.P. $\lesssim \sim 3.95$	≥ 0.72	(4)
Cs_2	0.394	3.893	$3.59 \leq$ I.P. $\lesssim$ 3.821	0.61	(4)
Hg_2	0.07_8	10.43	$9.39 <$ I.P. $\lesssim$ 9.61	$0.91 \lesssim D_0^\circ \lesssim 1.11$	(4)

[a)] Using: $D_0^\circ(M_2^+) = D_0^\circ(M_2) + I.P.(M) - I.P.(M_2)$.

Empirical Correlations of Bond Energies and Comparison with Experiment

The use of empirical models of bonding has been invaluable for the interpretation of the experimental dissociation energies of diatomic intermetallic molecules as well as for the prediction of the bond energies of new molecules. In the course of our work, conducted for over a decade, we have extended the applicability of the Pauling model of a polar single bond (31) and have developed new models such as the empirical valence bond model for certain multiple bonded transition metal molecules (32,33) and the atomic cell model (34).

According to the Pauling model, the bond energy, D(A-B) of a diatomic molecule, AB, may be expressed by the relation:

$$D(A\text{-}B) = \tfrac{1}{2}[D(A\text{-}A) + D(B\text{-}B)] + 96\ (X_A - X_B)^2 \quad (\text{in kJ mol}^{-1})$$

where D(A-A) and D(B-B) are the "single bond" energies of the component elements A and B, and X_A and X_B the respective electronegativities. The range of applicability of this model to diatomic intermetallic compounds has been significantly extended during the last decade due to the experimental determination of the bond energies of many new diatomic metal molecules. The Pauling model can also be used for triatomic intermetallic molecules (35). In this case it is applied to each individual heteronuclear bond and the contributions of the various bonds in the molecule are added to yield the atomization energy. An illustration of the application of the Pauling model for diatomic and triatomic intermetallic compounds with gold is given in Table 8 of Reference (2). Except for the barium-gold compounds, the agreement between calculated and experimental values is satisfactory. In principle the model can be used to calculate the bond energies of many intermetallic molecules (36). Limitations of the model appear in connection with the molecules of alkali and alkaline earth compounds with group IB metals. The model gives bond energies which are too high for these molecules. Similar limitations have been previously noted by Pauling in connection with alkali and alkaline earth hydrides (31). The barium-gold compounds in Table 8, Ref. (2) are an example. Another example are many lithides (see Table V in Ref. (23). The model is also not expected to apply to multiple bonded molecules as had been predicted by Brewer (37) for combinations between group IV transition metals and platinum metals. In Brewer's treatment of the corresponding condensed alloys (38) the platinum metal atom, which contains filled d-orbitals, transfers some of the d-electrons to the group IV transition metal that contains vacant d-orbitals, thus providing on either atom an increased number of unpaired d-electrons. This increased number of unpaired electrons leads to efficient multiple bonding, which accounts for the high stability observed for the corresponding alloys (38).

In our mass spectrometric work we have tested and confirmed the high stability of corresponding diatomic gaseous molecules (39) and subsequently shown that actinide, rare earth, and group V transition metals also form such stable diatomic molecules with platinum metals with a bond order of more than one. This has led to an empirical valence bond model (32,33) for calculating the dissociation energies of such intermetallic compounds. In this model, the bond energies per electron are taken to be the same as the corresponding values calculated by Brewer (38) for the various kinds of s, p or d valence electrons on the basis of the cohesive energies and valence electron configurations of pure transition metals. Then, knowing the number of valence state bonding electrons for each given pair of atoms in the molecules and the valence state promotion energies, the dissociation energies can be estimated.

The empirical valence bond model has shown good predicting power if one defines the bond multiplicity to be two for compounds such as CePd or BaPd and three for compounds such as LaRh, YIr, or RhTh. The limitations of the model could also be shown (40) when applied to assumed quadruple bond formation, as in RuV or ThRu or to double bond formation as in CePt or ThPt. The case of ThRu (41) shows that quadruple bonding is approached where suitable valence states are available, but not fully achieved, apparently because of the directional requirements of the bonding in a diatomic molecule. The examples of CePt and ThPt (40) show the calculated values that have been based on an assumed double bond to be too low. Platinum has no suitable valence state for triple bond formation, but apparently forms triple bonds with other atoms when the latter has a suitable valence state. In spite of certain limitations, which can be identified, the model can account for or predict the dissociation energies of about 200 intermetallic molecules with multiple bonds (40). It has been complementary to the Pauling model. Comparison of calculated and predicted values have been given elsewhere (2,10,40).

The atomic cell model proposed by Miedema and Gingerich (34) for calculating enthalpies of formation of diatomic intermetallic molecules is more general than the empirical valence bond model or the Pauling model in its application to intermetallic diatomic molecules. It is based on a model description by Miedema and associates (42) that was originally designed to account for the heat of formation of solid and liquid metal alloys. In simplified form, if the metal atoms are of approximately equal size the dissociation energy, D(A-B) may be expressed by

$$D(A-B) = \tfrac{1}{2}[D(A-A) + D(B-B)] + P(\Delta\phi^*)^2 - Q(n_{ws}^{1/3})^2$$

where P and Q are empirical constants, ϕ^* is the electronegativity parameter, which is very similar to the work function of pure metals, ϕ, and n_{ws} is a semi-empirical parameter representing the electron density at the boundary of a Wigner-Seitz atomic cell in the solid pure metal. This relation may be compared to the Pauling relation for a polar single bond (31) discussed above. For metal atoms of different size a correction term needs to be added. Other special cases are discussed in the original paper (34). Miedema has since presented the calculated values for the dissociation energies of all diatomic transition metal intermetallic molecules (43). A comparison of experimental values for intermetallic diatomic molecules with gold with the corresponding value calculated by the Pauling model and by the atomic cell model has been given in Table 6 of Reference (2). Table 7 of Reference (2) shows a comparison between experimental dissociation energies with values calculated by the atomic cell model and the empirical valence bond model. Table 9 of Reference (2) takes Miedema's refinements (43) of the atomic cell model into account in these comparisons.

Effect of Estimated Molecular and Electronic Structure on Third Law Reaction Enthalpies

For the third law evaluations of the reaction enthalpies from mass spectrometric equilibrium measurements the Gibbs free energy functions of the reactants are needed. Likewise the enthalpy functions are needed in order to correct the second law reaction enthalpies obtained at the average temperature of measurement to the desired reference temperature. These thermal functions can be calculated according to standard statistical thermodynamic methods from known or estimated molecular parameters. Sources containing known molecular parameters or papers describing empirical correlations for estimating them have been listed elsewhere (10). Here the effects of assumptions concerning the variation of assumed geometries for triatomic metal molecules or the variation in electronic structure of diatomic metal molecules are illustrated using examples of current interest.

The first example is taken from the study of triatomic Cu_3, Au_3 and Ag_3 (12). The evaluation of the third law enthalpies of the reaction $M_3(g) = 3M(g)$ has been based on three different possible geometries: linear ($D_{\infty h}$), bent (C_{2v}), and triangular (D_{3h}). The effect of these various assumptions in geometry on the third law reaction enthalpies (atomization energies) is illustrated in Table VI. The current interest in

Table VI. Effect of variation in the assumed geometry for $M_3(g)$ on the enthalpy change, ΔH°_0, for the reaction $M_3(g) = 3\ M(g)$ (M = Cu,Ag,Au). Values are in kJ mol^{-1}.[a)]

Molecule	No. of data points	Temp-range (K)	linear $D_{\infty h}$	equil. triangle D_{3h}	bent C_{2v}
Cu_3	6	1801-1897	293.5 ± 4.7	294.2 ± 4.6	282.8 ± 4.8
Ag_3	13	1402-1599	253.3 ± 5.1	254.9 ± 4.9	245.4 ± 5.1
Au_3	8	1780-1985	367.1 ± 7.2	366.1 ± 6.9	354.2 ± 7.3

[a)] Taken from reference (12).

the effect of various possible geometries in triatomic metal molecules is documented by its discussion in the presentations by Dixon, Gole, Schumacher, and Moskovits during this Symposium. Apparently any of these three geometries may be favored in one specific molecule or the other. Recent theoretical calculations appear to indicate that the energy differences between these three structures are surprisingly small. The fluxional nature of the geometry was first noticed in ab initio calculations for Li_3 by Gerber and Schumacher (44) and further stressed by Dixon and Gole in this symposium for many other triatomic Group I metals.

The Pb_2 molecule is given as an example for which a more accurate third law value of the dissociation energy could be obtained from equilibrium measurements on the basis of recent experimental spectroscopic data and improved theoretical estimates. Gingerich et al. (14) have studied this molecule by Knudsen effusion mass spectrometry. They assumed an electronic ground state multiplicity of 3 analogous to the lighter analog Si_2 which has a $^3\Sigma_g^-$ ground state (4). Recently Pitzer (22) drew attention to the fact that for molecules with very heavy atoms the relativistic effects yield large spin-orbit splittings which in some cases reduce the electronic partition function significantly. Spectroscopic data from Bondybey and English (45) confirm the ground state of Pb_2 as 0_g^+, a single state. Pitzer estimates the energy of excitation to the 1_g state (the other component of $^3\Sigma_g^-$) as about 7800 cm^{-1}. Thus at the experimental temperature of 1000 K, this state will be essentially unpopulated. In his revision of the reported experimental third law dissociation energy (see Table VII) Pitzer also considers the

Table VII: Revision of the dissociation energy of $Pb_2(g)$.[a)]

Approximate temperature	1000 K
Original ΔH_0^o of diss. (3rd law)[b)]	74±1 kJ mol^{-1}
Revised ΔH_0^o of diss. (3rd law)[a)]	82.5±1 kJ mol^{-1}
ΔH_0^o of diss. (2nd law) unchanged[b)]	84±2 kJ mol^{-1}
Selected value[a)]	83±1 kJ mol^{-1}

a) Ref. (22)

b) Ref. (14)

new value of the vibration frequency (45). As can be seen from the table the second and third law values are now in excellent agreement leading to a selected value, $D_0^o(Pb_2)$ = 83±1 kJ mol^{-1} (23) as compared with that of 79±6 kJ mol^{-1} originally reported (14). This illustration also demonstrates the considerable value of obtaining good second law data.

The final example illustrates the tremendeous effect the electronic contribution to the partition function can have on the third law dissociation energy of certain diatomic transition metals. As examples the molecules Ni_2 and Fe_2 are chosen. For these molecules Kant and collaborators have measured the equilibrium concentration of the dimer above liquid metals and derived the dissociation energies using estimated molecular parameters (46, 47). Recently the ground state vibrational frequencies have been measured by matrix isolation spectroscopy as ω_e= 381 cm^{-1} for Ni_2 (48) and ω_e= 299.6 cm^{-1} for Fe_2 (49). These values compared with the corresponding estimates of Kant et al. (46, 47) of 469 and 204 cm^{-1}, respectively, and their change have a comparatively small influence on the dissociation energies of the respective molecules. A much more significant correction results from the large number of low lying electronic states found in ab initio calculations for Ni_2 by Shim et al. (50), and Upton and Goddard (51) who independently predict 30 electronic states with relative energies between zero and 1.2 eV. For Fe_2 Shim (52) predicts even more, namely 112 low lying electronic states between 0 and 0.6 eV. Taking these low lying states into account leads to a significant downward correction of third law values obtained by Knudsen cell mass spectrometric experiments for the dissociation energy. This is illustrated in Table VIII.

Table VIII: Revision of third law dissociation energies of $Ni_2(g)$ and $Fe_2(g)$

Molecule	Ni_2	Fe_2	
Approximate temperature	2100 K	2000 K	
Original ΔH_0^o of diss. (3rd law)	220±21[a)]	126±21[b)]	kJ mol^{-1}
Revised ΔH_0^o of diss. (3rd law)	166[c)]	57[c)]	kJ mol^{-1}

a) Ref. (46)
b) Ref. (47)
c) Ref. (52)

The corresponding 2nd law value for $D_0^o(Ni_2)$ changes from 195±29 kJ mol^{-1} given by Kant (46) to 197 kJ mol^{-1}. For Fe_2 no reliable 2nd law value is available. In the discussion following this presentation, Professor Moskovits cited spectroscopic evidence that the dissociation energy of Fe_2 must be larger than the revised value of 57 kJ mol^{-1} given in Table VIII. Consideration of spin-orbit coupling effects in the theoretical calculations would be expected to lower the electronic contribution to the partition function and consequently raise the dissociation energy of Fe_2 somewhat. A reliable value cannot be given at present for the dissociation energy of Fe_2, but it will unlikely be higher than 80±25 kJ mol^{-1}.

The examples given in this chapter illustrate the significance of recent experimental and theoretical molecular parameters on the reported dissociation energies that had previously been based on equilibrium measurements and on estimated molecular parameters. This additional information will permit the calculation of improved dissociation energies on the basis of the same experimental equilibrium data as typically obtained from Knudsen effusion mass spectrometry.

Quite large downward corrections are also expected for other transition metal molecules due to the now expected metal like band electronic structure of other such molecules. Diatomic chromium is a recent example (53). The dissociation energies of Mo_2, Nb_2 and Pt_2 (2) that have been recently measured in our laboratory and for which we had considered no (Mo_2, Nb_2) or only a small (Pt_2) electronic contribution to the partition function are likely to be downward corrected, as has already been suggested by Brewer and Winn (54) for Mo_2 and Nb_2. Such corrections would bring the experimental values for these molecules closer to those predicted by various empirical models (54, 55).

We have shown that the Knudsen effusion method is currently the most widely applicable method to obtain the atomization energies of diatomic metals and metallic clusters. For diatomic intermetallic molecules various empirical models of bonding have been illustrated that permit the quite reliable prediction of the dissociation energies of previously undetermined intermetallic molecules, especially those involving the transition metals. These models thus bridge the gap between experimental measurements and semi-empirical and ab initio calculations which cannot as yet provide reliable predictions for dissociation energies. The empirical models offer comparatively little insight into the nature of bonding, although the relations between the molecular state and the condensed state are noteworthy. Semi-empirical and ab initio calculations often provide the experimentalist with useful predictions for molecular and electronic structure and, as their sophistication progresses, give an increasingly deeper insight into the nature of bonding. The recent progress in the experimental characterization of metal molecules by various experimental techniques and in theoretical calculations has been rewarding and we have illustrated its direct influence on improving the values for atomization energies from equilibrium data already at hand.

The study of diatomic metals and small metallic clusters has been presented here as a vivid and current example of the effects of the mutually beneficial and illuminating interaction between experimentalists and theoreticians - an interaction that has contributed greatly to the accelerated progress of this area of scientific endeavor.

Acknowledgements

The author thanks Drs. Irene Shim and Klaus Hilpert for the permission of using not yet published data. He also expresses his appreciation for the support by the National Science Foundation and the Robert A. Welch Foundation.

Literature Cited

1. Faraday Discussions of The Royal Society of Chemistry, Faraday Symposium 14, "Diatomic Metals and Metallic Clusters," 1980.
2. Gingerich, K. A. Faraday Symp. Chem. Soc. 1980, 14, 109.
3. Gaydon, A. G. "Dissociation Energies and Spectra of Diatomic Molecules"; 3rd Ed.; Chapman and Hall, London, 1968.
4. Huber, K. P.; Herzberg, G. "Molecular Spectra and Molecular Structure IV. Constants of Diatomic Molecules"; Van Nostrand Reinhold, New York, 1979.
5. Herrmann, A.; Schumacher, E.; Wöste, L. J. Chem. Phys. 1978, 68, 2327.
6. Inghram, M. G.; Drowart, J. in "High Temperature Technology"; McGraw-Hill Book Company, Inc., New York, 1960, pp. 219-240.
7. Grimley, R. T. in "Characterization of High Temperature Vapors"; Margrave, J. L., Ed.;Wiley-Interscience Publishers, New York, 1967, pp. 195-243.
8. Drowart, J.; Goldfinger, P. Angew. Chem. 1967, 79, 589; Angew. Chem. Intern. Edn. 1967, 6, 581.
9. Drowart, J. in Proc. Int. School Mass Spectrom. Lublijana, Yugoslavia, J. Marsel, Ed.; August 19-29, 1969, pp. 187-242.
10. Gingerich, K. A. in "Current Topics of Materials Science"; Kaldis, E., Ed.; North Holland Publishing Co., Amsterdam, 1980, Vol. 6, pp. 345-462.
11. Wu, C. H. J. Chem. Phys. 1976, 65, 3181.
12. Hilpert, K.; Gingerich, K. A. Ber. Bunsenges. Phys. Chem. 1980, 84, 739.
13. Gingerich, K. A.; Desideri, A.; Cocke, D. L. J. Chem. Phys. 1975, 62, 731.
14. Gingerich, K. A.; Cocke, D. L.; Miller, F. J. Chem. Phys. 1976, 64, 4027.
15. Wagner, L. C.; Grimley, R. T. Chem. Phys. Letters 1974, 29, 594.
16. Kordis, J.; Gingerich, K. A.; Seyse, R. J. J. Chem. Phys. 1974, 61, 5115.
17. Certificate of Analysis; Stand. Ref. Material 745; Gold, Natl. Bureau Stand. 1969.
18. Hilpert, K. presented at Faraday Symposium 14 (Ref. 1) and private communication, 1980.
19. Winans, J. G.; Heitz, M. P. Z. Physik 1952, 133, 291; 1953, 135, 406.
20. Busse, B.; Weil, K. G. Ber. Bunsenges. Phys. Chem. 1981, 85, 309.
21. Riekert, G., Doktorarbeit (Ph.D. Thesis) Universität Stuttgart, 1980.
22. Pitzer, K. S. J. Chem. Phys. 1981, 74, 3078.
23. Neubert, A.; Ihle, H. R.; Gingerich, K. A. J. Chem. Phys. 1980, 73, 1406.

24. Neubert, A.; Ihle, H. R.; Gingerich, K. A. presented at the 6. Int. Conf. Thermodynamics, Merseburg, D.D.R., Aug. 26-29, 1980.
25. Gingerich, K. A.; Nappi, B. M. Inorg. Chem. 1981, 20, 522.
26. Gupta; Satish, K.; Nappi, Bianca M.; Gingerich, Karl A. Inorg. Chem. 1981, 20, 966.
27. Fox, R. E.; Hickam, W. M.; Kjeldaas Jr., T.; Grove, D. J. Phys. Rev. 1951, 84, 859.
28. Berkowitz, J. in "Advances in High Temperature Chemistry"; Eyring, L.; Ed.;Vol. 3, pp. 123-176.
29. Zmbov, K. F.; Wu, C. H.; Ihle, H. R. J. Chem. Phys. 1977, 67, 4603.
30. Christiansen, P. A.; Pitzer, K. S. J. Chem. Phys. 1981, 74, 1162.
31. Pauling, L. "The Nature of the Chemical Bond"; Cornell University Press, Ithaca, N.Y., 3rd Edn, 1960.
32. Gingerich, K. A. Chem. Phys. Letters 1973, 23, 270.
33. Gingerich, K. A. J.C.S. Faraday II 1974, 70, 471.
34. Miedema, A. R.; Gingerich, K. A. J. Phys. B. 1979, 12, 2255.
35. Gingerich, K. A. Chem. Phys. Letters 1972, 13, 262.
36. Gingerich, K. A. Chimia 1972, 26, 619.
37. Brewer, L. private communication, 1968.
38. Brewer, L. Science 1968, 161, 115.
39. Gingerich, K. A.; Cocke, D. L. Chem. Commun. 1972, 536.
40. Gingerich, K. A. Int. J. Quantum Chem. Sym. 1978, 12, 489.
41. Gingerich, K. A. Chem. Phys. Letters 1974, 25, 523.
42. Miedema, A. R.; Boom, R.; de Boer, F. R. Calphad 1977, 1, 341.
43. Miedema, A. R. Faraday Symp. Chem. Soc. 1980, 14, 136.
44. Gerber, W. H.; Schumacher, E. J. Chem. Phys. 1978, 69, 1692.
45. Bondybey, V. E.; English, J. H. J. Chem. Phys. 1977, 67, 3405.
46. Kant, A. J. Chem. Phys. 1964, 41, 1872; 1968, 49, 5144, (revised).
47. Lin, S. S.; Kant, A. J. Phys. Chem. 1969, 73, 2450.
48. Ahmed, F.; Nixon, E. R. J. Chem. Phys. 1979, 71, 3547.
49. Moskovits M.; DiLella, D. P. J. Chem. Phys. 1980, 73, 4917.
50. Shim, I.; Dahl, J. P.; Johanson, H. Int. J. Quant. Chem. 1979, 15, 311.
51. Upton, T. H.; Goddard III, W. A. J. Amer. Chem. Soc. 1978, 100, 5659.
52. Shim, I. private communication 1980.
53. Goodgame, M. M.; Goddard III, W. A. J. Phys. Chem. 1981, 85, 215.
54. Brewer, L.; Winn, J. S. Faraday Symp. Chem. Soc. 1980, 14, 127.
55. Miedema, R. A.; Gingerich, K. A. J. Phys. B. 1979, 12, 2081.

RECEIVED September 17, 1981.

10

Bound–Free Transitions in Weakly Bound Metal Aggregates

J. L. GOLE and G. J. GREEN

Georgia Institute of Technology, Department of Chemistry, Atlanta, GA 30332

Supersonic expansions have been used to form small metal aggregates, M_n ($2 \leqslant n \leqslant 4$). Emphasis is placed on the analysis of bound-free transitions in these small metal clusters. Discussion focuses on the characterization of variously produced sodium supersonic expansions and the analysis of laser induced atomic fluorescence resulting from the photodissociation of triatomic sodium clusters. We will consider: (1) the nature of observed "fluctuation" bands corresponding to bound-free transitions involving a repulsive excited state which dissociates to yield ^2P (Na-D line) sodium atoms and ground state, $^1\Sigma_g^+$, Na_2; (2) the possibility of controlled "hot band" formation through disruption of the free jet expansion allowing the mapping of vibronic structure in the ground electronic state of sodium trimer, and (3) the observation of very efficient energy transfer between sodium dimer and polyatomic sodium. The current studies are discussed in the light of recent experimental and theoretical characterization of Na_3. Preliminary studies of the reaction $Na_3 + Cl \rightarrow NaCl + Na_2$ are discussed as they pertain to the dissociation energy of Na_3.

Because of their probable importance to the understanding of the fundamental mechanisms of catalysis and numerous chemical conversions, the basic properties (geometry, bond strength, reactivity) of small metallic clusters M_n ($2 \leqslant n \leqslant 6$) have become the subject of intense theoretical and experimental study (1-36). Because experimental characterization is complicated, theory abounds and experimental studies are much less prevalent. Ligand-

0097-6156/82/0179-0125$07.00/0

free metallic clusters have been produced in the gas phase through "free jet" supersonic expansion of the pure element (18-21) or a seeded gas (helium)-metal mixture and through direct cluster deposition or diffusion controlled clustering in rare gas matrices (22-28).

In our laboratory, supersonic expansions of refractory metals are being characterized using a combination of laser induced fluorescence and mass spectroscopy. Specific emphasis is on the metals copper, sodium, silver, and antimony where we are attempting to produce and excite fluorescence from "cooled" small metal clusters, M_n ($2 \leq n \leq 4$). Although the production of cold clusters represents one of the primary goals in our efforts, we are also concerned with the controlled "Disruption" of the supersonic expansion in order to obtain "hot band" fluorescence. Collisional heating facilitates the characterization of a significant component of the vibronic structure for the ground electronic states of these "weakly" bound species. In the present discussion, we will concentrate primarily on a segment of this work which compliments the efforts of Professor Schumacher and his coworkers (37,38). We will focus on the characterization of bound-free transitions in the weakly bound sodium trimer molecule.

In the hope of obtaining optical spectra for sodium clusters, Na_n, several groups are investigating the supersonic expansion of pure sodium vapor (19-21,37), direct specific cluster deposition through Stern-Gerlach selection in rare gas matrices (38), and diffusion controlled clustering. The most significant results thus far have been obtained by Schumacher and coworkers who have studied the spectroscopy of Na_3 employing a combination of optical excitation and photoionization, clusters being ionized in a two photon process and detected mass spectrometrically. Vibronic structure attributed to Na_3 and assigned to an excited state progression has been detected in a spectral region centered at ~6250 Å. A stronger system of bands centered at ~6700 Å appears to result from the combination of both a short progression and weak sequence structure. The studies reported here are indeed quite complimentary to the results obtained for bound-bound transitions in two-photon ionization spectroscopy. Here we are concerned with the pumping of transitions at slightly higher energies and the characterization of the products of disssociation along a repulsive upper excited state potential.

Experimental

The apparatus used in the studies which are the primary focus of our discussion is shown schematically in Figure 1. The details of the apparatus are discussed elsewhere (1,39). Briefly, the output of an argon ion pumped dye laser is brought into a suitably equipped vacuum chamber. Here, the laser beam intersects a supersonically expanded metal beam produced through use of an appropriate oven system. The oven depicted in the figure is a

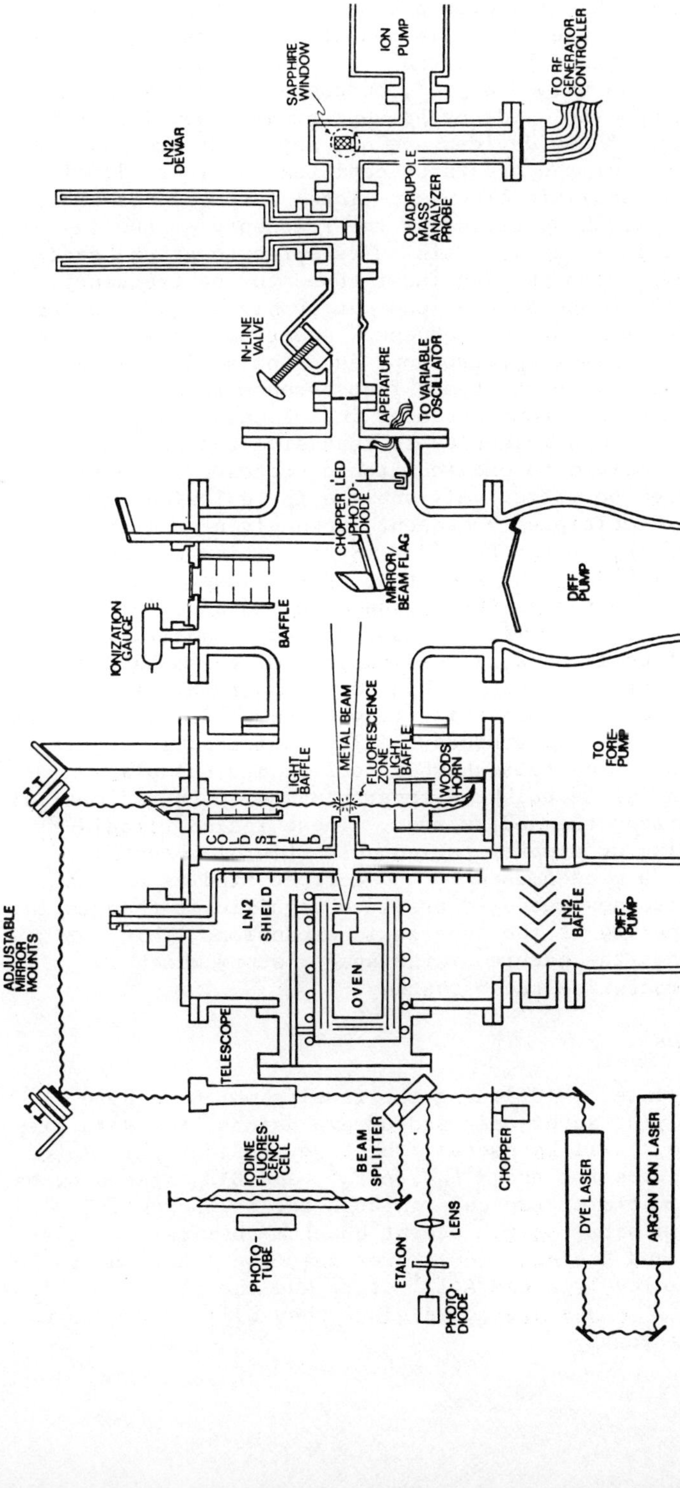

Figure 1. Diagram of apparatus for supersonic expansion of refractory metals.

double oven used for the supersonic expansion of sodium; it consists of a large backing oven and smaller frontal "nozzle" device. In general, the nozzle is run at a temperature approximately 50 to 100 degrees hotter than the backing oven. The fluorescence zone is viewed at 90°, on one side by a blank phototube and on the opposite side by an appropriate focusing system and spectrometer. These devices are not shown but are located in the back and front of the apparatus depicted in Figure 1. Two types of spectral analysis are used. Total fluorescence is collected by the blank phototube as the frequency of the incoming dye laser is scanned. This is referred to as an "excitation" spectrum. With the dye laser tuned to one frequency, the light from the fluorescence zone can be dispersed by a spectrometer yielding a "photoluminescence" spectrum. A variant of these two types of laser spectroscopy is associated with the monitoring of laser induced atomic fluorescence resulting from electronically excited atoms produced in molecular photodissociation. A laser pumps molecules to repulsive excited electronic states which dissociate to excited atomic states. A spectrometer is then used to selectively analyze the emission from individual atomic multiplet components in a given excited atomic state ($^2P_{3/2}$ and $^2P_{1/2}$ for the ^2P - ^2S transition corresponding to the Na D-line). Here the magnitude of the laser induced atomic emission is noted as the frequency of an incoming dye laser is scanned. Coincidental with the spectroscopic analysis of laser induced atomic fluorescence is the analysis of the mass distribution of the supersonic beam. This is accomplished either with electron impact or photoionization (light source focused through sapphire window) mass spectroscopy. Also indicated in Figure 1 are the positions of certain chamber cooling devices placed directly in front of (a bulkhead separating the oven and optical chambers) and around the sodium oven. These include liquid nitrogen shielding or a water cooled cold shield. Several cooling schemes have been used in the current studies and the variation of those devices used provides significant changes in the nature and purity of the supersonic expansion. This, of course, influences the nature of the spectroscopy which is observed to characterize these systems.

Sodium Expansions

For the purpose of outlining the results which we have obtained in studying supersonic sodium expansions, the data presented in Figure 2 will be useful. Here we consider calculated RKR potential curves for the $X^1\Sigma_g^+$, $A^1\Sigma_u^+$, and $B^1\Pi_u$ states of Na_2 and a rough schematic of two cuts through the Na_3 potential surface based, in part, on the recent quantum chemical calculations of Martin and Davidson (40). For Na_2, the location of the 33rd vibrational level in the $A^1\Sigma_u^+$ state and the 18th vibrational level in the $X^1\Sigma_g^+$ state are noted since they will be the focus of future discussion.

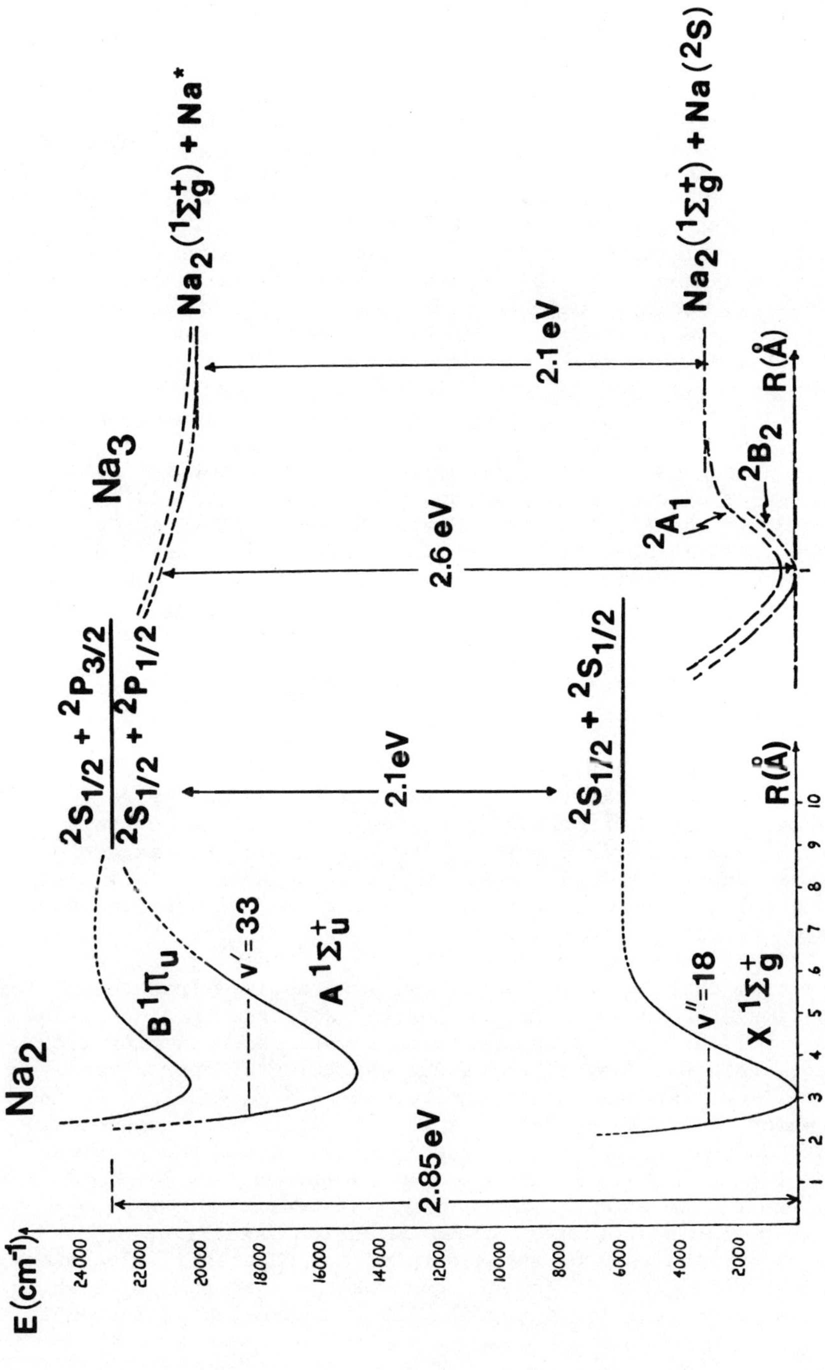

Figure 2. Potential curves for Na_2 generated using known spectroscopic data. Rough cuts in the ground and excited state Na_3 surfaces indicate energetics with respect to Na_2 for the bound–free systems studied.

We wish to use a supersonic expansion to generate cooled metallic species and to induce the formation of small polymeric metal clusters. In order to excite fluorescence from polymeric sodium compounds, spectral regions must be chosen with minimal atomic and diatomic sodium absorption. This requirement results because the oscillator strengths for sodium atom or dimer transitions are expected to exceed significantly those for any sodium polymer and thus spectra in an overlapping region will be dominated overwhelmingly by atomic or diatomic features. It we obtain the requisite cooling in supersonic expansion, Figure 2 suggests an approach whereby we can attempt to excite features from trimeric sodium. Several workers (1-3,41,42) have demonstrated that supersonically formed Na_2 is readily produced at rotational and vibrational temperatures such that the overwhelming molecular population is in the v" = 0 level of the ground electronic, $^1\Sigma_g{}^+$, state. Excitation of molecular species in the sodium expansion over the wavelength range 5500-6600 Å produces a long Na_2 progression corresponding to optical pumping from the v" = 0 level of the ground electronic state to levels v' = 10-33 in the $A^1\Sigma_u{}^+$ state. Given the appropriate laser dyes and optics, it is possible to access v' levels less than 10. The highest accessible level, v' = 33 has a Franck-Condon factor with respect to v" = 0 approaching 10^{-6}. A portion of this Na_2 progression, obtained as an excitation spectrum, including some pumping from v" = 1, is shown in Figure 3. In a system in which successful and "complete" cooling has been accomplished, no sodium dimer transitions can be pumped in the region from 18000 to 20000 cm^{-1}

Based on the calculations of Martin and Davidson, we attempted to excite fluorescence from Na_3 in the energy region lying between the $A^1\Sigma_u{}^+$ and $B^1\Pi_u$ states. The success of this venture is indicated in Figure 4. Here we depict the nature of the experimental excitation spectrum (top) obtained upon pumping (at an effective resolution of 0.5 cm^{-1}) to the highest accessible levels of the $A^1\Sigma_u{}^+$ state (from v" = 0, $X^1\Sigma_g{}^+$). Unfortunately, in addition to the features corresponding to pumping from v" = 0, the spectrum contains several virtually unresolved features. These can be simulated in a computer generated spectrum (Fig. 4b.) and they are believed to correspond primarily to laser pumping from ground state hot bands. The feature onsetting at approximately 18200 cm^{-1} corresponds to the excitation of $B^1\Pi_u$ - $X^1\Sigma_g{}^+$ fluorescence which would be inaccessible if the system were completely cooled. The experimental spectrum in Figure 4 and the progression in Figure 3 were obtained using a liquid nitrogen cooled "front" shield however chamber pumping speed was not sufficient to obtain a completely clean expansion (see Figure 1).

A closer analysis of the nature of our supersonic expansion is indicated in Figure 5 for a region of the spectrum at slightly lower energy to that in Figure 4. The top spectrum corresponds

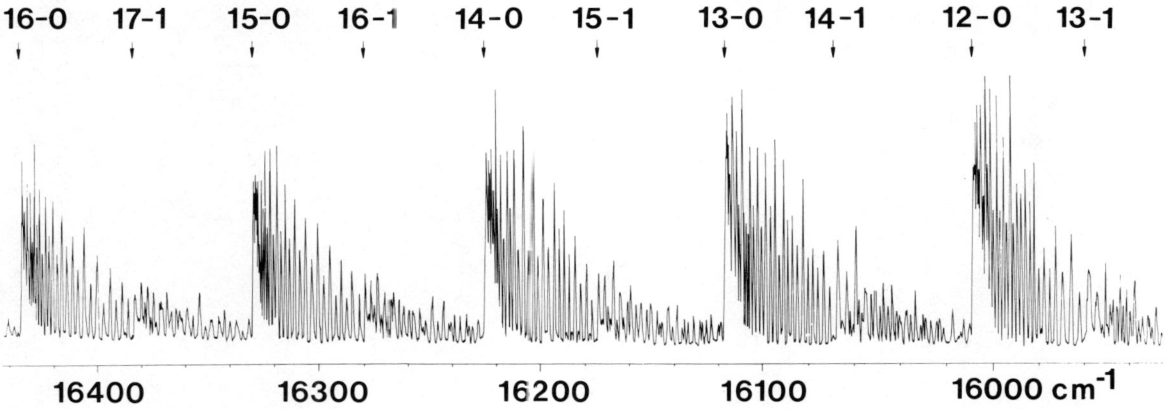

Figure 3. Excitation spectrum for supersonically expanded Na corresponding to the Na_2 $A^1\Sigma_u^+ - X^1\Sigma_g^+$ transition. Bands are denoted (v', v''). Several excited state (v') levels are pumped upon excitation from the lowest v'' = 0,1 levels of the ground electronic state. This spectrum correlates with $T_{Vib} \cong 50$ K, $T_{Rot} \cong 30$ K.

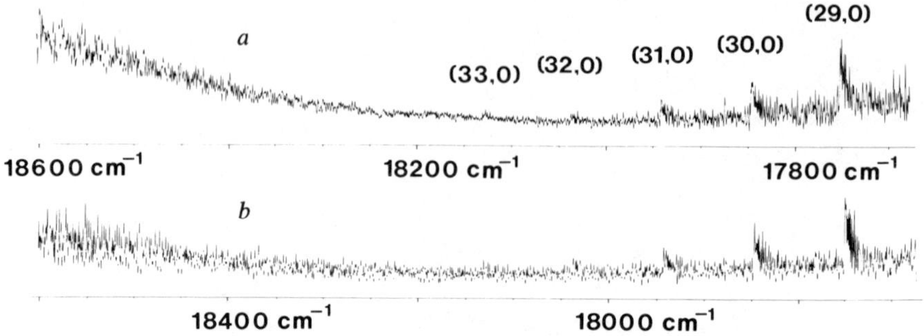

Figure 4. a: Excitation spectrum for supersonically expanded Na corresponding to pumping in the energy range 17,700–18,600 cm^{-1}. This spectrum shows the onset of $Na_2B^1\Pi_u - X^1\Sigma_g^+$ (~18,100 cm^{-1}) and the tail of $Na_2A^1\Sigma_u^+ - X^1\Sigma_g^+$ total fluorescence. b: Computer simulation of excitation spectrum.

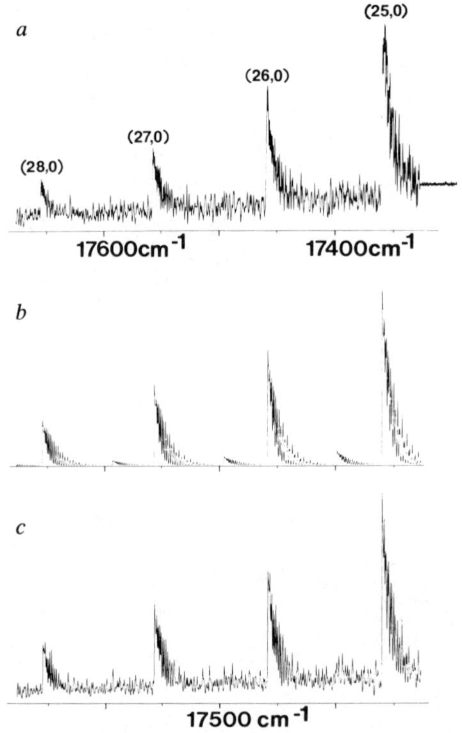

Figure 5. a: Excitation spectrum for supersonically expanded Na corresponding to the pumping of several vibrational levels in the $A^1\Sigma_u^+$ state. b: Computer simulation for an Na_2 vibrational temperature of 50 K and a rotational temperature of 30 K. c: Computer simulation: 99.75% of Na_2 molecules at $T_{Vib} = 50$ K, $T_{Rot} =$ 30 K; 0.25% at $T \approx 500$–1000 K (temperature range determined on basis of surface temperature measurement throughout oven expansion chamber).

to the experimentally observed excitation spectrum dominated by portion of the $A^1\Sigma_u^+ - X^1\Sigma_g^+$ v' progression from v" = 0. The central spectrum corresponds to a computer simulation of the A-X excitation spectrum for an assumed rotational temperature of 30 K and an assumed vibrational temperature of 50 K. The lower spectrum corresponds to a simulation in which 99.75% of the Na_2 molecules are at T_{Vib} = 50 K, T_{Rot} = 30 K and 0.25% are at a temperature ranging from 500 to 1000 K. The laser excitation technique has provided a sensitive probe for an extremely small component of hot sodium dimers, precluding the ready observation of a Na_3 excitation spectrum in the region 18000-20000 cm^{-1}.

In the course of similar experiments in which the liquid nitrogen baffle above is removed, an intriguing phenomena has been observed. We have been able to induce sodium D-line fluorescence upon single photon pumping at energies far lower than that required to simultaneously dissociate dimeric sodium and produce emission from excited atoms. The atomic fluorescence and its characterization will be the subject of our following discussion. Briefly, we have observed and characterized laser induced atomic fluorescence upon photodissociation of sodium trimers formed under a variety of conditions.

Laser Induced Atomic Fluorescence and Sodium Trimer Fluctuation Bands

The data presented in Figure 6 represents a portion of that obtained in scanning the frequency range 15200 - 19650 cm^{-1} at an effective resolution of 0.5 cm^{-1}. These spectra were obtained when the apparatus in Figure 1 was operated with a water-cooled bulkhead in the absence of liquid nitrogen cooling. The uppermost spectral scan corresponds to an excitation spectrum (the analog of Figure 4 taken at a considerably lower Na_2 $X^1\Sigma_g^+$, v"= 0 concentration) where total fluorescence is monitored as a function of the frequency of an exciting dye laser (Ar^+ ion pumped - Rhodamine 110). As in Figure 4, this spectrum depicts the onset of $B^1\Pi_u - X^1\Sigma_g^+$ fluorescence at ~18100 cm^{-1}. The onset of the B-X fluorescence is found to occur at 18100 $\pm$ 100 cm^{-1} for a large number of observed spectra. The feature extending from 17300 to 17900 cm^{-1} corresponds to the tail of the $A^1\Sigma_u^+ - X^1\Sigma_g^+$ excitation spectrum.

The second and third traces in Figure 6 depict the nature of the laser induced atomic fluorescence (LIAF) which we have observed under a variety of experimental conditions. Here atomic fluorescence is observed, again as the frequency of the dye laser is scanned. Upon "single photon" laser pumping at laser frequencies 500 cm^{-1} or more energetic than the components of the Na D-line ($^2P_{3/2}$ 16973, $^2P_{1/2}$ 16956 cm^{-1}) (43), we observe D-line fluorescence. Strikingly, this fluorescence corresponds primarily to the $^2P_{3/2}$ component of the 5890 Å Na* ($^2P - {}^2S$) emission. The emission from $^2P_{3/2}$ Na* is shown in spectrum 6b. whereas that from

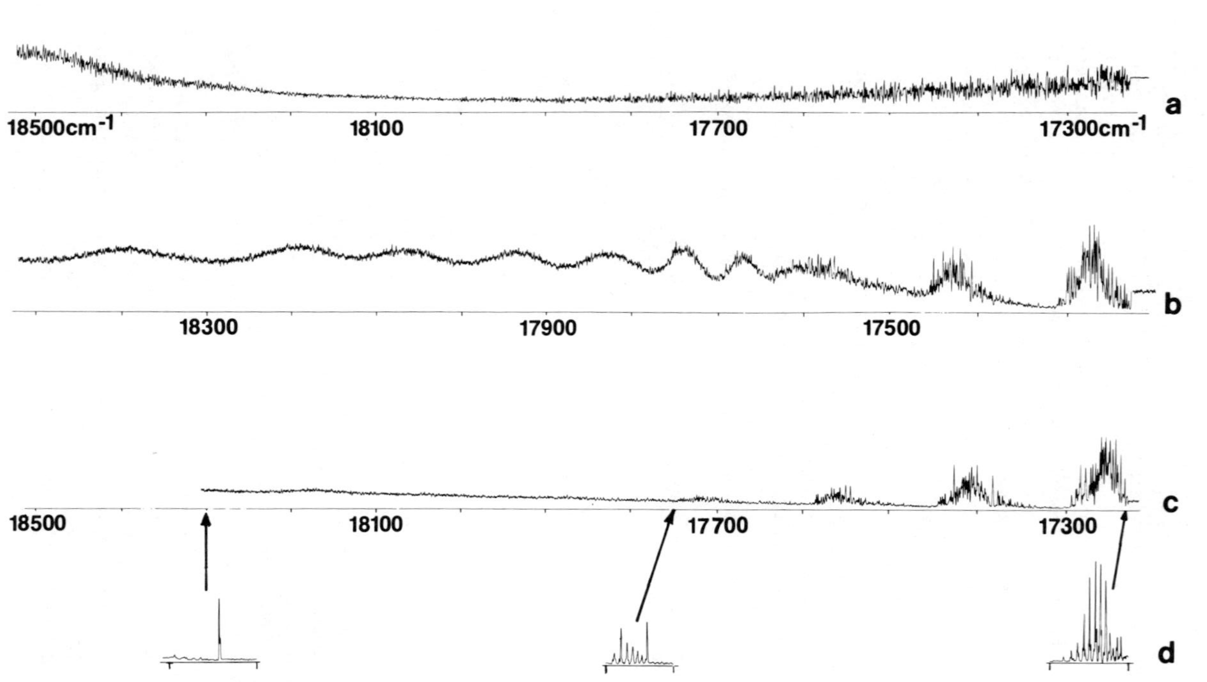

Figure 6. Spectra, as a function of dye laser excitation frequency, of supersonically expanded Na.

a: Excitation spectrum showing the onset of Na_2 $B^1\Pi_u - X^1\Sigma_g^+$ (~18000 cm^{-1}) fluorescence and the tail of $Na_2 A^1\Pi - X^1\Sigma_g^+$ total fluorescence. b: Laser induced atomic fluorescence from $^2P_{3/2}$ Na including fluctuation bands at ν(laser) > 17500 cm^{-1}. c: Minimal laser induced atomic fluorescence from $^2P_{1/2}$ Na*. d: fast photoluminescence scans indicating relative magnitudes of Na D-line and Na_2 A-X resonance fluorescence (tick marks indicate ν = 16 000 and 18 000 cm^{-1}). The lower frequency features appearing in scans b and c ν_{laser} < 17 600 cm^{-1}) correspond to Na_2 fluorescence coincidental with the D-line components.*

$^2P_{1/2}$ is depicted in spectrum 6c. Common to both spectra are groups of sharp features appearing at frequencies less than 17000 cm^{-1}. These features correspond to Na_2 fluorescence which is coincidental with the D-line components at the resolution necessary to resolve these individual ($^2P_{3/2}$, $^2P_{1/2}$) emissions. In other words, each individual line in these groupings corresponds to a small portion of the total Na_2 photoluminescence spectrum which would be obtained if the laser were tuned to the frequency coincidental with this line. Hence, one is sampling small portions of numerous photoluminescence spectra. This is further exemplified in the three photoluminescence spectra shown at the bottom of Figure 6 and taken at the laser frequencies indicated in the figure. The tick marks in the photoluminescence indicate 16000 and 18000 cm^{-1} respectively. For the lowest frequency scan, the photoluminescence is dominated by A-X Na_2 emission; at 17760 cm^{-1} the diatomic emission has considerably weakened and an atomic D-line feature is clearly evident; at 18300 cm^{-1}, the photoluminescence spectrum is clearly dominated by D-line emission.

The dissociation energy of Na_2 to two ^2S sodium atoms is 6022 cm^{-1} (<u>44</u>). Therefore the photodissociation of dimeric sodium to produce the onset of D-line fluorescence would require that Na_2 possess 5500 cm^{-1} of internal excitation ($X^1\Sigma_g^+$, $v'' \geqslant 45$). From the onset of the B-X fluorescence in Figures 4 and 6a and the known Franck-Condon factors for the B-X transition, it can be determined that the highest "hot band" vibrational quantum level populated in the ground electronic, $^1\Sigma_g^+$, state of Na_2 is $v'' = 18$. Only at frequencies below 18000 cm^{-1} will transitions involving levels $v'' \geqslant 19$ contribute significantly to absorption corresponding to the B-X system. Therefore, an onset at 18100 cm^{-1} precludes an Na_2 internal excitation in excess of 2740 cm^{-1}. Dimeric photodissociation cannot account for the observed atomic fluorescence; neither can a collisional process involving sodium atoms and electronically excited sodium dimers explain the experimental observations. Based upon recorded pressures in the fluorescence zone, known Na_2 radiative lifetimes (<u>45</u>), and available cross section measurements for Na_2*-Na intermolecular energy transfer (<u>46</u>), the D-line fluorescence in this region cannot result from a collisional process involving Na_2*. The atomic fluorescence features must be associated with the photodissociation of polymeric sodium.

The $^2P_{3/2}$ Na fluorescence varies with laser frequency, producing a series of fluctuation bands extending from 17500 to 19700 cm^{-1}. Figure 7 presents the continuation of the spectra in Figure 6; in order, an excitation spectrum dominated by the Na_2 B-X system, $^2P_{3/2}$ laser induced atomic fluorescence, $^2P_{1/2}$ fluorescence, and appropriate photoluminescence spectra which at higher frequencies correspond in large part to the B-X system. The spectra in Figure 7 furnish a further indication that the process $Na_2^* + Na \rightarrow Na_2 + Na^*$ cannot account for the observed $^2P_{3/2}$ laser induced atomic fluorescence. Whereas the B-X excitation spectrum varies by several hundred percent as a function of laser scanning frequency, no such variations are observed in the $^2P_{3/2}$ fluctuation bands. We have analyzed the $^2P_{3/2}$ fluctuation band spectrum simultaneous with our mass spectrometric detection of Na_3^+. The results of this effort indicate that the $^2P_{3/2}$ fluctuation bands are to be associated with sodium trimer. Based upon the calculations of Martin and Davidson, the dissociation energy D(Na_2-Na) of the trimer is ≈3000 cm^{-1} (40). This result, in conjunction with the onset of LIAF, suggests that the fluctuation bands must result from the pumping of a bound-free photodecomposition via excitation from vibrationally hot levels of a low-lying or ground state potential.

Several regions of the fluctuating atomic spectra shown in Figures 6 and 7 (resolution ≈0.5 cm^{-1}) have been scanned at considerably higher resolution (0.03 cm^{-1}). No sharp structure characterizes the broad peaks at frequencies higher than 17650 cm^{-1}. This result causes us to correlate the peaks in this figure with a discrete lower state and repulsive excited states. In other words it appears that we have observed the photodecomposition of Na_3 via excitation from a shallow "low-lying or ground state" potential to closely lying repulsive excited electronic states.

Further detailed studies in this (39) and other laboratories (47) indicate that it is possible to induce internal excitation in the products of supersonic expansion by inserting devices into the oven expansion chamber which restrict the sodium expansion creating a "cloud" in the region directly in front of the oven. The creation of this condition and hence the purposeful "poisoning" of the expansion is found to enhance the atomic fluorescence and the B-X spectrum depicted in Figures 6 and 7. It is significant that the formation of a "cloud" provides a route for internal excitation of the aggregates formed in expansion. By controlling the degree of cloud formation, we have been able to modify the degree and onset of the atomic fluorescence observed. It must be emphasized that there is no atomic fluorescence without laser excitation.

If we accept the calculated trimer dissociation energy of Martin and Davdison (40), there is an excellent correlation between the internal excitation observed for Na_2 and Na_3. We have found that the maximum v" hot band level in the ground state of diatomic sodium appears to be v" = 18. This internal excitation corresponds to ~2600 cm^{-1}. In order for sodium trimer photodissociative effects to onset 500 cm^{-1} to the blue of the Na D-line, Figure 2 demonstrates that sodium trimer must have up to 2500 cm^{-1} of internal excitation. Hence, it appears that collisional poisoning of cooled Na_2 and Na_3 has produced a very similar internal heating.

Given the fact that we have observed photodissociation from "hot bands" of the sodium trimer, it should be possible to alleviate this hot band structure and in this way obtain photodissociation spectra which result only from the pumping of bound-free transitions involving cooled sodium trimer.

Upon careful modification of the sodium expansions in order to produce a "clean" supersonic beam, the fluctuation bands shown in Figures 6 and 7 were eliminated, indicating that they do indeed correspond to hot bands. In view of this result, we attempted to excite laser induced atomic fluorescence from cooled sodium trimers by tuning our lasers further to the blue. This effort has been quite successful and we have obtained the spectra depicted in Figure 8. Here we indicate the fluorescence observed upon monitoring both the $^2P_{1/2}$(8a) and $^2P_{3/2}$(8b) components of the sodium D-line. Although there appears to be a very close matchup in the fluctuation band positions, the intensity distributions and relative intensities for the $^2P_{1/2}$ and $^2P_{3/2}$ fluctuation bands differ significantly. The ratio of relative intensities is ~2:1 $^2P_{3/2}$ to $^2P_{1/2}$, virtually in agreement with a statistical prediction. These results would seem to indicate pumping from the same lower discrete state levels to two closely lying but slightly differing repulsive states, one of which dissociates to $^2P_{1/2}$ and the other to $^2P_{3/2}$ sodium atoms (and ground state diatomic sodium). A slower scan of the $^2P_{3/2}$ photodissociation spectrum is presented in Figure 9 where we also indicate the coincidental sodium dimer B-X excitation spectrum. The frequencies corresponding to the peaks noted in Figure 9 are presented in Table I. Scans to 23000 cm^{-1} reveal no further strong atomic fluorescence features.

There is an outstanding correlation between the current studies, the two-photon photoionization work of Schumacher and coworkers (37,48), and detailed quantum mechanical calculations of Martin and Davidson (40), Gerber (48), and Gerber and Schumacher (49). Martin and Davidson (40) have performed an extensive configuration interaction study on the ground and lowest-lying states of sodium trimer. Gerber (48) has carried out extensive CI calculations on the ground and several excited states of Li_3. In addition, he has performed a full vibronic calculation on several of these states and his results may be correlated with experimental data now obtained for sodium trimer (48,49). The results of Martin and Davidson (40) and Gerber (48) are shown in Tables II and III.

To first order, the ground state of Na_3 corresponds to an equilateral triangle configuration (Table II) of symmetry $^2E'$ (40). This $^2E'$ configuration Jahn-Teller distorts into a $^2B_2$ obtuse angled and a $^2A_1$ acute angled configuration. The $^2B_2$ configuration corresponds to the ground state (minimum on the Na_3 surface) and lies approximately 0.6 kcal (200 cm^{-1}) below the $^2A_1$ acute angled configuration (40) which represents a saddle point on the Na_3 potential energy surface. There is another saddle point corresponding to a $^2\Sigma_u^+$ linear symmetric conformation which lies some 3 kcal/mole (1050 cm^{-1}) higher in energy than the $^2B_2$ conformation (40). Based upon the location of the $^2B_2$ state and the $^2A_1$ saddle point, the pseudorotation barrier

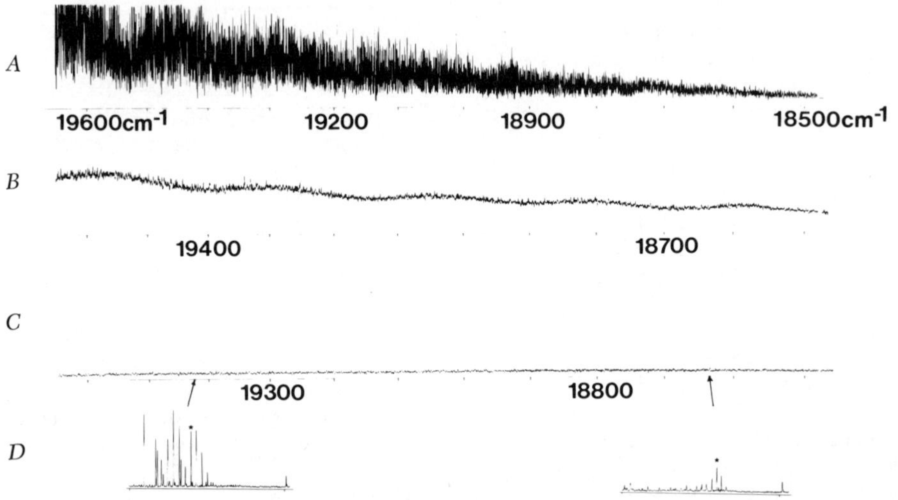

Figure 7. Continuation of spectra presented in Fig. 6. Key: a, excitation spectrum (Ar ion pumped coumarin 7) showing Na_2 $B^1\Pi_u - X^1\Sigma_g^+$ fluorescence; b, $^2P_{3/2}$ Na laser induced atomic fluorescence; c, $^2P_{1/2}$ laser induced atomic fluorescence; and d, Fast photoluminescence scans indicating relative magnitudes of Na D-line and Na_2 B-X fluorescence (tick marks indicate $\nu = 17,000$ and $21,000$ cm^{-1} and * indicates laser excitation frequency.*

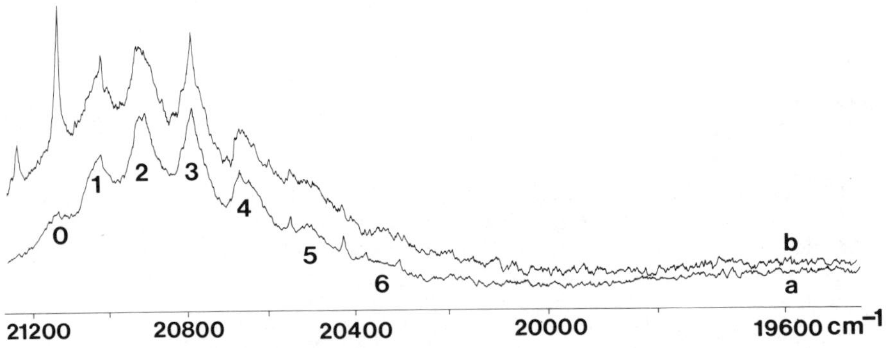

Figure 8. Laser induced atomic fluorescence (LIAF) from the photodissociation of "cold" Na_3 with an Ar ion pumped (coumarin 102) dye laser. Key: a, $^2P_{1/2}$; b, $^2P_{3/2}$ LIAF.

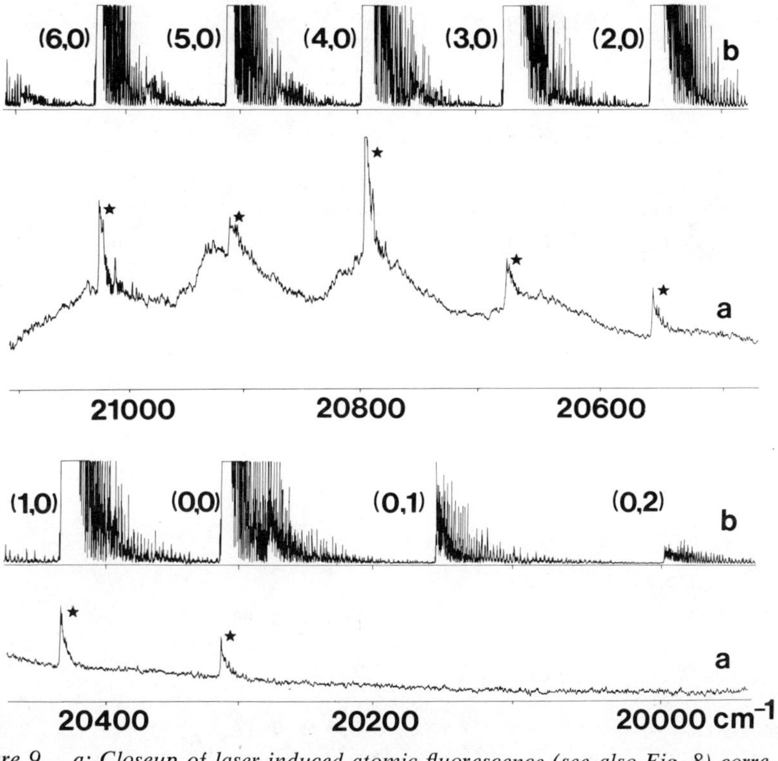

Figure 9. a: Closeup of laser induced atomic fluorescence (see also Fig. 8) corresponding to the $^2P_{3/2}$ component of the Na D-line. b: Na_2 $B^1\Pi_u$ — $X^1\Sigma_g^+$ excitation spectrum.

Table I
Peak Separations Sodium
Trimer Photodissociation Spectrum

Peak No.[a]	Frequency (cm^{-1})	Δ(Separations)	
0	21133 ± 5[b]		
		113 ± 10[b] ($\Delta_{0,1}$)	
1	21020 ± 4		
		99 ± 4 ($\Delta_{1,2'}$)	
2'	20921 ± 2		$\Delta_{1,2}$ = 110 ± 4
		122 ± 4 ($\Delta_{1,2''}$)	
2"	20899 ± 2		
		131 ± 3 ($\Delta_{2,3}$)	
3	20779 ± 3		
		133 ± 2 ($\Delta_{3,4}$)	
4	20647 ± 3		
		134 ± 3 ($\Delta_{4,5}$)	
5	20513 ± 4		
		133 ± 8 ($\Delta_{5,6}$)	
6	20380 ± 5		

a. From labeled peaks in Figures 8 and 9.

b. Uncertainties represent standard deviations in the measurements taken from a number of experimental spectra.

Table II*
Calculated Binding Energies (kcal/mole) Relative to Various Asymptotes

	Na_3 ($^2B_2$,obtuse)[a]	Na_3 ($^2A_1$,acute)[b]	Na_3 ($^2E'$,equilateral)[c]	Na_2 + Na ($^1\Sigma_g^+$ + ^2S)[d]
Na_3($^2B_2$, obtuse)[a]	0.0	0.6	1.6	8.5
Na_3($^2A_1$, acute)[b]		0.0	1.1	7.9
Na_3($^2E'$,equilateral)[c]			0.0	6.8
Na_2($^1\Sigma_g^+$) + Na(^2S)[d]				0.0

a. $\bar{R}$ = 6.658 au, Δ = .994 au, $\theta = \Pi/2$; the absolute energy in au was -485.50182.

b. $\bar{R}$ = 6.699 au, Δ = .749 au, $\theta = 3\Pi/2$.

c. $\bar{R}$ = 6.544 au, Δ = 0.

d. R_{Na-Na} = 5.819 au.

* From reference 40

in this system is only 200 cm^{-1}. Although the equilibrium bond angles for the $^2B_2$ and $^2A_1$ configurations differ substantially ($^2B_2$, 73°; $^2A_1$, 52°), Table II demonstrates that the equilibrium bond lengths for these two configurations ($^2B_2$, $\bar{R}$ = 6.66 au.; $^2A_1$, $\bar{R}$ = 6.70 au.) are quite similar. Hence for the purpose of considering photodissociation via the repulsive portion of an excited state, the rough drawing of the two potential surface cuts in Figure 2 is qualitatively representative. Based upon the data in Table II, it would not be unreasonable to observe photodissociation processes involving both the $^2B_2$ and $^2A_1$ configurations since, on the ground state pseudorotation potential surface, there are points at which the wavefunction is $^2B_2$ and other points at which it is $^2A_1$. The dynamically pseudorotating molecule may undergo a vertical transition leading to photodissociation while passing through a $^2A_1$ saddle point.

It is appropriate to correlate our results and those of Schumacher and coworkers with the excited state calculations presented in Table III. The two photon spectra generated by Schumacher and coworkers are presented in Figures 10 and 11. The 6250 Å system (Figure 10) appears to correspond to a long excited state vibronic progression, while the 6717 Å (Figure 11) system has been attributed to a short progression in conjunction with a sequence. Based upon our own results and the nature of the intensity distribution in the 6717 Å system we feel that the major feature corresponds to a (0,0) band and the bands (1) and (6) correspond to the (1,0) and (0,1) bands associated with one quanta of the symmetric stretch in the ground and excited states of the trimer. Further discussion of this point will follow.

The excited states to which we pump in the present experiments lie at higher energies than the bound states characterized by Schumacher and coworkers. We access potential surfaces, with an open channel to dissociation, whose absorption cross sections are expected to be at least comparable to and possibly greater than those for the bound-bound systems. This suggestion is based on the anticipation of a considerably larger Na 3p component mixing in these repulsive states. Transitions involving these states may therefore be closer to the analog of the strong 3s-3p transition in the sodium atom than are the transitions corresponding to the bound-bound systems. Based upon vibronic calculations on Li_3 and the available experimental data for bound-bound transitions in Na_3, Gerber (48) for the purpose of interpreting the Na_3 spectrum, has provided a tentative quantitative indication of the appropriate vibronic curves for the trimer. His results are presented in Figure 12. The notation for the curves $^2E'$ or $^2E''$ corresponds to that for D_{3h} geometry. The curves labeled $^2E'_{3/2}$ and $^2E'_{1/2}$ correspond to the $^2B_2$ and $^2A_1$ states which result from the distortion of the ground state equilateral triangle conformation. Here, one attempts to take into account spin orbit coupling and an avoided crossing at the D_{3h} geometry. As is apparent from Table III, Li_3 has an excited $^2E''$

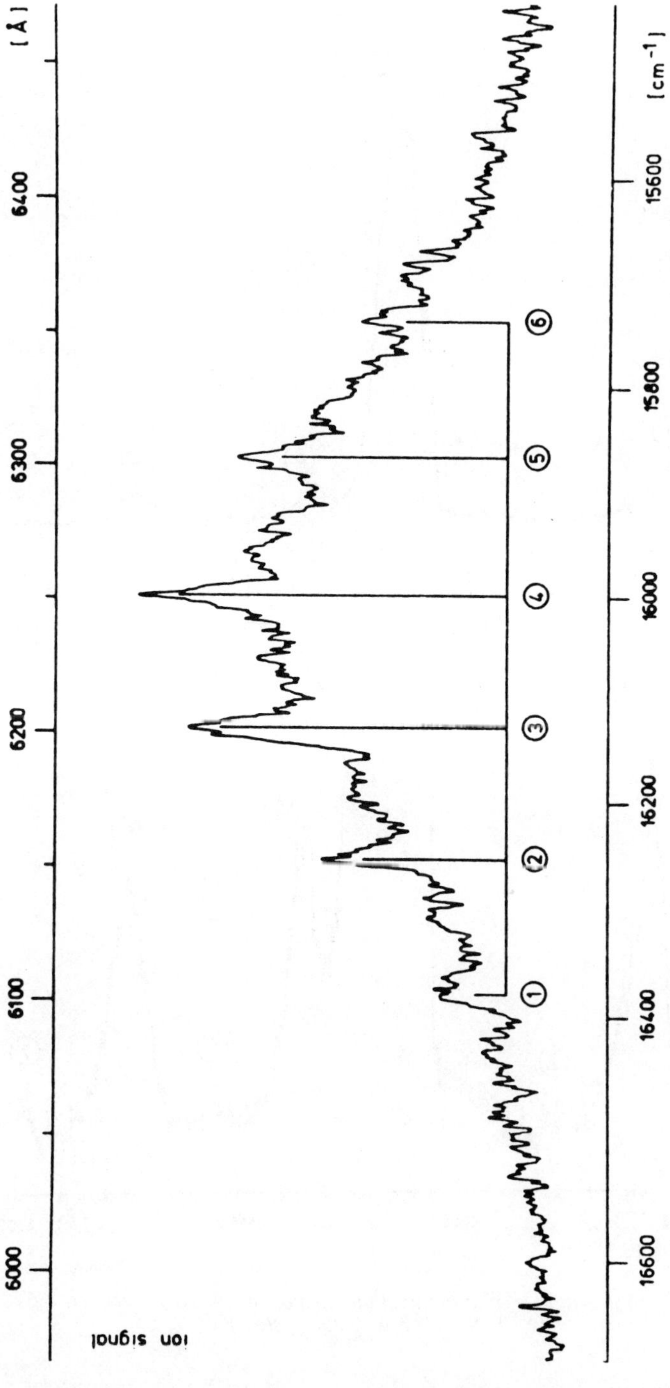

Figure 10. Two-photon photoionization spectrum of Na_3 between 6000 and 6500 Å (6250-Å band system) identifying a long excited state vibronic progression (37).

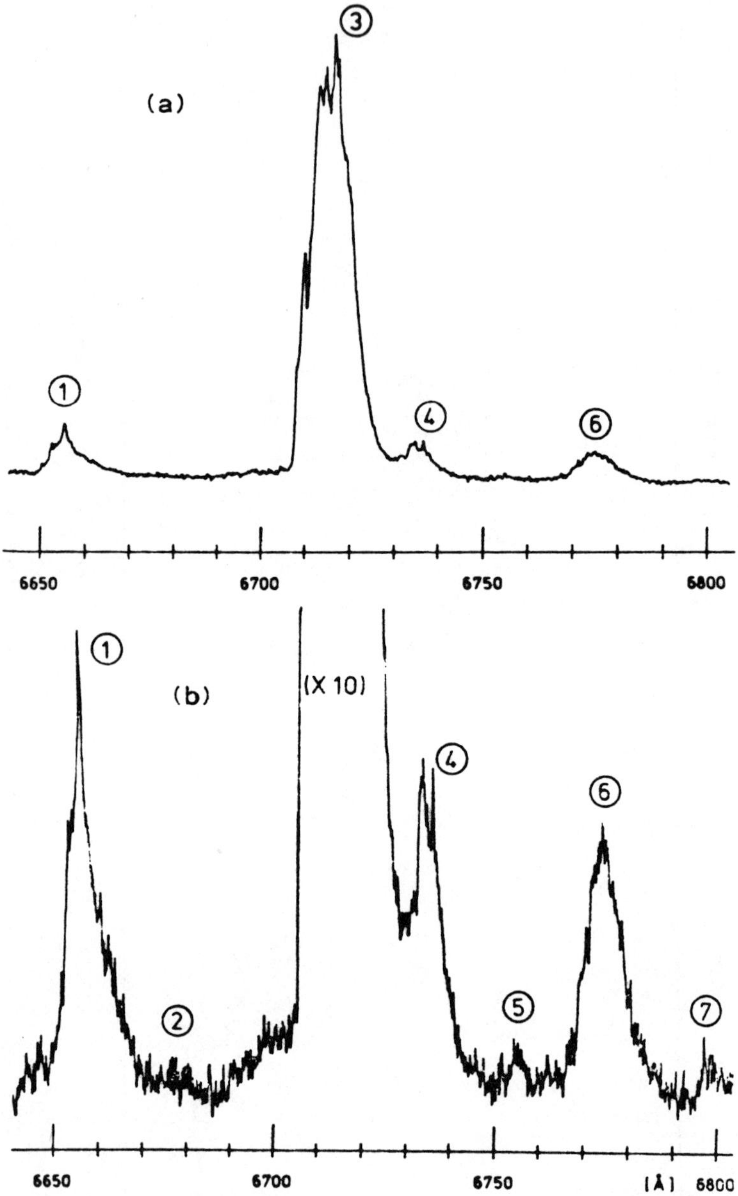

Figure 11. Two-photon photoionization spectrum of Na_3 between 6640 and 6810 Å (6717-Å band system) (37).

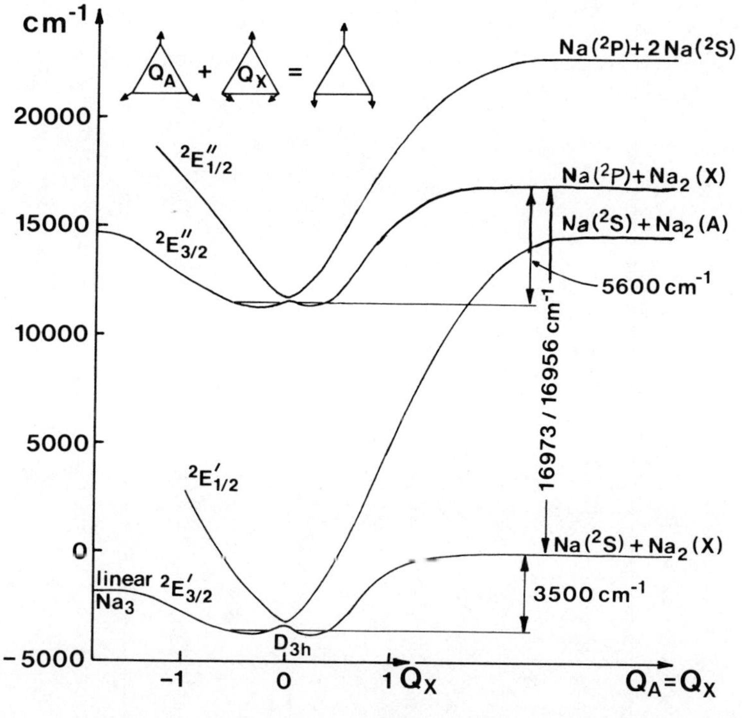

Figure 12. Proposed potential functions (cuts in the Na_3 surface) for the interpretation of the Na_3 spectrum (48).

Table III*
Li_3 Doublet-States

No.	Configuration	State	Energy (cm^{-1})	Li_3-Xe-Matrix Absorption
0	$(2a_1')^2(2e')^1$	$^2E'$	0	(ground state)
1	$(2a_1')^2(1a_2'')^1$	$^2A_2''$	5000	
2	$(2a_1')^2(3a_1')^1$	$^2A_1'$	9260	
3	$(2a_1')^2(3e')^1$	$^2E'$	13520	13890
4	$(2a_1')^2(1e'')^1$	$^2E''$	17550	18050
5†	$(2a_1')^2(4e')^1$	$^2E'$	≈ 20300	21050
6††	$(2a_1')^1(2e')^1(1a_2'')^1$	$^2E''$	≈ 23400	23310

†In gas dissociation into $Li(^2S) + Li_2(X)$; bond energy approx. 20 kcal mol^{-1}.
††Dissociation into $Li(^2S) + Li_2(^1\Pi_u)$, metastable; bond energy approx. -8 kcal mol^{-1}.

(Matrix experiments: T. Welker, T. P. Martin, J. Chem. Phys. 1979, 70, 56831.
*E. Schumacher privately communicated-used by permission.

configuration at 17550 cm^{-1}. The curves shown in Figure 12 correspond to the analog of this $^2E''$ state for Na_3. Again, an avoided crossing may lead to complicated potential curve structure. The states with which we are concerned dissociate to $Na(^2P)$ + Na $(X^1\Sigma_g^+)$. They would be represented by repulsive states in Figure 12 located at energies in excess of 17000 cm^{-1}.

It is appropriate to make further comparisons between the detailed quantum chemical calculations and experiment. From a fit of their generated potential surface, Martin and Davidson have determined force constants for the three normal modes of vibration of sodium trimer. These force constants lead to the fundamentals 147 cm^{-1} for the symmetric stretch, 84 cm^{-1} for the asymmetric stretch, and 89 cm^{-1} for the bending mode. Because the sodium trimer surface is extremely flat, the trimer being very "floppy", the error bounds on the calculated normal mode frequencies must be on the order of $\pm$ 20 cm^{-1}. The pseudo-rotation interaction requires that the bending and asymmetric stretching modes couple, necessarily producing complicated spectral effects. However, the manifold of symmetric stretch levels should be only mildly effected provided that Fermi resonances do not play an important role. Thus if several trimer ground state vibronic levels are populated, we may expect a complicated spectral pattern from which might emerge peaks associated with the symmetric stretching mode. Figure 9 and the data in Table I may confirm our beliefs. The peak separations for the highest energy features (lowest levels in the ground state) appear to be on the order of 100 to 110 cm^{-1}. It would not be unreasonable to associate these features with significant populations in the bending and asymmetric stretching manifolds. This is only a tentative speculation at present since the region at frequencies in excess of 20800 cm^{-1} is rather overlapped, however, higher resolution scans should soon yield more information. The 130-136 cm^{-1} separations for the bands 2 through 6 (Figures 8,9) are intriguing. In analyzing their 6717 Å band system, Herrmann et al. (37) conclude that they have observed a short excited state progression. However, the portion of their spectrum assigned to the progression consists of a very intense peak surrounded by two relatively weak features of comparable intensity. We would like to suggest that this central peak corresponds to a (0,0) band and that the side peaks separated from the central feature by ~134 cm^{-1} correspond to (0,1) and (1,0) bands. This would indicate a symmetric stretch frequency, ~134 cm^{-1}, very similar to the symmetric stretch fundamental calculated by Martin and Davidson (40). In addition, based on extrapolation of the detailed vibronic calculations on Li_3, Gerber (48) has estimated a value of 137 cm^{-1} for the symmetric stretch frequency. If the symmetric stretch frequency for the ground electronic state of sodium trimer is ~135 cm^{-1} and the anharmonicity for the symmetric stretch manifold is reasonably small, the features 3, 4, 5, and 6 in Figures 8 and 9 may correspond, at least in part, to vibrationally hot levels in the symmetric stretch manifold. If, for example, we assume that at least a major portion of the intensity associated with the feature labeled 3 in Figures 8 and 9 can be ascribed to the third vibrational quantum level in the symmetric stretch manifold, the zeroth level should be at ~21180 cm^{-1} in reasonable agreement with the observation of weak structure in this region. Of course the confirmation of absolute quantum level numbering will await further study. There is one additional point which should be emphasized about the spectrum in Figure 8. The broad nature of the observed peaks, especially 4, 5, and 6 may reflect the effect of pseudorotation among the ground obtuse angled $^2B_2$ and low-lying acute angled $^2A_1$ configurations.

The intensities of the $^2P_{3/2}$ and $^2P_{1/2}$ fluctuation bands in Figure 8 are in the ratio 2 to 1, virtually a statistical distribution; however, the relative intensity distributions for the $^2P_{1/2}$ and $^2P_{3/2}$ fluctuation bands differ notably. Much more pronounced differences are observed for the "hot bands" depicted in Figures 6 and 7 where the $^2P_{3/2}/^2P_{1/2}$ ratio varies with exciting frequency and in many instances approaches 8/1. This difference in intensity ratio must reflect very different geometries for the lower discrete states from which pumping occurs in bound-free transition. Again such a result would appear to correlate well with the theoretical analysis of the sodium trimer surface. Martin and Davidson predict that a linear symmetric $^2\Sigma_u^+$ conformation lies only 1050 cm^{-1} above the ground $^2B_2$ conformation. Gerber (Figure 12) notes the very low-lying "$^2E'_{1/2}$" surface dissociating to $Na(^2S) + (Na_2(^1\Sigma_u^+)$. Either of these possible configurations presents a substantial change in geometry relative to the ground state of the trimer. The "hot band" structure (Figures 6 and 7) terminates at 19700 cm^{-1}, laser induced atomic fluorescence being absent to 20100 cm^{-1}. This dip in the atomic fluorescence intensity is apparent in Figure 8. The cutoff in fluorescence intensity leads us to believe that the "hot band" structure is not associated with high vibrational levels of the $\tilde{X}\ ^2B_2$ Jahn-Teller conformation. The laser induced atomic fluorescence for cooled sodium trimer terminates at 21200 cm^{-1}. If we assume that the 19700 and 21200 cm^{-1} terminations correspond to pumping from the zeroth levels of the two "states" involved, the separation, which is 1500 cm^{-1}, is in reasonable agreement with the calculated $^2B_2$-$^2\Sigma_u^+$ separation.

Further studies indicate that the phenomena characterized here can be extended to the study of other bound-free transitions, furnishing a means of mapping the repulsive states which characterize small metal clusters (51). This may have significant importance for the characterization of photocatalytic processes. Further it appears that one can make use of the shallow nature of the bonding potentials characterizing many metal clusters and supersonic expansions in general to obtain "hot band" structure close to the dissociation limit of the ground state of the cluster. It should also be possible to bridge the middle ground between efficiently cooled and highly perturbed aggregate expansions. Indeed, we are now formulating experiments which involve the gradual "poisoning" of the supersonic expansion so as to build up "hot band" structure in a controlled manner.

There are two further points of interest in Figures 8 and 9. From the short wavelength (high frequency) limit of the laser fluorescence spectrum, we obtain an upper bound to the Na_3 dissociation energy. This value, 4250 cm^{-1} (21200-16950 cm^{-1}) is in good agreement with the work of Martin and Davidson (40). In Figure 9, one notes several sharp features superimposed on the broad structure which results from direct photodissociation of the trimer. These sharp features correlate precisely with strong Na_2 B-X absorption (note the excitation spectrum), and it was first thought that they corresponded to Na_2* emission coincidental with the Na-D-line, (in analogy to our observations of similar behavior in the "hot band" region). Further experiments, however, have demonstrated that this is not the case. The sharp features denoted with an asterisk correspond to atomic fluorescence. The source of these atomic fluorescence features is definitely an energy transfer process.

There are two possibilities which may explain the observed energy transfer. Although we have eliminated Na_2*-Na collisions as possible sources of the "hot band" laser induced atomic fluorescence observed for Na_3 photodissociation, the Franck-Condon factors for the observed low v' bands in the B-X system of Na_2 (Figure 9) exceed those for high vibrational levels of the (v',0) progression in the A-X system by four orders of magnitude. Therefore, the possibility of Na* formation via Na_2-Na collisions is considerably heightened. There are, however, problems with this interpretation. The intensity of the sharp atomic fluorescence features associated with the B-X bands does not correlate well with the Franck-Condon factors for the B-X system and the relative ease with which a given (v',0) transition is pumped. This can be readily seen from Figure 9. In addition, any attempts to associate the observed intensities with a reasonable energy gap model involving the formation of vibrationally excited Na_2 and $^2P_{3/2}$ sodium atoms cannot readily explain the varying intensities which characterize the sharp fluorescence features. Although not depicted here, this situation is further complicated by notably different energy transfer features which are associated with the $^2P_{1/2}$ laser induced atomic fluorescence spectrum (52).

Another possibility, consistent with observation involves the energy transfer processes

$$Na_2^* + Na_3 \rightarrow Na_3^* + Na_2$$

$$Na_3^* \rightarrow Na^* + Na_2$$

$$Na^* \rightarrow Na + h\nu$$

In the current study and similar studies reported by Schumacher and coworkers, the ratio of sodium atoms to sodium trimers is thought to be between 10 and 40 to one as a function of experimental conditions (a more exact estimate will require more certain measurement of the cross sections for electron impact and/or photoionization cross sections). The difference in monomer and trimer abundance can easily be overcome by the density of effective states in Na_3 which significantly exceeds those in atomic sodium. Because of the increased density of states, it would not be surprising if the cross section for Na_2^*-Na_3 energy transfer far exceeded that for Na_2^*-Na. The energy transfer phenomena is currently the subject of further investigation in our laboratory.

Further Confirmation of the Na_3 Dissociation Energy

In order to augment the upper bound for the Na_3 dissociation energy obtained in the studies outlined above, we have undertaken an experiment involving sodium trimer oxidation. The experiment involves the study of the reaction

$$Na_3 + Cl \rightarrow NaCl + Na_2^*$$

If sodium trimers react with chlorine atoms (53), the process to produce sodium chloride and dimeric sodium is sufficiently exothermic so as to lead to the production of excited states of the dimer. There is reason to believe that the Na_3-Cl reaction will proceed via a relatively long-lived complex and that a reasonable component of the reaction exoergicity will be transferred to the dimer (54). The short wavelength limit of the observed Na_2 chemiluminescence then provides an estimate of the $Na_3 \rightarrow Na_2 + Na$ dissociation energy. Several experiments have now been performed and we have definitely observed Na_2 chemiluminescence from both the A and B states which appears to correlate with the Na_3-Cl reaction. Although a detailed analysis of the Na_2 emission is not complete, the short wavelength limit of the Na_2 chemiluminescence yields a lower bound of 2300 cm^{-1} for the Na_2-Na dissociation energy. Hence our lower (2300 cm^{-1}) and upper (4250 cm^{-1}) bound estimates of the Na_3 dissociation energy bracket the calculated value (3000 cm^{-1}) of Martin and Davidson (40).

Acknowledgement

It is a pleasure to acknowledge the support of this research by the National Science Foundation. Helpful discussions with Ernst Schumacher, Ernest Davidson, and W. C. Stwalley are gratefully acknowledged.

Literature Cited

Theory (a sampling containing in themselves references to other work)

1. Richtsmeier, S.; Gole, J. L.; Dixon, D. A. Proc. Natl. Acad. Sci. 1980, 77, 5611.

2. Schaeffer, H. F., III Acc. Chem. Res. 1977, 10, 287-293.
3. Bauschlicher, C., Jr.; Bagus, P.; Schaeffer, H. F., III IBM J. Res. Dev. 1978, 22, 213-220.
4. Companion, A. L.; Steible, D. J.; Starshak, A. J. J. Chem. Phys. 1968, 49, 3637-3640.
5. Companion, A. L. Chem. Phys. Lett. 1978, 56, 500-502.
6. Pickup, B. T. Proc. R. Soc. London Ser. A 1973, 333, 69-87.
7. Gelb, A.; Jordan, K. D.; Silbey, R. Chem. Phys. 1975, 9, 175-182.
8. Davies, D. W.; del Conde, G. Mol. Phys. 1977, 33, 1813-1814.
9. Bagus, P. S.; del Conde, G.; Davies, D. W. Faraday Spec. Discuss. Chem. Soc. 1977, 62, 321.
10. Kendrick, J.; Hillier, I. H. Mol. Phys. 1977, 33, 635-640.
11. Gole, J. L.; Childs, R.; Dixon, D. A.; Eades, R. A. J. Chem. Phys. 1980, 72, 6368-6575.
12. Fripiat, J. G.; Chow, K. T.; Boudart, M.; Diamond, J. R.; Johnson, K. H. J. Mol. Catal. 1975, 1, 59-72.
13. Anderson, A. B. J. Chem. Phys. 1977, 66, 5108-5111.
14. Baetzold, R. C. J. Chem. Phys. 1971, 55, 4355-4370.
15. Baetzold, R. C. Adv. Catal. 1976, 25, 1-55.
16. Goddard, W. A.; Walch, S. P.; Kappe, A. K.; Upton, T. H.; Melius, C. F. J. Vac. Sci. Technol. 1977, 14, 416-418.
17. Bachman, C.; Demuynk, J.; Veillard, A. Gazz. Chim. Ital. 1978, 108, 389-391.

Experiment (a sampling containing in themselves references to other work)

18. Preuss, D. R.; Pace, S. A.; Gole, J. L. J. Chem. Phys. 1979, 71, 3553-3560.
19. Herrmann, A.; Schumacher, E.; Wöste, L. J. Chem. Phys. 1978, 68, 2327-2336.
20. Herrmann, A.; Leutwyler, S.; Schumacher, E.; Wöste, L. Helv. Chim. Acta 1978, 61, 453-487.
21. Leutwyler, S.; Schumacher, E. Chimia 1975, 31, 475-478.
22. Moskovits, M.; Ozin, G. A."Cryochemistry"; Wiley, New York, 1976.
23. Ozin, G. A. Catal. Rev. Sci. Eng. 1977, 16, 191-230.
24. Ozin, G. A.; Huber, H. Inorg. Chem. 1978, 17, 155-163.
25. Moskovits, M.; Hulse, J. E. J. Chem. Phys. 1977, 67, 4271-4278.
26. Lindsay, D. M.; Herschbach, D. R.; Kwiram, A. L. Mol. Phys. 1976, 32, 1199-1213.
27. Schulze, W.; Becker, H. U.; Minkwitz, R.; Manzel, K. Chem. Phys. Lett. 1978, 55, 59-61.
28. Ozin, G. A.; Huber, H.; Mitchell, S. A. Inorg. Chem. 1979, 18, 2932-2934.

Related work is discussed in

29. Robinson, A. L. Science 1974, 185, 772-774.
30. Demitras, G. C.; Muetterties, E. L. J. Am. Chem. Soc. 1977, 99, 2796-2797.
31. Sinfelt, J. H. Acc. Chem. Res. 1977, 10, 15-20.
32. Muetterties, E. L. Science 1977, 196, 839-848.
33. Muetterties, E. L. Bull. Soc. Chim. Belg. 1975, 84, 959-986.

34. Muetterties, E. L. Bull. Soc. Chim. Belg. 1976, 85, 451-470.
35. Muetterties, E. L.; Rhodin, R. N.; Band, E.; Brucker, C. F.; Pretzer, W. R. Chem. Rev. 1979, 79, 91-137.
36. Band, E.; Muetterties, E. L. Chem. Rev. 1978, 78, 639-658.
37. Herrmann, A.; Hofmann, M.; Leytwyler, S.; Schumacher, E.; Wöste, L. Chem. Phys. Lett. 1979, 62, 216.
38. See previous discussion by Schumacher and coworkers in this volume.
39. Green, G. J.; Pace, S. A.; Preuss, D. R.; Gole, J. L. "The Characterization of Supersonic Metal Vapor Expansions"; to be submitted to Review of Scientific Instruments.
40. Martin, R. L.; Davidson, E. R. Mol. Phys. 1978, 35, 1713-1729.
41. For the initial demonstration of cooling using laser fluorescence, see Sinha, M. P.; Schultz, A.; Zare, R. N. J. Chem. Phys. 1973, 50, 549. For very recent work see Herman, A.; Schumacher, E.; Wöste, L. J. Chem. Phys. 1978, 68, 2327. NO_2 work: Smalley, R. E.; Ramakrishna, B. L.; Levy, D. H.; Wharton, L. J. Chem. Phys. 1974, 61, 4363; ibid. 1975, 63, 4977. S-tetrazine work: Smalley, R. E.; Wharton, L.; Levy, D.; Chandler, D. W. J. Mol. Spectrosc. 1978, 68, 2487.
42. See Levy, Donald H. "Laser Spectroscopy of Cold Gas Phase Molecules", Ann. Rev. Phys. Chem. 1980, 31, 197.
43. Moore, C. E. "Atomic Energy Levels"; NBS Circular 457, 1949.
44. Verma, K. K.; Vu, T.; Stwalley, W. C. J. Molec. Spectrosc. 1981, 74, 131.
45. Demtröder, W.; Stetzenbach, W.; Stock, M.; Witt, J. J. Mol. Spectrosc. 1976, 61, 382; Ducas, T. W.; Littman, M. G.; Zimmerman, M. L.; Kleppner, D. J. Chem. Phys. 1976, 65, 842; Callender, R. H.; Gersten, J. I.; Leigh, R. W.; Yang, J. L. Phys. Rev. A 1976, 14, 1672; Stevens, W. J.; Hessel, M. M.; Bertoncini, P. J.; Wahl, A. C. J. Chem. Phys. 1977, 66, 1477. See also references to earlier work in these articles.
46. Lam, L. K.; Fujimato, T.; Gallagher, A. C.; Hessel, M. M. J. Chem. Phys. 1978, 68, 3553. Kraulinya, E. K.; Kopeikina, E. K.; Janson, M. L. Chem. Phys. Lett. 1976, 39, 565.
47. Oldenborg, R. C. (Los Alamos Scientific Laboratory) and Smalley R. E. (Rice University) private communication.
48. Gerber, W. H. "Theorie des dynamischen Jahn-Teller-Effekts in Li_3 und Untersuchung von Lithium-Molekularstrahlen"; Ph.D. Thesis, Bern University, 1980.
49. Gerber, W. H.; Schumaker, E. J. Chem. Phys. 1978, 69, 1692-1703.
50. Schumacher, E., private communication. This data used by permission of E. Schumacher.
51. Such efforts are currently being undertaken in our laboratory with collaboration on the part of Prof. J. Tellinghuisen.
52. Gole, J. L.; Green, G. J.; Pace, S. A.; Preuss, D. R. in preparation.
53. Crumley, W. T.; Gole, J. L.; Dixon, D. A., work in progress.
54. Struve, W. S.; Krenos, J. R.; McFadden, D. L.; Herschbach, D. R. J. Chem. Phys. 1975, 62, 404.

RECEIVED August 26, 1981.

11

Resonance Raman Spectroscopy of Small Metal Clusters

M. MOSKOVITS and D. P. DI LELLA

University of Toronto, Department of Chemistry and Erindale College, Toronto, Ontario, Canada M5S 1A1

The resonance Raman spectra of Fe_2, NiFe, V_2, Ti_2 and Ni_3 in solid, rare-gas matrices are reviewed. Spectroscopic constants for these were ($\omega_e, \omega_e x_e$, cm^{-1}): Fe_2 (300.26, 1.45), NiFe (320.0, 1.32), V_2 (508.0, 3.3) and Ti_2 (407.9, 1.08). A Bernstein-LeRoy analysis was performed on the data for Fe_2, yielding a bond dissociation energy of 1.2 eV, close to that reported by Kant. All of the transition metal diatomics studied showed an inordinate ability to remain vibrationally excited when the first few vibrational levels were populated radiatively. This gave rise to intense antistokes emissions.

Interest in the electronic properties and structures of small transition metal diatomics and clusters has increased rapidly during the past decade, largely as a result of the connection, real or suggested, that exists between these entities and surface chemistry.

In this paper we will present some examples of the application of resonance Raman spectroscopy to the study of transition metal diatomics. The application of Raman spectroscopy to matrix-isolated metal clusters was first reported by Schulze et al. (1). Having observed only a single line in the Raman spectrum of Ag_3, Schulze concluded that the molecule was linear since a bent triatomic and an equilateral triangular geometry would have, in principle, 3 and 2 Raman-active modes. The evidence, however, is not conclusive since many C_{2v} molecules have very weak asymmetric stretches in the Raman (2) (for example, the ν_3 mode of O_3 is undetectable in the Raman (3)). Moreover, the bend (ν_2) of Ag_3 is expected to be a very low-frequency mode, perhaps lower than one can feasibly detect in a matrix Raman experiment.

Resonance Raman (RR) spectroscopy has two advantages and a disadvantage over ordinary Raman. The disadvantage is the unequal enhancement of all the fundamentals of a molecule in its RR spec-

0097-6156/82/0179-0153$05.00/0

trum. Normally it is totally symmetric modes which are enhanced (4). One advantage lies in the fact that several members of a progression are often seen in RR spectroscopy. Hence the anharmonicity and an estimate of the bond dissociation energy (at least in the case of diatomics) may be obtained. Secondly, the isotopic fine structure, if any exists, may be resolved quite easily in the high harmonics, making the identification and, in the case of polyatomics, the structure of the carrier determinable.

Experimental

The apparatus has been described in detail elsewhere (5). Briefly, metal atoms vaporized by electrically heating a metal ribbon filament directly were cocondensed with rare gas on a polished aluminum surface cooled to 11°K by means of an Air Products Displex refrigerator. The aluminum substrate which was contained in a vacuum chamber whose local pressure with the refrigerator on was approximately 10^{-8} torr could be viewed through a Pyrex or quartz bell surrounding it. The metal vapor stream was partially collimated by transit through the cylindrical neck of the flange which connected the metal vapor source to the sample area. Only the upper half of the substrate intercepted the metal stream and the matrix deposited there was colored. There was a distinct boundary between the colored part of the sample and the lower part, which was white. The thickness of the rare gas deposition was nearly constant over the substrate. For a particular sample, the concentration of the colored part was nearly constant, judging from the spectra obtained when the laser sampled various parts of it. The spectra obtained when the laser was focused on the white part of the sample showed that it also received some metal flux. RR spectra obtained when the laser was focused on the white part of the sample near the color boundary were nearly as strong as that on the colored part proper. This characteristic results from reabsorption of some of the emitted light by the colored part of the matrix. The intensity of the features in the RR spectra obtained from the white part of the sample quickly and steadily decreased as the laser was moved away from the color boundary. Parts of the sample with widely varying metal concentrations could therefore be studied in the same experiment. Mixed Fe/Ni deposits were generated by stringing two filaments, one of each metal, side by side across the water-cooled electrodes of the metal vapor furnace. An Fe/Ni sample had the same color as a sample with Fe alone.

Emission spectra were excited using either a Control model 554A or a Spectraphysics model 165 argon ion laser. A Spectraphysics model 375 dye laser (Rhodamine 6G) was also used. The emitted light was dispersed with a Spex 1401B double monochromator with 1800 lines/mm holographic gratings. A cooled Hamamatsu R955 multialkali photomultiplier detected the signal, followed by conventional photon counting.

Diatomics

Fe_2. Portions of the RR spectrum of Fe_2 (6) are shown in Figure 1. Analysis of the vibrational fine structure, employing the relation

$$G(v) = \omega_e v - \omega_e X_e(v^2+v) \tag{1}$$

yields the constants $\omega_e = 300.26$ and $\omega_e X_e = 1.45\ cm^{-1}$. These data were obtained from the progression of the more abundant species $^{56}Fe_2$. In addition, five members of a second progression belonging to $^{54}Fe^{56}Fe$ were observed. The vibrational constants imply a dissociation energy of 1.9 eV if the formula pertinent to a Morse oscillator, $D_e = \omega_e^2/(4\omega_e X_e)$, is used. This value is larger than the quantity reported by Kant (7), 1.3 eV, and much larger than that reported by Gingerich (8), 0.5 eV, obtained by re-evaluating Kant's data on the assumption that Fe_2 possesses a large number of low-lying, electronically excited states within one-half an eV of the ground state. A lower value is obtained on using an analysis due to LeRoy and Bernstein (9). By taking the contribution to the interatomic potential to be a single inverse power near the dissociation limit, one obtains an expression for the vibrational terms of the form

$$G(v) = D_e - X_0(n)(v_D-v)^{2n/(n-2)} \tag{2}$$

In eqn. 2 D_e is the dissociation energy, n the power of the asymptotically dominant inverse-power term in the potential, v_D the effective, non-integer value of the vibrational quantum number at dissociation and $X_0(n)$ is a constant of the form

$$X(n) = \overline{X}_0(n)/(\mu^n C_n^{\ 2})^{1/(n-2)} \tag{3}$$

In eqn. 3, μ and C_n are, respectively, the reduced mass of the diatomic and the coefficient of the r^{-n} term in the near-dissociation potential. $\overline{X}_0(n)$ is a constant tabulated by LeRoy (10).

LeRoy has recently enlarged the analysis to allow lower vibrational terms to be used (10). Among the various "corrections" to eqn. 3 proposed is one of the form

$$G(v) = D_e - X_0(n)(v_D-v)^{2n/(n-2)}\{1 + a_1(v_D-v) + a_2(v_D-v)^2 + \ldots\} \tag{4}$$

The constants in eqn. 4 have been previously defined except for a_1, a_2, etc., which are arbitrary, adjustable parameters. The $^{56}Fe_2$ data of Fig. 1 were fit to eqn. 4 using a Fletcher-Powell program to minimize the sums of residuals. For the $^5D_4$ ground state of the Fe atom the proper choice of n is 5 (11). C_5 was calculated to be $-3.1775 \times 10^4\ cm^{-1}\ Å^5$ using the formulas and

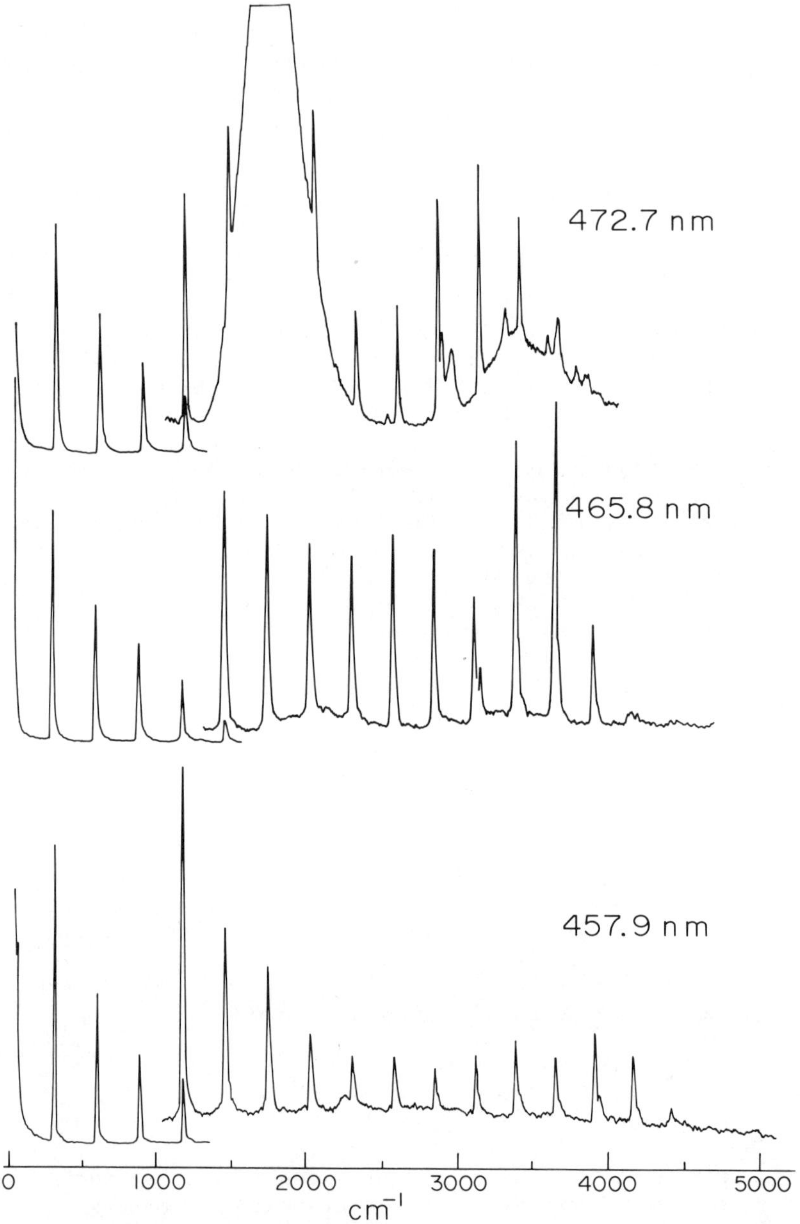

Figure 1. Spectra obtained from the red–orange part of the sample with three laser excitations for Fe codeposited with Ar at 11 K. The resonance Raman progression is assigned to Fe_2 and all fluorescence features to FeO (6).

data tabulated by Chang (11) and assuming a mean square radius for Fe of 1.4948 Å^2 (12). The ground state of Fe_2 was taken to be $^7\Sigma_g$ as reported by Montano (13) on the basis of a Mössbauer study. The parameters in eqn. 4 were thereby determined to be: D_e = 9735.97 cm^{-1}, v_D = 63.83, a_1 = -0.017367 and a_2 = 9.1583 × 10^{-5}. The value for the dissociation energy (1.2 eV) obtained in the analysis is in much better agreement with that reported by Kant than that based on the Morse oscillator. One should still regard this result as somewhat tentative as may be gauged from Fig. 2 which shows that the available vibrational terms are too few to distinguish between a linear Birge-Sponer plot for which $D_e = \omega_e^2/(4\omega_e X_e)$ and the $\Delta G(v)$ versus v curve obtained from eqn. 4. However, it is clear that Gingerich's estimate of 0.5 eV for D_e of Fe_2 is too low since we see spectral emissions (Fig. 1) to vibrational states of ground state Fe_2 which are higher than 4000 cm^{-1} above the ground vibrational state and which are still at a rather harmonic portion of the ground state potential, implying that they are not near the dissociation limit.

In addition to the progression observed on the Stokes side of the spectrum, an inordinately intense antistokes progression was obtained with Fe_2, as with the other diatomics discussed below.

The antistokes lines were too intense for them to arise simply from transitions from thermally populated v" = 1, 2, 3, etc., levels. We propose that these states are populated radiatively by the Stokes resonance Raman process. Thus the antistokes spectrum arises only from consecutive, two-photon (or multiple-photon) absorption processes, while the Stokes spectrum has contributions from both single- and two-photon absorptions. In the limit of very low laser powers the ratio of the intensity of an antistokes to a Stokes line should be proportional to the laser power. Moreover for the intensity of the antistokes spectrum to be so large, radiative depopulation of excited ground state vibrational states must compete effectively with thermal relaxation to the ground state. This implies very long lifetimes for the first few vibrationally excited states of the ground state of Fe_2 isolated in solid argon. Very long vibrational lifetimes have been previously reported for a number of matrix-isolated diatomics (14).

The dependence of the Stokes and antistokes intensities on incident laser power was considered in ref. 6 where it was shown that the intensity, S_{oj}, of the component corresponding to the transition $j \leftarrow o$ depends on the laser power, I, as

$$S_{oj} = k_{oj} \frac{aI + b_o I^2 + c_o I^3}{a + bI + cI^2} \qquad (5)$$

while for $j \leftarrow i$ transitions where $i \neq 0$ (this includes all antistokes transitions when $j < i$) the intensity S_{ij} is given by

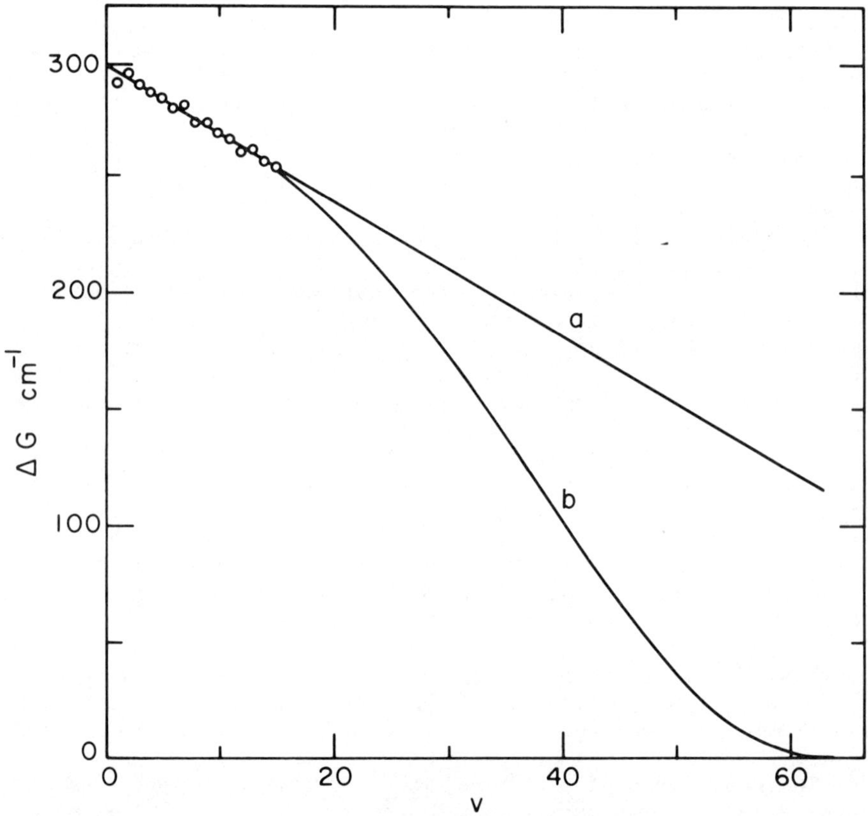

Figure 2. $G(v + 1) - G(v)$ vs. v for the Fe_2 data of Fig. 1. Key: a, Birge–Sponer extrapolation using $\Delta G(v) = 300.26 - 2.90(v + 1)$ cm^{-1}; and b, $\Delta G(v)$ obtained from Eq. 4.

$$S_{ij} = k_{ij} \frac{b_i I^2 + c_i I^3}{a + bI + cI^2} \tag{6}$$

In (5) and (6) b_o and $b_i \leq b$ and c_o and $c_i \leq c$.

The three terms in the denominators of (5) and (6) (and the numerator of (1)) come about, respectively, from the population (or depopulation) of a oth or i^{th} level following the absorption of one, two and three successive photons. The termination of the numerators of (5) and (6) at the third order arises from the fact that in reference 6 the authors considered only the first three vibrational levels of the electronic ground state to be significantly populated and truncated the solution with the v" = 2 level. Since three or fewer members of the antistokes progression are seen this assumption seems to suffice. When more vibrational levels are significantly populated, equations (5) and (6) must be enlarged to contain terms of higher order.

Equations (5) and (6) indicate, however, that the intensities associated with $j \leftarrow o$ transitions will first increase proportionally with I when I is small and ultimately increase with a smaller slope where I is large. $j \leftarrow i$ transitions on the other hand will gain intensity as I^2 for low incident laser powers while at high laser powers they will become proportional to the laser power. Hence both types of signals will show a saturation-like behavior as a function of I. The ratio of any S_{ij}, which includes all antistokes components, to any S_{oj} which forms the "main" RR progression (the progression which ordinarily exists when excited vibrational states are not populated) has the form

$$\frac{S_{ij}}{S_{o\ell}} = \frac{k_{ij}}{k_{i\ell}} \frac{b_i I + c_i I^2}{a + b_o I + c_o I^2} \tag{7}$$

This quantity rises linearly with I for small I and achieves a constant value for I large.

As a function of laser power, the ratio of antistokes to Stokes line intensities was found to follow eqn. (7) well.

NiFe. When Ni and Fe were condensed together one obtained in addition to the progression attributed to Fe_2 another, weaker progression (6) with spectroscopic constants ω_e = 320.0 and $\omega_e X_e$ = 1.32 cm^{-1}. This we assign to NiFe.

A fluorescence progression of $^{58}Ni_2$ has recently been reported by Nixon (15) for which ω_e was determined to be 380.9 cm^{-1}. The force constants for Fe_2, FeNi and Ni_2 are therefore 1.48, 1.72 and 2.48 mdyn/Å respectively, the middle value being roughly the mean of the other two.

$\underline{V_2}$. V_2 isolated in solid argon shows three resonance Raman (Fig. 3) and several fluorescence spectra (16). These have been assigned, respectively, to RR in the ground state, RR of V_2 in a radiatively populated, electronically excited state which is situated roughly 1860 cm^{-1} above the ground state (T_{00}) and an electronic RR process joining the two states. The ground state vibrational constants of V_2 were found to be ω_e = 537.5 cm^{-1} and ω_eX_e = 4.3 cm^{-1}. Those of the low-lying excited state were ω_e = 508.0 and ω_eX_e = 3.3 cm^{-1}. Every line of the electronic resonance Raman spectrum could be accounted for in terms of those four constants.

As in Fe_2, an inordinately intense antistokes progression was observed as well as sequence components in both the Stokes and antistokes progressions. The ratios of the antistokes lines to the non $j \leftarrow o$ sequence components are expected to follow the laser intensity in accordance with eqn. 7. This is shown in Fig. 4 for the ratios S(-528)/S(528) and S(521)/S(528). (The numbers in parentheses refer to the Raman frequencies of the components whose intensities are used.) In this figure the solid curves are best fits using equation 7 and the four adjustable parameters obtained when the latter is written in the form

$$S_{ij}/S_{o\ell} = (1+DI)/(A+BI+CI^2) \tag{8}$$

(A, B, C and D may be related to $(k_{ij}/k_{o\ell})$, b_i, c_i, a, b_o and c_o by inspection), while the points are experimentally derived.

The ratio S(501)/S(528) also fit equation 8 (Fig. 4), implying that the progression obtained with V_2 in the low-lying, electronically excited state, of which the 501 cm^{-1} line is the first component of the first member, arises entirely from levels which have been populated as a result of laser irradiation.

Each component of the ground state RR progression of V_2 was observed to consist of at least three subcomponents whose relative intensities do not depend greatly on the exciting line power (Fig. 5). Several possible explanations come to mind to account for the subcomponents. If the V_2 molecule lies in a divacancy (two neighboring substitutional sites) in argon and if it interacts weakly with the argon then the local "molecule" comprising the V_2 with its nearest neighbor argon atoms has a formal symmetry which is not cubic. The ground state of V_2 has been variously reported by several authors. Cooper and Hare report a $^1\Sigma_g$ low-spin configuration (17) based on an extended Hückel calculation in which only d-electrons were considered. Ford et al. (18) predict a $^9\Sigma_u^-(s\sigma_g^2 d\sigma_g^1 d\pi_u^2 d\delta_u^2 d\delta_g^2 d\pi_g^1)$ high-spin ground state while Harris and Jones (19) prefer a $^9\Sigma_u^-$ ground state with an $s\sigma_g^2 d\sigma_g^1 d\pi_u^2 d\delta_g^2 d\delta_u^2 d\sigma_u^1$ configuration. The last two authors also note that as a result of the small energy intervals between various V_2 levels originating mainly from 3d levels of V, a large number of molecular states exist which lie very close to the ground state. This, of course, is also true for other transition metal dimers with partially filled d-states. Consequently a small variation in the parameters

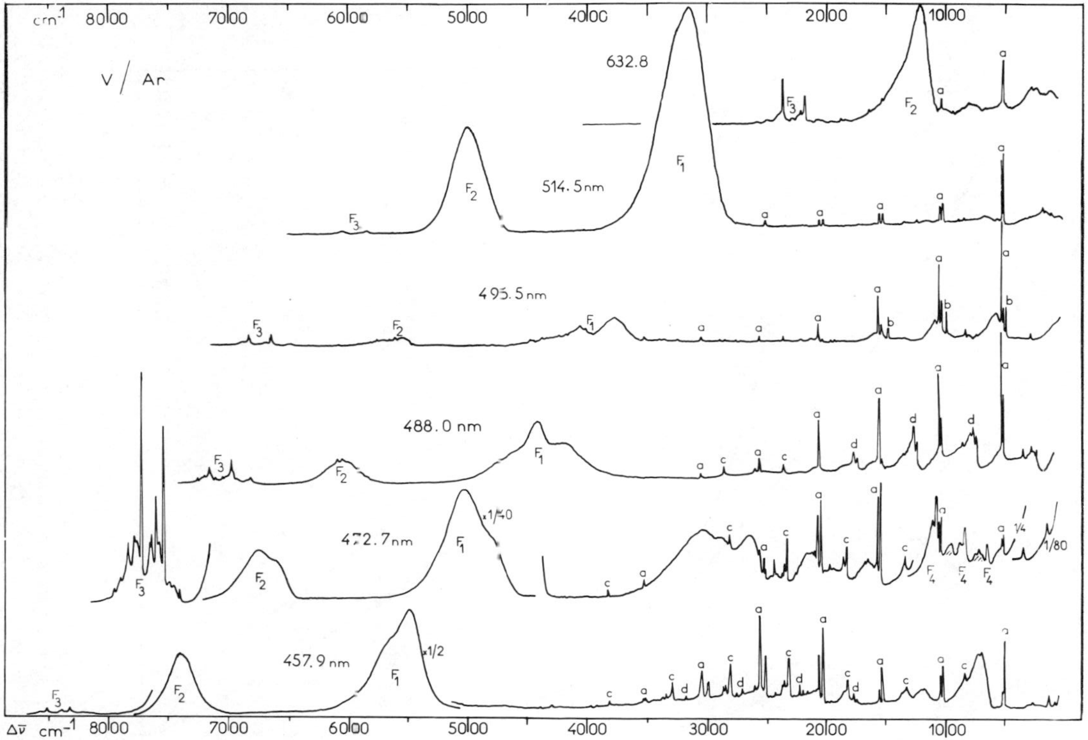

Figure 3. Resonance Raman and fluorescence spectra of V isolated in solid Ar matrix (15 K) for different exciting laser lines for a matrix with $V/Ar \sim 2 \times 10^{-3}$; spectral slits 4 cm^{-1}; laser power at the sample was 20 mW for the 6348-nm line and 5 mW for all others. Key: a, ground state resonance Raman spectrum; b, excited state resonance Raman spectrum; c, electronic resonance Raman; and F1 to F4, fluorescences (16).

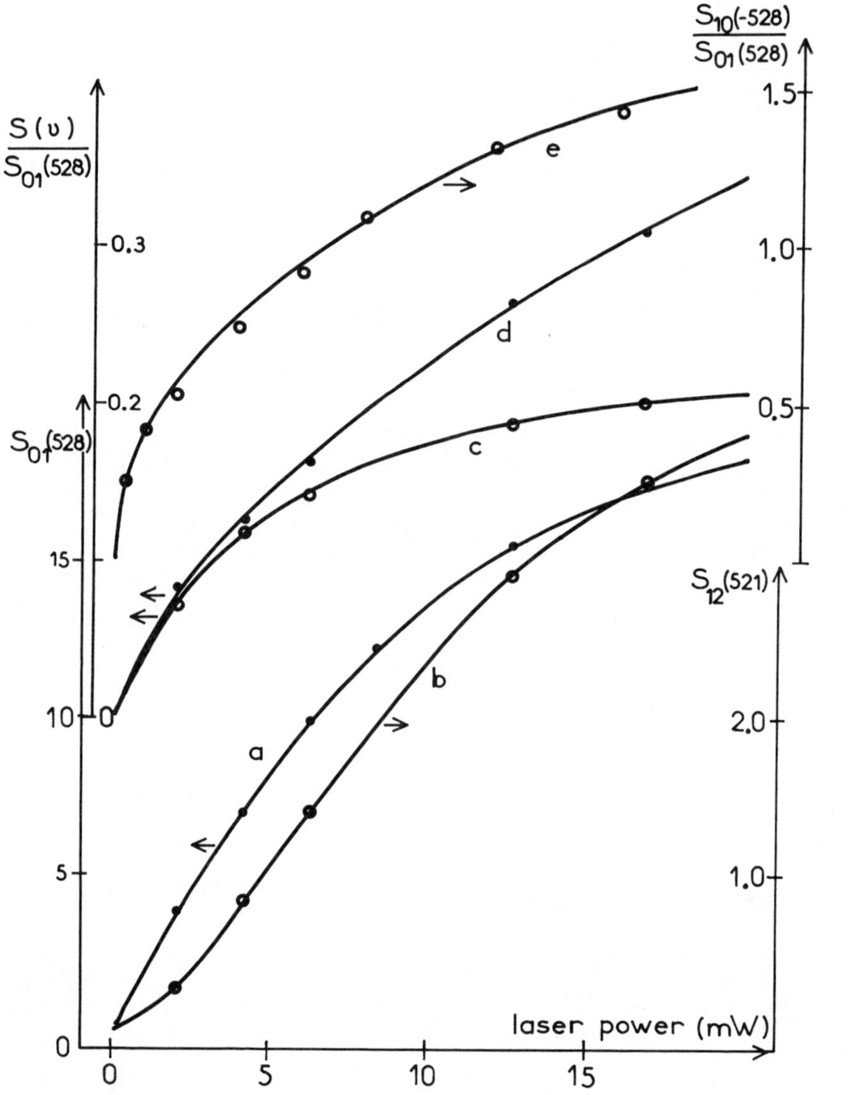

Figure 4. a: Intensity of $S_{01}(528)$. b: Intensity of $S_{12}(521)$. c: Ratio of intensity $S_{12}(521)$ to $S_{01}(528)$. d: Ratio of intensity $S_{0'1'}(501)$ to $S_{01}(528)$. a–d: Excited with 496.5 nm; e: Ratio of intensity of the first antistokes line $S_{10}(-528)$ to first Stokes $S_{01}(528)$ line excited with 514.5 nm. All as a function of incident laser power. The points are experimental. The lines drawn for c, d and e are best fits to Eq. 4. Parameters A, B, C and D equal to: 20.95, 3.71, 0, and 0 for c; for d, $A = 16.44$, $B = 6.31$, $C = 0.0706$, and $D = 0.0936$; for e, $A = 0.791$, $B = 2.64$, $C = 0.169$, and $D = 0.420$ (16).

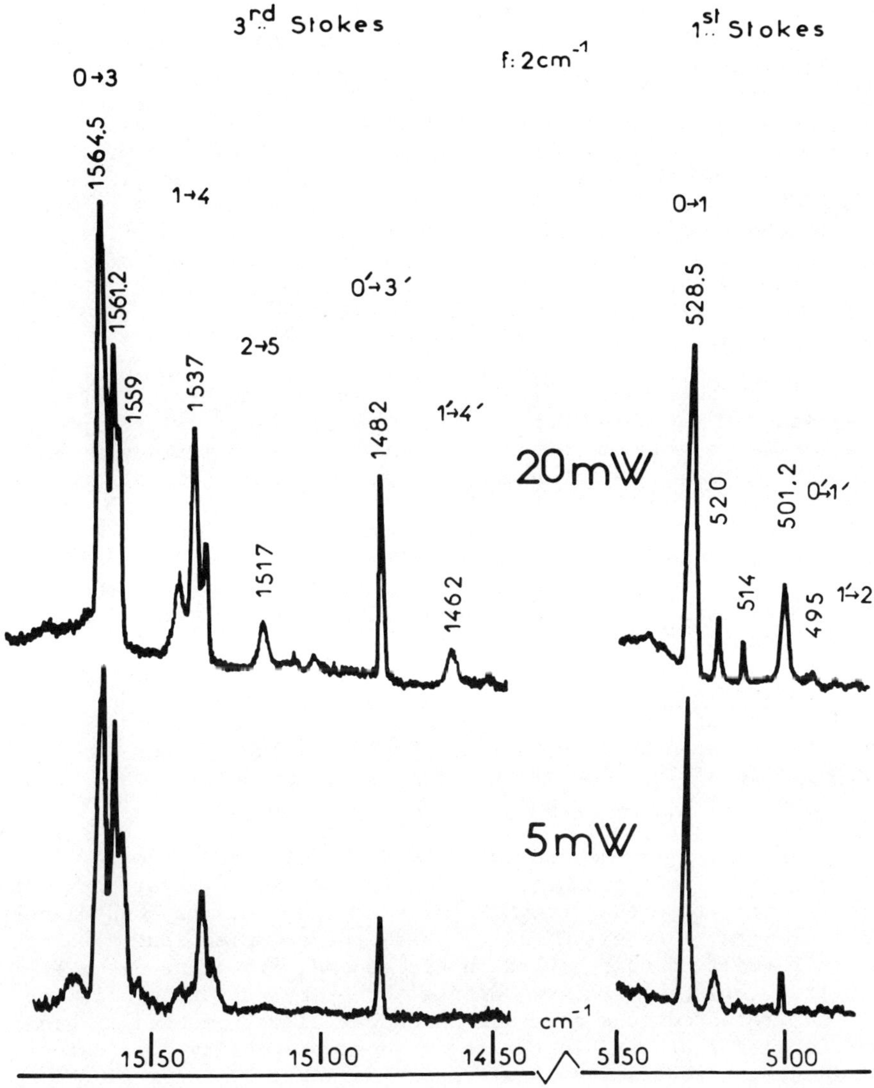

Figure 5. Portions of a high-resolution spectrum of V_2 isolated in an Ar matrix excited with 496.5-nm Ar^+ laser radiation, showing the effect of laser power on the first and third members of the Stokes GSRR and RRES progressions. Stokes RRES are indicated with primes (16).

of the quantum mechanical calculation may alter the ordering of these low-lying states, causing another state to be the ground state. Harris and Jones suggest that for V_2 a $^9\Delta_u$ state ($s\sigma_g^2 d\sigma_g^1 d\pi_u^2 d\delta_g^2 d\delta_u^1 d\delta_u^1 s\delta_u^1$) is another likely candidate as the ground state. Clearly, the symmetry of the correct ground state cannot be stated with confidence. It appears, however, that a state of high, odd spin multiplicity is most likely.

Focusing on a possible $^9\Delta$ ground state, we note that interactions with the matrix would remove the spatial degeneracy, yielding two states. We therefore hazard to guess that the ground state of V_2 is a $^9\Delta_u^-$ state whose spatial degeneracy was removed through the aforementioned interaction with the matrix while the excited state A is a state of symmetry $^9\Sigma_u^-$. The $^9\Sigma_u^- \leftarrow {}^9\Delta_u^-$ transition is, of course, Raman-allowed since they are of the same parity and their direct product transforms as the product of two displacements.

<u>Ti_2</u>. With titanium (<u>16</u>) in solid argon one obtains intense progressions only when exciting with red or orange laser light. As with the previous molecules both Stokes and antistokes progressions were observed and assigned to Ti_2, whose vibrational constants were determined to be ω_e = 407.9 and $\omega_e x_e$ = 1.08 cm^{-1}. With five abundant isotopes each of the Stokes and antistokes vibrational components split into a bouquet of lines. Superimposed on these were the sequence components arising from the radiative population of excited vibrational states each with its isotopic companions. This is shown in Fig. 7.

<u>Bonding in First-Row Transition Metal Dimers</u>. In transition metal dimers of low spin configuration, formal bond orders up to six (for Cr) are possible while in high spin, bond orders up to three are possible. These consist of an $s\sigma$ bond (even when high-spin level filling is assumed the $s\sigma_g$ orbital is generally taken to lie sufficiently below the d-manifold so as to be doubly filled first) and one σ, two π and two δ bonds arising from the d-orbitals. The multiple bonds of d-origin do not always contribute significantly to either the M-M force constant or its bond dissociation energy. Anderson (<u>20</u>) showed, for example, that although Ni_2 is formally double-bonded in his scheme, promoting one electron from the $s\sigma_g$ to the $s\sigma_u$ orbital, thereby rupturing only the $s\sigma$ bond, resulted in a molecule which was almost unbound. This fact is also reflected in the force constants observed to date for the first-row transition metals. Estimating the $s\sigma$ contribution to the force constant by means of Siebert's formula (<u>21</u>; the formula $k(mdyn/Å) = 7.2 Z_A Z_B/(n_A n_B)$ was used for the single bond force constant of a diatomic molecule AB in which Z and n refer, respectively, to the atomic number and the principal quantum number appropriate for atoms A and B) we see (Table I) that the observed force constant for Cu_2, which contains only one $s\sigma$ bond, is well reproduced by the formula, that of Fe_2 is only slightly

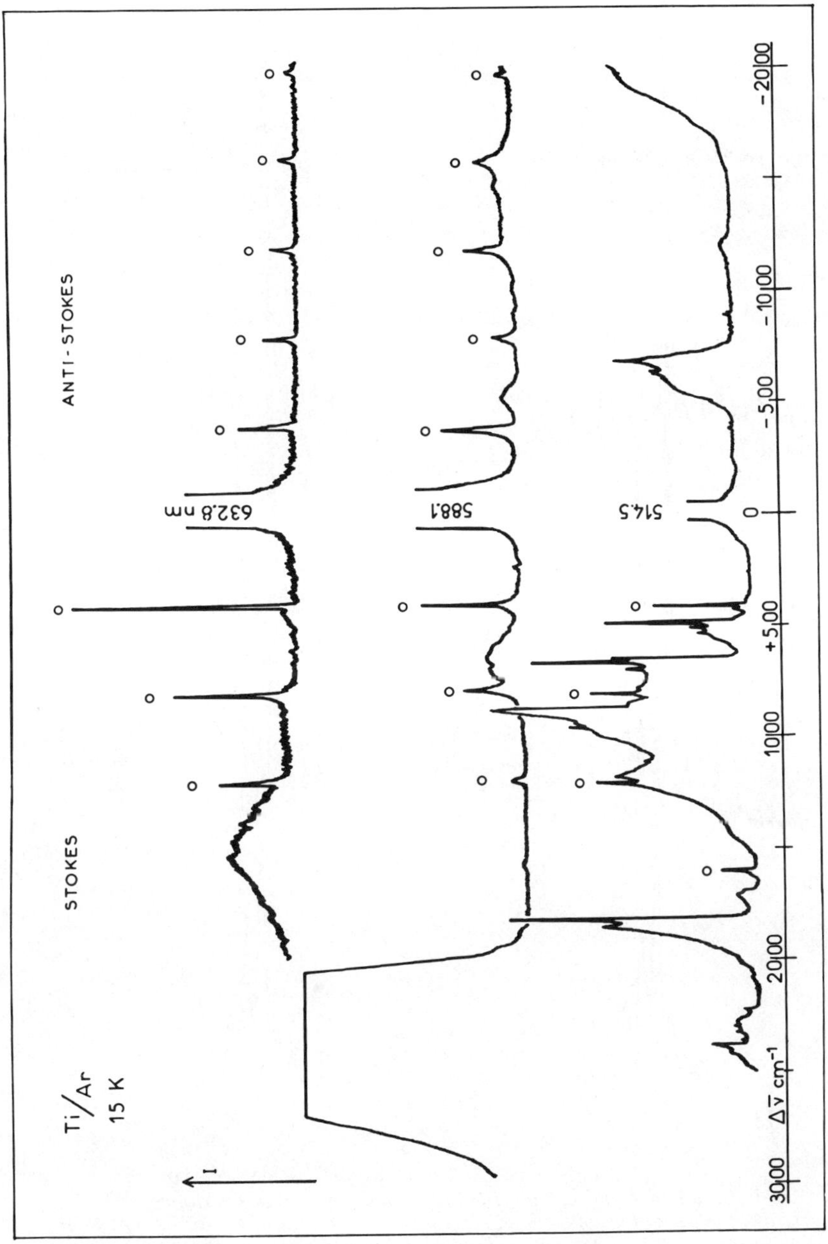

Figure 6. Resonance Raman scattering (○) of Ti_2 in Ar matrix at 15 K obtained with the three following excitations: 6328 nm (30 mW) He–Ne laser (6-cm^{-1} slits); 588.1 nm (10 mW) 6G dye laser excited with Ar^+ (5-cm^{-1} slits); 514.5 nm (10 mW) Ar^+ ion laser (5-cm^{-1} slits) (16).

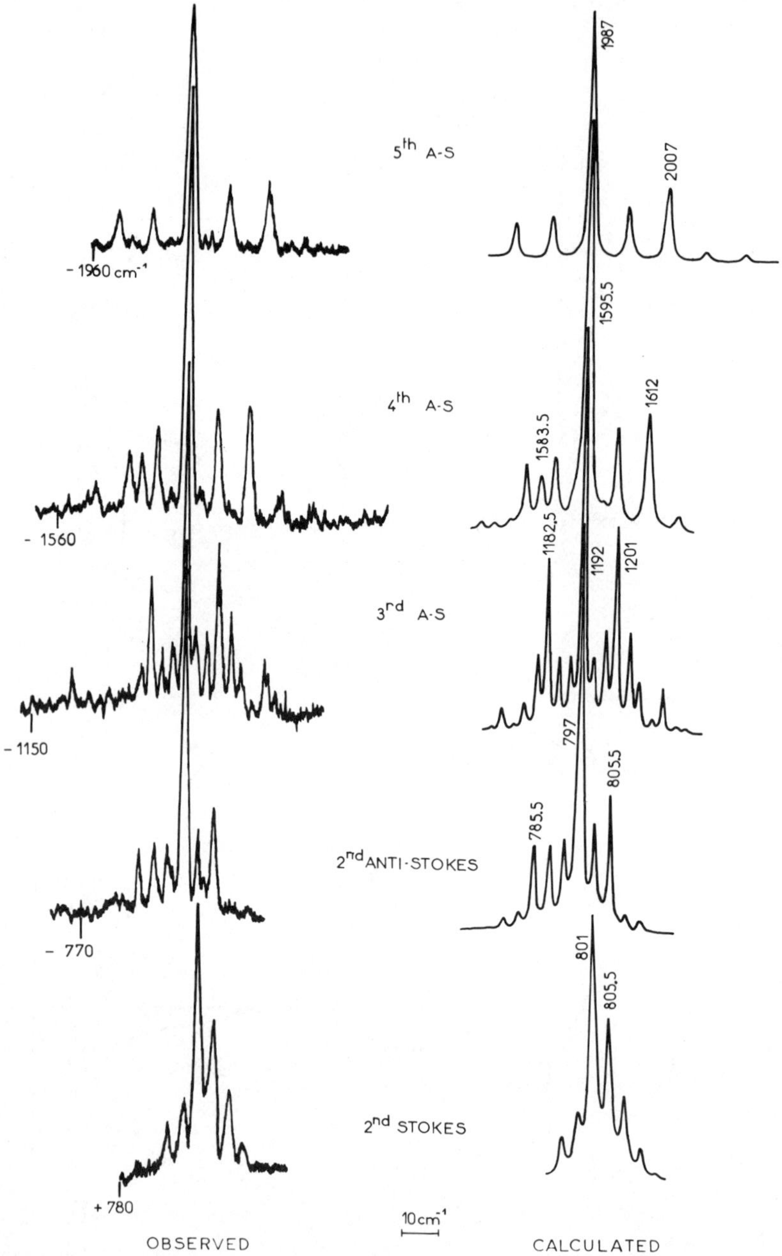

Figure 7. Experimental and calculated Stokes and antistokes overtones of Ti_2 in Ar matrix at 15 K recorded under high resolution (1-cm^{-1} slits) with 632.8-nm laser excitation. Frequencies shown on calculated lines indicate the location of each component used in the calculations, to reproduce the observed spectrum best (16).

TABLE I. Selected Properties of the First-Row Transition Metal Dimers

	Calculated ground state[a]	Calculated ω_e[a]	Observed ω_e (ref.)	k(mdyn/Å)	Single bond k(mdyn/Å)[c]	Bond order[a]	D_e (eV)[e]	dd[f] (eV)	σ-σ^*[f] (eV)
Sc_2	$^5\Sigma_u^-$	200	--	--	0.78	2	1.13	--	1.9
Ti_2	$^7\Sigma_u^+$	220	407.9 (*b*)	2.35	0.85	3	1.40	--	2.0
	$^5\Sigma_g^+$	630				*d*			
V_2	$^9\Sigma_u^-$	230	537.5 (*b*)	4.34	0.93	2	2.48	1.20	2.5
Cr_2	$^{11}\Sigma_u^+$	220	--	--	1.01	1	1.90	--	2.7
Mn_2	$^{11}\Pi_u$	220	--	--	1.10	1	0.33, 0.56	--	--
	$^{11}\Sigma_u^+$	210				1			
Fe_2	$^7\Delta_u$	390	300.3 (26)	1.48	1.19	3	1.3	0.33	2.9
Co_2	$^5\Delta_u$	360	--	--	1.28	2	1.72	--	--
Ni_2	$^3\Sigma_g^-$	320	380.9 (14)	2.48	1.38	2	2.37	0.94	3.0
Cu_2	$^1\Sigma_g^+$	280	266.1	1.31	1.48	1	1.9		3.5

[a] Taken from or based on the work of Harris and Jones (21). [b] Ref. 16. [c] Calculated from the formula of Siebert (21) which assumes only a single sσ bond. [d] The configuration was not reported for this state, hence the bond order could not be determined. [e] As reported by Cooper and Hare (17) based largely on the experimental work of Kant (7). [f] Calculated as described in the text from data given in the reference, dd refers to the span of the levels in the diatomic derived mainly from atomic d-orbitals while σ-σ^* is the observed $\sigma \rightarrow \sigma^*$ transition energy.

greater than the "single bond" force constant, those of Ni_2 and Ti_2 are greater than the single bond value while that of V_2 lies higher still above the "single bond" value. This trend mirrors that of the observed bond dissociation energies for the series. (A summary of the bond dissociation energies of first-row transition metal dimers is given by Cooper et al. in ref. 17, based mainly on the work of A. Kant (7). A more up-to-date summary has been prepared by K.A. Gingerich and is referred to in his contribution to this volume.)

The strength of a bond depends, among other factors, on the difference in energy between the bonding and antibonding orbitals arising from the same atomic orbitals. Our observations are consistent with the known fact that the energy difference between the $s\sigma_g$ and $s\sigma_u^*$ orbital is considerably greater than the span of the manifold of levels arising from the d-orbitals. They suggest, moreover, that the width of the d-manifold is greater in V_2 than in Ti_2 or Ni_2 whose d-manifolds are broader, in turn, than that of Fe_2. This is corroborated by the UV-visible absorption spectra of three of these dimers. Moskovits and Hulse (22) report that in the absorption spectrum of Ni_2 two structural absorptions at 528 and 377 nm arise respectively from a $d\pi_g^* \rightarrow p\pi_u$ and a $d\sigma_g \rightarrow p\pi_u$ transition. Since the $d\pi_g^*$ lies near the top of the d-manifold while the $d\sigma_g$ lies near the bottom, the difference in frequency (0.94 eV) between the ν_{00} frequencies of these two absorptions should give a measure of the width of the d-manifold. Both transitions end on the same level. A featureless band at 410 nm (3.0 eV) was assigned to the $s\sigma \rightarrow s\sigma^*$ transition. A similar procedure for Fe_2 (23) and V_2 (using the absorption spectrum reported in ref. 18 and identifying the high-energy $d \rightarrow \pi$ transition with the structured band partially obscured by the 368 nm atomic line; in the case of V_2 the high-energy $d \rightarrow \pi$ transition which comes at 370 nm is partly obscured by an atomic absorption at 368 nm) yields estimates of the d-manifold widths of 0.33 eV and 1.2 eV respectively. The d-manifold widths of V_2, Fe_2 and Ni_2 calculated above correlate well with the extent to which their force constants are greater than those due to the $s\sigma$ single bond alone. There is, in addition, an excellent correlation between the $s\sigma \rightarrow s\sigma^*$ energy intervals and the "single bond" force constants of the seven first transition row dimers for which data were available (Table I).

It is clear, however, that the formal bond order is not a good indicator of bond strength in transition metal systems without concurrent knowledge of the width of the d-manifold.

The ground state vibrational frequencies of the known transition metal diatomics are not calculated with uniform accuracy by any of the aforementioned groups (17, 19). This is especially true of V_2, which has been underestimated in the calculations of both Cooper and Hare (17) and Harris and Jones (19).

Triatomics

With triatomics one has the added dimension of structure to contend with. This should, in principle, be determinable from the isotopic fine structure of the resonance progression, assuming of course that the various isotopic counterparts of a molecule are simultaneously in resonance with the laser beam. While the broadening of vibrational levels due to interaction with the matrix tends to work in this direction, one cannot guarantee this condition in all cases.

Occasionally the magnitude of the frequency may suggest a geometry. So, for example, in a D_{3h} molecule the highest frequency is the totally symmetric one and hence the one likely to form the resonance Raman progression. A frequency higher than that of the diatomic would suggest this structure. In a linear triatomic, on the other hand, the totally symmetric vibration is the lowest frequency vibration. This fact may help guide one in choosing the correct geometry. All this depends, of course, on the assumption that the metal-metal force constant in the triatomic is not too different from that of the diatomic, an assumption which is not always realized. In addition one may have a triatomic which is Jahn-Teller or Renner-Teller unstable producing a fluxional or pseudo-rotating molecule even in a matrix environment, as has been suggested for Li_3 (24) and K_3 (25). In this case the concept of a vibrational state is no longer tenable and one must discuss the spectroscopy of the trimers in terms of vibronic states.

Ni_3. Figure 8 shows the RR spectrum attributed to Ni_3 (26). A high resolution spectrum of the third Stokes component is shown in Figure 9 together with the predicted isotopic fine structure calculated for various geometries assuming, first, that all isotopic molecules are equally excited and hence produce spectral components proportional in height to their natural abundances and, second, that the spacing between adjacent components of the progression shown in Figure 8 is the totally symmetric vibration. Details of the calculation are given in ref. 26. The results of the calculation indicate that the spectrum calculated for a C_{2v} molecule with an apex angle (ϕ) between 95° and 95.5° fits the observed spectrum best. The calculation was repeated with different ratios of f_{rr}/f_r, f_r being the Ni-Ni force constant and f_{rr} the interaction force constant between two Ni-Ni bonds. For f_{rr} approximately equal to zero the best fit was obtained for a C_{2v} molecule with ϕ = 90° whereas for $f_{rr} = 0.1\ f_r$, ϕ = 111° gave the best fit. Since f_{rr}/f_r ratios in neutral triatomics are rarely greater than 10% and even less often negative (27), we propose that Ni_3 is a C_{2v} molecule with an apex angle between 90 and 100°, favoring the former value. (A positive interaction stretching constant f_{rr} implies that stretching one bond causes the other to shrink in length and therefore strengthen. In a molecule such as

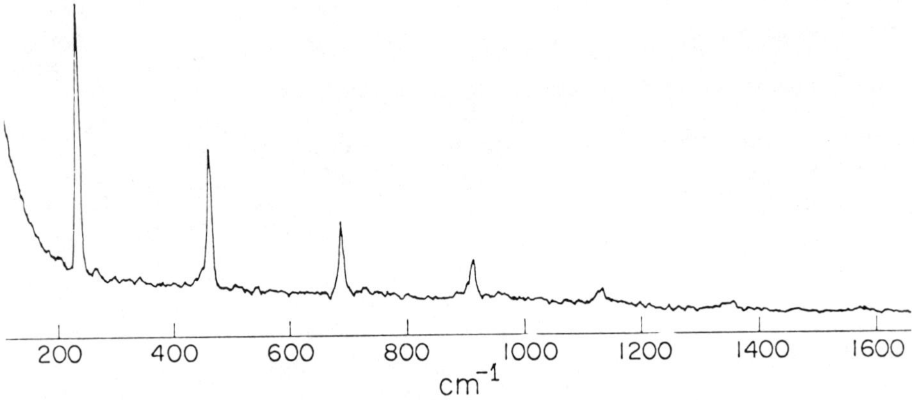

Figure 8. Raman spectrum of Ni-containing, solid Ar matrix excited with 4880 Å Ar-ion laser radiation (26).

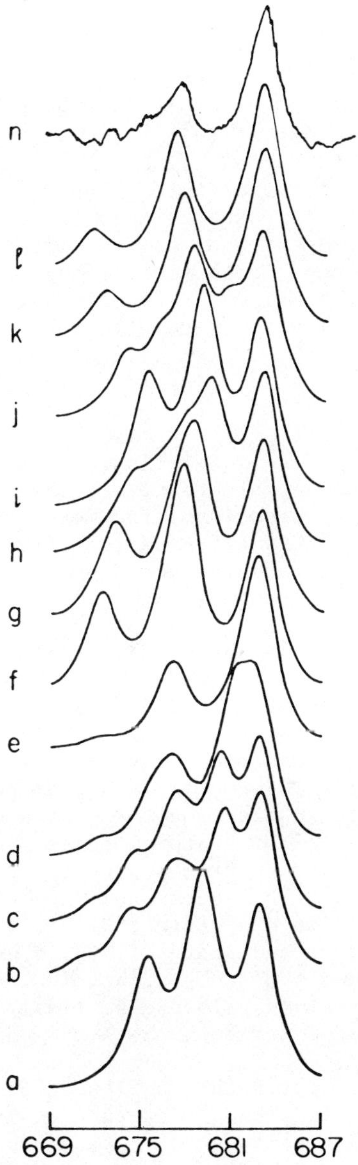

Figure 9. a–l *Calculated isotopic fine structure of the* $v = 3$ *component of the resonance Raman progression assuming the scatterer to be* Ni_3 *and that the normal mode is the symmetric stretching vibration and taking* f_{rr}/f_r *to be 0.05. Key:* a, D_{3h} *geometry;* b–i, C_{2v} *geometry; Apex angles, 80°, 90°, 95°, 95.5°, 96°, 96.5°, 100°, 120°, 140°, 160°, and 180°; and* n, *experimentally obtained fine structure on the* $v = 3$ *component excited with 4965-Å radiation* (26).

Ni_3, in which each individual bond is weaker than that in Ni_2 (as gauged by their relative force constants), the interaction force constant is almost certainly positive. Negative values of f_{rr} are encountered in KrF_2, H_2O and HCN, while unusually large values of f_{rr} are found in triatomic anions such as HCl_2^-, HF_2^-, Cl_3^-, $BrCl_2^-$ and other members of this class.

The geometry of Ni_3 deduced by the arguments given above is in apparent contradiction with that reported by Anderson using a semi-empirical molecular orbital calculation (28) and by Walch and Goddard (29), who suggest that Ni_3 is linear. Anderson finds, however, that the energy rise along the bending coordinate is so slight that even at $\phi = 120°$ it is only 0.1 eV higher than that calculated for the linear molecule (28). This may imply that the interaction with the Ar matrix, which is commonly greater than 0.1 eV, is a dominant force in determining the shape of the molecule. (An attempt to produce this Ni species in a krypton matrix in order to test this possibility failed because it is considerably more difficult to produce metal clusters in matrices formed from the larger rare gases.) Alternatively the distortion from a linear configuration or conformation may result from the Renner-Teller effect (30). This effect arises from the interaction between the electronic angular momentum and the angular momentum associated with the bending vibration, causing an electronic π or Δ state of a linear triatomic to split into A and B states of a C_{2v} molecule (30). It has been shown (31), however, that while the difference in the potential energy curves for the two new states contains both quadratic and quartic terms in the bending coordinate when the original electronic state is π, only quartic terms (i.e., only the anharmonicity) contribute in the case of Δ states. Consequently the Renner-Teller effect is expected to be much more pronounced in electronic π states than in Δ states. And, indeed, Renner-Teller distortion has been observed in many triatomic molecules formed from a parent π electronic state (30, 32) but only one report has been published of a substantial Renner-Teller effect in a Δ state (33).

The electronic ground state of Ni_3 is unknown. The united atom formalism allows $^1\Sigma_g^+$, $^3\Sigma_g^-$, $^1\pi_g$, $^3\pi_g$ and $^1\Delta_g$ states, while the separated atom method allows essentially any electronic state with Λ between 0 and 6 for both the d^8s^2 and d^9s^1 configurations of each Ni atom.

Anderson (28) reports the configuration of Ni_3 to be $1s\sigma_g^2$-$1s\sigma_u^2 1d\pi_u^4 1d\sigma_g^2 1d\delta_u^4 1d\pi_g^4 1d\sigma_u^2 1d\delta_g^4 2d\delta_u^3 2d\sigma_g^1 2d\pi_g^2$, which is obtained by filling the lowest seventeen orbitals of linear Ni_3 with the electrons in a high-spin configuration. The small energy spacing between the levels apparently dictates this choice. This ground state configuration allows spin quintets, triplets and singlets and suggests an electronic Δ_u state. (Had a low-spin configuration been chosen, the electronic ground state of Ni_3 would have been $^1\Sigma_g$.) Although one can rearrange the calculated levels somewhat to yield other electronic ground states, no reasonable configuration can be found which leads to a π ground state.

The "parent" ground state of Ni_3 would appear to be a Δ_u state of unknown spin multiplicity. This state has the possibility of distorting as a result of the Renner-Teller effect, although the distortion does not lead to a great deal of stabilization. In the gas phase the molecule is most likely fluxional with a linear average geometry. The matrix affects the stabilization of the molecule in a bent configuration. This possibility was also suggested in the context of the dynamical Jahn-Teller molecule Li_3 (24).

If Ni_3 occupies three neighboring vacancies (a trivacancy) in the matrix, assuming the ease of distortion suggested both by Anderson's calculations and the Renner-Teller argument, the bent configuration might come about as a result of the molecule's adjusting to the matrix environment in a manner least likely to cause crystalline strain. The four types of trivacancy which exist in a face-center-cubic crystal have linear, 120°-bent, 90°-bent and equilateral triangular geometries. Their relative abundances are in the ratio 1:4:2:2 (23). The D_{3h} geometry is unlikely to be sufficiently stabilized by the matrix if the "parent" molecule is linear. Consequently one has a six-to-one preponderance of bent molecules with apical angles of 90° and 120°, over the linear conformation. These angles are not too dissimilar to the value of ϕ determined by the isotopic fine structure analysis previously discussed. It is also possible that the matrix structure around a Ni_3 may deform slightly, compromising between argon crystal strain and Ni_3 stability. This may allow molecules in 90° and 120° trivacancy to assume a more uniform geometry with an apical angle between 90° and 120°. Linear molecules, if present at all, would not be discernible in the spectrum because of their reduced concentration.

Formally Ni_3 would be expected to have either an A_2 or a B_2 ground state in the matrix, these being the states to which Δ_u resolves in C_{2v}. The excited state which causes the resonance Raman is likely of similar symmetry, naturally of the same spin, and probably of opposite parity in its linear correlate (i.e., a Δ_g state). These conjectures are based on the similarity in the frequency of the ν_1 vibration in the two states. Moreover, the fact that a progression in ν_2 is not seen suggests that the apex angle is also similar in the two states. The molecule in its excited state is therefore subject to the same distortions discussed above. Several transitions (referred to the linear parent molecule) are likely candidates for the excited state, among them $d\pi_g \rightarrow p\pi_u$, $d\sigma_g \rightarrow p\sigma_u$. In each case neither the symmetry of the state nor the bond strengths which are primarily derived from s-orbital interactions are substantially affected.

Acknowledgements

We wish to thank NSERC and Imperial Oil for financial support. Discussions with Michel Tranquille, Thérèse Mejean, Robert Lipson and Kathleen Taylor are also gratefully acknowledged. The assistance of Professor R. J. LeRoy in performing the calculation of the dissociation energy of Fe_2 is noted with thanks.

Literature Cited

1. Schulze, W; Becker, H.U. Chem. Phys. Lett. 1978, 35, 177.
2. Beattie, I.R.; Gall, M.J. J. Chem. Soc. A 1971, 3569.
3. Selig, H.; Claassen, H.H. Is. J. Chem. 1968, 6, 499.
4. Kiefer, W. Appl. Spectrosc. 1974, 28, 115.
5. DiLella, D.P.; Gohin, A.; Lipson, R.; McBreen, P; Moskovits, M. J. Chem. Phys. 1980, 73, 4282.
6. Moskovits, M.; DiLella, D.P. J. Chem. Phys. 1980, 73, 4917.
7. Lim, S.S.; Kant, A. J. Phys. Chem. 1969, 73, 2450.
8. Gingerich, K.A. This volume, Chap. 000.
9. LeRoy, R.J.; Bernstein, R.B. J. Chem. Phys. 1970, 52, 3869; LeRoy, R.J. Ibid. 1980, 73, 6003.
10. LeRoy, R.J. Chem. Phys. Lett. 1980, 71, 544.
11. Chang, T.Y. Rev. Mod. Phys. 1967, 39, 911.
12. Clementi, E. "Tables of Atomic Functions," IBM J. Res. Develop. 1965, 9, 2.
13. Montano, P.A. Farad. Symp. Chem. Soc. 1980, 14.
14. Legay, F. "Chemical and Biological Applications of Lasers," C.B. Moore, ed., Academic, New York, 1977, 2, 43.
15. Ahmed, F.; Nixon, E.R. J. Chem. Phys. 1979, 71, 3547.
16. Cossé, C.; Fouassier, M.; Mèjean, T.; Tranquille, M.; DiLella, D.P.; Moskovits, M. J. Chem. Phys. 1980, 73, 6076.
17. Cooper, W.F.; Clarke, G.A.; Hare, C.R. J. Phys. Chem. 1972,
18. Ford, T.A.; Huber, H.; Klotzbücher, W.; Kündig, E.P.; Moskovits, M; Ozin, G.A. J. Chem. Phys. 1977, 66, 524.
19. Harris, J.; Jones, R.O. J. Chem. Phys. 1979, 70, 830.
20. Anderson, A.B. J. Chem. Phys. 1977, 66, 5108.
21. Siebert, H. "Anwendungen der Schwingungsspektroscopie in der Anorganischen Chemie"; Springer: Berlin, 1966.
22. Moskovits, M.; Hulse, J.E. J. Chem. Phys. 1977, 66, 3988.
23. a) Data of Hulse, J.E. Ph.D. Thesis, University Microfilms, Ann Arbor, Michigan, 1978. b) DeVore, T.C.; Ewing, A.; Franzen, H.F.; Calder, V. Chem. Phys. Lett. 1975, 35, 78.
24. Gerber, W.H.; Schumacher, E. J. Chem. Phys. 1978, 69, 1692.
25. Lindsay, D.M. This volume, Chap. 000.
26. Moskovits, M.; DiLella, D.P. J. Chem. Phys. 1980, 72, 2267.
27. Jones, L.H. "Inorganic Vibrational Spectroscopy, Vol. 1"; Marcel Dekker: New York, 1971.
28. Anderson, A.B. J. Chem. Phys. 1977, 66, 5108 and 1976, 64, 4046 and private communication.
29. Walch, S.P.; Goddard III, W.A. Surf. Sci. 1978, 72, 645.

30. Herzberg, G. "Molecular Spectra and Structure, Vol. III"; Van Nostrand Reinhold: New York, 1966.
31. Pople, J.A.; Longuet-Higgins, H.C. Mol. Phys. 1958, 1, 372.
32. Dressler, K.; Ramsey, O.A. J. Chem. Phys. 1957, 27, 971.
33. Merer, A.J.; Travis, O.N. Canad. J. Chem. 1965, 43, 1795.

RECEIVED September 21, 1981.

12

Electronic Structure of Small Group IA and IB Metal Clusters

STEVEN C. RICHTSMEIER, ROBERT A. EADES, and DAVID A. DIXON
University of Minnesota, Department of Chemistry, Minneapolis, MN 55455

JAMES L. GOLE
Georgia Institute of Technology, Department of Chemistry, Atlanta, GA 30332

The electronic structures of Group IA and IB metal clusters have been determined using two theoretical methods: ab initio molecular orbital theory and the semi-empirical diatomics-in-molecules (DIM) method. Electron affinities for Li_2, Li_3, Li_5, Na_2, and Na_3 are determined and various methods for calculating the electron affinity from ab initio wavefunctions are considered. Surfaces for Li_3 and Li_3^- are discussed in detail with respect to electron attachment and detachment. Similar considerations are outlined for Na_3. The structures of the Group IA trimers, Li_3, Na_3, K_3, Rb_3 and Cs_3 have been calculated using the DIM method. The trimers are all bound with respect to dissociation to atom plus diatomic having C_{2v} structures, a $^2B_2$ (obtuse angle) geometry being the most stable form. A $^2A_1$ state lies slightly higher in energy. A general discussion of the potential energy surface for the trimers is presented. Further DIM calculations on the Group IB trimers (Cu_3, Ag_3, Au_3) are outlined and their similarity to the Group IA trimers is emphasized. Vibrational frequencies for the $^2B_2$ and $^2A_1$ configurations are presented (uncorrected for the effect of higher potential surfaces) in the hope that they will aid the search for trimer spectroscopic features in both matrix isolation and gas phase spectroscopic studies.

The interplay between theory and experiment is of importance in high-temperature chemistry. The electronic structures of high-temperature molecules are being studied in detail with numerous experimental and theoretical techniques. Molecular

0097-6156/82/0179-0177$12.25/0

properties including ground state geometries (1), potential energy curves (2) and surfaces (1a), excitation energies (3), ionization potentials (4), electron affinities (4,5) and vibrational intensities (6) are being determined using theoretical methods for a wide range of high temperature species. Such theoretical studies are useful in providing a guide for the interpretation of experimental results, particularly in cases where novel species are encountered. This is especially true for metal clusters, many of whose structures are thought to be fluxional and for which little structural chemistry is yet known, even though they are of great practical importance (7).

A wide range of theoretical methods has been applied to the study of the structure of small metal clusters. The extremes are represented on the one hand by semi-empirical molecular orbital (Extended Hückel) (8) and valence bond methods (Diatomics-in-Molecules) (9) and on the other hand by rigorous *ab initio* calculations with large basis sets and extensive configuration interaction (CI) (10). A number of approaches lying between these two extremes have been employed including the X-α method (11), approximate molecular orbital methods such as CNDO (12) and PRDDO (13) and Hartree-Fock *ab initio* molecular orbital theory with moderate CI.

In applying any of the theoretical techniques mentioned, it is important that one be cognizant of the degree of sophistication needed to provide useful descriptive information which will aid the interpretation of experiment. Here certain constraints are operative; molecules and/or clusters which are the most amenable to experiment may be inaccessible to rigorous *ab initio* calculations by virtue of the atoms from which they are constituted. While rigorous calculations may be performed on small lithium or sodium clusters, such efforts are not yet feasible for potassium, rubidium, or cesium and one must resort to a meaningful semi-empirical theory. In studying the heavier metal clusters using semi-empirical techniques one would hope to employ a significant component of available experimental information as an aid to the evaluation of quantities inherent in the theory (e.g. coulomb and exchange integrals). Experimental data should enter into and gauge the calculations.

For this reason, we have chosen to apply the diatomics-in-molecules method to the study of the heavier Group IA trimers K_3, Rb_3, and Cs_3 and the Group IB trimers Cu_3, Ag_3, and Au_3. In this chapter, we will discuss the application of both *ab initio* and DIM methods to the study of the structure of small metal clusters, specifically clusters of the Group IA and Group IB atoms. These are, by choice, examples taken from our work but we have endeavored to comment on the general applicability of the various techniques.

Aspects of Theoretical Methods

Ab-Initio Calculations. *Ab-initio* molecular orbital theory has been widely discussed. Konowalow and Rosenkrantz, in this volume (14), have outlined the elegant multiconfiguration self-consistent (MCSCF) method (15) while Kurtz and Jordan (16) have presented detailed studies of metal atom-water interactions and alkali halide dimers. Here, we outline SCF+CI calculations on a number of small metal clusters (4,17,18). Our focus in this work has been on the determination of electron affinities for the alkali dimers and trimers. A variety of *ab initio* calculations have been employed in these studies. For example, if we wish to determine the adiabatic electron affinity (EA) for a given molecule, we must calculate the energy of the anion at its optimum geometry and the energy of the appropriate neutral at its optimum geometry. The difference in these energies yields a value for the EA uncorrected for vibrational effects. The calculation of electron affinities and, especially the molecular structure of the anion, should be done using extended basis sets including additional diffuse functions in order to properly describe the more diffuse nature of the anion charge cloud. Such diffuse functions are also required for determining the polarizability of the molecule and for determining the potential describing the adiabatic interaction of an electron with a molecule (19). An extended basis set may even be required for describing the potential energy surface of the neutral (1a,4). This is especially true for highly fluxional molecules where very different geometries have similar energies (see the discussion below). The calculations should be extended beyond the SCF level using methods which account for electron correlation, e.g. configuration interaction or MCSCF techniques (20).

In certain instances it may be appropriate to determine the ground state wavefunction using the generalized valence-bond (GVB) technique (21). This is especially true for closed shell anions where the pair of electrons in the highest occupied molecular orbital (HOMO) may be described by the wavefunction

$$\Psi_{GVB}(\mathrm{HOMO}) = [\phi_1(1)\phi_2(2) + \phi_1(2)\phi_2(1)](\alpha\beta-\beta\alpha) \quad (1)$$

instead of the restricted Hartree-Fock form

$$\Psi_{RHF}(\mathrm{HOMO}) = [\phi_1(1)\phi_1(2)](\alpha\beta-\beta\alpha) \quad (2)$$

The splitting of the HOMO may provide important flexibility if the anion is best described by the combination of a "neutral" core and a very diffuse electron.

There are a number of slightly more approximate methods for determining the electron affinity (EA) based on the restricted Hartree-Fock (RHF) and spin-unrestricted Hartree-Fock (UHF) methods. For closed shell anions, molecules which dissociate to an electron plus a radical in a doublet electronic state, one can employ Koopmans' theorem (22) to estimate the binding energy of the electron. This approach is most applicable for systems characterized by moderate to large electron affinities. Previous calculations of the electron affinities of GeH_3 and SiH_3 (23) using Koopmans' theorem have yielded good agreement with experimental estimates. For doublet open shell anions, which dissociate to an electron plus a closed shell neutral, the EA can be calculated using an SCF energy difference obtained from the difference between an RHF calculation on the neutral and a UHF calculation on the radical anion. This approach has previously been applied to the determination of EA's for Li_2, LiNa and Na_2 with some success (5).

The *ab initio* calculations described below were performed with the HONDO (Versions 3 and 5) (24) and MOLE programs (25). All calculations employed extended basis sets constructed from contracted Gaussian type orbitals. Basis sets were, in general, taken from the work of Huzinaga (26) and Dunning and Hay (27). They are described in detail elsewhere (4,18).

The Diatomics-in-Molecules Approach. The simple version of the DIM method that we employ is based on the Heitler-London approximation (28). In spirit, it is similar to the London-Eyring approach except that we use accurate diatomic potential curves (29a) rather than an approximate form for the diatomic triplet curve (e.g. the Sato parameter for LEPS surfaces) (29b). To exemplify, the energies for the doublet states of an A_3 system, where each A atom is in a ^2S state, are given by the following expression (the London equation)

$$E_{\pm} = J_{12}+J_{13}+J_{23}\pm 2^{-1/2}[(K_{12}-K_{13})^2+(K_{12}-K_{23})^2+(K_{13}-K_{23})^2]^{1/2} \quad (3)$$

(The same energy expression is also employed if all of the atoms are different.) The terms J_{ij} and K_{ij} are the Coulomb and exchange integrals between atoms i and j. The integrals J_{ij} and K_{ij} are evaluated from the empirical expressions

$$J_{ij} = 1/2[{}^1\Sigma_{ij}(R_{ij}) + {}^3\Sigma_{ij}(R_{ij})] \quad (4)$$

$$K_{ij} = 1/2[{}^1\Sigma_{ij}(R_{ij}) - {}^3\Sigma_{ij}(R_{ij})] \quad (5)$$

where ${}^1\Sigma_{ij}(R_{ij})$ is the potential curve for the ground electronic state of the diatomic at the distance R_{ij} between atoms i and j and ${}^3\Sigma_{ij}(R_{ij})$ is the lowest triplet curve; in general, the triplet curve is nearly repulsive but may be characterized by a small attractive well. These two curves in the asympototic limit correlate with the ground ^2S states of the atoms i and j. This representation of the J_{ij} and K_{ij} integrals follows from the simplest form of the Heitler-London approximation for the energy of a diatomic molecule formed from two ^2S atoms:

$$E(^1\Sigma) = (J + K)/(1 + S^2) \tag{6}$$

$$E(^3\Sigma) = (J - K)/(1 - S^2) \tag{7}$$

where S is the overlap integral between the two S orbitals. In the approximation we have employed, the value of S is set equal to zero. For four atoms an energy expression similar to that given above (equation (3)) is employed since only two Rumer (30) diagrams are needed to describe the molecule. For five and six atom molecules there are five Rumer diagrams (30) which must be combined to give the energy of the five appropriate singlet states. The energies of the singlet states for six atoms are given by the eigenvalues of the 5 x 5 matrix

$$E = \sum_{i>j}^{6} J_{ij}\,\underset{\sim}{1} - \underset{\sim}{K} \tag{3a}$$

where K is a matrix whose elements are given by Gelb, Jordan and Silbey (31). This expression, in slightly different form, was derived many years ago by Eyring et al. (32).

An important advantage of the DIM method as outlined above is that, for a maximum of eight atoms, the method is computationally efficient; even for ten atoms, a moderate size 42 x 42 matrix need be diagonalized. The diagonalization step requires little time using efficient matrix diagonalization routines such as EIGEN or GIVENS. Thus, extensive studies of the potential energy surface can be made. For clusters with more than 10 atoms, the method is considerably more difficult to apply as a result of the valence bond branching diagram. For a twelve atom cluster, a 132 x 132 matrix must be constructed and then diagonalized. If excited states are included in the DIM calculation (33), very large matrices must be diagonalized even for a three-atom system.

Perhaps the most significant advantage of the DIM method is that the empirical curves used in evaluating the Coulomb and exchange integrals are "exact" for the diatomic. Thus they include both correlation and relativistic effects between all atomic pairs. This latter correction is very important for heavy atoms (e.g. gold (Au)) (34). Although the singlet curves are well-established from experiment (35) or theory, the triplet curves are, in general, not well known. This deficiency is being rapidly remedied by new experimental and theoretical studies on metal diatomics. For example, a combination of theoretical (36) and experimental work (37) has now provided an excellent representation of the $^3\Sigma_u^+$ curve for Li_2.

In this chapter, we will focus on the application of DIM to small clusters of the Group IA (alkali) and Group IB (coinage) metals. The Group IA atoms are well represented by a ^2S ground state. The first excited ^2P state lies at considerably higher energy for sodium and potassium decreasing to 1.39 eV for cesium (38). Thus, the DIM method should be well-suited for studying clusters of sodium and potassium and of moderate utility for the study of rubidium and cesium. The application of DIM to clusters of Group IB atoms will be valid only if the ground atomic state $d^{10}s^1$ is well-separated from the first excited state, d^9s^2 (for Ag, $d^{10}p^1$ is the first excited level ≈ isoenergetic with d^9s^2). The excitation energies (38) are 1.39 eV, 3.75 eV and 1.14 eV for Cu, Ag and Au, respectively. Thus, we anticipate that DIM should be quite applicable to the study of silver clusters and should be of moderate applicability to the characterization of copper and gold clusters (see following discussions).

The parameters for the curves employed in these studies are given in Table I. The singlet curves are represented by a Morse potential

$$^1\Sigma_{ij} = D_1\{\exp[-2\beta(r_{ij} - r_e) - 2\exp[-\beta(r_{ij} - r_e)]\} \qquad (8)$$

with the parameters taken from experiment(35). The triplet curves are represented by Lennard-Jones 6-n potentials (39)

$$^3\Sigma_{ij} = 4D_3[(\sigma/r_{ij})^n - (\sigma/r_{ij})^6] \qquad (9)$$

with n = 8 for the Group IA metals and n = 12 for the Group IB metals.

The triplet curves for the Group IB metals were derived from the pseudopotential calculations of Ermler et al. (2). The triplet curve for Au_2 was fit to a Lennard-Jones 6-n potential and the value of n was optimized to 12. The experimental and theoretical equilibrium distances for the ground state of Au_2 were found to differ slightly. Therefore, the position of the triplet minimum was scaled slightly using the position of the experimental singlet minimum. The difference between the singlet and triplet minima was assumed to be a constant and the position of the triplet minimum was then set with respect to the experimental singlet minimum. The calculated singlet curves for Cu_2 and Ag_2 do not agree as well with experiment as does the Au_2 singlet curve. The theoretical triplet curves for Cu_2 and Ag_2 were thus scaled with respect to the experimental singlet curve as described above for Au_2.

The triplet curves for the Group IA diatomics were synthesized from molecular beam scattering data. The values of σ were obtained from the glory scattering experiments of Helbing and Rothe (40) on alkali-alkali pairs interacting in the $^3\Sigma$ state. We have used their values for σ which were obtained with the constraint of the van der Waals coefficient (Set A, Ref. 40). The values of the well depth were taken from the spin exchange experiments of Pritchard et al. (41). These authors found the well

Table 1. Potential Curve Parameters for DIM Calculations[a]

Atom	D_1(eV)	$\beta(au^{-1})$	r_e(au)	D_3(eV)	σ(au)
Li[b]	1.0677	0.45706	5.0493	0.02721	6.939
Na	0.7298	0.45312	5.8166	0.02721	7.2633
K	0.5197	0.40426	7.4134	0.02721	7.7573
Rb	0.4935	0.3814	7.8121	0.02721	7.986
Cs	0.3966	0.39029	8.2921	0.02721	8.379
Cu	1.98	0.7616	4.195	0.2349	4.149
Ag	1.63	0.7924	4.724	0.0513	5.507
Au	2.24	0.9087	4.668	0.16245	4.843

[a]See text for details of curves and references.

[b]Triplet curve parameters from Ref. 36 are σ = 6.406 au and D_3 = 0.0362 eV. The actual calculations were carried out using a Morse curve with parameters: r_e = 8.0 au., β = 0.45991 $au.^{-1}$ and D_e = 0.0362 eV.

depth to be virtually invariant with respect to scattering partner. Thus the well depths for the alkali $^3\Sigma$ states were all chosen to have the same value. The only comparison which we can make with accurate ab initio curves is with the $^3\Sigma_u^+$ curve calculated by Konowalow (36) and confirmed by the experimental work of Stwalley et al. (37). As demonstrated in Table I, the ab initio curve yields a deeper well depth and a smaller value of σ (the value of $^3\Sigma(r_{ij})$ for which the energy is zero). The curve constructed from the scattering results is not in serious disagreement. A detailed comparison of the determined structural parameters for the trimer using both cruves will be presented in later discussion.

Results

Ab-initio Studies

A. Surface for Li_3 (Na_3). In order to approach the calculation of electron affinities, we first discuss the nature of the Li_3 (1a,4) and Na_3 (3) surfaces. In considering the nature of these species, we emphasize the complications which are entailed in the description of their molecular structure. A logical structure for a metal trimer, M_3, where M has a ^2S ground state is an equilateral triangle. The simplest molecular orbital theory then yields a valence electron configuration...$a^2e'^1$ corresponding to a $^2E'$ state which must Jahn-Teller distort (42). The trimer can distort along the bending component of the degenerate "e" vibration (which has a_1 symmetry for a C_{2v} point group) to give either an obtuse triangular (Θ>60°) $^2B_2$ electronic state or an acute triangular (Θ<60°) $^2A_1$ electronic state. The presence of a conical intersection (43) (the crossing of the $^2A_1$ and $^2B_2$ surfaces at 60°) requires the introduction of a vector potential term in the nuclear Hamiltonian in order that the Born-Oppenheimer wavefunction be single-valued. This phenomena has been referred to as the molecular Aharonov-Bohm (MAB) effect (44). The effect is significant whenever the nuclear wavefunction is appreciable along a closed path about a conical intersection. The presence of the MAB effect requires a pronounced change in the method of evaluation of possible vibrational energy levels especially if the $^2A_1$ and $^2B_2$ states are connected by a low barrier. For a further discussion of the global topology of triatomic potential energy surfaces the reader is also referred to Davidson (45).

The surface for Li_3 has been explored by a number of workers at the SCF-CI level (1a, 4, 46, 47) (see Table 2 for a summary of results). These researchers have found that the molecule is bound with respect to monomer plus dimer by 9-10 kcal/mol and with respect to three Li atoms by ~34 kcal/mol. The Li_3 molecule has three low energy forms corresponding to $^2B_2(C_{2v})$, $^2A_1(C_{2v})$, and $^2\Sigma_u^+(D_{\infty h})$ electronic states. The $^2\Sigma_u^+$ state is simply the linear form corresponding to the $^2B_2$ state. The $^2B_2$

Table 2. Ab initio Geometries and Energies for Li_3 and Na_3

Binding energy[a]	R_e[b]	θ_e[c]	Binding energy[a]	R_e[b]	θ_e[c]	Ref.
$Li_3(^2B_2)$			$Li_3(^2A_1)$			
34.0	2.77	71	33.3	2.73	54	1a
33.6	2.84	73	33.4	2.79	54	44
29.4	2.80	68	29.4	2.70	52	43
28.7	2.96	74	--	--	--	4
$Na_3(^2B_2)$			$Na_3(^2A_1)$			
26.2	3.31	73	25.6	3.22	51.5	3

[a]Energy in kcal/mol relative to dissociation to three atoms.

[b]Equilibrium bond length in Å. 1 a.u. = 0.529177Å.

[c]Equilibrium bond angle in degrees.

state lies lowest in energy; the $^2A_1$ state has a slightly higher but comparable energy. Various calculations indicate that the linear form lies between 1 and 6 kcal/mole above the $^2B_2$ state. Gole et al. (4), in fact, find the linear form to be at a secondary minimum on the potential energy surface rather than at a saddle point. A comparison of the different calculations demonstrates that various basis sets treat different geometries at different levels of approximation. Apparently, it is difficult to obtain a moderate sized basis set which is equally appropriate for both the linear and highly bent forms. In highly fluxional molecules such as the alkali trimers, where small energy differences between structures of vastly different geometry are present, significant care must be taken in choosing the basis set. It should also be noted that these molecules are virtually unbound at the SCF level and that binding is obtained only when correlation effects are considered, e.g., with the use of CI techniques.

Martin and Davidson (3) have examined the structure of the sodium trimer at the SCF-CI level and find results that are very similar to those which have been obtained for Li_3. Na_3 has an optimum C_{2v} geometry with $^2B_2$ symmetry, a $^2A_1$ configuration corresponding to a saddle point lying only 0.6 kcal/mol higher in energy. The linear form lies only 3 kcal/mol above the minimum $^2B_2$ geometry. The Na_3 molecule is bound with respect to the dissociation limit Na_2 + Na by 8.5 kcal/mol.

B. Potential Surfaces for Li_3^+ and Li_3^-. The potential energy surfaces for the anion and cation of Li_3 (4) are significantly simpler in form than that of the neutral. At the D_{3h} geometry, the anion has an a^2e^2 configuration, the minimum corresponding to a triplet coupled state. The singlet state, $^1E'$, arising from the a^2e^2 configuration must Jahn-Teller distort. Indeed, the distortion is quite large and the anion is found to have a $D_{\infty h}$ geometry ($^1\Sigma_g^+$) (see Fig. 1a). The $^1\Sigma_g^+$ state is, in fact, the lowest lying state of Li_3^-. In contrast, the removal of an electron from $Li_3(D_{3h})$ leads to an a^2e^0 electron configuration for the cation and the D_{3h} geometry is the most stable structure (see Fig. 1b).

C. Electron Affinities and Ionization Potentials.

1. SCF-CI results for lithium trimer. Various values for the electron affinity and ionization potential are given in Table 3 and in Fig. 1. The exact definition of vertical and adiabatic energy increments is complicated by the fluxional nature of the Li_3 surface and the presence of a low-lying triplet surface for Li_3^-. As noted previously, the adiabatic energy is defined as the energy difference between the minimum energy geometries of the ion and the neutral. A secondary adiabatic quantity is defined with relation to the linear minimum on the Li_3 surface. A vertical energy is defined as the energy difference

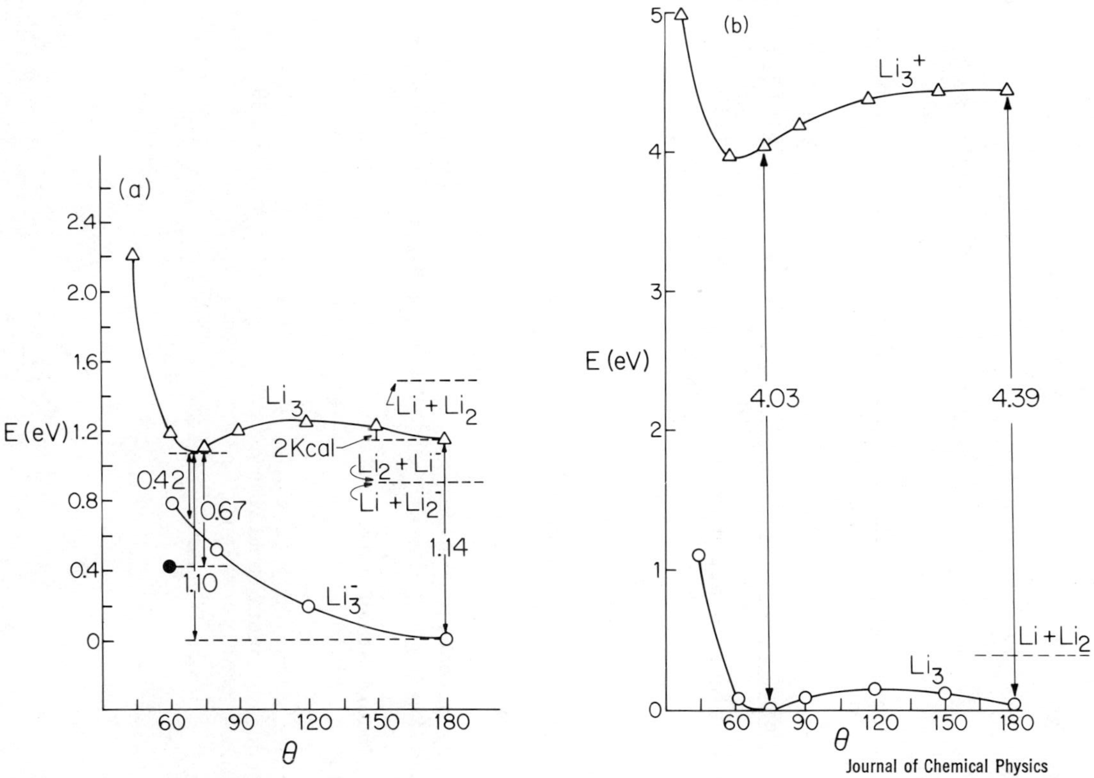

Figure 1. Potential energy surfaces as a function of bond angle for Li_3, Li_3^- and Li_3^+ obtained from SCF–CI calculations at a bond distance of 5.6 a.u. a: Cut in surfaces for Li_3 and Li_3^-. Key: △, Li_3; ○, Li_3^- $^1A_1$; and ●, Li_3^- $^3A_2'(D_{3h})$. Various vertical electron affinities are represented. b: Cut in surfaces for Li_3 and Li_3^+. Key: ○, Li_3 and △, Li_3^+. Various vertical ionization potentials are represented. Dissociation to Li + Li_2^+ at 5.24 eV (4).

Table 3. Summary of Electron Binding Energies in eV for Li_3

Quantity	ΔE(SCF)	ΔE(Cl)
Adiabatic electron affinity	0.46	1.10
Secondary adiabatic electron affinity	0.59	1.14
Vertical detachment energy (6.0 a.u.)	0.66	1.22
Neutral vertical electron affinity	0.30	0.67
Adiabatic ionization potential	3.64	3.95
Secondary adiabatic ionization potential	3.47	3.91
Vertical ionization potential (neutral minimum)	3.74	4.03
Vertical ionization potential (secondary neutral minimum linear configuration)	4.59	4.39

at a given geometry of the neutral or the ion. Table 3 demonstrates that correlation effects play a significant role in the calculation of the electron affinity. At the CI level, the adiabatic electron affinity is 1.10 eV while the vertical detachment energy is 1.22 eV (6.0 au.). The vertical attachment energy for the addition of a triplet coupled electron is only 0.67 eV and corresponds to formation of a $^3A_2'$ anion rather than a singlet species.

The adiabatic and vertical ionization potentials for Li_3 are very similar, both being approximately 3.95 eV. This results because of the similar geometries for the $^2B_2$ state of $Li_3(C_{2v})$ and the $^1A_1$ state of $Li_3^+(D_{3h})$. It must be noted, however, that the vertical ionization process for the removal of an electron from linear Li_3 to give linear Li_3^+ leads to the higher ionization potential, 4.39 eV. If both C_{2v} and $D_{\infty h}$ forms are present in an experiment, a complicated threshold dependence for the ionization process will be observed (4). The results for Li_3 exemplify that the fluxional nature of a small metal cluster may complicate the experimental determination of electron affinities and ionization potentials.

2. Further calculation of electron affinities for lithium and sodium clusters. The electron affinities of Li_2, Li_3, Li_5, Na_2 and Na_3 have been calculated (16) using the more approximate methods discussed previously. Since Li_2^- and Na_2^- are open shell anions, the EA's for Li_2 and Na_2 are calculated using the difference between the UHF energy of M_2^- and the RHF energy of M_2. Because the anions of Li_3, Li_5, and Na_3 are closed shell species, the EA's for their neutral clusters can be evaluated using Koopmans' theorem. Geometries for Li_2^- and Li_3^- were taken from previously optimized values. The structure determined for Na_3^- was geometry optimized using the gradient method (48) in $D_{\infty h}$ symmetry following the results found for Li_3^-. The geometry for Na_2^- was estimated using the ratio $R_e(Li_2^-)/R_e(Li_2)$ to obtain $R_e(Na_2^-)/R_e(Na_2)$. The Li_5^- geometry was presumed to be D_{5h} with a bond length of 6.0 au.

The various EA's determined using the above precipition are given in Table 4 where they are compared with other values. The present electron affinity for Li_2 is the highest value yet calculated. This is probably the result of the quality of the basis set which, at present, is the largest employed to describe the molecule. There is good agreement with our values for $EA(Li_2)$ and $EA(Na_2)$ and those of Shepard et al. (5) which were determined in the same fashion. As a further comparison, we have also given the values of the EA estimated from an application of Koopmans' theorem to UHF wavefunctions for Li_2^- and Na_2^-. Koopmans' theorem rigorously applies only to closed shell RHF calculations and, as expected, the current result demonstrates that the EA is seriously overestimated. (The actual EA for Li_2 is estimated to be between 0.50-0.65 eV.) (15). These results should also be compared to the SCF-CI calculation of Dixon et al. (15)

Table 4. Approximate Electron Affinities for Alkali Metal Clusters in eV

Molecule	Method[a]	EA(ev)[b]	Ref.
Li_2	KT(UHF)	0.92(V)	This work
Li_2	ΔE(UHF-RHF)	0.49(A)	This work
Li_2	ΔE(UHF-RHF)	0.46(A)	5
Li_2	ΔE(SCF-CI)	0.45(A)	15
Li_3	KT(RHF)	1.00(V)	4
Li_3	ΔE(SCF-CI)	1.10(A)	4
Li_5	KT(RHF)	1.26(V)	4
Na_2	KT(UHF)	0.85(V)	This work
Na_2	ΔE(UHF-RHF)	0.43(A)	This work
Na_2	ΔE(UHF-RHF)	0.42(A)	5
Na_3	KT(RHF)	0.93(V)	This work

[a]KT = Koopmanns' theorem, UHF = spin unrestricted Hartree-Fock, RHF = restricted Hartree-Fock, SCF-CI = Self consistent field with configuration interaction, $\Delta E(\text{UHF-RHF}) = E(M_n^-(\text{UHF}))-E(M_n(\text{RHF}))$, $\Delta E(\text{SCF-CI}) = E(M_n^-(\text{SCF-CI}))-E(M_n(\text{SCF-CI}))$.

[b]Electron affinity. V = vertical, A = adiabatic.

who, in their study, employed a basis set of Slater type orbitals somewhat smaller than that needed to provide sufficient flexibility in the valence space describing the anion. Consequently, their determined value for the EA was too low. The electron affinity has also been determined using the equations-of-motion method (49,50), however the predicted value is too large because the basis set used in these calculations is heavily biased in favor of the anion.

The value for EA(M_n) determined by Koopmans' theorem for M_n^- (closed shell) corresponds to a vertical detachment energy at the optimum structure of M_n^-. For the alkali trimers this does not introduce large errors since the energies of the linear and C_{2v} forms are quite similar. The value obtained at the SCF-CI level (4) is 1.14 eV for Li_3 while the application of Koopmans' theorem yields 1.00 eV. These values are in good agreement. The value obtained from Koopmans' theorem for Na_3 is 0.93 eV, similar to that found for Li_3. We thus expect the adiabatic EA for Na_3 to lie between 1.0 and 1.1 eV. As discussed previously, the interpretation of electron affinities for the alkali trimers is complicated by the fluxional nature of their potential energy surfaces. Significantly, more complication arises for slightly larger clusters as a result of the large number of possible geometrical isomers.

In studying the Li_5 cluster and its negative ion, a number of structures should be investigated. A few studies have been completed. The vertical detachment energy for the negative ion calculated for a D_{5h} geometry is 1.26 eV. DIM calculations (51) predict that the D_{5h} neutral structure, which must Jahn-Teller distort, is significantly less stable than the more compact pseudo-trigonal-bipyramidal structure. Detachment from this trigonal-bipyramidal structure would yield a somewhat lower adiabatic EA if the D_{5h} geometry is optimum for Li_5^-.

3. General comments on electron affinities and ionization potentials. From the investigations carried out thus far, it does appear that the electron affinity increases with increasing odd numbers of metal atoms in a cluster. Experimentally, it is well known that ionization potentials (IP) show an even-odd dependence on cluster size, where even n clusters have a higher IP than odd n clusters (52). For electron affinities the opposite effect probably exists with even n clusters having lower EA's than odd clusters. These results can be explained in terms of simple orbital models.

The ionization potential is determined by removing an electron from the highest occupied molecular orbital (HOMO). For an even n cluster, in a singlet state, the electron which is removed will see a greater nuclear charge than will an electron in the HOMO of an odd n cluster in a doublet state. This results because the nuclear charge for a closed shell orbital is only partially screened by the other electron in that orbital. In contrast, an odd electron in what is probably a more diffuse HOMO sees an effective nuclear charge of one since the other closed shell electrons screen the remaining nuclear charge. For similar reasons, the addition of an electron to form an anion will lead to an opposite effect. The addition of an electron to an open shell HOMO allows this electron to experience a larger nuclear charge as a result of the partial screening effect of the other electron. This, of course, is a more significant effect than the addition of an electron to the LUMO of a closed shell molecule (an even n singlet cluster). In this latter case, the closed shell electrons screen most of the nuclear charge so that the added electron is more loosely bound, resulting in a lower electron affinity.

Calculations Based on the Diatomics-in-Molecules Method

A. Comparison with ab-initio results - Li_3 and Na_3. In order to demonstrate the quality of the results which we have obtained using the DIM method, we have compared our results for Li_3 and Na_3 with various *ab initio* calculations (see Table 2 and Table 5). We have also compared the results obtained with our synthesized triplet curve for Li_2 with those obtained using the more accurate curve determined by Konowalow (36). Our value for the atomization energy to three atoms is in good agreement with the best *ab initio* value obtained by Gerber and Schumacher (1a). The bond angle which we determine using DIM is too large and the bond distance is too long. However, because the surface is very flat, the agreement is reasonable. The comparison is important since it is well known that the simple DIM method seriously overestimates the stability of the strongly bent molecule H_4 (53) (D_{4h} symmetry). The results which we obtain using the more accurate triplet curve of Konowalow show a binding energy 1.7 kcal/mol larger than the *ab initio* results and a bond length that is in excellent agreement with the *ab initio* values. The bond angle increases slightly and is still too large when compared with more accurate calculations. The shorter bond length found using the *ab initio* triplet curve can be attributed to the smaller value of the σ parameter in the Lennard-Jones potential; the *ab initio* triplet curve is less repulsive in this region than our constructed triplet curve. The larger binding energy found for the trimer reflects the smaller value of σ obtained from the *ab initio* triplet curve. This can easily be demonstrated using simplified expressions derived from equation (1). In terms of the potential energy curves for a C_{2v} geometry with atom 2 at the apex (29a):

$$E(^2B_2) = 3/2\,^1\Sigma_g^+(1\text{-}2) + 1/2\,^3\Sigma_u^+(1\text{-}2) + {}^3\Sigma_u^+(1\text{-}3) = E_+ \quad (10)$$

$$E(^2A_1) = 1/2\,^1\Sigma_g^+(1\text{-}2) + {}^1\Sigma_g^+(1\text{-}3) + 3/2\,^3\Sigma_u^+(1\text{-}2) = E_- \quad (11)$$

Table 5. The DIM Structures of Pure Alkali Metal Trimers

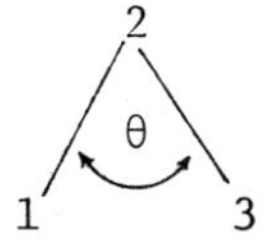

Alkali	R_{12}(a.u.)	R_{13}(a.u.)	θ	BE^a(kcal/mol)	R_{12}(a.u.)	R_{13}(a.u.)	θ	BE^a(kcal/mol)
	$E_+(^2B_2)$				$E_-(^2A_1)$			
Li	5.56	8.01	92.1	33.0	6.44	5.06	46.3	33.3
Li[b]	5.30	7.94	97.0	35.3	6.26	5.09	48.0	34.2
Na	6.16	8.39	85.8	23.7	6.97	5.81	49.3	23.4
K	7.51	8.96	73.2	18.1	7.91	7.41	55.9	17.9
Rb	7.89	9.22	71.5	17.3	8.25	7.81	56.5	17.1
Cs	8.37	9.67	70.6	14.0	8.71	8.30	56.9	13.8

[a]Binding energies with respect to dissociation to three atoms.

[b]Results using the triplet curve from Ref. 36.

An increase in the well depth of the $^3\Sigma_u^+$ curve will lead to an increase in the binding energy of the $^2B_2$ state since the $^3\Sigma_u^+(1\text{-}2)$ term is repulsive ($r<\sigma$) and the $^3\Sigma_u^+(1\text{-}3)$ term is attractive ($r>\sigma$). In a further comparison between the DIM and ab initio results, we focus on the structure and energy of the $^2A_1$ state of Li_3. The bond angle determined for this state is slightly smaller than that found using ab initio calculations; however, the bond length (R_{13}) is again in good agreement. The binding energy is also in good agreement and DIM predicts that the $^2A_1$ state lies only 1.1 kcal/mol above the $^2B_2$ state. This should be compared with a value of 0.7 kcal/mol determined from the ab initio results of Gerber and Schumacher (1a). The DIM method predicts that the linear conformation is very close in energy to the optimum geometry. Although the determined bond distance is in good agreement with ab initio results, the linear geometry is predicted to be too stable at the DIM level. Very similar comparisons can be made for Na_3. The bond length and binding energy of the $^2B_2$ state are in good agreement with the ab initio results of Martin and Davidson (3) while the bond angle is again predicted to be too large. The structure and energy of the $^2A_1$ configuration are in excellent agreement with the ab initio results. Again, the energy of the linear form evaluated at the DIM level is too close to that of the $^2B_2$ configuration. These comparisons demonstrate that for Li_3 and Na_3 the DIM method yields results in reasonable agreement with ab initio studies. Therefore, we feel that it can be applied to the study of the structures and energetics of other alkali metal clusters and will yield reasonable and useful results.

B. Parameters Governing Predicted Stabilities. The important diatomic terms governing the stabilities of the $^2B_2(E_+)$ and $^2A_1(E_-)$ states are indicated in equations 10 and 11. The triplet curves for the alkali dimers have small wells, their σ values being close to the value of r_e for the singlet curve. They are not strongly repulsive in the region where the attractive singlet interaction is large. The singlet interaction will therefore tend to be the dominant term in the sum describing the binding energies of these molecules. These properties will lead to the prediction of molecular clusters which are strongly bound. In contrast, the simplest three-electron triatomic, H_3, has an energetically unstable structure with respect to the asymptotic limit, atom plus diatomic. The triplet curve for H_2 is strongly repulsive in the region where the singlet curve is strongly attractive. The cancellation of energies is such that the repulsive triplet effects are dominant. This leads to the prediction of an unstable H_3 species. This argument is similar to that initially employed by Cashion and Herschbach in discussing H_3 (29a) and later by Dixon, Stevens and Herschbach (54) in considering H_6.

C. General Form of Potential Energy Surface. The general form of the potential energy surface for the alkali trimers is depicted in Fig. 2 for the Li_3 molecule. There are two low-lying states, $^2B_2$ and $^2A_1$, resulting from a distortion in C_{2v} symmetry, which cross at a D_{3h} conical intersection ($^2E'$). These $^2B_2$ and $^2A_1$ states are labelled as E_+ and E_- (Fig. 2a) respectively since they are derived from equation (3) giving equations 10 and 11 respectively. The $^2B_2(E_+)$ state can easily distort to the linear form bending past 180° to reach another conical intersection.

The $^2A_1$ and $^2B_2$ states can be connected by an asymmetric motion placing the three atoms in a C_s symmetry configuration and allowing the two states to mix. This is also represented in Fig. 2. An expanded view of the conical intersection is given in Fig. 2b. There are two other identical sets of C_{2v} distortions located at 120° and 240° on the hypersurface relative to the first C_{2v} distortion. These three distortions all intersect at their conical intersections. Because of the fluxional nature of the $^2B_2$ state and the low energy motion required to generate the linear form, the hypersurface is more complicated than that for other molecules (trimethylenemethane) which exhibit similar behavior (55). We note that our curves for E_+ and E_- are very similar to those obtained by Gerber and Schumacher (1a) in their fit of the Li_3 surface. The structures of the alkali trimers are summarized in Table 5. For comparison, experimental results for the dimer are given in Table 2.

D. Periodicity of M_3 Molecular Structure. The general behavior for M_3 alkali species is similar to that found previously in our discussion of Li_3 and Na_3. The lowest energy structure corresponds to the $^2B_2$ state (C_{2v}, obtuse angle) with the $^2A_1$ configuration at slightly higher energy. The $^2\Sigma_u^+$ linear structure is also of comparable energy to the $^2B_2$ state. The bond angle for the $^2B_2$ state decreases with the increasing size of the alkali atom, approaching a value of 60°. The bond lengths, r_{12} and r_{23}, (labelled with atom 2 as the unique atom), are slightly longer than those found for the diatomic. The bond angle for the $^2A_1$ configuration increases with the increasing size of the alkali atom also approaching 60°. The trimers are all bound with respect to the dissociation limit, atom plus diatomic. We find that the binding energy decreases with increasing atomic number; however, it is perhaps more relevant to compare this binding energy to the binding energy of the diatomic. The lowest ratios are found for Li_3 and Na_3 while the ratios for K_3, Rb_3 and Cs_3 are approximately 0.5.

The structure of the $^2A_1$ configuration is interesting in that the bond length r_{13} is virtually that of the dimer. It is apparent that this state looks very much like an atom bound to a diatomic. The binding is, of course, significantly greater than a typical van der Waal's interaction. This may result, in part, from the large polarizabilities of the alkali atom and of the alkali diatomic. In addition, there may also be a possible small admixture of the ionic configurations $M^-M_2^+$ and $M^+M_2^-$.

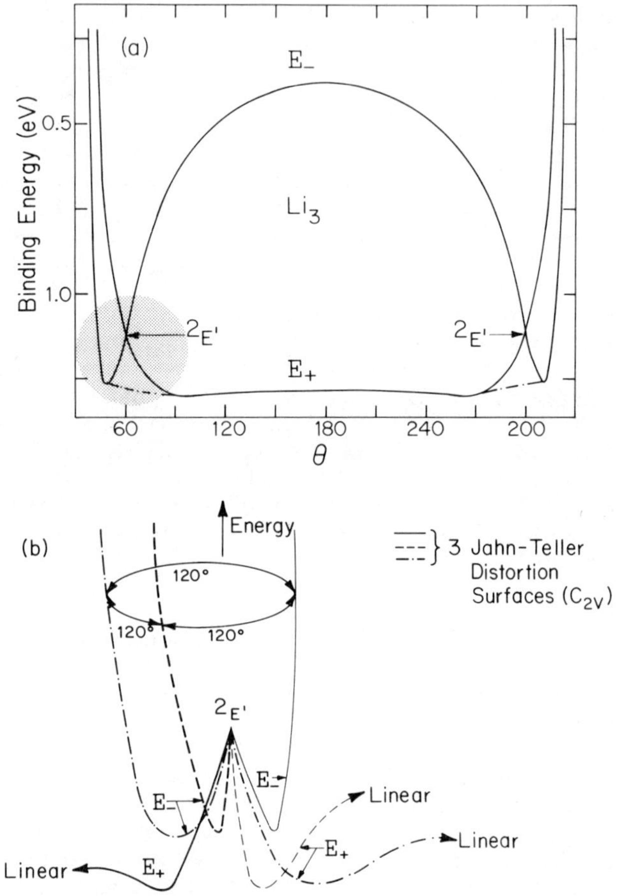

Figure 2. Scheme of Li_3 potential energy surface obtained from DIM calculations.

a: Full surface representing one C_{2v} distortion as a function of bond angle. The bond distances are taken from the optimized E_+ ($r = 5.3$) and E_- ($r = 6.26$) structures. The binding energy corresponds to dissociation to three atoms. Two surfaces cross at 60° giving a $^2E'$ state. DIM states E_+ and E_- correspond to the $^2B_2$ and $^2A_1$ electronic states. The · – · shows the general form of the asymmetric stretch distortion in C_s symmetry which yields a low energy path between the $^2B_2$ and $^2A_1$ states.

b: Expanded view of the $^2E'$ intersection region showing the full hypersurface. The three equivalent C_{2v} distortions are shown by three sets of curves. At values of $\theta < 60°$, the E_- state ($^2A_1$) is dominant and for values of $\theta > 60°$, the E_+ state ($^2B_2$) is dominant. The $^2B_2$ states all lead to linear geometries, $\theta = 180°$.

E. Comparison with Previous Calculations. A number of previous semi-empirical calculations on alkali trimers have been reported and are summarized by Companion. Companion (56) has used the DIM method to study Li_3 and finds results essentially identical to those reported here except that the bond angle for the $^2B_2$ state is slightly larger. Pickup (57), in studying Li_3, derived an elegant approach to generate the DIM matrices. Employing a triplet curve with a smaller well depth, Pickup found slightly smaller binding energies. Whitehead and Grice (58) have also examined Li_3 and Na_3 (also $NaLi_2$ and $LiNa_2$) with the DIM method. These authors employed a triplet curve with a very shallow well, finding the linear forms to be more stable than the $^2B_2$ C_{2v} structures for both Li_3 and Na_3. They also significantly underestimate the binding energy of the $^2B_2$ structure. In examining the $^2A_1$ state, formed by allowing an atom to approach a diatomic, they find a relative energy lower than that of the $^2B_2$ state for both Li_3 and Na_3, in contrast to the present DIM and ab initio results. We believe the current results (and those of Companion for Li_3) are more reliable since a more accurate representation of the triplet curve has been employed. Two pseudopotential calculations have been carried out for some of the alkali trimers. Hart and Goodfriend (59) report no binding energies and assume a linear geometry. The bond lengths determined in these calculations show moderate agreement with our results. Pickup and Byers-Brown (60), using a description close to the SCF level of approximation with a small "s" basis set, found no evidence that K_3 and Na_3 are bound with respect to monomer plus dimer. This is not unreasonable if we consider the previous discussion of ab initio calculations on the Li_3 surface where correlation effects are shown to account for most of the binding energy.

F. Extension of DIM to the Group IB Trimers. The Group IB and IA trimers are quite analogous. Previously we have reported the results of DIM calculations on the structure of the Group IB trimers with a focus on the $^2B_2$ surface (61). We report here new features of the Group IB surfaces and compare calculated binding energies with the recent experimental work of Hilpert and Gingerich (62). In addition, we have also explored the $^2A_1$ surface of the trimers.

The bending potentials for the $^2A_1$ and $^2B_2$ states of the pure Group IB trimers are shown in Figure 3. These potentials are determined using optimized $^2B_2$ and $^2A_1$ bond lengths obtained at the optimum bond angle. The energy of the $^2A_1$ state is comparable to that of the $^2B_2$ state for all of the trimers (see Table 6). In fact, for Ag_3, the $^2A_1$ geometry is predicted to be slightly more stable than the $^2B_2$ structure (0.3 kcal/mol). We calculate that the $^2A_1$ structures for Cu_3 and Au_3 have bond angles close to 55° and determine a bond angle for $Ag_3(^2A_1)$ of 51°. This trend is

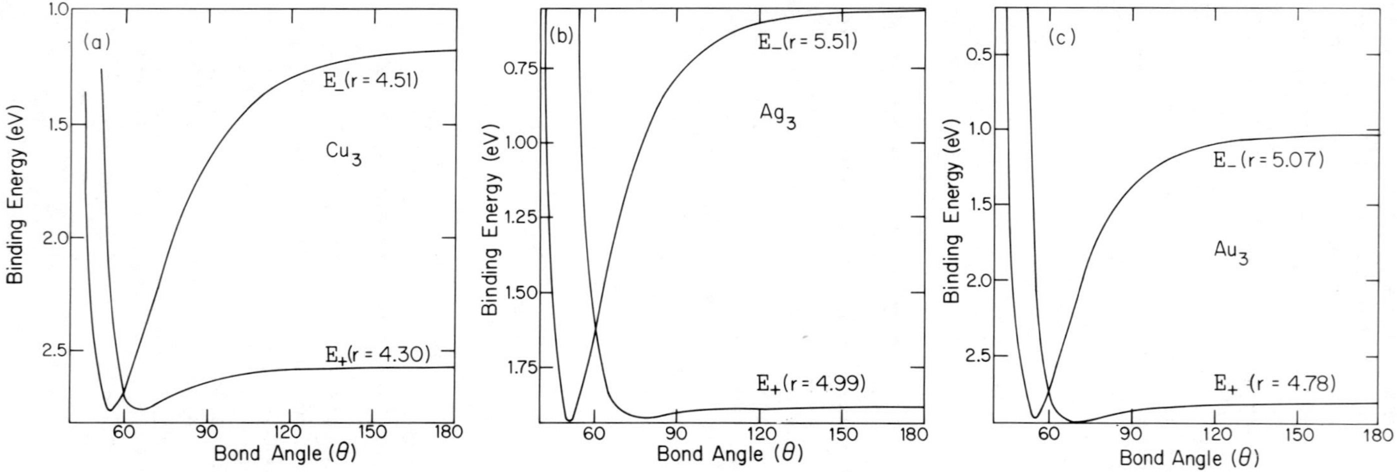

Figure 3. DIM potential energy surfaces as a function of bond angle for the pure Group IB trimers. The bond distances are taken from the optimized E_+ ($^2B_2$) and E_- ($^2A_2$) structures. The binding energy corresponds to dissociation to three atoms. The state crossing at 60° is due to the presence of a $^2E'$ state at this geometry. Key: a, Cu_3; b, Ag_3; and c, Au_3.

Table 6. The DIM Structures of Pure Group IB Metal Trimers

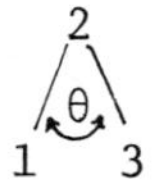

Metal	R_{12} (au)	R_{13} (au)	θ	BE (kcal/mol)[a]	R_{12} (au)	R_{13} (au)	θ	BE (kcal/mol)[a]	BE (kcal/mol)[a,b]	BE (kcal/mol)[a,b]
		$E_+(^2B_2)$				$E_-(^2A_1)$			$\exp(D_{\infty h})$	$\exp(C_{2v})$
Cu	4.30	4.66	65.6	75.2	4.51	4.19	55.4	75.2	70.3	68.7
Ag	4.99	6.19	76.6	52.1	5.57	4.73	50.8	52.4	60.5	58.9
Au	4.78	5.44	69.3	79.8	5.07	4.67	54.8	79.2	87.7	84.8

[a]Binding energies with respect to dissociation tc three atoms.

[b]Experimental results from Ref. 62.

reversed from the values of θ found for the $^2B_2$ structure where θ is smaller for Cu_3 and Au_3 than it is for Ag_3. Hilpert and Gingerich (62) have generated the Group IB trimers in a Knudsen cell, and, using mass spectrometry, have measured the atomization energies which are given in Table 6. The calculated values are in good agreement with the experimental quantities in view of our neglect of vibrational effects. Hilpert and Gingerich (62) have interpreted their results assuming a linear structure for the metal trimer in contrast to our finding of an optimum highly bent structure. The use of a $C_{2}v$ structure lowers the experimentally determined binding energies by a few kcal/mol which, for Ag_3 and Au_3, yields closer agreement between theory and experiment.

G. Calculation of Vibrational Frequencies. We have calculated vibrational frequencies for the isolated Group IB $^2A_1$ structures; these are tabulated in Table 7 and compared with the previously determined quantities for the $^2B_2$ state. The symmetric stretch frequencies are similar for the $^2B_2$ and $^2A_1$ structures and are significantly higher than the stretching frequency of the pure diatomic. The bending frequencies for the $^2A_1$ state considerably exceed the values found for the $^2B_2$ state. The value of ω_3, the asymmetric stretch is even lower in the $^2A_1$ state than the $^2B_2$ state. In fact, a real value of ω_3 for Au_3 cannot be determined, the $^2A_1$ structure representing a saddle point on the trimer surface. The asymmetric stretch is qualitatively the mode that connects the $^2A_1$ and $^2B_2$ states via a C_s structure. There is no crossing of the potential curves in the conversion between the states since both states belong to the same representation in C_s symmetry. This implies that our treatment of the full vibrational motion is only qualitative because we have not included the effects of the conical intersection, i.e., the vector potential term required by the molecular Aharonov-Bohm effect has not been included in our wave functions describing the nuclear motion. We expect these structures to be highly fluxional and pseudorotating about a 60° bond angle at moderate temperatures. If the molecule is frozen in a rare gas matrix at low temperatures, e.g., at 5°K, the predicted vibrational frequencies for the $^2A_1$ and $^2B_2$ states are relevant. Under these experimental conditions the molecules should not have access to the full region about the conical intersection (60°). The small value of the asymmetric stretch demonstrates that low temperatures are required to isolate certain forms. ESR spectral studies on the similar Group IA trimers Na_3 (63) and K_3 (64) show that isolated C_{2v} structures are observed at temperatures less than or equal to 5°K. Based on our semi-quantitative comparison of the alkali and Group IB surfaces it should be possible to isolate non-fluxional forms of the Group IB trimers in low temperature matrices.

The predominant feature of the Group IB trimer surface is the low barrier connecting the $^2B_2$ and $^2A_1$ structures about the conical intersection. This has significant ramifications

Table 7. Vibrational Frequencies for Group IB Trimers

Frequency (cm^{-1})[a]	Cu	Ag	Au
$\omega_1(^2B_2)$	445	264	276
$\omega_2(^2B_2)$	138	36	57
$\omega_3(^2B_2)$	163	57	143
$\omega_1(^2A_1)$	450	273	290
$\omega_2(^2A_1)$	237	168	168
$\omega_3(^2A_1)$	102	49	–[b]
$\omega_1(^2E')$[c]	460	275	285

[a] ω_1 = symmetric stretch, ω_2 = bend, ω_3 = asymmetric stretch.

[b] Imaginary frequency.

[c] Symmetric stretch at D_{3h} geometry.

with regard to the spectral features expected for these molecules and will play a key role if they have sufficient internal energy to access the different states in the region of the conical intersection. It is interesting to note that the energy of the upper surface is relatively high compared to the lowest-lying surface except in the region about $\theta = 60°$. The conical intersection actually moderates the two energies so that they intersect at $\theta = 60°$. As an example, the energy of the $^2B_2$ state for Au_3 is ~25 kcal/mol above the energy of the $^2A_1$ state in the region of the $^2A_1$ minimum.

Summary

The general structural feature of the Group IA and Group IB trimers is the fluxional nature of the potential energy surface characterizing these molecules. Although specific structural forms may be isolated in matrices at low temperatures, most gas phase trimers (except those produced in very cold supersonic expansions) will probably be quite fluxional and pseudorotate between the various $^2A_1$ and $^2B_2$ geometries. This adds complication to the treatment of the vibrational levels and the spectroscopy of these molecules. These small metal trimers appear to be quite simple at first glance yet their structural features suggest that they are quite complicated and will offer new insights about bound but highly fluxional molecules. They represent a new but extremely important form of chemical species.

Acknowledgement

This work was supported, in part, by National Science Foundation Grants CHE-7905985 (David A. Dixon) and CHE-7909075 (James L. Gole). David A. Dixon is an Alfred P. Sloan Foundation Fellow (1977-78), a Camille and Henry Dreyfus Teacher Scholar (1978-1983), and a Dupont Young Faculty Grantee (1978).

Literature Cited

1. a) Gerber, W. H.; Schümacher, E. J. Chem. Phys. 1978, 69, 1692.
   b) Dunning, T. H., Jr. J. Chem. Phys. 1980, 73, 2304.
2. a) Konowalow, D. D.; Olson, M. L. J. Chem. Phys. 1979, 71, 459.
   b) Ermler, W. C.; Lee, Y. S.; Pitzer, K. S. J. Chem. Phys. 1979, 70, 293.
3. Martin, R. L.; Davidson, E. R. Mol. Phys. 1978, 35, 1713.
4. Gole, J. L.; Childs, R. H.; Dixon, D. A.; Eades, R. A. J. Chem. Phys. 1980, 72, 6368.
5. Shepard, R.; Jordan, K. D.; Simons, J. J. Chem. Phys. 1978, 69, 1788.

6. Eades, R. A.; Scanlon, K.; Overend, J.; Dixon, D. A. unpublished results.
7. a) Sinfelt, J. H. Acc. Chem. Res. 1977, 10, 15.
   b) Robinson, A. L. Science, 1974, 185, 772.
   c) Muetterties, E. L.; Rhodin, T. N.; Band, E.; Bruker, C. F.; Pretzer, W. R. Chem. Rev. 1979, 79, 91.
   d) Stwalley, W. C.; Koch, M. E. Optical Engineering 1980, 19, 71.
8. Hoffman, R.; Lipscomb, W. N. J. Chem. Phys. 1962, 36, 3489.
9. a) Kunz, P. J. "Atom-Molecule Collision Theory"; Ed. R. B. Bernstein, Plenum, New York, 1979; p. 79
   b) Tully, J. C. "Modern Theoretical Chemistry: Semi-Empirical Methods of Electronic Structure Calculations"; Ed. G. A. Segal, Plenum, New York, 1977; Vol. 7B, p. 173.
10. Schaeffer, H. F., Ed.; "Modern Theoretical Chemistry. Methods of Electronic Structure Theory"; Plenum, New York, 1977; p. 111.
11. Johnson, K. H. Annu. Rev. Phys. Chem. 1975, 26, 39.
12. Pople, J. A.; Beveridge, D. L. "Approximate Molecular Orbital Theory"; McGraw-Hill, New York, 1970.
13. Halgren, T. A.; Lipscomb, W. N. J. Chem. Phys. 1973, 58, 1569.
14. See Konowalow, D. D.; Rosenkrantz, M. E. "The Electronic Structure and Spectra of Light Alkali Diatomic Molecules and their Molecular Spectra"; ACS Symp. Series, 1981.
15. See for example Wahl, A. C.; Das, G. "Modern Theoretical Chemistry Methods of Electronic Structure Theory"; Ed. H. F. Schaefer, III, Plenum, New York, 1977; Chap. 3.
16. See Jordan, K. D.; Kurtz, H. A. "Theory of Metal Atom-Water Interactions and Alkali Halide Dimers"; ACS Symp. Series, 1981.
17. Dixon, D. A.; Gole, J. L.; Jordan, K. D. J. Chem. Phys. 1977, 66, 567.
18. Eades, R. A.; Dixon, D. A.; Gole, J. L. unpublished results.
19. Dixon, D. A.; Eades, R. A.; Truhlar, D. G. J. Phys. 1979, B 12, 2741.
20. Shavitt I. "Modern Theoretical Chemistry. Methods of Electronic Structure Theory"; Ed. H. F. Schaeffer, III, Plenum, New York, 1977; Chap. 6.
21. Goddard, W. A., III; Dunning, T. H., Jr.; Hunt, W. J.; Hay, P. J. Acc. Chem. Res. 1973, 6, 368.
22. Koopmans, T. Physica (Utrecht) 1934, 1, 104.
23. Eades, R. A.; Dixon, D. A. J. Chem. Phys. 1980, 72, 3309.
24. a) Dupuis, M.; Rys, J.; King, H. F. J. Chem. Phys. 1976, 65, 111.
    b) King, H. F.; Dupuis, M.; Rys, J. Nat. Resour. Comput. Chem. Software Cat. 1980, Vol. 1, Prog. No. QH02 (HONDO).
25. Rothenberg, S.; Kollman, P.; Schwartz, M. E.; Hayes, E. F.; Allen, L. C. Int. J. Quant. Chem. Symp. 1970, 3, 715.
26. Huzinaga, S. J. Chem. Phys. 1965, 42, 1293.

27. Dunning, T. H., Jr.; Hay, P. J. "Modern Theoretical Chemistry. Methods of Electronic Structure Theory"; Ed. H. F. Schaeffer, III, Plenum, New York, 1977; Chap. 1.
28. Glasstone, S.; Laidler, K. J.; Eyring, H. "The Theory of Rate Processes"; McGraw-Hill, New York, 1941.
29. a) Cashion, J. K.; Herschbach, D. R. J. Chem. Phys. 1964, 40, 2358.
b) Sato, S. J. Chem. Phys. 1955, 23, 592, 2465.
30. Rumer, G. Göttingen Nachr. 1932, 377.
31. Gelb, A.; Jordan, K. D.; Silbey, R. Chem. Phys. 1975, 9, 175.
32. Taylor, H. S.; Eyring, H.; Sherman, A. J. Chem. Phys. 1933, 1, 68.
33. a) Issacson, A. D.; Muckerman, J. T. J. Chem. Phys. 1980, 73, 1729.
b) Faist, M. B.; Muckerman, J. T. 1979, 71, 225, 233.
34. Pitzer, K. S. Accts. Chem. Res., 1979, 12, 271.
35. Huber, K. P.; Herzberg, G. "Molecular Spectra and Molecular Structure IV. Constants of Diatomic Molecules"; Van Nostrand Reinhold, New York, 1979.
36. Olson, M. L.; Konowalow, D. D. Chem. Phys. 1977, 21, 333 and J. Chem. Phys. 1979, 71, 450.
37. Koch, M. E.; Stwalley, W. C.; Collins, C. B. Phys. Rev. Lett. 1979, 42, 1052.
38. Moore, C. E. "Atomic Energy Levels"; NBS Circular No. 467 (Government, Washington, D.C.) Vols. 1 and 2.
39. Hirschfelder, J.O.; Curtiss, C. F.; Bird, R. B. "Molecular Theory of Gases and Liquids"; Wiley, New York, 1954.
40. Helbing, R. K. B.; Rothe, E. W. J. Chem. Phys. 1968, 48, 3945.
41. a) Pritchard, D.; Carter, G., Chu, F. Y.; Kleppner, D. Phys. Rev. A 1970, 2, 1922.
b) Pritchard, D.; Chu, F. Y. Phys. Rev. A 1970, 2, 1932.
42. Jahn, H. A.; Teller, E. Proc. Roy. Soc. London, Ser. A 1937, 161, 220.
43. Mead, C. A.; Truhlar, D. G. J. Chem. Phys. 1979, 70, 2284.
44. Mead, C. A. Chem. Phys. 1980, 49, 23.
45. Davidson, E. R. J. Am. Chem. Soc. 1977, 99, 397.
46. Kendrick, J.; Hillier, I. H. Mol. Phys. 1977, 33, 635.
47. Bagus, P. S.; del Conde, G.; Davies, D. W. Faraday Discuss. Chem. Soc. 1977, 62, 321.
48. Pulay, P. "Modern Theoretical Chemistry. Applications of Electronic Structure Theory"; Ed. H. F. Schaeffer, III, Plenum, New York, 1977; Chap. 4.
49. Simons, J. Annu. Rev. Phys. Chem. 1977, 28, 15.
50. Anderson, E.; Simons, J. J. Chem. Phys. 1976, 64, 4548.
51. Richtsmeier, S.; Dixon, D. A.; Gole, J. L. unpublished results.
52. Herrmann, A.; Schümacher, E.; Wöste, L. J. Chem. Phys. 1978, 68, 2327.
53. Eaker, C. W.; Parr, C. A. J. Chem. Phys. 1976, 65, 5155.

54. Dixon, D. A.; Stevens, R. M.; Herschbach, D. R. Faraday Discuss. J. Chem. Soc. 1977, 62, 110.
55. Davidson, E. R.; Borden, W. T. J. Am. Chem. Soc. 1977, 99, 2054.
56. Companion, A. L. Chem. Phys. Lett. 1978, 56, 500.
57. Pickup, B. T. Proc. Roy. Soc. (London) A 1973, 333, 69.
58. Whitehead, J. C.; Grice, R. Mol Phys. 1973, 26, 267.
59. Hart, G. A.; Goodfriend, P. L. Mol. Phys. 1975, 29, 1109.
60. Pickup, B. T.; Byers Brown, W. Mol. Phys. 1972, 23, 1189.
61. Richtsmeier, S. C.; Gole, J. L.; Dixon, D. A. Proc. Nat. Acad. Sci., USA 1980, 77, 5611.
62. Hilpert, K.; Gingerich, K. A. Ber. Bunsen ges. Phys. Chem. 1980, 84, 739.
63. Lindsay, D. M.; Herschbach, D. R.; Kwiram, A. L. Mol. Phys. 1976, 32, 1199; 1980, 39, 529.
64. Thompson, G. A.; Lindsay, D. M. J. Chem. Phys. 1981, 74, 959.

RECEIVED August 26, 1981.

13

Statistical Mechanical Prediction of Entropies and Free Energy Functions for Small Clusters of Atoms

D. J. FRURIP and M. BLANDER

Argonne National Laboratory, Chemical Engineering Division, Argonne, IL 60439

C. CHATILLON

Laboratorie de Thermodynamique et Physico–Chimie Metallurgiques,
E.N.S.E.E.G., B.P. 44,38401, Saint–Martin–D'Heres, France

Previously successful correlations of thermodynamic data for various simple gaseous molecules are extended to larger molecules. In the statistical mechanical theory one calculates the nonelectronic entropy and free energy function with good precision using as input only the interatomic distance and atomic masses. The general class of molecules MX_n ($1 \leq n \leq 6$) appears to follow the predictions of the theory. When known thermodynamic data for small homonuclear clusters of atoms, X_n ($n \leq 8$), are tested the resulting correlations with the MX_{n-1} data are relatively good, further confirming the insensitivity of the entropy to the exact form of the pair potential.

The study of small, homonuclear clusters of atoms is important in understanding nucleation because such clusters are intermediates in the formation of bulk condensed phases. The dynamic process of condensation from a gas must initially involve the formation of tiny aggregates of the new phase. This can be illustrated by the reaction sequence $A(g) \rightarrow A_2(g) \rightarrow A_3(g) \rightarrow \ldots \rightarrow A(\ell)$. One of the major weak points in the present day understanding of such nucleation phenomena is the unknown thermodynamic properties of clusters. Certainly, the common practice of treating a 2-200 atom cluster as a tiny piece of the bulk with a large surface is inexact. There is a need for precise thermodynamic data on atomic and molecular clusters to better define nucleation kinetics.

In the classical theory of nucleation (1) for condensing vapors, there exists a "critical size cluster" through which the passage of clusters smaller than critical to clusters larger than critical, is rate limiting. According to this theory, the exact

0097-6156/82/0179-0207$05.00/0

size of this critical size nucleus is determined by a maximum in the curve of the free energy, ΔG_n°, versus cluster size. Depending on conditions the apparent critical size cluster may, in fact, be as small as the dimer, A_2. Blander and Katz (2) have pointed out some of the ambiguities in the conventional macroscopic theory for ΔG_n° in terms of bulk liquid surface tensions and densities. For smaller size clusters ($n \lesssim 20$), a molecular approach to the nucleation problem is clearly preferred. In this regard, Bauer and Frurip have recently used a somewhat empirical approach to determine the cluster entropies of condensing iron (3).

In this paper we will apply our dimensional model (4) to the problem of determining standard entropies for small homonuclear gas clusters. The model can also be used to determine free energy functions. Our approach will be statistical mechanical in nature and less empirical than the entropy correlation by Bauer and Frurip. In section II, the dimensional model is reviewed and its application to molecules of the type MX_n $2 \leq n \leq 6$ is discussed. In section III, the model is tested for homonuclear clusters, A_n, by comparing literature data with corresponding data on MX_{n-1} type molecules. In section IV the results are discussed.

The Dimensional Model

The calculation of entropies for gaseous species generally requires detailed knowledge of geometry, bond distances and vibrational frequencies. We have developed a statistical mechanical model which allows an estimation of the entropy of an unknown molecule using as input only the atomic masses and interatomic distances. Details of the development of the model have been given in previous publications (4). A brief summary of the important assumptions and equations is given below.

Fundamentally, the dimensional model involves an investigation into the properties of the classical configurational integral. For a molecule C_cA_a, with a characteristic size parameter d the configurational integral Z can be written:

$$Z = \frac{1}{a!c!} \int \cdots \int_V \exp(-U/kT) d\tau^c d\tau^a \qquad (1)$$

where U is the total potential energy. The development of the model is pursued by comparison of Z for a given molecule in a volume V with the integral Z_o for another salt D_cB_a with a size parameter d_o in a volume V_o. For each configuration of the atoms in C_cA_a in the volume V, there exists a geometrically similar configuration of D_cB_a in the volume V_o such that V and V_o are related by the equation:

$$\frac{V_o}{V} = \left(\frac{d_o}{d}\right)^3 \qquad (2)$$

Thus, we can scale the value of the configurational integral by this dimensionless ratio. This scaling can be accomplished exactly for coulombic pair potentials of a particular form. However, it was shown (4a) that the results should not differ greatly for other types of pair potentials.

The final result for the dimensional model is an equation relating the entropy (excluding electronic contributions) to the interatomic distance and atomic masses, m.

$$\frac{S_T^\circ}{R} = \sigma_o + \ln\left[d^{3(a+c-1)}(m_A^a m_C^c)^{3/2}\right] \qquad (3)$$

The term, σ_o, is a universal constant related to theoretical integrals which are dependent only on the stoichiometry and temperature. There is an analogous expression for the free energy function $-(G_T^\circ-H_{298}^\circ)/RT$, with a different constant, γ_o. We have found that a convenient method for testing the validity of the dimensional model is through plots of S_T°/R or $\frac{-(G_T^\circ-H_{298}^\circ)}{RT}$ versus $\ln[d^{3(a+c-1)}(m_A^a m_C^c)^{3/2}]$. At constant temperature Eq. (3) predicts a linear relationship with unit slope and intercepts σ_o or γ_o. Such plots are shown in references 4a and 4b for a range of stoichiometries. These include the alkali halides, alkali halide dimers, diatomic oxides, nitrides, and sulfides, and molecules of the type MX_2. Good agreement with the model is obtained with many species, including those which are non-ionically bonded.

Tests of the dimensional model have been extended to larger and more complex species. For example, the series of molecules of the general stoichiometry MX_n, $2\leq n\leq 6$, appear to correlate well with the model (5). Figure 1 is representative of the type of correlation found for the MX_n species. Shown are the MX_3 and MX_6 entropy plots at 1000 K. The solid lines are least square fits of the data points under the constraint of unit slope. The data were taken mainly from the JANAF tables and supplements (6). Points plotted as open circles indicate that most of the spectroscopic input for these molecules (*i.e.*, vibration frequencies, etc.) used in the quantum statistical calculation were estimated. On the average, these "estimated" points deviate more from the least squares line than the filled circles. Note also the paucity of data for the larger MX_6 species. Table I lists the results of all such plots for the MX_n species.

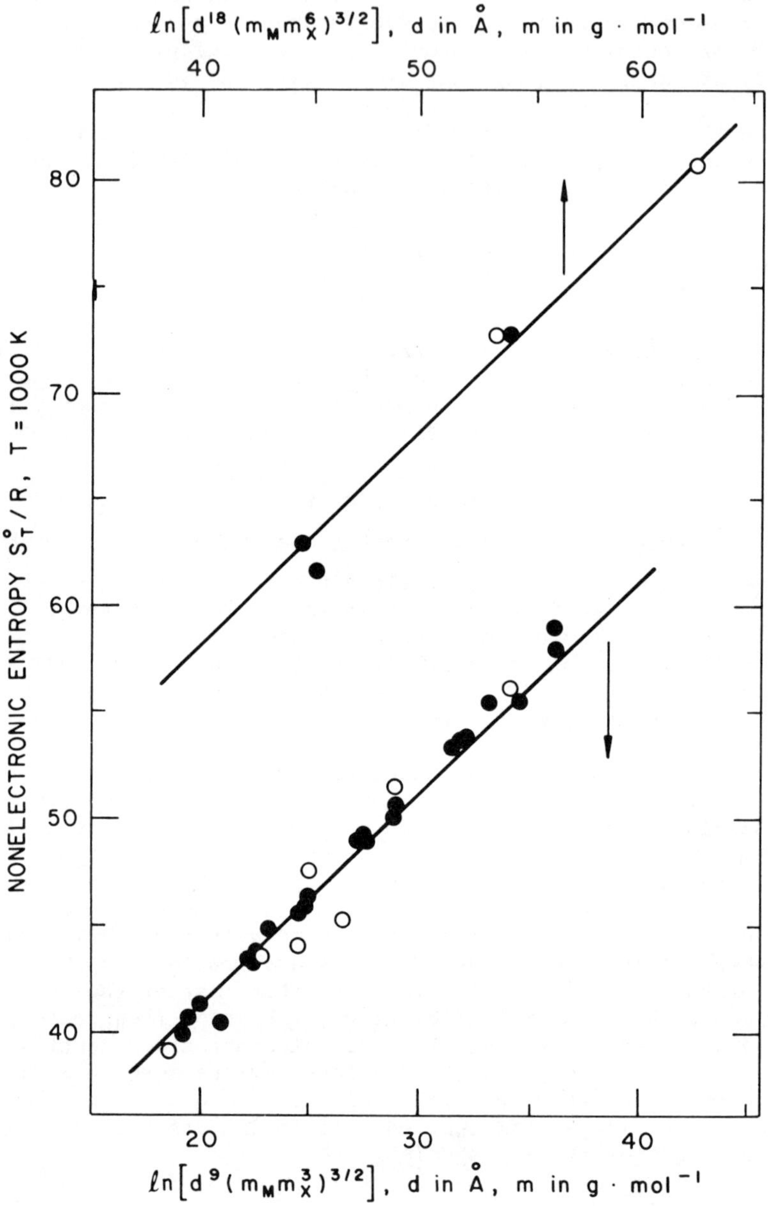

Figure 1. Representative tests of the correlation of the dimensionless, nonelectronic entropy with ln $[d^{3(a+c-1)}(m_A{}^a m_C{}^c)^{3/2}]$ for gaseous MX_3 (lower scale) and MX_6 (upper scale) species at 1000 K (6). Key: ○, vibrational frequencies and bond distances estimated; and ——, least squares fits of data with the constraint of unit slope.

Table I. Dimensional Analysis Fitting Results[a] for Gaseous MX_n Species ($1 \leq n \leq 6$) at 1000 K.

	No. of molecules tested	σ_o at T = 1000 K
MX	13	20.4
MX_2	76	21.0 (linear) 21.9 (nonlinear)
MX_3	24	21.1
MX_4	17	20.5
MX_5	11	19.9
MX_6	5	17.9

[a]Plots showing the MX (alkali halides) and MX_2 correlations are given in ref. 3; the MX_3 plot is shown in Fig. 1; the MX_4, MX_5, and MX_6 plots are unpublished (5). Most of the data for the MX_3-MX_6 tests were taken from the JANAF tables (6).

Applications for Homonuclear Species

The applicability of the dimensional model to ionic and nonionic molecules containing as many as seven atoms is gratifying. This demonstrates the insensitivity of the entropy (as determined by the configurational integral) to the exact form of the pair potential. This encouraged us to attempt similar correlations with homonuclear clusters. We will test whether the X_n species follow the prescriptions of the dimensional model for the corresponding MX_{n-1} species. For convenience, we will restrict ourselves to the entropy at 1000 K. The correlations at other temperatures and for the free energy functions are similar.

Dimeric Clusters. By far, the largest body of thermodynamic information exists for vapor phase homonuclear diatomics. Our source of data was the JANAF tables and supplements.[6] This compilation is by no means exhaustive but there are certainly sufficient X_2 type molecules listed (20) to provide a suitable test of the model. Due to the prefactor $(a!c!)^{-1}$ in the configurational integral (Eq. 1), an amount Rln2 was added to the entropies of the X_2 species to allow for a meaningful comparison with the MX species. Fig. 2 exhibits the test of the model for 19 dimers at 1000 K. (H_2 was not included due to quantum effects). The solid line is the best fit line of unit slope for the alkali halides as taken from reference 4a. The actual alkali halide data points would deviate from this line to a considerably smaller extent (± 0.1 dimensionless entropy units) than the dimers. Still, for the dimers, one sees an adequate correlation with the predictions of the model. The largest deviations from the alkali halide line occur with Pb_2 and Cs_2. Both of these molecules have vibrational frequencies and interatomic distances which were estimated. Another interesting observation is that dimers of the alkali metals consistently lie above the alkali halide curve by 0.5 units. The dashed line is a line of unit slope drawn through these points. Although the predictions of the model are good for groups of unrelated molecules, it seems to be even better for correlations of molecules containing atoms within the same group.

Trimeric Clusters. Thermodynamic data (or sufficient spectroscopic data to allow a thermodynamic calculation) for several homonuclear trimeric species exist in the literature. Data for the species C_3, N_3, Si_3, and O_3 are found in the JANAF tables (6); S_3 and Se_3 are in Mills, (7) Sb_3 and Pb_3 from Gingerich (8) and coworkers; As_3 in Hultgren et. al. (9); Li_3 in Companion et. al. (10,11); and Bi_3 from Rovner et.al.(12). Except for the species O_3 and C_3 most of the spectroscopic input for these clusters have been estimated. In the entropy data presented in Mills and Hultgren (i.e., As_3, S_3, and Se_3), it is not clear whether or not electronic effects are included. Figure 3 presents the scaled entropy ($S^\circ_T/R + \ln 3$) versus the appropriate logarithmic term for the 11 trimeric species. The uncertainty bars on Pb_3 (9) are due to the range of estimations by Gingerich in which various structures were assumed (C_{2v}, $D_{\infty h}$, D_{3h}). The solid lines in the plot are the least squares fits to the corresponding MX_2 species as taken from reference 4a. The upper line is for the nonlinear MX_2 and the lower line is for linear MX_2. As in the homonuclear dimer plot (Fig. 2) we see that the homonuclear trimeric species are adequately represented by the MX_{n-1} results. It is unfortunate that both of the relatively well characterized trimers (O_3 and C_3) are in the low mass range for this precludes a meaningful extension to higher masses.

Tetramers. Fig. 4 shows the dimensional model test for 8 homonuclear tetramers. The species tested are (references in parentheses): C_4, P_4 (6); As_4, Sb_4 (9); S_4, Se_4 (7); Bi_4 (12); and Li_4 (11). Of these species the only molecule which has been investigated experimentally and for which all spectroscopic constants have been measured is P_4. The solid line is the least squares fit result for the MX_3 species (see Fig. 1). One notices a larger amount of scatter about the MX_{n-1} line than with the

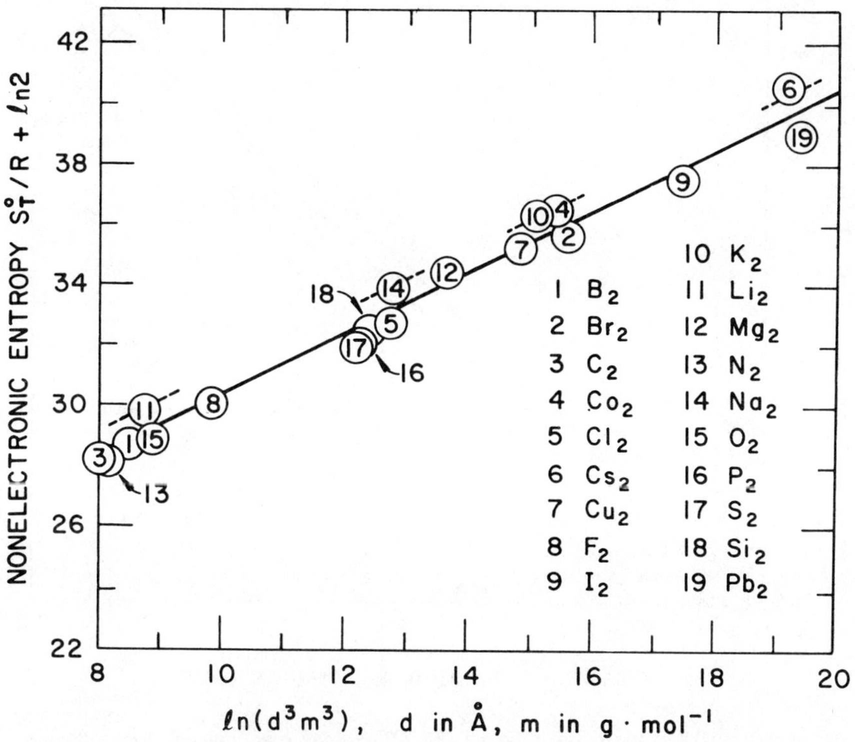

Figure 2. Test of dimensional analysis for gaseous homonuclear diatomics at 1000 K. Key: ——, alkali halide; and – – –, unit slope though diatomic alkali metals (6, 10, 11, 14).

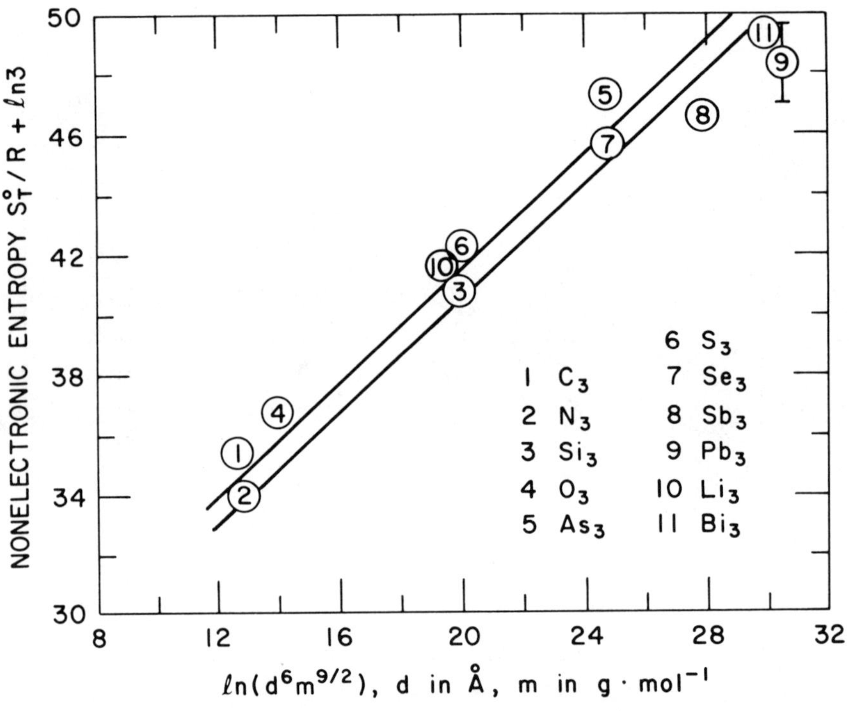

Figure 3. Test of dimensional model for 11 gaseous homonuclear triatomics at 1000 K. Key: ——, results of $MX_2(g)$ correlation (3).

dimers or trimers. Even though the deviations are larger, the general correlation with the model appear to hold up. The data for the species S_4, Se_4 and Li_4 (points marked 5, 6 and 8 respectively) probably contain larger uncertainties than the other species and these are marked with dashed circles. In the S_4 and Se_4 cases, the method of estimation was not discussed in reference (7) and these points may in fact include some electronic contributions. In a thermodynamic data compilation by Glushko,[13] a detailed calculation of $S_4(g)$ is presented containing some experimental data. This result is shown as the point labeled 5' in Fig. 4 and one sees that this lies very close to the MX_3 line. In the Li_4 case (the point labeled 8), the entropy calculation was based entirely on ab initio theoretical calculations and thus may also contain substantial uncertainties. Considering the remaining points, the dimensional model appears to do an adequate job in correlating the tetramer entropies.

Larger Homonuclear Clusters. As one might expect, the scarcity of data on homonuclear polymers increases with cluster size. To date we have only been able to locate published entropy data on four pentamers: S_5, Se_5, and Te_5 from Mills' compilation (7) and C_5 from the JANAF tables (6). Furthermore, only data on the hexamers and heptamers of S and Se exist (7). All of these data have been estimated by various means. These entropy data are plotted in Fig. 5 against the appropriate form of the logarithmic term in Eq. 3. Please note that the dimensionless entropy scale is shifted for each polymer type to facilitate comparison. Also plotted in Fig.5 are the least squares fits of unit slope from the corresponding dimensional model tests for the MX_{n-1} species (Table 1). These are shown as the solid lines in Fig. 5 and represent the results for MX_4 on the pentamer plot, MX_5 on the hexamer plot, and MX_6 on the heptamer plot.

Although it is difficult to draw any firm conclusions due to the lack of data, one sees a general compliance with the dimensional model. The deviations from the line are certainly less than the uncertainties in the estimations. Note that both heptamers lie above the MX_6 line by 4-5 dimensionless units. This might be explained partially by the low value of σ_o(17.9) for the MX_6 data (Fig. 1, Table 1). Most of the other σ_o values (at 1000 K) for the other MX_n species lie close to 20.

Discussion and Conclusions

The results of the testing of the dimensional theory for homonuclear clusters, X_n ($n \leq 7$), indicates a general agreement with the model. Unfortunately an accurate and meaningful comparison of the data on clusters larger than trimers is severely limited by the lack of accurate experimental spectroscopic data. The uncertainties involved in the standard estimation procedures are

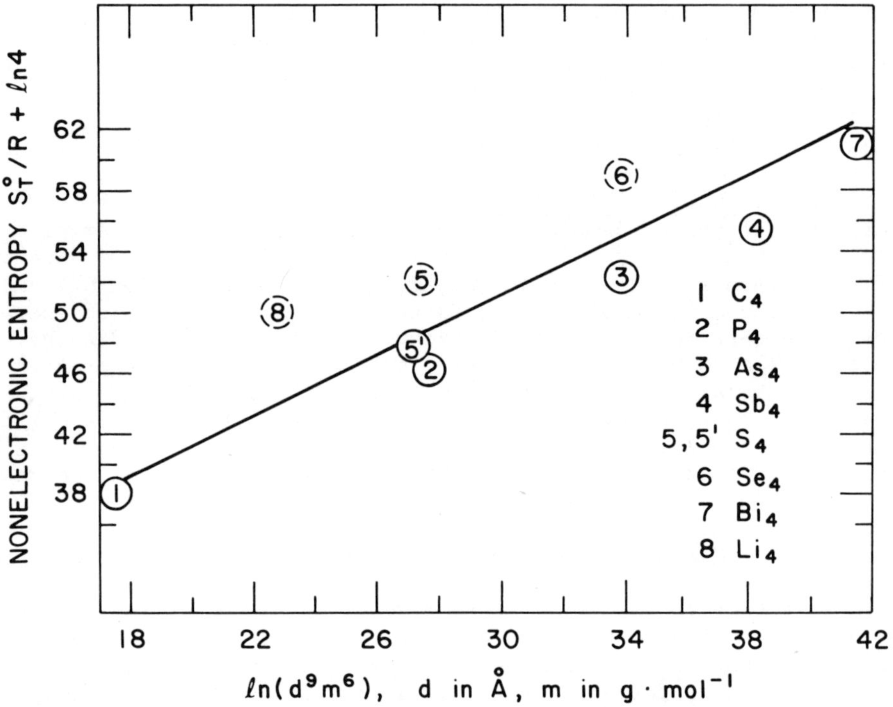

Figure 4. Test of dimensional model for 7 gaseous tetramers at 1000 K. Key: ——, results of $MX_3(g)$ correlation (Fig. 1); and ( ), uncertain estimations.

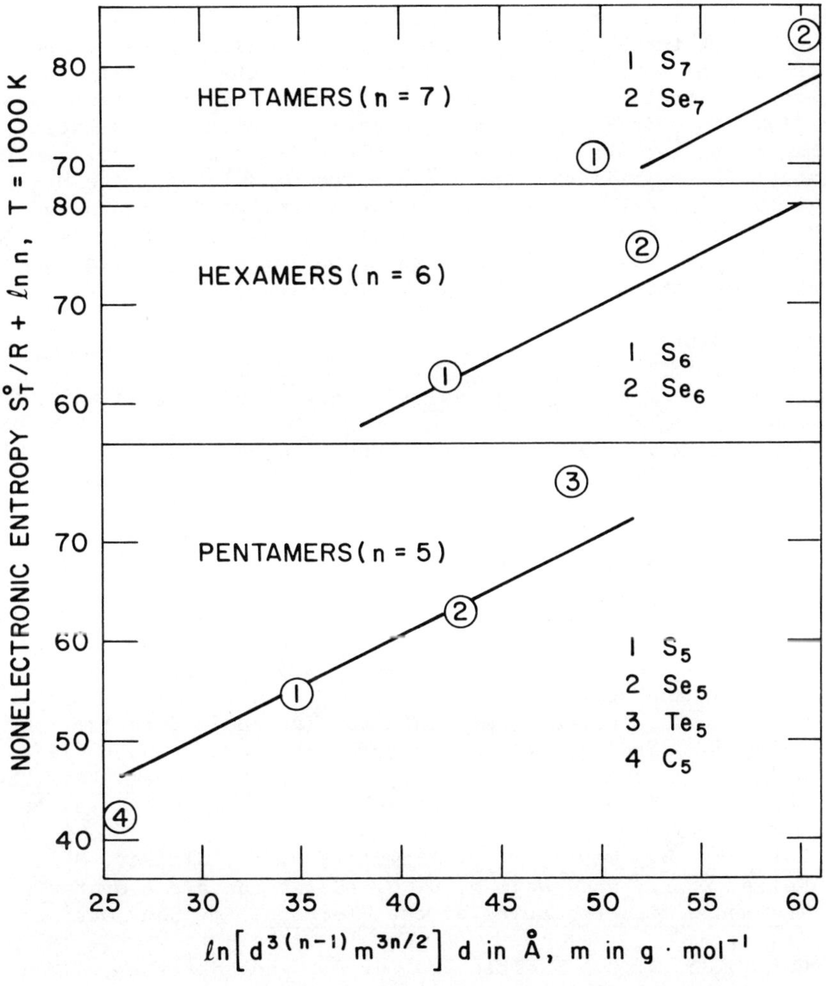

Figure 5. A test of the dimensional model for homonuclear polymers (X_n) containing 5 (pentamers), 6 (hexamers), and 7 (heptamers) atoms. Key: ——, least squares fit results for corresponding MX_{n-1} species (Table I).

usually as large as the deviations from the dimensional model predictions.

The comparison of the X_n thermodynamic data with the corresponding MX_{n-1} data also appears to suggest a viable method for making predictions from the dimensional model. In every case tested except the X_7 species, the MX_{n-1} line lies close to the points for the X_n polymers. We believe that the disparity in the X_7 case is probably due to the lack of enough $MX_6(g)$ data points (Fig. 1). Note in Table I that for every species tested except the MX_6 molecules the σ_o values at 1000 K lie close to 20. Interestingly, if we artificially increase the σ_o value for the MX_6 from 17.9 to 20, the discrepancy between the two X_7 points and the MX_6 line is reduced considerably (Fig. 5). Ultimately we plan to numerically evaluate the configuration integral with various potential functions for homonuclear clusters.

Literature Cited

1. McDonald, J. E. Am. J. Phys., 30, 870 (1962) and 31, 31 (1963).
2. Blander, M.; Katz, J. J. Stat. Phys., 4, 55 (1972).
3. Bauer, S. H.; Frurip, D. J., J. Phys. Chem., 81, 1015 (1977).
4a. Frurip, D. J.; Blander, M. J. Chem. Phys. 73, 509 (1980).
4b. Frurip, D. J.; Blander, M. Proc. 10th Mat. Res. Symposium: Characterization of High Temp. Vapors, NBS Pub. 561, 1597 (1979).
5. Frurip, D. J.; Chatillon, C.; and Blander, M. in preparation.
6. Stull, D. R.; Prophet, H. JANAF Thermochemical Tables, Natl. Stand. Ref. Data. Ser., Nat. Bur. Stand. 37 (1971) 2nd edition; and supplements.
7. Mills, K. C. Thermodynamic Data for Inorganic Sulphides, Selenides, and Tellurides, Butterworth & Co., London (1974).
8. Gingerich, K. A.; Cocke, D. L.; Miller, F. J. Chem. Phys. 64, 4027 (1976); Kordis, J; Gingerich, K. A., J. Chem Phys. 58, 5141 (1973).
9. Hultgren, R.; Desai, P. D.; Hawkins, D. T.; Gleiser, M.; Kelley,K. K.; and Wagman, D. D. Selected Values of the Thermodynamic Properties of the Elements, Am. Soc. Met. (1973)
10. Companion, A. C.; Steible Jr., D. J.; and Starshuk, A. J. J. Chem. Phys. 49, 3637 (1968).
11. H. Beckman, J. Koutecky, P. Bofschwina, and W. Meyer, Chem. Phys. Lett. 67, 119 (1979).
12. L. Rovner, A. Drowart, and J. Drowart, Trans. Farad. Soc. 63, 2906 (1967).
13. V. P. Glushko, Thermodynamic Properties of Individual Substances, Nauk, Moscow (1978).

RECEIVED September 1, 1981.

14

Laser Fluorescence Spectroscopy of Molecular Molybdenum Isolated in Rare Gas Matrices

M. J. PELLIN, T. FOOSNAES, and D. M. GRUEN

Argonne National Laboratory, Argonne, IL 60439

Characterization of the vibronic structure of the homonuclear molybdenum diatom in argon matrix supports has been accomplished by laser fluorescence spectroscopy. A ground state vibrational frequency of 475($\pm$4.5) cm^{-1} was determined for Mo_2 isolated in either of two matrix sites whose presence is manifested by spectrally shifted emission. The presence of two trapping sites gives rise to complicated overlapping absorption spectra which have been separated and elucidated by excitation spectroscopy. As well Raman spectra of these matrices show a peak at 476 cm^{-1}. The measured emission lifetimes are long and generally not single exponential. In argon matrices 2.1 msec decay times were measured from one site and a tri-exponential decay with components of 2.1 msec, 3.93 msec, and 6.3 msec were measured. The length of these lifetimes and the large spectral shift of the emission relative to the excitation energy leads to the conclusion that emission does not emanate from the initially excited, optically allowed level detected in absorption. Rather emission must occur from one of the dense manifold of higher spin multiplicity levels which lie at slightly lower energies than the $^1\Sigma_u$ level coupled by laser radiation to the $^1\Sigma_g$ ground state.

0097-6156/82/0179-0219$05.00/0

The Mo_2 diatom has recently been the subject of intense theoretical (1) and experimental investigation (1, 3, 7-10). This attention is due to the uniqueness of the molybdenum to molybdenum bond which arises from the 5d and 1s electrons available on each molybdenum atom for bonding. Recent calculations suggest that the Mo_2 bond is indeed "sextuple" with two of the d-electron bonds contributing only a small fraction to the overall bond strength (5, 6). These calculations are in part confirmed by the bond orders and bond lengths of several organo-metallic compounds which contain molybdenum to molybdenum bonds. $Mo_2Cl_8^{-4}$ (11) has a bond order of 3 and a bond length, r_e, of 2.1 Å. The purpose of this investigation was to determine the force constant of the naked metal cluster by both laser fluorescence spectroscopy and raman spectroscopy.

The laser fluorescence apparatus is displayed in Fig. 1. In general gated photon counting techniques were used to detect the long lived, weak emission from Mo_2 matrices. The argon matrices were grown on a sapphire plate attached to a closed cycle refrigerator which maintained the plate at 14°K. The metal which was seeded into the matrix under carefully controlled conditions was produced by a hollow cathode source (10). Matrix raman was accomplished with a standard Spex Ramanor system, pumped by various argon-ion laser lines with 50-m watts of laser intensity.

Results and Analysis

Matrices produced had absorption spectra similar to that shown in Fig. 2 when fluorescence experiments were run. For raman experiments Mo_2 absorption strengths of 2-3 OD were produced. The absorption spectra are perplexing in that peak shapes and spacings are widely varying.

Depending on excitation wavelength either of two emission spectra may be observed (Fig. 3). The twin peaks at 690.5 nm and 692.2 nm represent long lived emission resulting from excitation of the sapphire matrix plate and are useful fiducial markers. The fluorescence spectrum in the 720 nm - 860 nm spectral region can be assigned to the Mo_2 molecule. Excitations at 522.9, 533.6, 540.5 or 543.3 nm produced spectra essentially the same as that shown for 517.9 nm excitation (Fig. 2) and will be referred to as "red" site emission. Excitation at 507.8 nm and 498.2 nm give emission spectra similar to that shown for 502.2 nm excitation (Fig. 2) and will be referred to as "blue" site emission. The two vibrational progressions are similar in peak spacing ($\Delta\nu$ = 473.2 ± 5 cm^{-1} with 514.0 nm excitation and $\Delta\nu$ = 477.9 cm^{-1} ± 6 cm^{-1} with 502.2 nm excitation) and in Franck-Condon factors.

The two emission spectra can be compared in several ways. The lifetime of the "red" site emission is not represented by a single exponential decay and may be characterized by a triple exponential decay of τ = 2.1 msec, 3.93 msec, and 6.3 msec. Blue site emission has a single 2.1 msec decay time. The peak shapes show in all

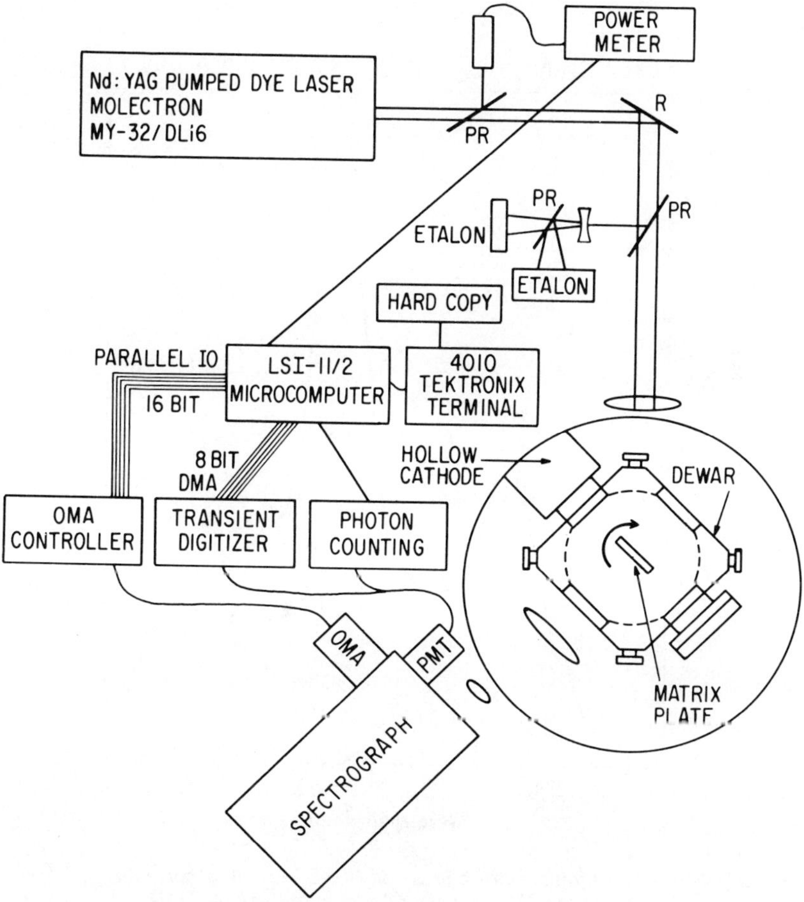

Figure 1. Laser fluorescence apparatus.

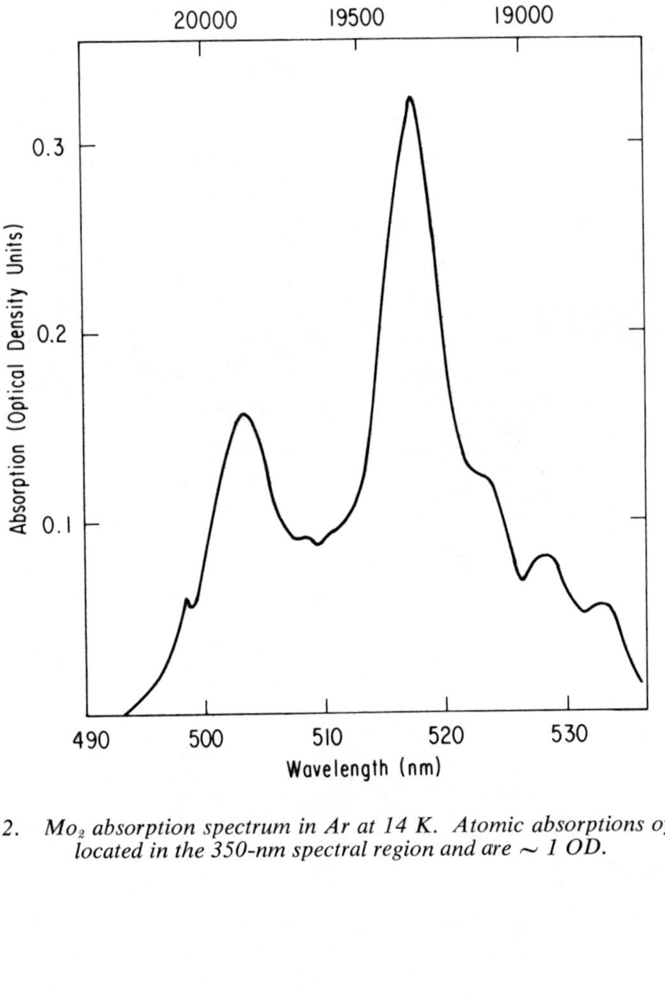

Figure 2. Mo_2 absorption spectrum in Ar at 14 K. Atomic absorptions of Mo are located in the 350-nm spectral region and are ~ 1 OD.

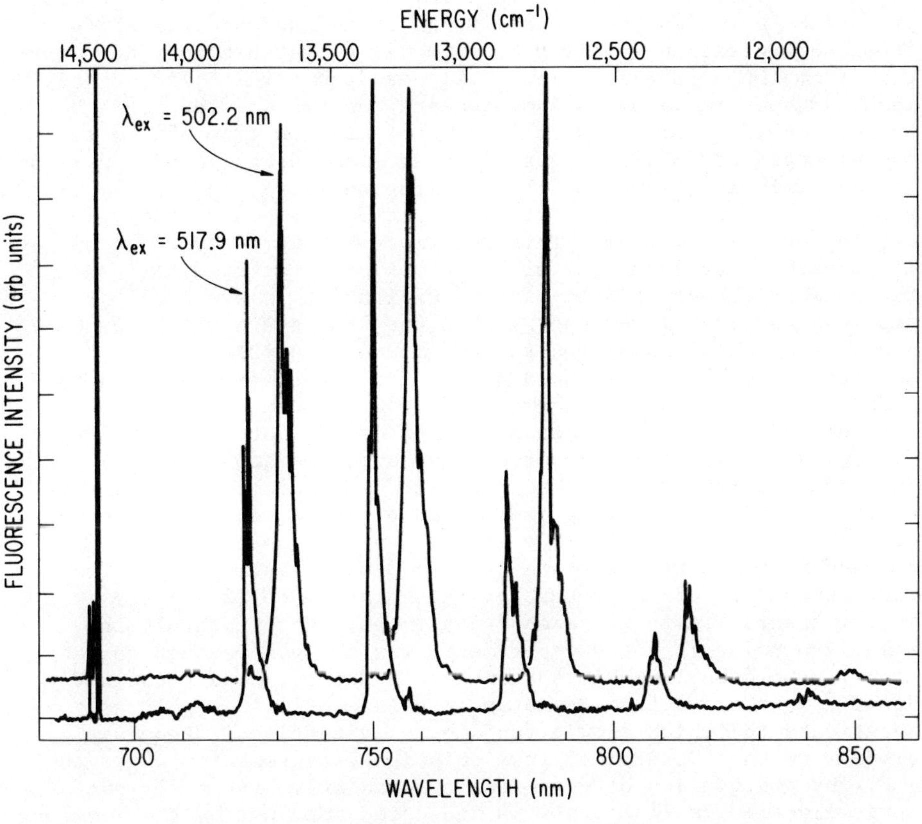

Figure 3. Fluorescence spectrum of Mo_2 in Ar at 14 K. The upper curve is found with 502.2-nm, the lower with 517.9-nm excitation (λ_{ex}).

cases that the red site has a smaller zero phonon line relative to its phonon band than does the blue site. Excitation spectra provide a third comparison (Fig. 4). It seems likely that the two emissions occur from Mo_2 trapped in two separate matrix sites and that the complicated absorption band is the superposition of these two sites. In general the excitation spectra should mirror the absorption spectra of each site (Fig. 4).

The emission is interesting not just because of the two trapping sites but also because of its lifetime and spectral shift from the excitation wavelength. Clearly emission cannot be occurring from the optically allowed $^1\Sigma_u^+$ excited level. Indeed only weak and spectrally broad luminescence can be detected near the excitation line ($\Delta\omega$ = 300 cm^{-1}) and no emission from 550 nm to 690 nm could be found. Figure 5 details an energy level model for the Mo_2 emission sequence. The excitation energy is rapidly lost from the excited $^1\Sigma_u^+$ excited level and decays non-radiatively with a rate ($K_{20} + K_{12}$). This rate must be extremely rapid since it competes very favorably with the optically allowed $^1\Sigma_u^+ \rightarrow {}^1\Sigma_g^+$ fluorescent channel. Some of this excitation is eventually trapped in a level at lower energy than the $^1\Sigma_u^+$ state which has only a weakly allowed transition to the ground $^1\Sigma_g^+$ state but whose non-radiative channel is so slow as to allow detection along the radiative channel.

Ground state repopulation studies demonstrate that the observed emission has as its lower energy level the ground ($^1\Sigma_g^+$) state. In these experiments the absorbance of a Mo_2 doped Ar matrix was monitored at 517 following excitation of these red absorption site diatoms. At t = o there is a large bleaching as population is removed from the ground state by the 30 nsec laser pulse. The time dependence of the return of this absorbance (2.2 $\pm$ 0.3 msec) is commensurate with emission decay times. Unfortunately the signal to noise in these experiments was not good enough to determine if the decay was bi-exponential in nature. Additional evidence can be drawn from Raman spectra (see below) as well as gas phase data (8) which show a fundamental vibrational frequency similar to that determined from emission measurements.

The postulation of very fast non-radiative rates for the optically allowed $^1\Sigma_u^+$ state is indicated not just by the lack of direct line emission but also by the lack of a resonance Raman effect in Mo_2 matrices. Figure 6 shows Raman spectra in the 450 cm^{-1} frequency region as produced by several exciting lines. At these low intensities each line produces some broad luminescent features which varied from exciting line to exciting line. One sharp feature was present at each exciting line and is displayed in Fig. 6. This peak though weak can be assigned in conjunction with the fluorescent data as due to Mo_2. The most striking feature of the Raman signal is the lack of resonant enhancement of the signal in changing excitation from 488 nm to 514 nm. One can write the intensity of a Raman scattering event as (12):

$$I \approx \left[1/h[\sum_I (< \nu o/p|I>< I/p|\nu 1 >)(\Delta\omega_I = 1\Gamma_L)^{-1}] \right]^2 \quad (1)$$

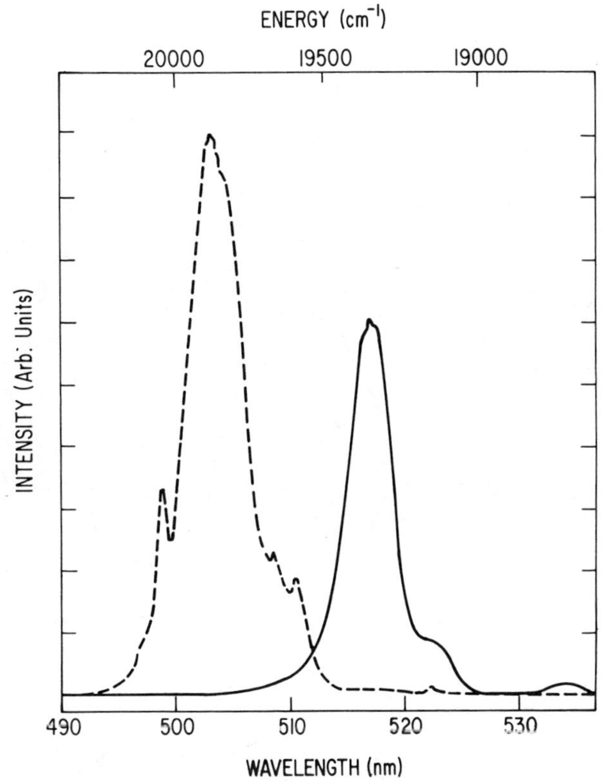

Figure 4. Excitation spectrum of Mo_2 in Ar at 14 K. The – – – is found by monitoring emission at 7561 Å and scanning the laser wavelength as shown. The —— is similar but with 7505-Å monitor.

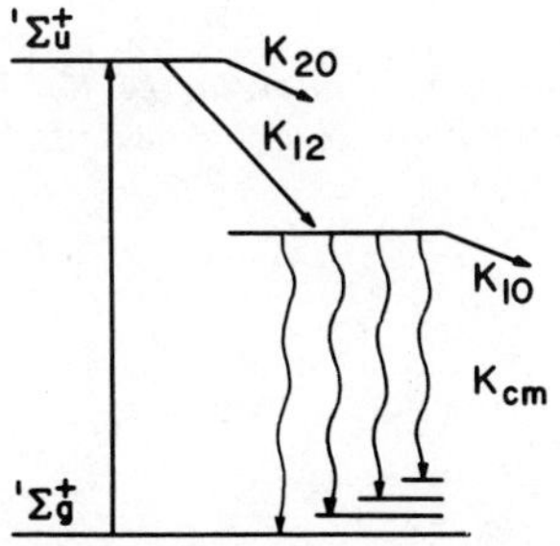

Figure 5. Energy level diagram for the Mo_2 excitation–emission sequence. Initially the diatomic is excited by laser excitation to the $^1\Sigma_u^+$ state, Mo_2 then decays nonradiatively with rate $K_{20} + K_{12}$. Some fraction is trapped in an intermediate emitting level which has the radiative and nonradiative rates K_{cm} and K_{10}, respectively.

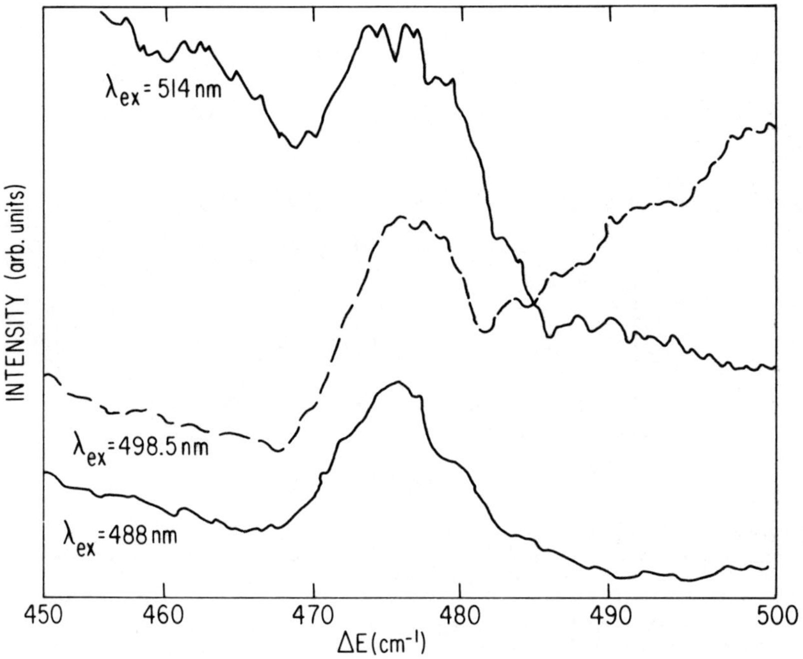

Figure 6. Raman spectrum of Mo_2 in Ar at 14 K. Three lines are for 488.0-, 496.5-, and 514-nm excitation. Note the lack of resonant enhancement.

where $|\nu o\rangle$, $|\nu\ I\rangle$, and $|\nu 1\rangle$ are the wavefunctions of $\nu = o$, $\nu = 1$ and of the intermediate level and $\Delta\omega_I$ the frequency difference between the excitation and the electronic level I. For a discrete resonance and small damping Γ_L the Raman intensity will be dominated by a single level and show considerable enhancement. However in this case the $^1\Sigma_u^+$ level fluorescent lifetime is drastically shortened by matrix interactions allowing large non-radiative rates. Thus Γ^1 is so large that the $^1\Sigma_u^+$ state imparts negligible Raman enhancement.

In conclusion Raman and laser fluorescence experiments have been carried out on Mo_2 isolated in Ar matrices. These studies indicate an $\omega_e'' \approx 475\ cm^{-1}$.

Acknowledgement

Work performed under the auspices of the Office of Basic Energy Sciences, Division of Materials Sciences, U. S. Department of Energy. The submitted manuscript has been authored by a contractor of the U.S. Government under contract No.: W-31-109-ENG-38.

Literature Cited

1. Klotzbucher, W.; Ozin, G. A. Inorg. Chem. 1977, 16, 984.
2. Norman, J. G.; Kori, H. J.; Gray, H. B.; Trogler, W. C. Inorg. Chem. 1977, 16, 987.
3. Klotzbucher, W.; Ozin, G. A.; Norman, J. G.; Kolari, H. J. Inorg. Chem. 1977, 16, 2871.
4. Norman, J. G.; Ryan, P. B. J. Comp. Chem. 1980, 1, 59.
5. Bursten, B. E.; Cotton, F. A.; Hall, M. B. J. Amer. Chem. Soc. 1980, 102, 6348.
6. Atha, P. M.; Hillier, I. H.; Guest, M. F. Chem. Phys. Lett. 1980, 75, 84.
7. Bates, J. K.; Gruen, D. M. J. Mol. Spectroscopy 1979, 78, 284.
8. Efremov, Y. M.; Samoilova, A. N.; Kozhukhuvsky, V. B.; Gurvich, L. V. J. Mol. Spectroscopy 1978, 73, 430.
9. Becker, K.; Shurgers, M. Z. Naturforch 1971, A26, 2072.
10. Pellin, M. J.; Foosnaes, T.; Gruen, D. M. J. Chem. Phys. In Press.
11. Brencic, J. V.; Cotton, F. A. Inorg. Chem. 1969, 8, 7.
12. Bondebey, V. E.; English, J. H. Chem. Phys. Lett. 1978, 60, 69.

RECEIVED August 26, 1981.

15

Magnetic Circular Dichroism of Metal Atoms and Dimers: Copper and Lead

KYLE ZERINGUE, JALAL SHAKHS EMAMPOUR, and MARTIN VALA
University of Florida, Department of Chemistry, Gainesville, FL 32611

Analysis of the absorption and MCD spectra of copper in krypton matrices provides strong evidence that the $^2P \leftarrow ^2S$ Cu atom transition at 310 nm is split by simultaneous Jahn-Teller and spin orbit interactions. The order of the s.o. components is inverted compared to the gas phase. Similar inversions are noted for higher-lying ^2P transitions. One band is assigned to a Cu dimer transition.

The $^3P_1 \leftarrow ^3P_0$ Pb atom transition is confirmed by the appearance of a positive MCD $\mathcal{A}$ term. Two Pb_2 transitions at 500nm and 244nm are assigned to $0_u^+ \leftarrow 0_g^+$ transitions.

Introduction. A considerable body of literature has developed recently on the spectroscopy and chemistry of metal atoms, dimers and higher aggregates in low temperature matrices (1-4). This interest has arisen for a number of reasons. First, the electronic structures of metal dimers and clusters are important to an understanding of the processes of heterogeneous catalysis (5), nucleation(6), chemisorption (7), and photography (8). Second, optically pumped dimers, including some metals, have recently been shown (9, 10) to lase in both CW and pulsed modes. And, finally, a large number of novel chemical compounds have been synthesized (1, 4) by codepositing a metal vapor prepared in a high temperature furnace with a reactive partner on a cryogenic surface.

In this paper we report on our magnetic circular dichroism (MCD) study of matrix-isolated copper and lead atoms and their dimers. The primary advantage of the matrix isolation technique lies in its ability to trap otherwise reactive and difficult-to-study species. Because of the low temperatures used it also ensures that any absorptive transitions start from the ground electronic state. This characteristic has proven extremely useful in the assignment of the spectra of high temperature molecules (11) where gas phase studies can often lead to ambiguous results. The rotational structure present in gaseous investigations has, of course, been effectively utilized in the analysis of elect-

0097-6156/82/0179-0229$05.00/0

ronic and vibronic transitions. In combination the two methods have provided much needed information on the electronic states of high temperature species. To assign matrix bands, however, great reliance has been placed on a simple correlation between gaseous and matrix bands, i.e., on their relative intensities and frequencies. This procedure, developed by Gruen and co-workers (12), has provided a good overall understanding of the spectra of matrix-isolated atoms, but has been unable to provide insight on finer details, such as the shifts in relative intensity going from gas to matrix phases or the triplet appearance of the $^2P \leftarrow ^2S$ transition in Cu or alkali metals isolated in rare gas matrices. We show below that MCD of matrix isolated species can provide some of these answers.

MCD is the measurement of the difference in the absorption of left- and right- circularly polarized light by a sample placed in a longitudinal magnetic field: $\Delta A = A_L - A_R$, where A_L is the absorbance of left-circularly polarized light. There are three possible origins for measureable ΔA's (13);(1) $\mathcal{A}$ terms arise from a degeneracy in either ground or excited electronic states and have a distinctive derivative shape where the zero- crossing point coincides with the absorption band maximum: (2) $\mathcal{B}$ terms are the result of magnetic field mixing of (usually) nearlying states; (3) $\mathcal{C}$ terms arise from the Boltzmann distribution in a degenerate ground state. Both $\mathcal{B}$ and $\mathcal{C}$ terms have a single-signed signal with a band shape similar to the absorption band shape. The latter may be differentiated from the former by its temperature dependence. Generally, for low temperature matrices, species with degenerate ground states display $\mathcal{C}$ term activity while species with nondegenerate ground states and degenerate excited states show $\mathcal{A}$ term signals. Transitions between nondegenerate states exhibit a zero MCD signal or a weak $\mathcal{B}$ term.

Experimental. Our matrix isolation absorption and MCD apparatus has been described previously (14). The Knudsen cell (Ta) was filled with either Cu or Pb powder (Spex) and degassed thoroughly at elevated temperatures. Resistive heating of the cell (to 1100°C for Cu and 875° for Pb) produced a metal beam which was directed at a CaF_2 window held at 20°K (Displex cryostat) and co-condensed with the isolant krypton gas. Absorption and MCD spectra were measured sequentially on the same matrix sample. Matrices were generally clear, but polycrystalline in nature. Depolarization effects were checked by comparing the CD spectrum of an aqueous solution of $[Co(en)_3]C\ell$ (d-tartrate) placed after the sample matrix with a similar run of the solution alone. This solution was also used to establish the sign of the induced circular dichroism: its absorption band at 469nm has a positively-signed CD and a negatively-signed MCD. The MCD signal intensity was checked as a function of magnetic field strength and found to be linear.

Results. Copper in Krypton Matrices. The absorption and MCD spectra obtained for Cu in a krypton matrix are shown in Figure 1. The oft-observed triplet structure of the 310nm absorption band is evident; its MCD displays (with increasing energy) positive, negative, and negative bands. From their temperature dependence (cf. Figure) it is clear that these are $\mathcal{C}$ terms. Another lower intensity triplet of bands is observed at 245nm. The MCD of this set is very different from the lower energy group. The first band (251nm) shows only a weak, temperature independent signal (i.e. $\mathcal{B}$ term), while the other two display opposite $\mathcal{C}$ term behavior. Finally, two higher energy bands show positive $\mathcal{C}$ terms.

Lead in Krypton Matrices. Selected regions of the absorption and MCD spectra of lead in krypton are shown in Figures 2 and 3. In the U.V. region (Figure 2) two strong (263.5 and 246.9nm) and one weak (232.7nm) band are observed. The relative intensities of these three peaks were observed to change under different deposition conditions (sample window temperature and Knudsen cell temperature). Only the lowest energy band at 263.5 nm shows an MCD signal: a temperature independent, positive $\mathcal{A}$ term. In the visible range (Figure 3) a highly structured absorption band commences at 508.2 nm. This band shows no MCD under our experimental conditions. Four broad, higher energy bands (469nm, 448nm, 419nm and 400nm) also show no MCD. With longer deposition times these bands did show some MCD attributable to very weak $\mathcal{B}$ terms. With deposition conditions similar to those used to obtain the Figure 3 spectrum we also observed a broad, symmetrical band at 612.9nm with an intensity comparable to the 510nm band. The relative intensities of these two bands varied somewhat depending on deposition conditions.

Discussion. Copper in Krypton. The absorption spectra of copper atoms isolated in rare gas matrices have been extensively studied (15-25) and the triplet of bands at 310nm attributed to a number of different causes. These include (1) spin orbit splitting and static axial site distortion (17), (2) multiple matrix sites (18), (3) exciplex formation between the metal and a single matrix atom (19), (4) long range metal-metal interactions (20), and (5) Jahn-Teller (JT) effect resulting from matrix cage atom vibrations on the excited metal (21,22,23). The MCD of Cu atoms in the rare gas matrices has recently been reported (24,25) and the results interpreted as consistent with either the distorted site or JT hypotheses (39).

Kasai and McLeod (26) have measured the ESR spectrum of Cu atoms in krypton and have shown that there is no evidence for multiple trapping sites and that the Cu atoms reside in an octahedral site. For Cu atoms in an octahedral site, spin orbit interaction splits the ^2P state into $^2P_{3/2}$ and $^2P_{1/2}$ components. Further splitting of the $^2P_{3/2}$ state may occur via the JT effect

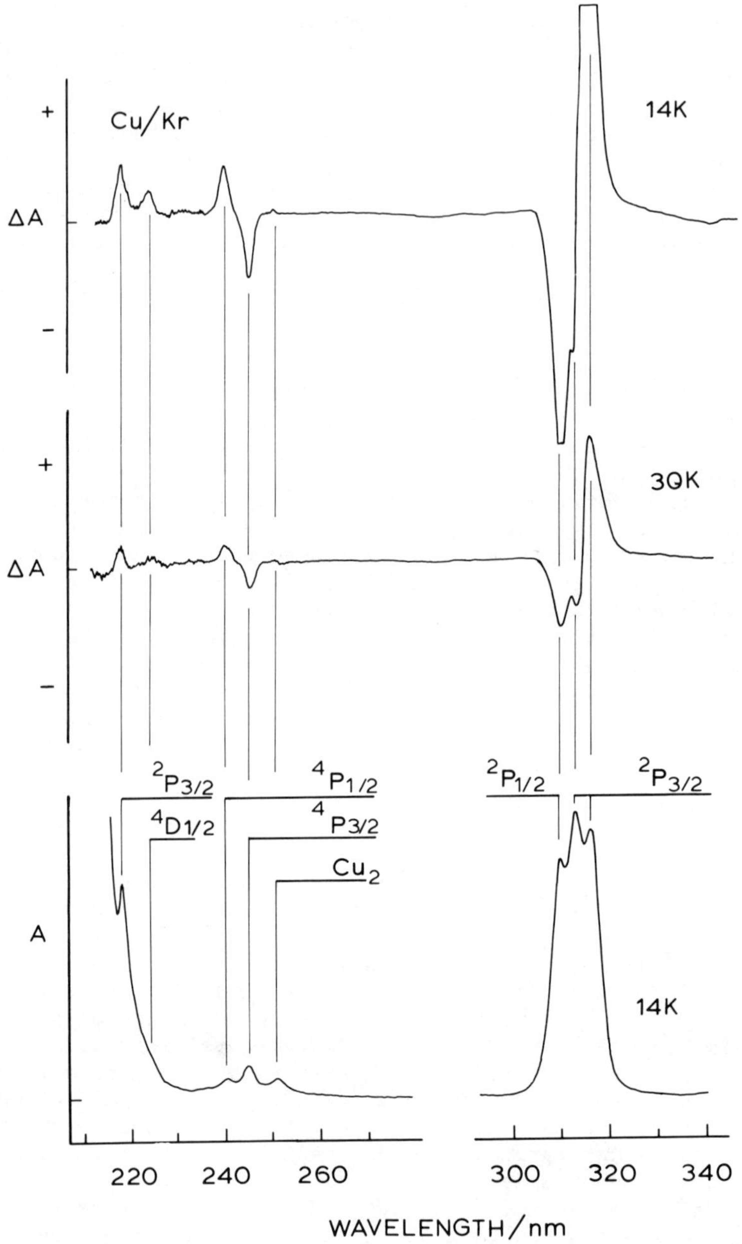

Figure 1. Top: MCD spectrum of Cu in a Kr matrix at 14 K. Middle: MCD spectrum of Cu in a Kr matrix at 30 K. Bottom: Absorption spectrum of Cu in a Kr matrix at 14 K. The top and middle panels are from the same matrix sample with identical gain settings. The left hand portion (210–270 nm) of the bottom panel is from the same matrix. The right portion (300–340 nm) is on a more dilute sample matrix.

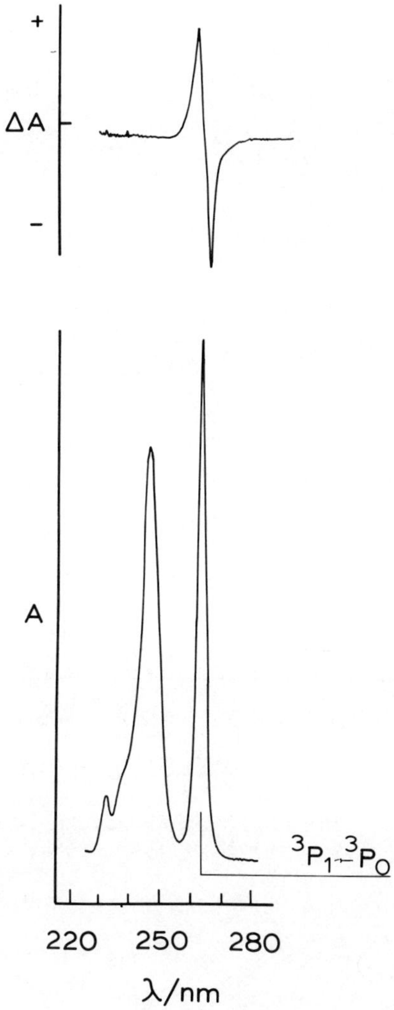

Figure 2. MCD spectrum (top) and absorption spectrum (bottom) of Pb in a Kr matrix at 14 K in the UV range.

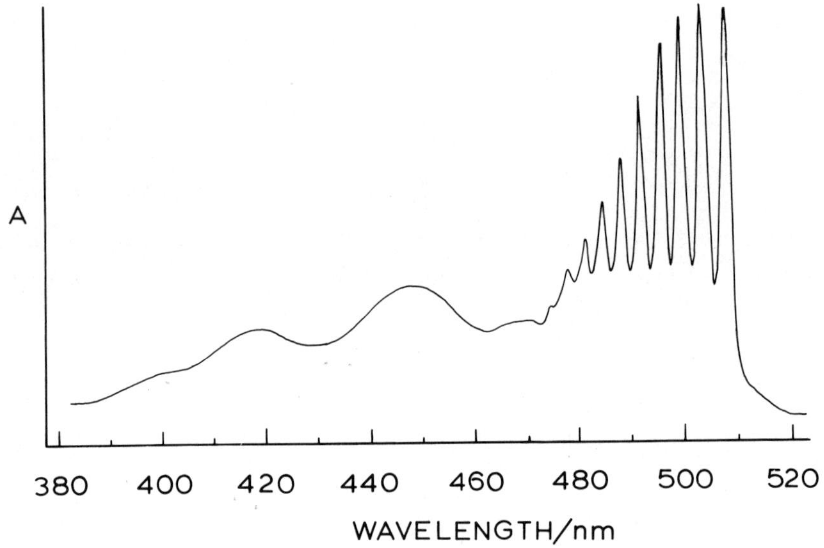

Figure 3. Absorption spectrum of Pb in a Kr matrix at 14 K in the visible range. No observable MCD was noted.

(the $^2P_{1/2}$ remains unaffected). In the JT-distorted excited state the spin-orbit components become: $^2E\ E'\ ^2P_{3/2}$, $^2EE''\ ^2P_{3/2}$, and $^2A_2E'\ ^2P_{1/2}$ (the symbols denote D_3 distorted point group, D_3^* double group, and free atom spin-orbit states). We have previously shown (28) that the predicted MCD C_o terms to these states from the $^2A_1E'\ ^2S_{1/2}$ ground state are positive, negative, and negative, respectively. These predictions are consistent with the experimental observations only if the $^2EE'\ ^2P_{3/2}$ level lies lowest, i.e. the $^2P_{3/2}$ level lies below the $^2P_{1/2}$ level. This is opposite to the order in the gas phase (29). The reversal in s.o. component ordering could be attributed to the overlap of the Cu and cage matrix atoms. This effect has already been invoked by Kasai and McLeod (26) to explain the dependence of the ground state g factors on matrix atom type. However, it will probably prove necessary to include the effect of the JT-induced distortion together with the s.o. interaction to fully understand this reversal.

The triplet of bands at 245nm is much weaker than the set at 310nm. In the gas phase, several spin-forbidden levels ($^4P_{3/2}$, $^4P_{1/2}$) lie just above the ^2P components. Using the Wigner-Eckart theorem and Piepho's tables of vector coupling coefficients (27) for the O^* group, we predict that $\mathcal{C}_o/\mathcal{D}_o$ for the $^4P_{3/2}(^4T_1U'_{3/2}$; notation, O_h and O^* symbols) transition is -1, and for the $^4P_{1/2}(^4T_1E')$ transition is +2. Of the three peaks observed in absorption only the upper two display an MCD $\mathcal{C}$ term, with the highest energy one positive. We therefore assign that band (240nm) to the $^4P_{1/2} \leftarrow {}^2S_{1/2}$ transition and the lower (245nm) one to $^4P_{3/2} \leftarrow {}^2S_{1/2}$. The lowest energy band (251nm) shows only a weak $\mathcal{B}$ term and thus cannot be a Cu atom band. We attribute it to a Cu_2 $^1\Sigma_u^+ \leftarrow {}^1\Sigma_g^+$ band since the Cu_2 ground state ($^1\Sigma_g^+$) will not give rise to a $\mathcal{C}$ term. Note that the order of the ^4P levels is not inverted in the matrix compared to the gas phase and that (with the resolution and field strength used in the present measurements) there is no evidence for a JT splitting in this excited state.

In the gas phase, the next three bands are with increasing energy) $^4D_{1/2}$, $^2P_{1/2}$, and $^2P_{3/2}$. The MCD in this higher energy region (230-210nm) displays two clear, positive $\mathcal{C}$ terms. Calculation of the $\mathcal{C}_o$ term for the $^4D_{1/2} \leftarrow {}^2S_{1/2}$ transition gives a positively-signed quantity. We assign the lower energy band (224nm) to this transition. The next band (218nm) is assigned to the $^2EE'$ component of the $^2P_{3/2}$ state, since, as discussed above, only this component of the ^2P state will exhibit a positive $\mathcal{C}$ term. The $^2P_{1/2}$ component presumably lies above our observable energy range. This assignment accounts for all the Cu atom transitions in the u.v. region and is consistent with the assignment of the 310nm band components.

Armstrong, Grinter, and McCombie (25) have previously discussed the MCD of the Cu $^2P \leftarrow {}^2S$ transition in rare gas matrices.

Our experimental results for Cu/Kr samples are similar to theirs (30). However, these authors propose an assignment entirely different from ours. They suggest that the lowest energy band is $^2P_{1/2,1/2}$ while the upper two are $^2P_{3/2,3/2} < {}^2P_{3/2,1/2}$ (within the distorted site model). These designations depend on the magnitudes and signs of the distorted crystal field parameter and the spin orbit coupling parameter. We believe that Kasai and McLeod's ESR work (26) shows unequivocally that the Cu cage site is octahedral and that an interpretation of the optical spectrum based on a ground state distorted site is not warranted. Our interpretation of simultaneous JT and spin-orbit interaction in the excited ^2P state reconciles both the ESR conclusions and the optical spectral observation of three bands in the $^2P \leftarrow {}^2S$ matrix transition. Work is continuing on other similar systems and other matrices (Ar, Xe) and will be reported at a later time. Details of the C term calculations and moment analyses of the $^2P \leftarrow {}^2S$ band will also be reported later.

Lead in Krypton. Brewer and Chang (BC) (31) were the first to study the absorption spectra of Pb and Pb_2 isolated in rare gas matrices. They discovered two dimer bands (denoted A←X, 508nm and E←X, 244nm) and assigned the $^3P_1 \leftarrow {}^3P_o$ atom transition to a band at 261.4nm (Kr). Earlier, Shawhan (32) and then Weniger (33) had investigated the spectrum of Pb_2 vapor. Teichman and Nixon (34) reported on the absorption and laser-induced emission of Pb_2 in neon, argon, and krypton matrices. Bondybey and English (BE) (35) revised Weniger's analysis of the 500nm dimer band, determined a value of $112cm^{-1}$ for the vibrational frequency of the ground state, and provided evidence for six states of Pb_2 below the 500nm one. They also suggested that transitions to the latter state be assigned as $F0_u^+ \leftarrow X0_g^+$. Prior to this no symmetry assignments for any Pb_2 transitions had been proposed.

In this paper we confirm BE's assignment of the 500nm band provide a similar assignment for the 244nm band first observed by BC, and corroborate the $^3P_1 \leftarrow {}^3P_o$ designation for the 261.4nm band. The ground state of atomic Pb is ^3P_o(29), with the $^3P_1$ multiplet component lying at $7819cm^{-1}$. The first allowed gas phase transition in the u.v. region occurs at $35,287cm^{-1}$ ($^3P_1 6p7s \leftarrow {}^3P_o 6p^2$). A positive A term is expected for such a transition. The 261.4nm band, assigned by BC to $^3P_1 \leftarrow {}^3P_o$, is observed to have a positive A term (cf. Figure 2), thus confirming this assignment. The gas-to-matrix shift is large, $+2957cm^{-1}$.

For Pb_2, the appearance of no observable MCD for the 500nm band or for the 244nm band is strong evidence for the assignment of the Pb_2 ground state to 0_g^+. Two ground state ^3P_o atoms can combine to produce such a state (36). C terms would have been expected for any other degenerate symmetries. The absence of A terms in these transitions leads us to conclude that both their excited states are 0_u^+.

Stranz and Khanna (SK) recently reported (38) on the resonance Raman of small Pb clusters in xenon matrices. Laser excitation into a band analogous to our Pb/Kr band at 469nm gave vibrational frequencies which were interpreted as arising from a Pb_3 species with D_{3h} geometry. Such a molecule could have an orbitally degenerate ground or excited state which might give rise to $\mathcal{C}$ or $\mathcal{A}$ term MCD activity, although JT distortions might be expected to reduce the degeneracy. The observation (cf. Figure 3) of no MCD bands (or very weak $\mathcal{B}$ terms upon further deposition) provides no support for or against SK's suggestion. If this band is due to the Pb_3 species, it does however provide evidence that the ground and excited states of Pb_3 involved in this transition are orbitally and spin nondegenerate.

Acknowledgements. We gratefully acknowledge the National Science Foundation (grant CHE-77-00988) for its support of this research.

Literature Cited.

1. Blackborow, J.R.; Young, D.; "Metal Vapor Synthesis in Organometallic Chemistry"; Springer-Verlag, Berlin, 1979.
2. Power, W.J.; Ozin, G.A.; in Adv. Inorg. Chem. and Radiochem., Vol. 23, eds. H.J. Emeleus and A.G. Sharpe, Academic Press, 1980.
3. "Cryochemistry," eds. M. Moskovits and G.A. Ozin, Wiley-Interscience, New York, 1976.
4. Klabunde, K.; "Chemistry of Free Atoms and Radicals," Academic Press, New York, 1980.
5. "The Physical Basis of Heterogeneous Catalysis," eds. E. Drauglis and R.I. Jaffee, Plenum Press, New York, 1975.
6. Abraham, F.F.; "Homogeneous Nucleation Theory," Academic Press, New York, 1974.
7. Ugo, R.; Catal. Rev. 11, 225 (1975).
8. Hamilton, J.F.; Logel, P.C.; Photo. Sci. Eng., 1974, 18, 507.
9. Wellegehausen, B.; IEEE J. Quan. Elec., 1979, QE15, 1108.
10. Kóffend, J.B.; Field, R.W.; Guyer, D.R.; Leone, S.; in Laser Spectroscopy III, eds. J.L. Hall and J.L. Carlsten, Springer-Verlag, Berlin, 1977.
11. Weltner, W.; in Adv. High Temp. Chem., Vol. 2, 1969.
12. Carstens, D.; Brashear, W.; Eslinger, D.; Gruen, D., Appl. Spec., 1972, 26, 184.
13. Stephens, P.J.; in Adv. Chem. Phys., Vol. 35, eds. I. Prigogine and S. Rice Wiley, New York, 1976.
14. Brittain, R.; Powell, D.; Voigtman, E.; Vala, M., Rev. Sci. Instrum., 1980, 51, 905.
15. Brewer, L.; King, B.; J. Chem Phys. 1970, 53. 3981.
16. Forstmann, F.; Kolb, D.; Leutloff, D.; Schulze, W., J. Chem. Phys., 1977, 66, 2806.
17. Gruen, D.; Gaudioso, S.; McBeth, R.; Lerner, J.; J. Chem. Phys., 1974, 60, 89.
18. Meyer, B.; "Low Temperature Spectroscopy," American Elsevier, New York, 1971.

19. Moskovits, M.; Hulse, J.; J. Chem. Phys., 1977, 67, 4271.
20. Andrews, L.; Pimentel, G.C.; J. Chem. Phys. 1967, 47, 2905.
21. Belyaeva, A.A.; Predtechenskii, Yu. B.; Shcherba, L.D.; Opt. Spec. 1973, 34, 21.
22. Mowery, R.L.; Miller, J.C.; Krausz, E.; Schatz, P.N.; Jacobs, S.M.; Andrews, L. J. Chem. Phys. 1979, 70, 3920.
23. Ozin, G.A.; Fara. Symp. Chem. Soc. 1980, 14, 7.
24. Grinter, R.; Armstrong, S.; Jayasooriya, U.; McCombie, J.; Norris, D.; Springall, J.; Fara. Symp. Chem. Soc. 1980, 14, 94.
25. Armstrong, S.; Grinter, R.; McCombie, J., J. Chem. Soc., Fara. Trans. 2, 1981, 77, 123.
26. Kasai, P.H.; McLeod, D.; J. Chem. Phys. 1971, 55, 566.
27. Piepho, S.; in Recent Adv. in Group Theory and Their Appl. to Spec., ed. J. Donini, Plenum Press, New York, 1979.
28. Zeringue, K.; ShakhsEmampour, J.; Vala, M., to be submitted.
29. Moore, C.E.; "Atomic Energy Levels," Nat'l Bureau of Stnds. Circ. 467, Vol. 1, 1949; Vol. II, 1952; Vol. III, 1958.
30. These authors follow an older convention for the signs of the MCD signals than the one (13) commonly in use now. In the older convention a calculated negative C_0 term corresponded to a positive observed C term.
31. Brewer, L.; Chang, C.; J. Chem. Phys., 1972, 56, 1728.
32. Shawhan, E.N., Phys. Rev. 1935, 48, 343.
33. Weniger, S., J. Phys. (Paris) 1967, 28, 595.
34. Teichman, R.; Nixon, E.R., J. Mol. Spec. 1976, 59 299.
35. Bondybey, V.E.; English, J.H., J. Chem. Phys. 1977, 67, 3405.
36. Stranz, D.D.; Khanna, R.K., J. Chem. Phys. 1981, 74 2116.

RECEIVED August 26, 1981.

BONDING AND INTERACTIONS OF ALKALI HYDRIDES AND HALIDES

16

Ionic–Covalent Interactions in Alkali Hydrides

SZE–CHENG YANG

University of Rhode Island, Department of Chemistry, Kingston, RI 02881

WILLIAM C. STWALLEY

University of Iowa, Iowa Laser Facility and Departments of Chemistry and Physics, Iowa City, IA 52242

The recently available spectroscopic data and the RKR potentials of the alkali hydrides allow us to determine the "experimental" values of the parameters relevant to the transition probability of the charge transfer processes. In the Landau-Zener model these parameters are the energy gap between the $A^1\Sigma^+$ and $X^1\Sigma^+$ adiabatic potentials at the avoided crossing distance and the coupling matrix elements. In this paper the coupling matrix elements are evaluated in a two-state ionic-covalent interaction model. The systematic trends found in the alkali hydride series for their $X^1\Sigma^+$ potentials are presented. This leads to a simple model for the ionic potentials.

Crawford and Jorgensen (1) in their early spectroscopic measurements of the LiH A-X transitions discovered an anomaly in the spectroscopic constants, $B_{v'}$ and $\Delta G_{v'+\frac{1}{2}}$ of the $A^1\Sigma^+$ state. Both $B_{v'}$ and $\Delta G_{v'+\frac{1}{2}}$ first increase as a function of the v' to a maximum before they revert to the usual monotonic decreasing behavior at higher v'. Mulliken later offered a qualitative explanation in terms of the avoided crossing of the ionic and covalent potential curves (2). The concept has been very useful for understanding not only the optical spectra of alkali hydrides but also a large number of important processes: charge transfer reactions, electronic to vibrational (E $\rightarrow$ V) energy transfer, chemi-ionization, the harpooning mechanism of the chemical reactions, predissociation, ion pair formation and ion-ion recombination.

The optical spectra of alkali halides provide another example of the effect of ionic-covalent curve crossing on spectra. The qualitative features are explained nicely by the theory of non-adiabatic transitions due to Berry (3). Unfortunately the spectra cannot be resolved and identified to yield quantitative data relevant to the ionic-covalent interaction. For alkali halides these interactions are more amenable to atomic beam scattering experiments (4). In contrast the optical spectra of

0097-6156/82/0179-0241$05.00/0

alkali hydrides are simple and assignable. The accuracy of the spectroscopic measurements and the abundance of details allow us to extract information relevant to the dynamics of the charge transfer process.

New spectroscopic measurements (5-11) of the past few years allow the construction of the $X^1\Sigma^+$ state RKR potentials (5-13) very close to the avoided crossing region. Since the electronic character of the alkali hydrides changes as the internuclear distance is changed adiabatically across the avoided crossing region, the potential curves are expected to bend sharply in that region. The shape and the energy gap of the curves contain quantitative information about the ionic-covalent interaction.

In this study we first examine the systematics that exist in the RKR potentials of the $X^1\Sigma^+$ state to establish that the potential curves are strongly ionic for $R \ll R_c$ (R_c is the distance of the pseudocrossing point). Next we evaluate an essentially experimental value of the parameters relevant to the cross section for the charge transfer process in the Landau-Zener approximation. We also construct a model ionic potential which can be used to describe the charge transfer process in the ionic region.

Systematics of the $X^1\Sigma^+$ RKR Potentials

Table I shows some of the parameters of the RKR potentials.

Table I. Parameters obtained from the optical spectra of the alkali hydrides.

	IP-EA (cm^{-1})	$X^1\Sigma^+$		$A^1\Sigma^+$	
		D_e (cm^{-1})	R_e (Å)	D_e (cm^{-1})	R_e (Å)
LiH	37403.3	20286.6 ± 0.5	1.5957	8681.6 ± 0.5	2.5839
NaH	35366.0	15900 ± 500	1.8872	10137 ± 500	3.1934
KH	28926.8	14500 ± 500	2.2424	8732 ± 500	3.7629
RbH	27606.1	14240 ± 500	2.3673	8598 ± 500	3.8642
CsH	25323.7	14500 ± 500	2.4938	7286 ± 500	3.9823

The parameters given in Table I are based on the following: IP = ionization potential of the alkali atom (14); EA = electron affinity of the hydrogen atom (15); D_e = bonding energy of the potential well (16); R_e = internuclear distance at the minimum of the potential curve.

When we plot the energy of the X state potential minimum relative to the energy of the ion pairs as a function of R_e^{-1}, the inverse of the internuclear distance at the minimum, we get a straight line correlation among the alkali hydrides (see Figure 1):

$$V(R_e) = -(IP - EA - D_e) = -\frac{80468}{R_e} - 7709$$

where units are in cm^{-1} and Å for V and R_e. Figure 2 expresses

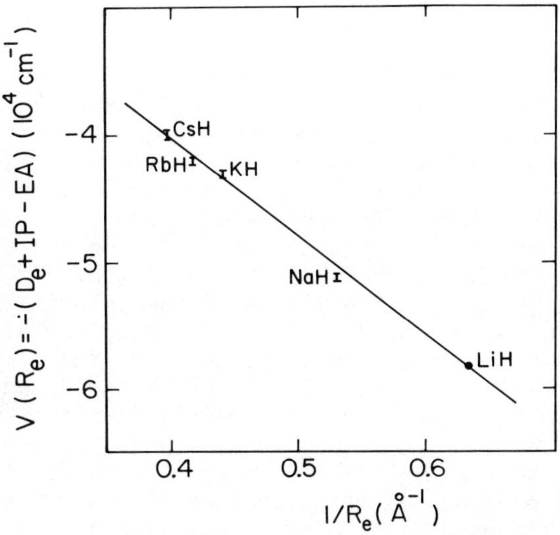

Figure 1. Correlation between the energy of the potential minima relative to the ion pair dissociation limit and the reciprocal of the equilibrium bond distance for the $X^1\Sigma^+$ state alkali hydrides.

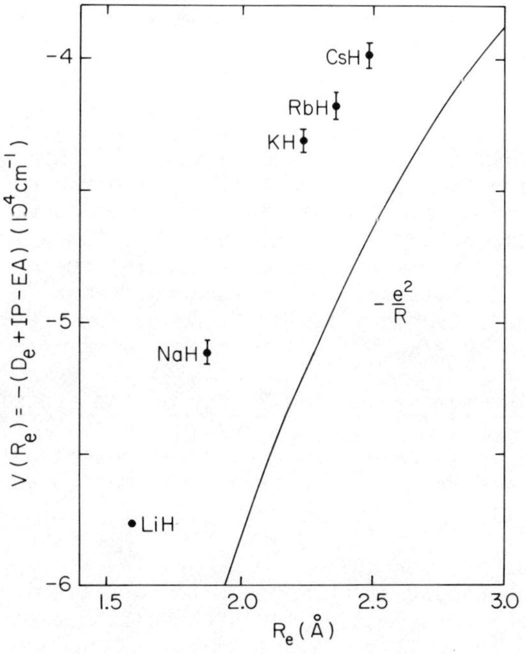

Figure 2. Energy of the potential minima relative to the free ions vs. the equilibrium bond distance. The curve $V(R) = -(e^2/R)$ is the Coulombic point charge potential.

this correlation in another manner. We can see that the $V(R_e)$ curve runs parallel to the coulombic potential, $V_{coul}(R) = -e^2/R$. This simply reflects the fact that the ground electronic state is primarily ionic, i.e. M^+H^-. The fact that $V(R_e)$ points stay smoothly on a curve above $V_{coul}(R)$ suggests that the repulsive exchange interaction also follows the trends in this series. It is interesting to note that the systematic trend is obtained by considering the ionic character alone even though the adiabatic curves dissociate into neutral atoms.

There is another striking similarity among the alkali hydride systems. Figure 3 shows a reduced plot of the X state RKR turning points (5 - 13). The minima of the potential wells are taken as the zeroes of energy and the internuclear distance R is scaled by the equilibrium bond distance R_e of the alkali hydride. The $X^1\Sigma^+$ potentials fall into almost the same reduced form by this simple scaling. Note that the vertical axis is not scaled. The dissociation limits of the molecules are indicated in the figure. It is remarkable that a common reduced potential for all the alkali hydrides is maintained up to an energy very close to their significantly different dissociation limits. However if we temporarily ignore the avoided crossing and pretend that the molecules dissociate into their ion pairs, then the common reduced form is not so surprising because they represent less than the bottom 30% of the ionic potential well. The implication is that the molecules are dominantly ionic at all internuclear distances shorter than those in the avoided crossing region. Moreover, the energy of the $X^1\Sigma^+$ potential curve in the avoided crossing region is quite close to the adiabatic dissociation limit of the molecules. The reduced potential curves are expected to deviate from each other eventually upon entering the avoided crossing region. The curves will then bend sharply to their different dissociation limits. The RKR potential of NaH (8) is an example which already shows some deviation from the common reduced form at the higher vibrational levels with outer turning points in the avoided crossing region. However, if the avoided crossing region is ignored, the $X^1\Sigma^+$ potentials of all the alkali hydrides appear to be well represented by a single function of R/R_e only.

Coupling Matrix Element and the Energy of the Crossing Point

In the above we have argued that the potential curve in the avoided crossing region contains information about the charge transfer process:

$$M + H \rightarrow M^+ + H^-$$

We consider the Landau-Zener model (17) for calculating the charge transfer cross section for high collision energies:

$$Q_{ZL}(E) = 4\pi g\, R_c{}^2\, (1 - H_{ii}(R_c)/E)\, G(\lambda)$$

$$G(\lambda) = \int_1^\infty \exp(-\lambda x)\,[1 - \exp(-\lambda x)]\, x^{-3} dx$$

$$\lambda = \frac{2\pi\, T_{ic}^{\,2} \times \left| \dfrac{dH_{ii}(R)}{dR} - \dfrac{dH_{cc}(R)}{dR} \right|_{R=R_c}}{\{hv\,[1 - H_{ii}(R_c)/E]\}^{\frac{1}{2}}}$$

where Q_{ZL} is the total cross section for charge transfer reaction, E is the collision energy, $H_{ii}(R)$ and $H_{cc}(R)$ are the ionic and covalent curves, R_c is the crossing distance, g is a symmetry factor, v is the relative velocity of the collision and T_{ic} is the coupling matrix element. In the two-state configuration approximation (18 - 20)

$$T_{ic} = |H_{ic}(R_c) - H_{ii}(R_c)\, S|$$

where H_{ic} and S are the off diagonal Hamiltonian matrix element and the overlap integral between the ionic and covalent representations, respectively.

Our intention is to use as much of the experimental RKR potentials as possible to evaluate the relevant quantities T_{ic} and $H_{ii}(R_c)$.

Figure 4 shows the known RKR turning points for both the X and A states of the alkali hydrides. The internuclear distance is scaled by the equilibrium bond distance in the $X^1\Sigma^+$ state. It can be seen that the A state potentials (6, 7, 10, 12, 13) are known accurately in the avoided crossing region. The X state RKR potentials are known to internuclear distances very close to the crossing region, but not into it. However, with a very short extrapolation the $X^1\Sigma^+$ RKR potential can be extended to the crossing distance. We adopted the following procedure to extrapolate the X state potential curves of NaH, KH, RbH and CsH based on the available theoretical X state potential curves. The energies of the theoretical potential curves (21 - 23) are scaled to make them smoothly fit with the outermost of the existing RKR turning points to extrapolate the RKR potentials into the crossing region. The extrapolation of the NaH potential is based on the theoretical curves of Olson and Liu (21). The extrapolations of KH and RbH are based on the theoretical curves of Stevens and Hiskes (22). The theoretical curve of Stallcop and Laskowski (23) is scaled to extrapolate the RKR curves for CsH. The uncertainty is estimated to be $\lesssim 300\ cm^{-1}$ based on the deviation of the theoretical curves from the known portions of the RKR curves. LiH is unique because the dissociation energy is accurately determined (16). The hybrid potential for LiH (24) is believed to be highly reliable. This is further supported by the high quality ab initio calculations on LiH by Partridge and Langhoff (25) and by Docken and Hinze (26). The former agrees with the hybrid potential within a fraction of a percent. We simply use the reported hybrid potential (24) for LiH in the avoided

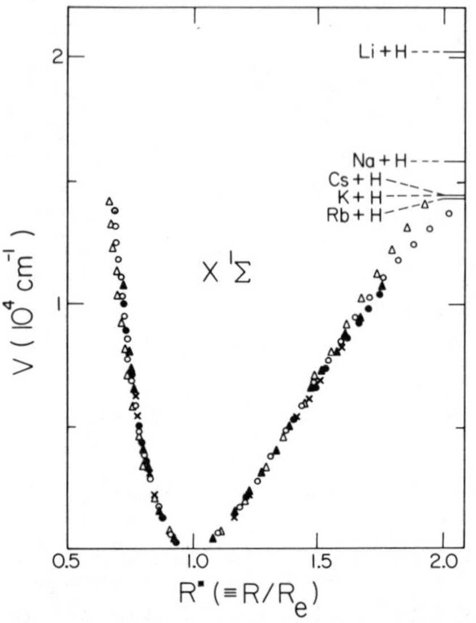

Figure 3. Potential energy curves as a function of reduced distances (R/R_e) for the $X^1\Sigma^+$ state of the alkali hydrides. Key: △, LiH; ○, NaH; ▲, KH; ×, RbH; and ●, CsH.

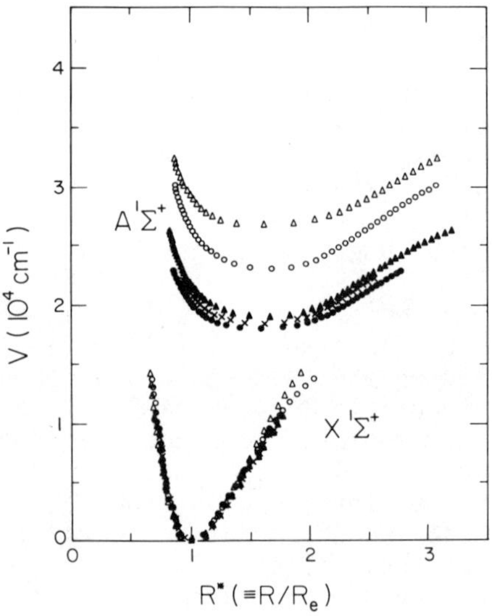

Figure 4. RKR turning points of the $X^1\Sigma^+$ and $A^1\Sigma^+$ states of the alkali hydrides plotted vs. the reduced internuclear distance R/R_e. Key: see Fig. 3.

crossing region. We determine the crossing distance, R_C, and the adiabatic energy gap, $\Delta V(R_C)$, at the crossing distance as follows: First we define the crossing distance as the internuclear distance where the energy gap between the A and X state potentials is a minimum. It is also at this point where the two potential curves have the same slope, i.e.

$$\left.\frac{dV_A(R)}{dR}\right|_{R=R_c} = \left.\frac{dV_X(R)}{dR}\right|_{R=R_c}.$$

The R_c's are determined graphically and listed in Table II.

Table II. The energy (in cm^{-1}) of the adiabatic potentials at the crossing distance with respect to the ion pair dissociation limit $M^+ + H^-$.

	R_c(Å)	$V_A(R_c)$	$V_X(R_c)$
LiH	3.90	-29326.3	-38796.3
NaH	4.08	-27250.3	-36758
KH	4.70	-23424	-29959
RbH	4.86	-22633.5	-28814
CsH	5.15	-21323.7	-26423.7

The corresponding energies of the X and A potentials are also listed in this table.

In the two state configuration mixing approximation of Grice and Herschbach (18), Adelman and Herschbach (19) and Janev and Radulović (20), the coupling matrix element can be related to the energy of adiabatic curves at the crossing distance:

$$T_{ic} = \frac{1 - S^2}{2} (V_A(R_c) - V_X(R_c))$$

$$H_{ii}(R_c) = \tfrac{1}{2}[V_A(R_c) + V_X(R_c)] - \frac{S}{1 - S^2} T_{ic}$$

The overlap integral S has been evaluated by Grice and Herschbach (18) for the crossing distances (listed in parentheses, column 2 of Table III) slightly different from our semi-

Table III. Parameters for the ionic-covalent interaction in the two-state approximation.

	R_c(Å)	$S(R_c)$	$\lvert H_{ic}-H_{ii}S\rvert$ (cm^{-1})	$H_{ii}(R_c)$ (cm^{-1})
LiH	3.90 (3.92)	0.34	4187.6	-35671.3
NaH	4.08 (4.00)	0.33	4236.2	-33572.9
KH	4.70 (4.66)	0.30	2973.4	-27871.5
RbH	4.86 (4.82)	0.29	2830.4	-26630
CsH	5.15 (5.15)	0.27	2225	-24562

experimental values. Since the differences in the R_c's are small, we use their value of S listed in column 3 of Table III to evaluate the semiempirical values of T_{ic} and $H_{ii}(R_c)$ which are listed in columns 4 and 5 of Table III.

Previously the coupling matrix elements were calculated by Janev and Radulović (20) employing the asymptotic wave functions in a two electron approximation. In Table IV we compare these

Table IV. Comparison of $|H_{ic}-H_{ii}S|$

	Janev and Radulović[a]		This work	
	$R_c(a_o)$	$\frac{1}{2}\Delta(cm^{-1})$	$R_c(a_o)$	$\lvert H_{ic}-H_{ii}S\rvert(cm^{-1})$
LiH	7.40	2970	7.37	4188
NaH	7.66	3065	7.71	4236
KH	8.86	2415	8.87	2973
RbH	9.11	2359	9.19	2830
CsH	9.74	2065	9.73	2225

[a] Reference 20.

values with our essentially experimental values. There are differences in the magnitudes of the coupling matrix elements, especially for the lighter alkali hydrides. Note that both treatments are in the framework of a two-state model, but we feel our empirically based values should be preferred for calculation of $Q_{LZ}(E)$.

A semiempirical correlation among coupling matrix elements and crossing distances for many molecules was examined by Olson, Smith and Bauer (27) and Olson (28). The coupling matrix elements calculated from their semiempirical formula agree with our values to within a factor of two, which is not surprising for such an approximate correlation.

The Modeling of an Ionic Potential

The systematics found in the alkali hydride $X^1\Sigma^+$ potentials suggest that it might be possible to model a simple ionic potential to reflect the behavior of the whole series of alkali hydrides. Here we construct a "practical" diabatic curve which reflects the physical properties, e.g. the dipole moment function. We expect our diabatic curve to follow the ionic part of the $X^1\Sigma^+$ adiabatic potential and to begin to deviate in the avoided crossing region.

We have chosen to model the diabatic ionic curve ($H_{ii}(R)$) as a truncated Rittner potential with a turning-off function for the polarization term:

$$H_{ii}(R) = A\ \exp[-R/\rho] - \frac{e^2}{R} - \frac{(\alpha_+ + \alpha_-)}{2R^4} f(R) \qquad (1)$$

where $f(R) = 1 - \exp[-(R/\Upsilon)^6]$. α_+ and α_- are the ion polarizabilities (31, 32), Υ the distance parameter where the charge overlap diminution of the polarization term becomes significant, and A and ρ are parameters for the repulsion to be fitted from RKR turning points.

At large internuclear distances ($R >> \Upsilon$), where both the exchange and charge overlap effects are negligible, the turning-off function f(R) approaches unity. There the potential is the usual R^{-1} power series expansion. In the intermediate range ($R \simeq \Upsilon$), the overlap of charges causes the polarization term to be reduced. At short internuclear distances ($R \simeq \rho$), the repulsive exchange interaction becomes important.

We now justify the introduction of the turning-off function f(R). In a study of H_2^+, Kreek and Meath (29) showed that the coefficients of the usual R^{-1} expansion for long range forces are reduced at an internuclear distance of intermediate range due to charge overlap effects. This effect is taken as the basis for the introduction of the turning-off function f(R). The functional form of f(R) was suggested by Numerich and Truhlar (30). The charge overlap effect is expected to be important in the alkali hydrides because of the very diffuse electron distribution in H^- (the polarizability of H^- is 30.5 Å^3 (31)!).

The parameter Υ is essentially fixed by the requirement that the ionic curve pass through the crossing point. At R_C the exchange repulsive term is small and

$$H_{ii}(R_c) \simeq -\frac{e^2}{R_c} - \frac{\alpha_+ + \alpha_-}{2R_c^4} \exp\{1 - (\frac{R}{\Upsilon})^6\} \qquad (2)$$

The parameters A and ρ are determined by a nonlinear least square fit of equation (1) to the RKR inner turning points of each of the alkali hydrides. This fitting procedure is justified by the fact that the magnitude of the dipole moment functions of LiH (25, 26) and NaH (33) at these internuclear distances are very close to those of opposite point charges a distance R apart. Based on our definition for the "practical" diabatic curves, we require that the ionic curve agree with the inner wall of the adiabatic curve. The fitted parameters A and ρ are listed in Table V.

Table V. Parameters for the ionic curve.

	$\alpha_+ + \alpha_-$ (Å^3)	Υ(Å)	A(cm^{-1})	ρ(Å)
LiH	30.55	3.6607	1081730	0.3826
NaH	30.70	3.7810	1335760	0.4030
KH	31.36	4.2320	1497650	0.4460
RbH	31.94	4.4364	1558760	0.4577
CsH	32.96	4.9033	1930060	0.4530

Figure 5 illustrates the essentially experimental hybrid potentials and the fitted ionic curve of LiH. Note that the ionic curve starts to deviate from the $X^1\Sigma^+$ potential in the avoided crossing region and runs parallel to the ionic portion of the $A^1\Sigma^+$ potential slightly outside the crossing region and crosses the $A^1\Sigma^+$ curve at roughly the position of the second crossing distance of the $A^1\Sigma^+$ state. The relation between the coupling matrix element T_{ic} and the energy gap R_c in the two-state interaction approximation is also indicated there.

Figure 6 illustrates the RKR turning points and the ionic curves of NaH, KH, RbH and CsH. The ionic curves all seem to be satisfactory in view of the very simple model we have employed.

Note that in our two-state model, the energy of the crossing points are above the dissociation limit of the $X^1\Sigma^+$ potential. Thus the $H_{cc}(R)$ is repulsive at the crossing distance in this two-state model. This may imply either the two-state approximation is not sufficiently accurate or the interpretation of $H_{cc}(R)$ is not simple. In fact Garrett *et al*. (41) and Numerich and Truhlar (30) found that the interaction is not well localized at the crossing distance and that the interaction with a third state (although small) is not negligible. For the same reasons, the Landau-Zener approximation for the charge transfer reaction is not expected to be accurate for low energy (thermal) collisions (41, 42) (although it is expected to be adequate for medium and high energy collisions).

There are other measurable physical properties which reflect the change of electronic character as a function of the internuclear distance R in the avoided crossing region. Examples can be found in the measurements of the vibrationally resolved dipole moments of the $A^1\Sigma^+$ state of LiH and NaH (34) to complement earlier $X^1\Sigma^+$ results (35). Also, measurements on the lifetimes (36, 37) and the fluorescence intensity patterns (38, 39) provide information about the transition moment functions. Further experimental measurements when coupled with the theoretical calculations of the dipole moments and the transition moments (25, 26) and the radiative lifetimes (40) will eventually give a more detailed picture of the ionic-covalent interactions, and should also more clearly establish the limitations of a two-state model.

Finally we note that similar scaling procedures to that shown here for the alkali hydrides are discussed extensively for the alkali dimers in this volume (43).

Conclusions

New spectroscopic measurements of the alkali hydrides have provided information related to the dynamical charge transfer process for these systems. We have examined the RKR potentials derived from these spectra. Several striking regularities for the $X^1\Sigma^+$ potentials are presented along with an interpretation based on a simple model of ionic potentials for internuclear distances shorter than the crossing distance R_c.

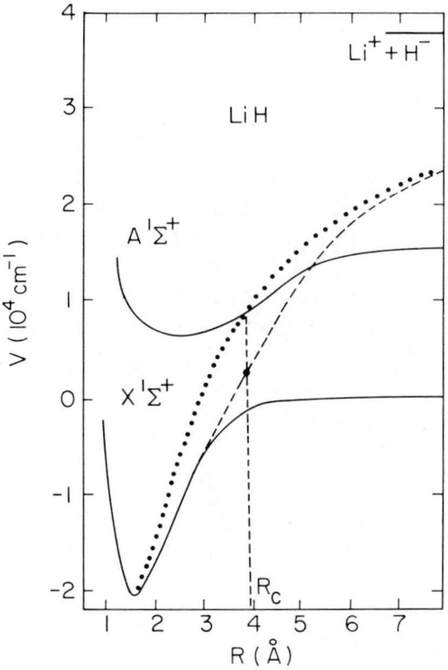

Figure 5. LiH potential energy curves. The $A^1\Sigma^+$ and $X^1\Sigma^+$ curves are the hybrid potentials of Ref. 24. Key: – – –, ionic curve of equation 1; · · ·, ionic curve without R^{-4} polarization term; and vertical – – –, R_c.

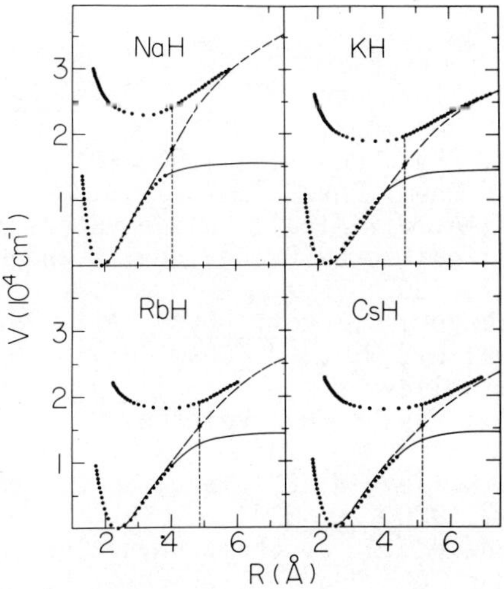

Figure 6. Potential energy curves for the $X^1\Sigma^+$ and $A^1\Sigma^+$ states of NaH, KH, RbH, and CsH. Key: · · ·, RKR turning points; ——, extrapolations of $X^1\Sigma^+$ states based on ab initio calculations; – – –, ionic curve of Eq. 1; and vertical – – –, R_c.

In the avoided crossing region the experimental RKR potential of the $A^1\Sigma^+$ state and the "essentially experimental" potential of the $X^1\Sigma^+$ state are used to determine the crossing distance R_c and the coupling matrix element T_{ic} in the two state approximation. These quantities are relevant to the evaluation of the total charge transfer cross section at high energy (e.g. in the Landau-Zener model).

In this study we used the scaled theoretical potential curves of Olson and Liu (21), Stevens, Karo and Hiskes (22), and Laskowski and Stallcop (23) to make a short extrapolation of the $X^1\Sigma^+$ RKR potentials into the avoided crossing region. Experimental measurements to obtain strictly experimental RKR potentials for this region are in progress. New optical measurements on the dipole moments (34, 35), the transition moments (38, 39) and radiative lifetimes (36, 37) of the alkali hydrides are also becoming available. This type of information will soon provide additional details about the ionic-covalent interactions in these molecules.

Acknowledgements

The authors wish to thank W. Stevens, A. M. Karo, J. R. Hiskes, B. Laskowski, J. Stallcop, H. Partridge and S. R. Langhoff for making their work available prior to publication. S. C. Yang would like to thank William Meath and David Freeman for their valuable suggestions and discussions. Work at Rhode Island was supported by the Research Corporation; work at Iowa was supported by the National Science Foundation and the National Aeronautics and Space Administration.

Literature Cited

1. Crawford, F. H.; Jorgensen, T., Jr. Phys. Rev. 1935, 47, 358; Phys. Rev. 1935, 47, 932.
2. Mulliken, R. S. Phys. Rev. 1936, 50, 1017, 1028.
3. Berry, R. S. J. Chem. Phys. 1957, 27, 1288.
4. Los, J.; Kleyn, A. W. "Alkali Halide Vapors"; Academic Press: New York, 1979; edited by P. Davidovits and D. L. McFadden; Chapter 8, p 275-330.
5. Ennen, G.; Ottinger, C. Chem. Phys. Lett. 1975, 36, 16.
6. Li, K. C.; Stwalley, W. C. J. Chem. Phys. 1979, 70, 1736; Orth, F. B.; Stwalley, W. C. J. Mol. Spectrosc. 1979, 76, 17; Stwalley, W. C.; Way, K. R.; Velasco, R. J. Chem. Phys. 1974, 60, 3611.
7. Orth, F. B.; Stwalley, W. C.; Yang, S. C.; Hsieh, Y. K. J. Mol. Spectrosc. 1980, 79, 314.
8. Giroud, M.; Nedelec, O. J. Chem. Phys. 1980, 73, 415; Baltayan, P.; Jourdan, A.; Nedelec, O. Phys. Lett. 1975, 36, 16.
9. Cruse, J. A.; Zare, R. N. J. Chem. Phys. 1974, 60, 1182.

10. Hsieh, Y. K.; Yang, S. C.; Tam, A. C.; Verma, K. K.; Stwalley, W. C. J. Mol. Spectrosc. 1980, 83, 311
11. Tam, A. C.; Happer, W. J. Chem. Phys. 1976, 64, 2156.
12. Yang, S. C.; Hsieh, Y. K.; Verma, K. K.; Stwalley, W. C. J. Mol. Spectrosc. 1980, 83, 304.
13. Hsieh, Y. K.; Yang, S. C.; Tam, A. C.; Stwalley, W. C. J. Chem. Phys. 1978, 68, 1448.
14. Moore, C. E. "Atomic Energy Levels"; Washington, D. C.: National Bureau of Standards, 1941; Volumes 1 and 2.
15. Pekeris, C. L. Phys. Rev. 1958, 112, 1649.
16. D_e for ^{7}LiH was precisely determined in the following paper: Vidal, C. R.; Stwalley, W. C. submitted to J. Chem. Phys. (based on Way, K. R.; Stwalley, W. C. J. Chem. Phys. 1974, 59, 5298). D_e's for the other alkali hydrides are estimated by scaling the theoretical potentials of References (21-23) to the existing RKR potentials of References (5-13).
17. Olson, R. E.; Peterson, J. R.; Moseley, T. J. Chem. Phys. 1970, 53, 3391; Janev, R. K. Adv. At. Mol. Phys. 1976, 12, 1.
18. Grice, R.; Herschbach, D. R. Mol. Phys. 1974, 27, 159.
19. Adelman, S. A.; Herschbach, D. R. Mol. Phys. 1977, 33, 793.
20. Janev, R. K.; Radulović, Z. M. Phys. Rev. A 1978, 17, 889.
21. Olson, R. E.; Liu, B. J. Chem. Phys. 1980, 73, 2817.
22. Stevens, W. J.; Karo, A. M.; Hiskes, J. R. J. Chem. Phys. 1981, 74, 3989.
23. Laskowski, B.; Stallcop, J. J. Chem. Phys. 1981, 74, 4883.
24. Stwalley, W. C.; Zemke, W. T.; Way, K.R.; Li, K. C.; Proctor, T. R. J. Chem. Phys. 1977, 66, 5412; 1977, 67, 4785.
25. Partridge, H.; Langhoff, S. R. J. Chem. Phys. 1981, 74, 2361.
26. Docken, K. K.; Hinze, J. J. Chem. Phys. 1972, 57, 4928, 4936.
27. Olson, R. E.; Smith, F. T.; Bauer, E. Appl. Optics 1977, 10, 1848.
28. Olson, R. E. Phys. Lett. 1981, 77A, 143.
29. Kreek, H.; Meath, W. J. J. Chem. Phys. 1969, 50, 2289.
30. Numerich, R. W.; Truhlar, D. G. J. Phys. Chem. 1975, 79, 2745; 1978, 82, 168.
31. Chung, K. T. Phys. Rev. A 1971, 4, 7; Adelman, S. A. Phys. Rev. A 1971, 5, 508.
32. Pauling, L. Proc. R. Soc. A 1972, 114, 19.
33. Sachs, E. S.; Hinze, J.; Sabelli, N. H. J. Chem. Phys. 1975, 62, 3367, 3377.
34. Dagdigian, P. J. J. Chem. Phys. 1980, 73, 2049; Brieger, M.; Hese, A.; Renn, A.; Sodeik, A. Chem. Phys. Lett. 1980, 76, 465; also to be published.
35. Wharton, L.; Gold, L. P.; Klemperer, W. J. Chem. Phys. 1962, 37, 214; Rothstein, E. J. Chem. Phys. 1969, 50, 1899.
36. Dagdigian, P. J. J. Chem. Phys. 1976, 64, 2609.
37. Wine, P. H.; Melton, L. A. J. Chem. Phys. 1976, 64, 2692; ACS Symp. Ser. 1977, 56, 167.

38. Yang, S. C.; Hsieh, Y. K.; Tam, A. C.; Zemke, W. T.; Verma, K. K.; Stwalley, W. C. J. Chem. Phys. (to be published).
39. Ennen, G.; Fiedler, B.; Ottinger, C. J. Chem. Phys. 1981, 75, 59.
40. Zemke, W. T.; Crooks, J. B.; Stwalley, W. C. J. Chem. Phys. 1978, 68, 4628.
41. Garrett, B. C.; Redmon, M. J.; Truhlar, D. G.; Melius, C. F. J. Chem. Phys. 1981, 74, 412.
42. Miller, W. H.; George, T. F. J. Chem. Phys. 1972, 56, 5637; see also Preston, R. K.; Sloane, C.; Miller, W. H. J. Chem. Phys. 1974, 60, 4961.
43. Konowalow, D. D.; Rosenkrantz, M. E. ACS Symp. Series, 1981, XX, xxxx [this volume].

RECEIVED September 3, 1981.

17

Cesium Hydride Formation from Cs(7P) and H_2 Interactions

B. SAYER, M. FERRAY, J. LOZINGOT, and J. BERLANDE

Centre d'Etudes Nucléaires de Saclay, Service de Physique Atomique,
91191 Gif-sur-Yvette, Cedex, France

Laser photoexcitation of cesium atoms to the 7P state in the presence of hydrogen gas leads to the formation of cesium hydride by a process which is shown to be indirect. The concentration of the reactants is studied under various experimental conditions using a laser fluorescence technique. The rate coefficient of CsH formation is found to be proportional to the Cs(6S) density suggesting a two step process with the formation of an intermediate species which reacts with ground state cesium atoms.

Several attempts have been made during the last few years to study the formation of alkali hydrides in the vapor phase when a mixture of alkali metal vapor and hydrogen is exposed to laser irradiation which photoexcites the alkali atom. The present work is devoted to the following overall reactions:

$$Cs(6S) + h\nu_L \rightarrow Cs(7P) \qquad (1)$$

$$Cs(7P) + H_2 \rightarrow CsH + H \qquad (2)$$

This process was reported for the first time by Tam and Happer (1). Using the lines of an argon ion laser they photoexcited the cesium atoms to the 7P state through the broadened 6S-7P lines and observed the formation of macroscopic crystals of cesium hydride ("laser snow"). More recently Picqué and coworkers (2) have reported the same observation using a tunable dye laser resonant with each of the fine structure components of the 6S-7P transition.

However, in spite of the important features apparent in these studies, the detailed mechanism of the Cs(7P) + H_2 reactive collision is still uncertain. As proposed by Tam and Happer (1), it is conceivable that either a direct process (equation 2) or a two-step process with the formation of an intermediate species might be responsible for the CsH formation.

0097-6156/82/0179-0255$05.00/0

It is the purpose of the present work to try to obtain insight into CsH formation and to estimate the efficiency of the process. For this reason, we performed a spectroscopic investigation of the relative concentrations of the reactants - Cs(7P), H_2 and CsH - under various experimental conditions in stationary as well as in transient regimes. A more detailed discussion of these results will be published in J. Chem. Phys.

Experimental Set-Up

A CW tunable dye laser (Stilbene 3) is pumped by the ultraviolet lines of an argon ion laser. The laser is focused in a cell filled with a mixture of cesium and hydrogen. The cell (10 cm long with an inside radius of 2.6 cm) is connected to a reservoir containing liquid cesium and to a pumping system or a hydrogen gas tank. It is placed in an oven heated independently from the reservoir. The laser beam enters the cell from below through a flat window and is focused at a point 1 cm above the window. The fluorescence light is observed at right angles to the laser beam by an optical system and a small monochromator followed by a RCA Quantacon photomultiplier, the signal of which is sent to an amplifier and a chart recorder when the cell is continuously irradiated or to a photon counting system and a multichannel analyser when transient effects are studied.

Preliminary Observations

The typical experimental conditions are: [Cs] = 0.8 - 8 x 10^{16} cm^{-3}, [H_2] = 8 - 80 x 10^{16} cm^{-3}, T = 630 - 670 K. The laser power in the cell is of the order of 20 mw. When the laser wavelength is tuned on the 6S-7$P_{1/2}$ cesium line (459.3 nm), we observe that the cesium atoms are photoexcited to the 7$P_{1/2}$ state and that excitation transfer to the 7$P_{3/2}$ state occurs because of inelastic collisions. These Cs(7$P_{3/2}$) atoms fluoresce at 455.5 nm. If we now shift the laser line by a few tens of GHz towards the short wavelength side, we observe in addition to the atomic fluorescence some doublets which are characteristic of the CsH ($A^1\Sigma^+ \rightarrow X^1\Sigma^+$) spectrum (3). This indicates that the laser is broad enough to photoexcite cesium atoms and the CsH molecules created by reaction (2).

The data on the energy of the rovibrational levels (4) and the intensities of the $A^1\Sigma^+$ (v' = 19) $\rightarrow$ $X^1\Sigma^+$ lines (3), lead us to the conclusion that the CsH transition responsible for the laser light absorption is $X^1\Sigma^+$ (v" = 0, J" = 11) $\rightarrow$ $A^1\Sigma^+$ (v' = 19, J' = 12). The CsH fluorescence therefore corresponds to the various transitions to the (v", J" = 11, 13) levels of the $X^1\Sigma^+$ state.

In contrast to the observations of Tam and Happer (1) and Picqué *et al*. (2), no macroscopic CsH crystals were observed. This is a result of our experimental conditions where the cesium densities and the laser power are lower than in previous experiments, leading to lower CsH concentrations, which do not permit crystal formation. Moreover, in the two previous studies, a CsH fluorescence intensity depending on the laser intensity I_L to the power $\frac{3}{2}$ was observed; this dependence is interpreted in terms of a destruction process mainly due to dimerization. In our experiment we find a fluorescence intensity which varies as I_L^2 indicating that this process is of minor importance in the loss of CsH molecules, and that diffusion out of the irradiated region is probably the dominant loss process.

Experimental Method

The first step consists of establishing the method for measuring the ratio between the Cs(7P) and CsH populations. The Cs(7P) concentration is related to the $7P_{3/2}$-$6S_{1/2}$ line intensity. In order to do this, we calculated the fraction γ of Cs(7P) atoms which are in the $7P_{3/2}$ sublevel and β the fraction of the $7P_{3/2}$-$6S_{1/2}$ line intensity which is not reabsorbed.

The $7P_{1/2}$ and $7P_{3/2}$ states have steady state populations determined by the following competitive contributions: (a) radiative transitions:

$$Cs(7P_{1/2,3/2}) \rightarrow Cs(7S, 5D, 6S) + h\nu \qquad (3)$$

(b) excitation transfer by collision with Cs (5) and H_2 (6):

$$Cs(7P_{1/2}) + (Cs \text{ or } H_2) \rightarrow Cs(7P_{3/2}) + (Cs \text{ or } H_2) \qquad (4)$$

and (c) quenching collisions with H_2 molecules (6):

$$Cs(7P_{1/2,3/2}) + H_2 \rightarrow Cs(6S) + H_2(v) \qquad (5)$$

where $H_2(v)$ is a vibrationally excited H_2 molecule. The fraction $\gamma = [Cs(7P_{3/2})]/[Cs(7P)]$ has been obtained from the radiative transition probabilities and the rate coefficients of reactions (4) and (5) (5, 6).

Reabsorption of the cesium fluorescence line has also been calculated from the cesium density and the line profile. Under our experimental conditions, line broadening by collisions with H_2 molecules is the predominant effect (50 MHz per Torr according to (7)).

The relationship between [Cs(7P)] and I_3, the intensity of the 455.3 nm fluorescence line, is then:

$$I_3 = G\, A_3\, \beta\, \gamma\, [Cs(7P)] \qquad (6)$$

where G is a factor which includes the contribution of both sensitivities and geometry and A_3 the Einstein coefficient of the transition.

In order to relate the CsH concentration to the CsH fluorescence intensity, we have assumed that the CsH $(X^1\Sigma^+)$ molecules produced, which undergo many collisions before absorbing a photon, are distributed over the rovibrational states according to the thermodynamic partition function at the gas temperature. α_p is the fraction of these molecules in the state (v" = 0, J" = 11). Let us consider the absorption of the CsH line:

$$\mathrm{CsH}\ [X^1\Sigma^+\ (v'' = 0,\ J'' = 11)] + h\nu_L \rightarrow \mathrm{CsH}\ [A^1\Sigma^+\ (v' = 19,\ J' = 12)] \quad (7)$$

The laser line is located at a wavelength between those of the CsH and $Cs(7P_{1/2})$ absorption lines such that the CsH fluorescence is a maximum, i.e. the product of the absorption by the lines is maximum (see Figure 1).

The photoexcitation rate is calculated:

$$\phi_{CsH} = I_L(\nu_{CsH})\ \alpha_p\ [\mathrm{CsH}] \int_0^\infty k(\nu)\ d\nu \quad (8)$$

where $I_L(\nu_{CsH})$ is the laser intensity at the wavelength of the CsH line. The integral over $k(\nu)$ is related to the Einstein A_a coefficient of the corresponding absorption transition.

By analogy with what is observed for Cs* atoms, we have considered that quenching of CsH* by H_2 molecules cannot be a priori neglected with respect to radiative depopulation and we have tentatively written:

$$[\mathrm{CsH}^*] = \phi_{CsH}/(\tau^{-1}_{CsH^*} + k_Q\ [H_2]) \quad (9)$$

where k_Q is the quenching rate coefficient which is unknown. The intensity I_F of the two unresolved fluorescence lines corresponding to the transitions:

$$\mathrm{CsH}\ [A^1\Sigma^+\ (v' = 19,\ J' = 12)] \rightarrow \mathrm{CsH}\ [X^1\Sigma^+\ (v'' = 4,\ J'' = 11,\ 13)] + h\nu_F \quad (10)$$

is then given by A_F [CsH*], where A_F is the transition probability of the observed pair of fluorescence lines. From equations (7), (9) and the relation which exists between k and A_a, the transition probability for the CsH absorption line, we have:

$$I_F = \frac{c^2}{8\pi\nu_{CsH}} \frac{g_s}{g_i} G' \frac{A_a\ A_F\ I_L(\nu_{CsH})\ \alpha_p\ [\mathrm{CsH}]}{\tau^{-1}_{CsH^*} + k_Q\ [H_2]} \quad (11)$$

where g_s/g_i is the ratio between the statistical weights of upper and lower levels of the absorbed transition and G' is a factor similar to G in equation (6). Recent calculations of CsH [$A^1\Sigma^+ \rightarrow X^1\Sigma^+$] transition probabilities (8) are used for the present calculation in spite of the slight difference between the J values of the transition considered here and those for which calculations were carried out. The radiative lifetime of the upper state (in fact $A^1\Sigma^+$ (v' = 19, J' = 10) instead of J' = 12) has also been calculated (8) (85 ns).

From equations (6) and (11), it is thus possible to relate [CsH]/[Cs(7P)] to I_F/I_3:

$$[CsH]/[Cs(7P)] = (BI_F/I_3) \cdot (1 + k_Q\,[H_2]\,\tau_{CsH*}) \qquad (12)$$

where B is a coefficient which can be calculated from the experimental conditions and does not depend on [Cs] and $[H_2]$ concentrations.

Experimental Results

In a first set of experiments, the hydrogen density was kept constant and that of cesium varied. The quantity BI_F/I_3 is plotted on Figure 2 as a function of [Cs] and we observe a linear variation.

Two interpretations of this variation are proposed: a CsH formation process requiring a ground state cesium atom in addition to Cs(7P) and H_2 molecules; a CsH destruction process inversely proportional to [Cs]. Collision controlled diffusion out of the irradiation region with cesium atoms could be such a process.

In a second set of experiments both [Cs] and $[H_2]$ have been varied. A slight increase of $B[Cs]^{-1}\,I_F/I_3$ with $[H_2]$ is observed (see Figure 3). As [CsH]/[Cs(7P)] should be proportional to $[H_2]$, we have plotted on the same graph the quantity $B[Cs]^{-1}\,(1 + k_Q\,[H_2]\,\tau_{CsH*})\,I_F/I_3$ with $k_Q = \sigma_Q <v>$ so as to have this quantity proportional to $[H_2]$. The approximate quenching cross section σ_Q of CsH* by H_2 that we find is equal to 4 $Å^2$, a value which is of the same order of magnitude as that of $Cs(6P_{1/2})$ (9).

Neither the precise nature of the CsH formation process nor the global rate of formation can be derived from the determination of $[CsH]\,([Cs(7P)]\,[H_2])^{-1}$ in a stationary regime because the loss rate for CsH molecules is unknown.

This fact has lead to investigations of transient laser irradiation. This is obtained by chopping the laser beam with a rotating wheel at a frequency of 1 kHz. It is observed that the rise time of I_F is of the order of a few hundred microseconds while I_3 follows closely the laser irradiation. It is possible to show that the initial dI_F/dt is related to the time constant of CsH formation, because CsH* molecules have a very short lifetime and because diffusion does not play an important role at the very beginning of the laser irradiation. Thus the risetime constant $\tau = I_F\,(dI_F/dt)^{-1}$ is assigned to $[CsH]\,(d[CsH]/dt)^{-1}$ and has been measured under various experimental conditions ([Cs], $[H_2]$). We have derived from these measurements the quantity $k = ([Cs(7P)]\,[H_2])^{-1}\,d[CsH]/dt = (\tau[H_2])^{-1}\,[CsH]/[Cs(7P)]$ which represents the rate of CsH formation according to the global reaction (2). This rate coefficient has been found to be independent of both [Cs(7P)] and

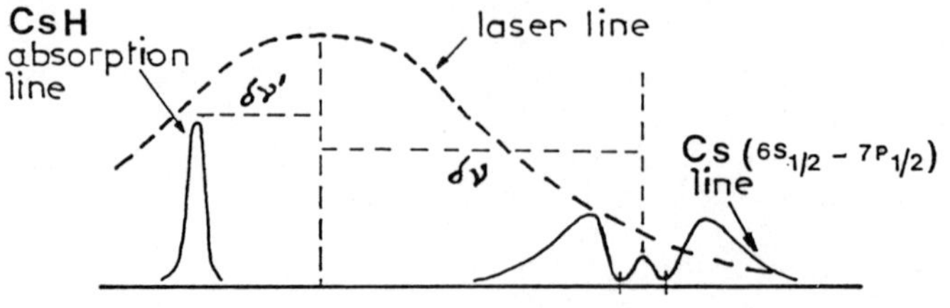

Figure 1. Relative positions of laser line and absorption lines in absorption region (7).

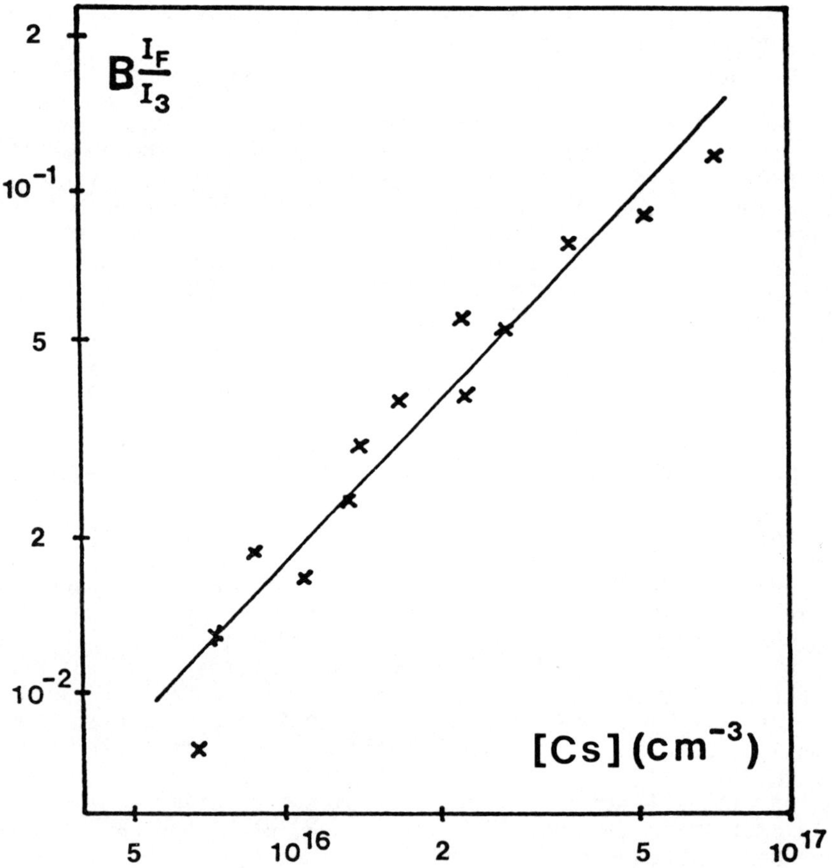

Figure 2. Variation of BI_F/I_3 with Cs density at a given $[H_2]$ ($\simeq 9 \times 10^{16}$ cm^{-3}). This quantity proportional to [CsH]/[Cs(7P)] (Eq. 12) (7).

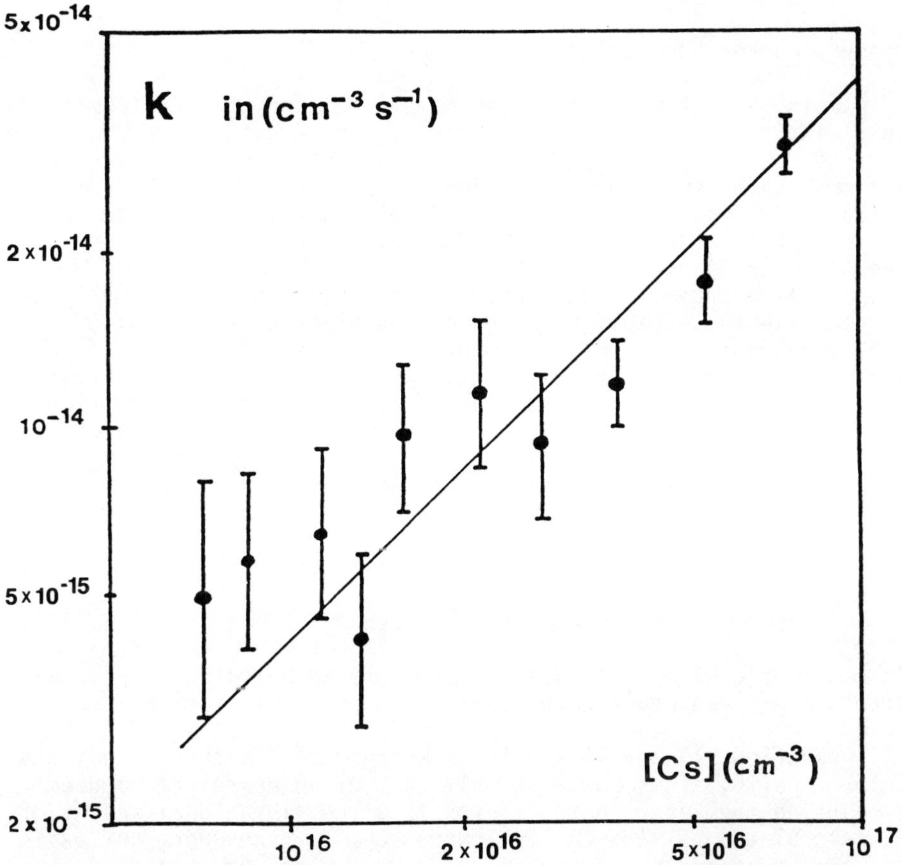

Journal of Chemical Physics

Figure 3. Experimental values of S fitted using various quenching cross sections of CsH by H_2. The upper straight line (○) shows the linear variation of a quantity proportional to [CsH]/[Cs(7P)] [Cs] with [H_2] (Eq. 12) (7).*

$[H_2]$, but approximately proportional to [Cs] as shown in Figure 4, where the experimental values of k corresponding to various $[H_2]$ have been plotted as a function of [Cs]. The best straight line corresponding to $k \propto [Cs]$ is presented.

This result indicates that the dependence of [CsH] on [Cs] found in the stationary regime originates from the CsH formation itself and not from the nature of the loss process as might be envisaged.

Discussion and Conclusion

Using a multimode laser, we have simultaneously photoexcited, in a mixture of cesium and hydrogen, cesium atoms to the $7P_{1/2}$ state and the resulting CsH molecules. The present experimental observations can be explained as follows:

- The Cs(7P) population is established by a balance between laser photoexcitation and quenching plus excitation transfer by collision between Cs and H_2 molecules.
- It is also probable that photoexcitation and quenching by H_2 are responsible for establishing the steady state CsH* population.
- By varying the cesium and hydrogen concentration, it was found that the experimental results were consistent with a CsH formation mechanism which can be written as a stepwise process, involving collisions with cesium atoms. This seems to eliminate the possibility of direct CsH formation by a Cs(7P) + H_2 process, a possible mechanism initially suggested by Tam and Happer (1). On the contrary the other mechanism proposed:

$$Cs(7P) + H_2 \rightarrow H_2(v) + Cs(6S) \qquad (13)$$

$$Cs(6S) + H_2(v) \rightarrow CsH + H \qquad (14)$$

is consistent with our results. It is to be noted that free hydrogen atoms can react with cesium atoms to form other CsH molecules.
- If reactions 13 and 14 are those which occur in the medium, the role of Cs(7P) atoms would be only that of an energy transducer converting the laser photon energy into the vibrational energy of H_2 molecules. We have the following scheme: the energy of excited cesium atoms is transfered to H_2 molecules, some of them having a vibrational energy high enough (v = 6) to react with ground state cesium atoms. It should be noted that these $H_2(v)$ molecules have a long lifetime as regards diffusion out of the observation region (10).
- Another mechanism which could explain the present experimental observations is the formation of a transient complex (11). In the present case this would correspond to:

$$Cs(7P) + H_2 \rightarrow (CsH_2)^* \qquad (15)$$

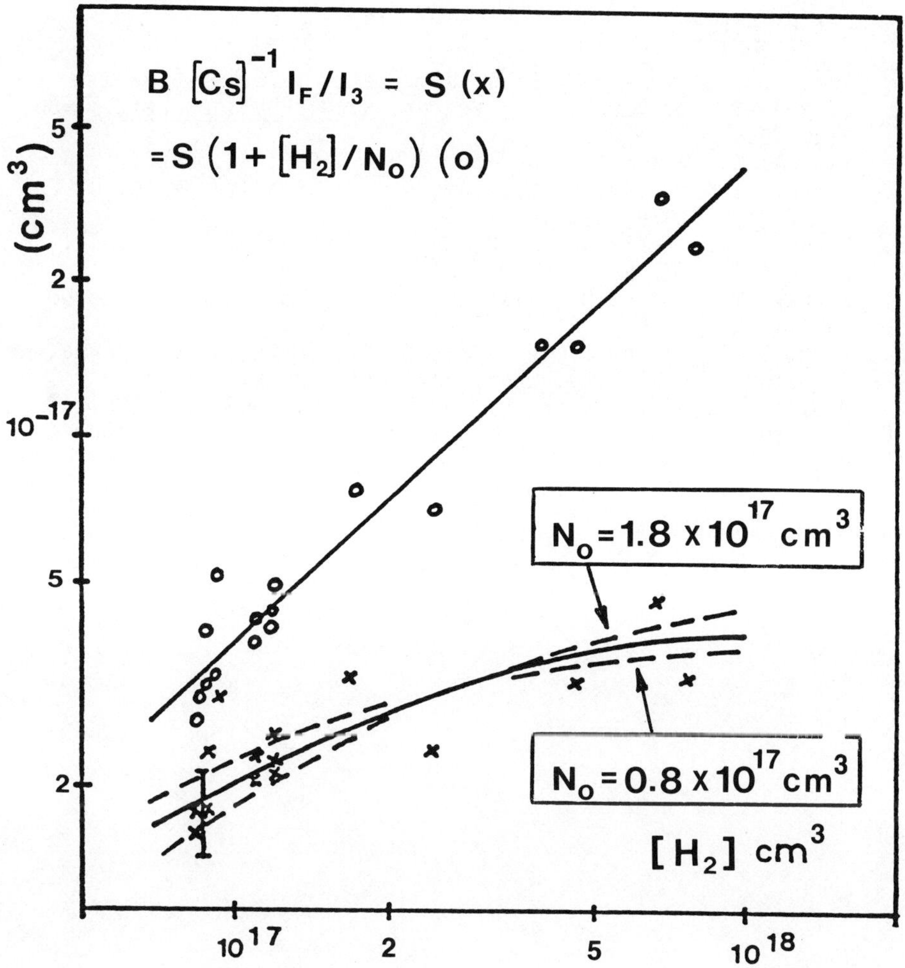

Journal of Chemical Physics

Figure 4. Variation of the rate coefficient of CsH formation with Cs density. Key: ——, $k \sim N_{Cs}$ (7).

$$(CsH_2)^* + Cs \rightarrow 2CsH \text{ or } CsH + Cs + H \qquad (16)$$

If $(CsH_2)^*$ has a long enough lifetime, this process can play a role. However, nothing is known about the possible existence of such a complex.

Literature Cited

1. Tam, A. C.; Happer, W. Phys. Rev. Lett. 1975, 35, 1630.
2. Picqué, J. L.; Vergès, J.; Vetter, R. J. Physique Lett. 1980, 41L, 305.
3. Tam, A. C.; Happer, W. J. Chem. Phys. 1976, 64, 2456.
4. Hsieh, Y. K.; Yang, S. C.; Tam, A. C.; Stwalley, W. C. J. Chem. Phys. 1978, 68, 1448.
5. Pace, P. W.; Atkinson, J. B. Can. J. Phys. 1974, 52, 1641.
6. Lukaszewicz, M. Bull. Acad. Polon. 1975, 23, 501.
7. Ferray, M.; Sayer, B.; Visticot, J. P.; Lozingot, J., to be published.
8. Yang, S. C.; Hsieh, Y. K.; Tam, A. C.; Zemke, W. T.; Verma, K. K.; Stwalley, W. C. J. Chem. Phys. (in press).
9. McGillis, D. A.; Krause, L. Can. J. Phys. 1968, 46, 1051.
10. Bernstein, R. B. "Atom-Molecule Collision Theory"; Plenum Press: New York, 1979.

RECEIVED August 26, 1981.

18

Thermochemistry of the Dimer Lithium Hydride Molecule $Li_2H_2(g)$

C. H. WU and H. R. IHLE

Kernforschungsanlage Jülich GmbH, Institut für Chemie, D-5170 Jülich, West Germany

A stable molecule Li_2H_2 has been identified experimentally and its stability determined.

The gas species over solutions of hydrogen in liquid lithium were detected by mass spectrometric analysis of the saturated vapor effusing from a Knudsen cell. From the measurements of the gaseous equilibria

$$Li_2H_2(g) + Li_2(g) = 2Li_2H(g),$$
$$Li_2H_2(g) + Li(g) = LiH_2(g) + Li_2(g),$$
$$Li_2H_2(g) + Li_2(g) = Li_3(g) + LiH_2(g) \text{ and}$$
$$Li_2H_2(g) + LiH(g) = Li_2H(g) + LiH_2(g),$$

an atomization energy $D^o_0(Li_2H_2) = 164.3 \pm 10$ kcal/mol and a heat of dimerization $\Delta H^o_0 = -(52.6 \pm 10)$ kcal/mol was obtained.

Much theoretical work has been carried out on the lithium hydride molecule, which has become the workbench of the theoretical chemist (1). Browne (2), and Fraga and Ransil (3) have given the binding energy for the LiH^+ ion by ab initio calculation; Companion (4) has applied the diatomic-in-molecule theory to the Li_2H and LiH_2 molecules and predicted the stabilities of these molecules. We have intensively studied the Li-H system by means of Knudsen effusion mass spectrometry, and identified all predicted molecules and ions as cited above (5), and reported the thermochemical properties of these gaseous species (6, 7, 8).

The existence of a stable molecule Li_2H_2 has not been demonstrated experimentally prior to this investigation. Tyndall and Companion (9) have studied the stability of Li_2H_2 by application of diatomics-in-molecules theory and given following heats of reaction:

$$2LiH(^1\Sigma^+) \rightarrow Li_2H_2\ (^1B_{1g}) \quad \Delta H^o_0 = -28.4 \text{ kcal/mol}$$

0097-6156/82/0179-0265$05.00/0

$$Li_2(^1\Sigma_g^+) + H_2(^1\Sigma_g^+) \rightarrow Li_2H_2\ (^1A_1) \quad \Delta H_0^o = -26.7 \text{ kcal/mol}$$

Previously we have reported the mass spectrometric determination of the ionization potentials and dissociation energies of the molecules LiH(6), Li_2H(7), LiH_2(8) together with the binding energies of their ions. In continuation of our study on the system Li-H, the equilibrium constants of four gas reations involving this molecule were measured, from which its atomization energy has been deduced.

Experimental

The apparatus used for this work has been described in detail earlier(6). Briefly, it consists of a quadrupole mass spectrometer by which the effusate from a heated Mo-Knudsen cell containing solutions of hydrogen in lithium is analyzed, and a highly accurate gas delivery system permits the admission of e.g. 2×10^{-9} moles H_2 with an accuracy of ± 1%.

In the temperature range of 800 to 1000 K and at a concentration of about 1000 ppm hydrogen in lithium, the following gas equilibria have been measured:

$$Li_2H_2(g) + Li_2(g) = 2Li_2H(g) \quad (1)$$

$$Li_2H_2(g) + Li(g) = LiH_2(g) + Li_2(g) \quad (2)$$

$$Li_2H_2(g) + Li_2(g) = Li_3(g) + LiH_2(g) \quad (3)$$

$$Li_2H_2(g) + LiH(g) = Li_2H(g) + LiH_2(g) \quad (4)$$

Identification of the ions was accomplished from their mass-to-charge ratio, shutterability, appearance potential and isotope abundance.

Results and Discussion

When a dilute solution of hydrogen in lithium, with a concentration of 1000 ppm, is heated in a molybdenum cell in the temperature range of 830 K to 1000 K, the ions Li^+, Li_2^+, Li_3^+, LiH^+, LiH_2^+, Li_2H^+ and $Li_2H_2^+$ are observed and the appearance potentials for these ions were found to be 5.5, 5, 4.5, 8, 6.5, 5 and 6 eV. respectively by the linear extrapolation method, with an uncertainty of ± 0.5 eV. These agree with the known ionization potentials of the corresponding neutral gaseous species given in the literature(10, 11, 6, 7, 8); thus, the appearance potentials suggest, that the ions observed are produced by ionization of the corresponding neutrals. To minimize fragmentation an ionizing electron energy of 2.5 eV above the respective appearance potentials was used during the measurements.

A typical set of relative ion currents measured at 936 K is given by:

$Li^+ = 3.25 \times 10^7$, $Li_2^+ = 6.90 \times 10^5$, $Li_3^+ = 2.95 \times 10^2$, $LiH^+ = 2.60 \times 10^2$, $Li_2H^+ = 3.60 \times 10^2$, $LiH_2^+ = 1.95 \times 10$ and $Li_2H_2^+ = 2.4 \times 10$.

Equilibrium Measurements and Atomization Energy of $Li_2H_2(g)$

The measurements of the gaseous isomolecular exchange equilibria (1-4) were carried out at temperatures between 830 and 1000 K, and the third law enthalpies of reactions (1-4) were calculated from the relation

$$\Delta H^o_0 = -RT \ln K - T\Delta[(G^o_T - H^o_0)/T] \quad (5)$$

where K, R and $\Delta[(G^o_T - H^o_0)/T]$ represent the equilibrium constant, the gas constant, and the change in the free energy function for the corresponding reactions.

The equilibrium constants for the reactions were calculated from the measured ion currents, using the relations

$$K_{p1} = \frac{I(Li_2H^+)^2}{I(Li_2^+)I(Li_2H_2^+)} \quad (6)$$

$$K_{p2} = \frac{I(LiH_2^+)I(Li_2^+)}{I(Li^+)I(Li_2H_2^+)} \quad (7)$$

$$K_{p3} = \frac{I(Li_3^+)I(LiH_2^+)}{I(Li_2^+)I(Li_2H_2^+)} \quad (8)$$

$$\text{and } K_{p4} = \frac{I(Li_2H^+)I(LiH_2^+)}{I(LiH^+)I(Li_2H_2^+)} \quad (9)$$

where I is the relative ion intensity.

The experimental values for the ionization cross sections and multiplier gains, which are necessary for the evaluation of the equilibrium constants of gas reactions from measured ion intensities, are not available. Therefore an assumption, that the ionization cross sections and multiplier gains for isomolecular reactions can cancel each other out has been made.

By using the estimated molecular parameters given by Tyndall and Companion for the D_{2h} diamond-shaped and C_{2v} dimer of LiH, one obtains the free energy function for the $Li_2H_2(g)$ molecule by the satistical mechanical method and the resulting numerical values of the free energy functions at 800, 900 and 1000 K for D_{2h} dimer are 56.11, 57.74 and 59.25; and those for C_{2v} dimer are 54.63, 56.29 and 57.82 cal mol^{-1} respectively.

The free energy functions for Li(g) and Li_2(g) were taken from the JANAF Tables, those for Li_3(g), Li_2H(g) and LiH_2(g) were taken from our previous work (11, 7, 8).

The measured equilibrium constants as functions of temperature are shown in Figure 1 and the third law results of reactions (1), (2), (3) and (4) are given in Tables I, II, III, IV.

The third law values of reactions (1), (2), (3) and (4) were combined with the following atomization energies:

$D_0^o(Li_2)$=25.5 ± 1.5 kcal/mol(11) $D_0^o(Li_2H)$=89.7 ± 5.0 kcal/mol(7)

$D_0^o(Li_3)$=45.5 ± 4.0 kcal/mol(11) $D_0^o(LiH)$ =55.86 kcal/mol(12)

$D_0^o(LiH_2)$ = 127.0 ± 7.0 kcal/mol(8)

This yields the corresponding heats of atomization:

$D_0^o(Li_2H_2)$ = 162.0 ± 5.2 kcal/mol (D_{2h} Dimer)

$D_0^o(Li_2H_2)$ = 163.4 ± 5.2 kcal/mol (C_{2v} Dimer),

from reaction (1)

$D_0^o(Li_2H_2)$ = 164.5 ± 7.1 kcal/mol (D_{2h} Dimer)

$D_0^o(Li_2H_2)$ = 165.9 ± 7.1 kcal/mol (C_{2v} Dimer),

from reaction (2)

$D_0^o(Li_2H_2)$ = 165.9 ± 8.2 kcal/mol (D_{2h} Dimer)

$D_0^o(Li_2H_2)$ = 167.2 ± 8.2 kcal/mol (C_{2v} Dimer),

from reaction (3) and

$D_0^o(Li_2H_2)$ = 161.9 ± 8.5 kcal/mol (D_{2h} Dimer)

$D_0^o(Li_2H_2)$ = 163.2 ± 8.5 kcal/mol (C_{2v} Dimer),

from reaction (4).

The second law values of the heat of reaction for the reactions (1), (2), (3) and (4) are ΔH_T^o = 4.8 ± 3.0 kcal/mol, ΔH_T^o = 15.6 ± 2.2 kcal/mol, ΔH_T^o = 12.6 ± 11.4 kcal/mol and ΔH_T^o = 2.1 ± 1.5 kcal/mol respectively. These values agree well with the third law values within the limits of error. However, the third law values are much more reliable than those from the second law. Therefore only the third law values have been used for the evaluation of the atomization energy of Li_2H_2.

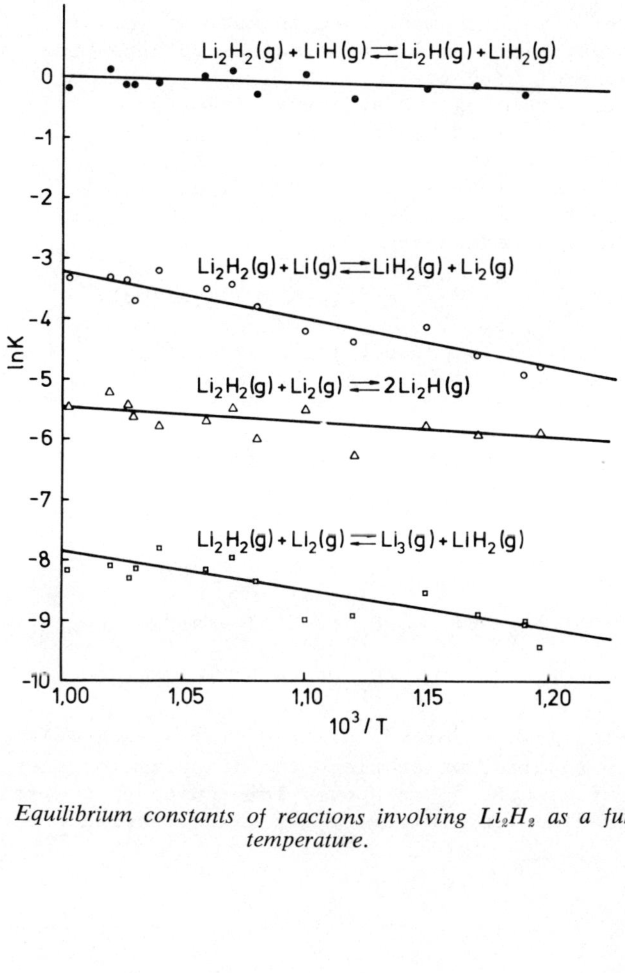

Figure 1. Equilibrium constants of reactions involving Li_2H_2 as a function of temperature.

It is seen that the agreement between the atomization energies derived from four independent isomolecular exchange reactions is excellent. This suggests that the measurements were conducted under equilibrium conditions in the Knudsen cell, and that there is no serious error in the free energy function for Li_2H_2(g) and in the previously determined atomization energies of Li_2H(g)(7), LiH_2(g)(8) and Li_3(g)(11). The average value of the atomization energy for Li_2H_2(g) is $D^o_0(Li_2H_2)$ = 164.3 kcal/mol. By using the same error treatment as mentioned in earlier works(7, 11), an estimated uncertainty of ±10 kcal/mol is obtained for the atomization energy of Li_2H_2(g).

Finally, combining the atomization energy of Li_2H_2(g), $\Delta H^o_0(Li_2H_2)$ = 164.3 ± 10 kcal/mol and LiH(g), ΔH^o_0(LiH) = 55.86 kcal/mol, with the heat of formation of Li(g), ΔH^o_0 = 38.5 ± 0.1 kcal/mol, and the dissociation energy $D^o_0(H_2)$ = 103.25 kcal/mol, yields the heat of formation $\Delta H^o_0(Li_2H_2)$ = 16.0 ± 10 kcal/mol, and the heat of dimerization

$$2LiH \rightarrow Li_2H_2 \qquad \Delta H^o_0 = -(52.6 \pm 10) \text{ kcal/mol.}$$

Kollman, Bender and Tothenberg(13) have predicted a heat of dimerization ΔH^o_0 = -47.2 kcal/mol by electronic structure calculation.

Milne and Cubicciotti(14) have applied ionic models to calculate the heats of dimerization for gaseous alkali halide dimer molecules, and the results are as follows:

$$2LiF \rightarrow Li_2F_2 \qquad \Delta H^o_0 = -57.6 \text{ kcal/mol}$$

$$2LiCl \rightarrow Li_2Cl_2 \qquad \Delta H^o_0 = -54.7 \text{ kcal/mol}$$

$$2LiBr \rightarrow Li_2Br_2 \qquad \Delta H^o_0 = -52.3 \text{ kcal/mol}$$

$$2LiI \rightarrow Li_2I_2 \qquad \Delta H^o_0 = -48.9 \text{ kcal/mol}$$

The calculated stability of Li_2H_2(13) agrees well with our experimental results, which place the dimerization energy of LiH close to that of LiBr given above. This supports the statement that the LiH dimer is similar to the alkali halide(9).

Table I: Third law enthalpies for the reaction

$$Li_2H_2(g) + Li_2(g) = 2\ Li_2H(g)$$

T (K)	$K_{p1}x10^3$	$-\Delta[(G^o_T-H^o_0)/T]$ (cal mol^{-1} K^{-1}) D_{2h} Dimer	C_{2v} Dimer	ΔH^o_0 (kcal mol^{-1}) D_{2h} Dimer	C_{2v} Dimer
836	2.70	-2.10	-0.63	8.07	9.29
869	2.90	-2.15	-0.70	8.22	9.48
890	1.68	-2.18	-0.72	9.35	10.66
850	2.52	-2.13	-0.66	8.29	9.54
908	4.07	-2.23	-0.78	7.91	9.22
924	2.23	-2.25	-0.81	9.13	10.46
936	4.03	-2.26	-0.82	8.14	9.48
946	3.37	-2.28	-0.84	8.54	9.91
962	2.99	-2.31	-0.87	8.89	10.27
967	3.45	-2.31	-0.88	8.66	10.04
840	5.62	-2.10	-0.63	6.88	8.12
855	1.41	-2.15	-0.68	5.40	6.67
968	4.53	-2.32	-0.88	8.13	9.53
978	5.30	-2.34	-0.90	7.89	9.30
998	3.91	-2.39	-0.93	8.61	10.06
			Av.	8.14±0.96	9.47±0.98

Table II: Third law enthalpies for the reaction

$$Li_2H_2(g) + Li(g) = LiH_2(g) + Li_2(g)$$

T (K)	$K_{p2}x10^2$	$-\Delta[G^o_T-H^o_0)/T]$ (cal mol^{-1} K^{-1}) D_{2h} Dimer	C_{2v} Dimer	ΔH^o_0 (kcal mol^{-1}) D_{2h} Dimer	C_{2v} Dimer
836	0.835	5.60	7.07	12.63	13.86
869	1.61	5.61	7.06	12.01	13.27
890	1.23	5.59	7.05	12.75	14.05
850	1.05	5.59	7.06	12.45	13.69
908	1.46	5.61	7.06	12.72	14.04
924	2.20	5.59	7.03	12.17	13.50
936	3.35	5.59	7.03	11.55	12.89
946	3.00	5.59	7.03	11.88	13.24
962	4.20	5.57	7.01	11.42	12.80
967	2.43	5.57	7.00	12.53	13.91
840	0.844	5.60	7.07	12.67	13.91
855	3.41	5.61	7.08	10.54	11.80
968	3.55	5.57	7.01	11.81	13.21
978	3.70	5.57	7.01	11.85	13.26
998	3.53	5.53	6.99	12.15	13.61
			Av.	12.08±0.60	13.40±0.60

Table III: Third law enthalpies for the reaction

$$Li_2H_2(g) + Li_2(g) = Li_3(g) + LiH_2(g)$$

T (K)	K_{p3}x10^4	$-\Delta[(G^o_T-H^o_0)/T]$ (cal mol^{-1} K^{-1})		ΔH^o_0 (kcal mol^{-1})	
		D_{2h} Dimer	C_{2v} Dimer	D_{2h} Dimer	C_{2v} Dimer
836	0.756	8.32	9.99	22.72	23.95
869	2.10	8.35	9.80	21.88	23.14
890	1.31	8.37	9.83	23.26	24.56
850	1.37	8.33	9.80	22.11	23.35
908	1.22	8.37	9.82	23.86	25.17
924	2.30	8.38	9.82	23.12	24.45
936	3.47	8.38	9.82	22.66	24.01
946	2.84	8.39	9.83	23.29	24.65
967	2.75	8.40	9.83	23.88	25.26
940	1.26	8.33	9.80	21.98	23.22
855	8.79	8.33	9.80	19.08	20.33
968	2.22	8.39	9.83	24.30	25.70
978	2.99	8.40	9.84	23.99	25.39
998	2.64	8.40	9.84	24.77	26.16
			Av.	22.92±1.41	24.24±1.46

Table IV: Third law enthalpies for the reaction

$$Li_2H_2(g) + LiH(g) = Li_2H(g) + LiH_2(g)$$

T (K)	K_{p4}	$-\Delta[(G^o_T-H^o_0)/T]$ (cal mol^{-1} K^{-1})		ΔH^o_0 (kcal mol^{-1})	
		D_{2h} Dimer	C_{2v} Dimer	D_{2h} Dimer	C_{2v} Dimer
836	1.16	1.09	2.56	0.66	1.89
869	0.836	1.08	2.53	1.24	2.51
890	0.667	1.06	2.52	1.66	2.95
850	0.875	1.08	2.55	1.14	2.39
908	1.09	1.05	2.50	0.80	2.11
924	0.687	1.03	2.47	1.64	2.97
936	1.12	1.02	2.46	0.74	2.09
946	0.993	1.01	2.45	0.97	2.33
962	1.35	0.99	2.43	0.39	1.76
967	0.865	0.99	2.42	1.25	2.62
840	0.730	1.10	2.57	2.45	2.68
855	1.08	1.08	2.55	0.79	2.05
968	0.872	0.99	2.43	1.23	2.62
978	1.11	0.98	2.42	0.76	2.16
998	0.822	0.94	2.40	1.33	2.78
			Av.	1.07±0.37	2.39±0.37

Acknowledgement

The authors wish to thank Mr. F. Fröschen for technical help.

Literature Cited

1. Stawlley, W. C.; Way, K. R. J. Chem. Phys. 1974, 60, 3611.
2. Browne, J. C. J. Chem. Phys. 1964, 41, 3495.
3. Fraga, S.; Ransil, B. J. J. Chem. Phys. 1961, 35, 669.
4. Companion, A. L. J. Chem. Phys. 1968, 48, 1186.
5. Ihle, H. R.; Wu, C. H. Adv. Mass Spectrometry 1978, 7A, 636.
6. Ihle, H. R.; Wu, C. H. J. Chem. Phys. 1975, 63, 1605.
7. Wu, C. H.; Ihle, H. R. J. Chem. Phys. 1977, 66, 4356.
8. Wu, C. H. J. Chem. Phys. 1979, 71(2), 783-7.
9. Tyndall, J. R.; Companion, A. L. J. Chem. Phys. 1970, 52, 2036.
10. Franklin, J. L.; Dillard, J. G.; Rosenstock, H. M.; Herron, J. T.; Draxl, K.; Field, F. H. Natl. Stand. Ref. Data 1969, Ser. 26.
11. Wu, C. H. J. Chem. Phys. 1976, 65, 3181.
12. Velasco, R. Can. J. Phys. 1957, 35, 1204.
13. Kollman, P.; Bender, C. F.; Rothenberg, S. J. Amer. Chem. Soc. 1972, 94, 8016.
14. Milne, T. A.; Cubicciotti, D. J. Chem. Phys. 1958, 29, 846.

RECEIVED September 25, 1981.

19

Photoionization and Photoelectron Spectroscopy of Alkali Halide Monomers, Dimers, and Trimers

J. BERKOWITZ, C. H. BATSON, and G. L. GOODMAN

Argonne National Laboratory, Argonne, IL 60439

The ultraviolet photoelectron spectra of diatomic alkali halide molecules are reviewed and interpreted. Data for lithium halide dimers, Li_2X_2, are presented and it is shown that the dimers have significantly larger ionization thresholds than the corresponding monomers. Some historical controversies regarding the presence of dimers and their ionization energies are clarified. Photoionization mass spectrometry is used to determine the adiabatic ionization potential of lithium chloride trimer, in order to probe the trend of I.P. with cluster size. The predictions of Hartree-Fock, Xα and ionic model calculations on this point are presented. Finally, the very weak stability of $M_2X_2^+$ and $M_3X_3^+$ and the high stability of M_2X^+ and $M_3X_2^+$ are discussed, and conjectures are made regarding the geometric structure of these entities.

It is now a well-recognized fact that the saturated vapors of alkali halides contain associated species (dimers, trimers, etc.) as well as the diatomic molecules. The first evidence of this behavior was an electron-impact mass spectrometric study by Ionov (1). He reported the observation of K_2I^+ and Na_2I^+, which he attributed to the process

$$M_2I_2 + \bar{e} \rightarrow M_2I^+ + I + 2\bar{e}, \text{ where } M = K,Na$$

He also indicated that similar data were obtained for K_2Cl^+, Na_2Cl^+ and Li_2Cl^+. Subsequently, Friedman (2) reported a similar, but more extensive study of lithium iodide vapor, which indicated that ∿30% of the ion current was from a dimer species.

A brief debate occurred in the middle 1950's, when Miller and Kusch (3) interpreted the velocity profile of alkali halide molecules effusing from an oven in terms of heavier species as

0097-6156/82/0179-0275$06.25/0

as well as diatomic molecules, while Klemperer and Rice (4, 5) were unable to detect the infrared spectra of the presumed heavier species and concluded that they contributed less than 15% to the vapor composition. In 1958, Berkowitz and Chupka (6) examined 13 alkali halides with a more sensitive mass spectrometer, and ruled out collisional processes (such as ion-molecule reactions) as possible mechanisms for the production of heavier ions. They were also able to observe ionic species attributable to trimers in all but one case, and to tetramers in four cases.

Their results are summarized in Table I. It can be seen that the degree of association is strongest for the lithium halides and diminishes monotonically toward the cesium halides.

The precise relative abundance of neutral monomers, dimers, trimers, etc. is difficult to deduce from mass spectrometric measurements, since the relevant ion intensities are influenced by ionization cross sections and fragmentation processes. Several schemes have been applied to overcome these problems. However, sublimation energies can be determined by measuring the temperature dependence of ion intensities, largely unaffected by the aforementioned problems. These measurements enable one to compute heats of formation, heats of dimerization, etc. for the neutral species. In addition, matrix-isolation techniques have yielded infrared active vibrational frequencies of dimers (and possibly one trimer) and geometrical structures of monomers and some dimers have been obtained from a combination of microwave spectra and electron diffraction.

The acquisition of such an extensive set of properties for this class of molecules has prompted a number of authors to explore the use of ionic models for systematically characterizing these properties. We have recently reviewed some of the successes and failures of these models,(7) including references to the various authors.

The newer technologies of photoelectron spectroscopy and photoionization mass spectrometry, when adapted to high temperature studies, have re-opened this topic for further investigation. In particular, photoelectron spectroscopy has provided information about the electronic structure of these species, a property largely ignored in the earlier investigations.

Photoelectron Spectroscopy—Its Application to the Alkali Halide Vapors

Photoelectron spectroscopy is the study of the electron kinetic energy spectrum produced upon photoionization of molecules with monochromatic radiation. We may write the basic photoelectric equation as

$$h\nu + AB \rightarrow AB^+ + \bar{e}$$

If AB^+ is produced in its ground state (electronic, vibrational

TABLE I

Relative Ion Intensities of Alkali Halide Vapors, Produced by Electron Impact[a]

	M^+	MX^+	M_2X^+	$M_3X_2{}^+$	$M_4X_3{}^+$
LiF	51.7	11.5	100.	7.2	0.038
LiCl	18.2	21.8	100.	4.3	0.14
LiBr	21.8	22.0	100.	3.9	0.11
NaF	100.	2.60	29.8	0.58	0.0058
NaCl	100.	59.0	71.3	0.59	-
NaBr	100.	70.7	64.8	1.01	-
NaI	85.6	100.	53.3	0.33	-
KF	100.	0.18	17.5	0.12	-
KCl	100.	16.8	21.4	0.058	-
KBr	100.	28.6	22.3	0.086	-
KI	100.	55.7	16.7	0.040	-
RbCl	100.	7.03	13.6	0.0061	-
CsCl	100.	1.16	7.15	<0.007	-

[a]Total pressures in all instances ca. 10^{-2} torr, using 75 eV electron energies.

and rotational) and the neutral molecule is in its ground state, then the kinetic energy of the electron will be $h\nu -$ I.P., where I.P. is the adiabatic first ionization potential of the molecule AB. In general, AB^+ can be found in a variety of excited electronic, vibrational and rotational states, and the kinetic energy of the electrons will be correspondingly reduced. Therefore, a measurement of the kinetic energy spectrum of the electrons generated can be interpreted in terms of the energy states of the molecular ion produced.

The monochromatic radiation used most frequently in studies of valence-shell photoelectron spectroscopy is the resonance line of He, at 584 Å ≡ 21.2 eV. Obviously, only energy states up to this energy are accessible with He I radiation. Some studies have been performed with He II radiation at 304 Å ≡ 40.8 eV, but they shall not concern us here.

Our earliest alkali halide studies were performed with cesium halides, since the prevailing evidence (Table I) indicated that the vapor would be predominantly monomeric, and hence simplest to interpret. The apparatus employed for these studies (Figure 1) consisted of a cylindrical mirror electron energy analyzer, a non-inductively wound oven for generating the cesium halide vapor, and a helium resonance lamp. The spectra we obtained (8) for the cesium halides are displayed in Figure 2. They reveal a clearly resolved doublet for CsI, a partially resolved doublet for CsBr, and broad single peaks for CsCl and CsF. We shall briefly reproduce here the arguments we used to interpret these spectra.

Let us focus initially on the ionization thresholds. Figure 3 is a representative potential energy diagram for the ground states of neutral CsX and the molecular ion CsX^+. On this figure, we have symbolized D_{oi} as the dissociation energy of the neutral molecule into ion pairs, E.A. (X) as the electron affinity of the halogen, I.P. (CsX) as the adiabatic first ionization potential and D_o (CsX^+) as the dissociation energy of the molecular ion. From Figure 3, we can note the identity

$$\text{I.P. (CsX)} + D_o\ (CsX^+) = D_{oi}\ (CsX) + \text{E.A. (X)}$$

Asymptotically, CsX^+ dissociates into $Cs^+ + X^o$, and the bond is essentially due to charge-induced dipole interaction $(\alpha e^2/2r^4)$ which leads to D_o (CsX^+) in the neighborhood of 0.1 — 0.2 eV. The ionic models are fairly good for calculating D_{oi} (CsX). A typical calculation (9) yields 5.55 eV, 4.77 eV, 4.55 eV and 4.29 eV for CsF, CsCl, CsBr and CsI, respectively. The values of E.A. (X) are (10) 3.399 eV, 3.615 eV, 3.364 eV and 3.061 eV for F, Cl, Br and I, in that order. With calculated values for D_o (CsX^+) we compute 8.85 eV, 8.18 eV, 7.71 eV and 7.15 eV for the adiabatic ionization potentials of CsF, CsCl, CsBr and CsI. From the thresholds in our experimental data, we deduce 8.80, 7.8_4, 7.4_6 and 7.1_0 eV for these values, in the same sequence.

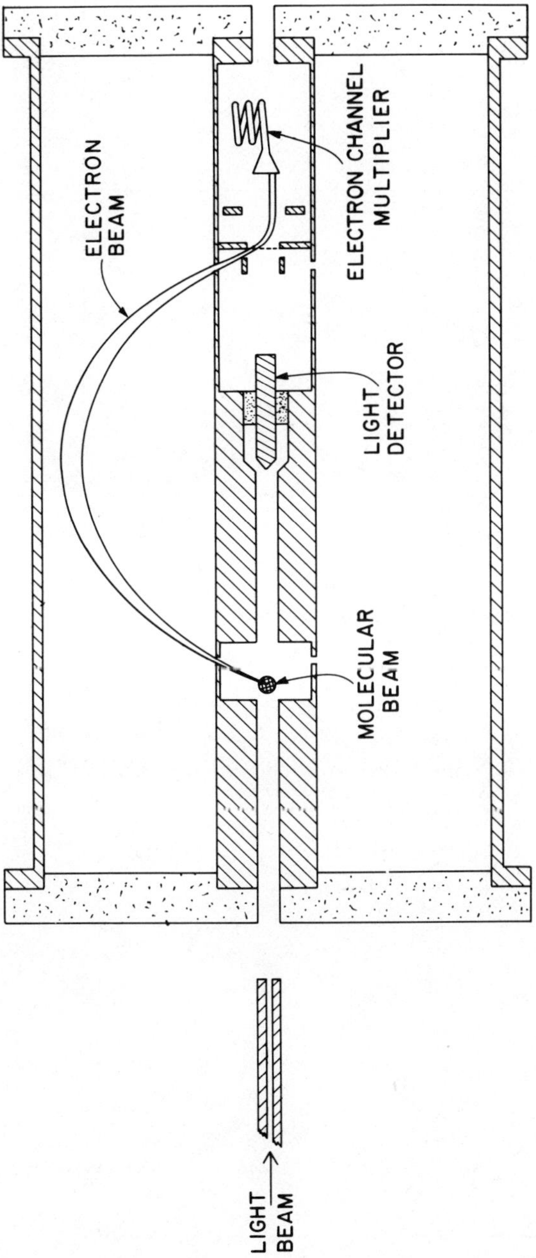

Journal of Chemical Physics

Figure 1. Cylindrical mirror electron energy analyzer (31).

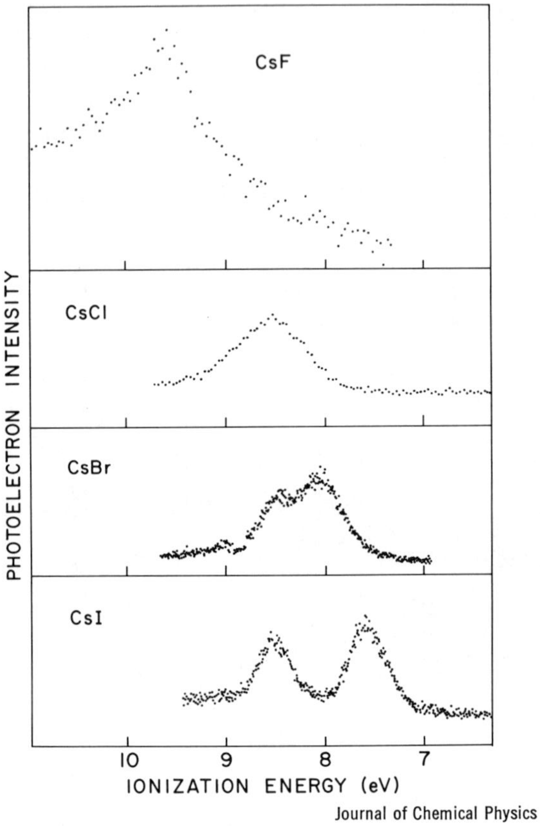

Figure 2. 584-Å photoelectron spectra of CsF, CsCl, CsBr, and CsI (8).

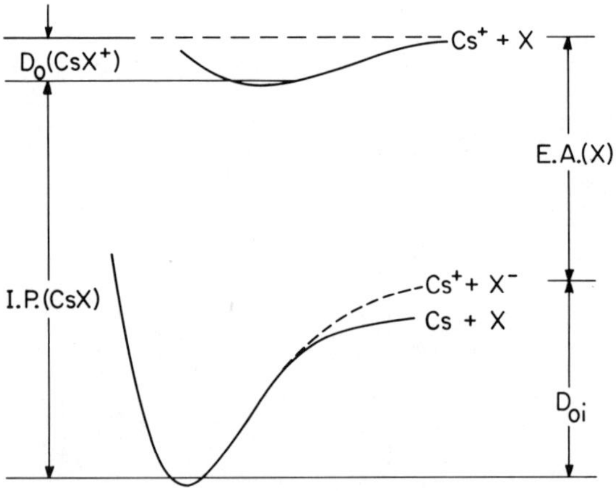

Figure 3. Schematic potential energy curves of CsX and CsX⁺ (19).

The agreement between calculated and experimental adiabatic ionization potentials is seen to be quite satisfactory. (The data of Potts, et al. (11) on CsCl, CsBr and CsI are in still better agreement with the calculations.) We can therefore apply the same simple scheme to the other alkali halides with some confidence.

Let us now try to understand the gross features of the spectra in Figure 2. A common approach to the interpretation of photoelectron spectra is a comparison of calculated orbital energies with observed spectra. According to Koopmans' theorem, (12) the vertical ionization potentials are the negative of the corresponding orbital energies, to first order. A rather extensive set of *ab initio* Hartree-Fock calculations on a number of alkali halides have been performed by Matcha. (13a—13f) These calculations demonstrated, inter alia, that a large degree of charge transfer occurred in the valence shell, from alkali s orbital to halogen p orbital, as might have been expected. The molecule in its ground state is then well-described as an ionic $M^+—X^-$ structure, as indeed ionic model calculations had successfully shown. The uppermost occupied orbitals, according to the *ab initio* calculations, are π and σ, both primarily centered on the halogen atoms and close in energy ($\lesssim$0.5 eV for all but the lithium halides). The next deeper molecular orbital will in most cases be metal p-like. For cesium, this level should be within the energy range accessible by 21.2 eV radiation, but for the lighter elements it would be too strongly bound.

If we ignore for the moment this deeper ionization, the 584 Å photoelectron spectrum of the alkali halides should be very simple — π and σ peaks, separated by a fraction of an electron volt, with the additional proviso that the π orbital is split due to spin-orbit interaction. The detailed structure of the photoelectron spectrum, and its interpretation, have however been the subject of some controversy. Experimentally, the dispute concerns the presence of one, two or three peaks in the valence region. Those who advocate three peaks (14) interpret them as $^2\Pi_{1/2}$, $^2\Pi_{3/2}$ and $^2\Pi^+_{1/2}$. Our approach (8) has been to solve the 2 × 2 secular equation resulting from inclusion of $\pi - \sigma$ and spin-orbit effects, and thereby to demonstrate that in most cases two of the three anticipated peaks will be inseparable, and in some cases all three will merge into a single, broad peak. The π —σ splittings for nine of the twenty alkali halides have been obtained directly from the *ab initio* calculations (13a—13f) and the others have been established from the trends observed. Table II summarizes these values. The spin-orbit parameters are dominated by the appropriate atomic halogen spin-orbit splitting for the valence p orbital. Slight departures from this spin-orbit constant can be estimated from the magnitudes of the contributions of other orbitals in the LCAO molecular wave function, and their spin-orbit parameters. Table III lists the spin-orbit coupling constants of the free halogen atoms, together

TABLE II

Separation of the π and σ Orbitals Derived from the Uppermost Occupied (Halogen p-like) Orbitals (electron volts)[a]

	F	Cl	Br	I
Li	0.591	0.754	0.770	(0.8)
Na	0.362	0.453	0.472	(0.5)
K	0.215	0.248	(0.27)	(0.3)
Rb	0.026	(0.1)	(0.15)	(0.19)
Cs	(0)	(0.03)	(0.05)	(0.1 —0.14)

[a]The π is least bound in all cases. Values have been obtained assuming Koopmans' theorem. The quantities in parentheses are estimated from the trends observed in the calculated values.

Source: Ref. 10.

TABLE III

Diagonal Spin-Orbit-Coupling Constants of the Uppermost Occupied π Orbitals of the Alkali Halides (electron volts)[a]

	F	Cl	Br	I
Atom	0.033	0.073	0.302	0.628
Li	0.028	0.063	0.263	0.530
Na	0.028	0.063	0.263	0.530
K	0.029	0.063	0.264	0.531
Rb	0.037	(0.068)	(0.273)	(0.546)
Cs	(0.043)	(0.073)	(0.298)	(0.628)

[a]The quantities in parentheses are estimated from the trends observed in the calculated values.

Source: Ref. 10.

with their modified values in the particular alkali halides. We shall not repeat the numerical details of the calculation, since they can be found elsewhere. (8, 15) However, a pictorial description is helpful. Taking as an example the case of CsBr, we begin with the estimated $\pi - \sigma$ separation and spin-orbit splitting, taken independently. (See Figure 4a.) Upon inclusion of the $\Omega - \Omega$ interaction, the two $\Omega = 1/2$ levels repel each other, leading to the final diagram (Figure 4b). In Figure 4b, the $^2\Pi_{3/2}$ and $^2\Sigma^+_{1/2}$ states are separated by only 0.032 eV, while $^2\Pi_{1/2}$ is separated from the mean of the other states by ∿0.45 eV. Both Potts, et al. (11) and Berkowitz, et al. (18) display two partially resolved peaks in their CsBr spectrum, the former reporting a splitting of 0.41 eV, the latter between 0.4 — 0.45 eV. In both experiments, the lower energy peak is more intense; the calculation indicates that it has two components, the higher energy peak only one. The inherent width of the peaks prevents separation of the two lower components for CsBr, although calculations on the other alkali halides (see Table IV) predict that three peaks should be observable in some cases.

Taking into account the peak widths, the corresponding calculation for CsI predicts a splitting into two well-resolved peaks, with a separation of ∿0.95 eV, and for CsCl and CsF, a single peak, in good agreement with the data of Potts, et al. (11) and Berkowitz, et al. (8) In 1973 we predicted the splitting patterns of all of the alkali halides, although our measurements at that time were restricted to the cesium halides. The splittings in most of the other alkali halides have been measured since. Table IV summarizes the calculated and experimental values. The agreement is seen to be very good. In particular, the calculations clearly show when one, two or three peaks should be observable.

The aforementioned discussion would appear to describe adequately the photoelectron spectra resulting from ionization of the uppermost occupied orbitals composed predominantly of the outer p halogen orbitals. The pattern is a consequence of the interplay between spin-orbit splitting and $\pi - \sigma$ splitting. The spin-orbit splitting is largely an atomic feature, whereas the $\pi - \sigma$ splitting is a measure of the departure from pure ionic bonding. It is largest for the lithium halides, which should therefore display some covalent character. However, Goodman et al. (14) and Timoshenko and Akopyan (16) present evidence for an additional peak to higher energy in almost every instance. Both of these groups identify the higher energy peak as the "missing" $^2\Sigma^+$ peak, Timoshenko and Akopyan go so far as to calculate the $^2\Pi_{3/2} - ^2\Sigma_{1/2}$ separation, and report fair agreement with experiment. We have examined this calculation in detail elsewhere, (15) and we believe that it is in error.

What, then, is the origin of the high energy peaks observed by both Goodman, et al. and by Timoshenko and Akopyan? We believed at the time (17) that the additional high energy peak

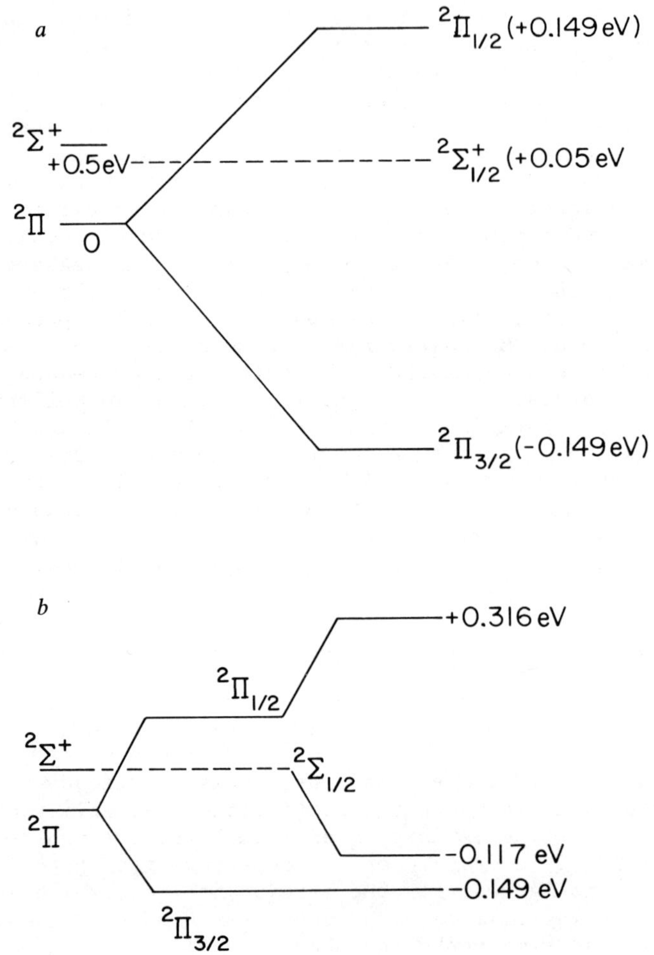

Journal of Chemical Physics

Figure 4. Energy splitting diagram for valence levels in $CsBr^+$ Key: a, Π–Σ and spin-orbit splittings taken separately; and b, final splitting, allowing for Ω–Ω interaction. (18)

TABLE IV

Comparison of Calculated[a] and Experimental[b] Splitting Patterns of Alkali Halide Monomers (in eV)

(Splittings are referenced in each instance to the $^2\Pi_{3/2}$ state)

		F		Cl		Br		I	
		$^2\Pi_{1/2}$	$^2\Sigma_{1/2}$	$^2\Pi_{1/2}$	$^2\Sigma_{1/2}$	$^2\Pi_{1/2}$	$^2\Sigma_{1/2}$	$^2\Pi_{1/2}$	$^2\Sigma_{1/2}$
Li	calcd.	0.027	0.606	0.060	0.788	0.213	0.951	0.337	1.238
	exptl.		(0.44)	0.06	0.74_5	0.14	1.00	0.27_5	1.27_5
Na	calcd.	0.027	0.377	0.058	0.489	0.183	0.684	0.254	1.041
	exptl.			t	0.46	t	0.65	0.22	1.18
K	calcd.	0.026	0.232	0.054	0.289	0.136	0.530	0.173	0.924
	exptl.				u	t	0.48	t	0.98
Rb	calcd.	0.012	0.070	0.043	0.160	0.087	0.472	0.116	0.892
	exptl.				u	t	0.45	t	0.97
Cs	calcd.	0.00	0.065	0.037	0.121	0.026	0.476	0.076	0.989
	exptl.		u		u	t	0.4 —0.45	t	0.95

[a] Calculated values are taken from ref. 8 (1973).

[b] Experimental values: lithium halides from ref. 18 (1979); sodium, potassium and rubidium halides from ref. 11 (1974); cesium halides from ref. 8 (1973).

[u] unresolved, only one peak observed.

[t] resolved into two, but not three components.

was due to dimer, an M_2X_2 species. Several years later we decided to re-examine this question with an improved apparatus. The new system (18) employed a hemispherical electron energy analyzer, rather than the cylindrical mirror, but that need not concern us. The oven assembly had provision for heating an upper and lower section (see Figure 5). This enabled us to alter the equilibrium concentrations of monomer and dimer. When the upper heater was off, the vapor had its maximum dimer composition. When it was on, the equilibrium could be shifted strongly in favor of monomer. In contrast to our early work with the cesium halides, where we sought to minimize the dimer contribution in order to simplify the spectrum, we now wanted samples which would maximize the dimer concentration. From the data given in Table I, we knew these to be the lithium halides.

In each of the following four figures (Figures 6—9), which pertain to LiF, LiCl, LiBr and LiI respectively, the uppermost curve (a) was obtained with the upper heater off, and the next lower curve (b) with the upper heater on. The data were obtained digitally, on a multichannel analyzer, and could be subtracted, point-by-point, on a computer. In the subtraction, it was necessary to choose a feature which was recognizable, common to both, and identifiable as a dimer peak. Such a feature should experience a reduction in relative intensity between spectrum (a) and spectrum (b). The two spectra would then be normalized at this peak, and the subtraction performed. In Figures 6—9, curve (c) displays this subtracted spectrum, which should be that of pure monomer. The splitting patterns from these spectra were already cited in Table IV, and were seen to be in very good accord with the predicted monomer splitting patterns. We were now in a position to determine the photoelectron spectra of "pure" dimers by subtracting the "pure" monomer spectrum (c) from the experimental spectrum (a) containing the largest dimer contribution. Again, a normalization was required, this time focussing on a feature that was clearly due to monomer. In Figures 6—8, curve (d) is the result of this subtraction, and represents the "pure" dimer spectrum. In Figure 9, we performed this subtraction in two stages, first with 50% of the normalized peak intensity subtracted (d) and then with 100% of the normalized peak intensity subtracted (e), in order to follow the relative intensities of the presumed dimer peaks, which should maintain the same relative intensities. Hence, in Figure 9, curve (e) represents the "pure" dimer spectrum.

Five peaks are readily apparent in Figure 9 (e). If the dimer valence region is predominantly a linear combination of two halogen outer p orbitals (12 electrons), we can expect six non-degenerate molecular orbitals. Presumably two are overlapped in the spectrum of Li_2I_2, whereas more are overlapped in the more compressed valence bands of Li_2F_2 (Figure 6a), Li_2Cl_2 (Figure 7d) and Li_2Br_2 (Figure 8d).

An obvious conclusion can be drawn by comparing curves (c)

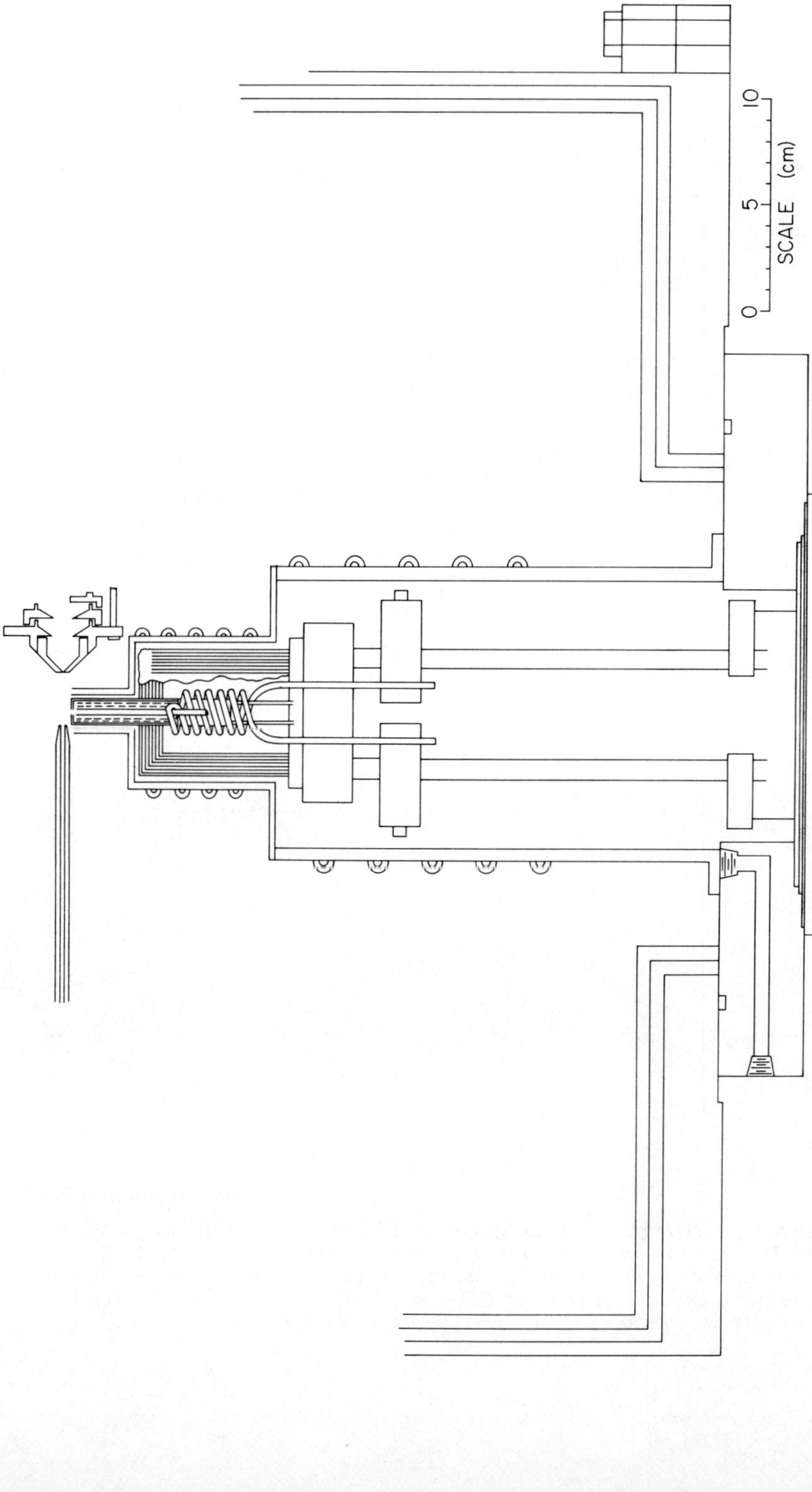

Journal of Chemical Physics

Figure 5. Juxtaposition of oven system, light source and electron energy analyzer. The energy analyzer is shown rotated by 90°, for clarity (18).

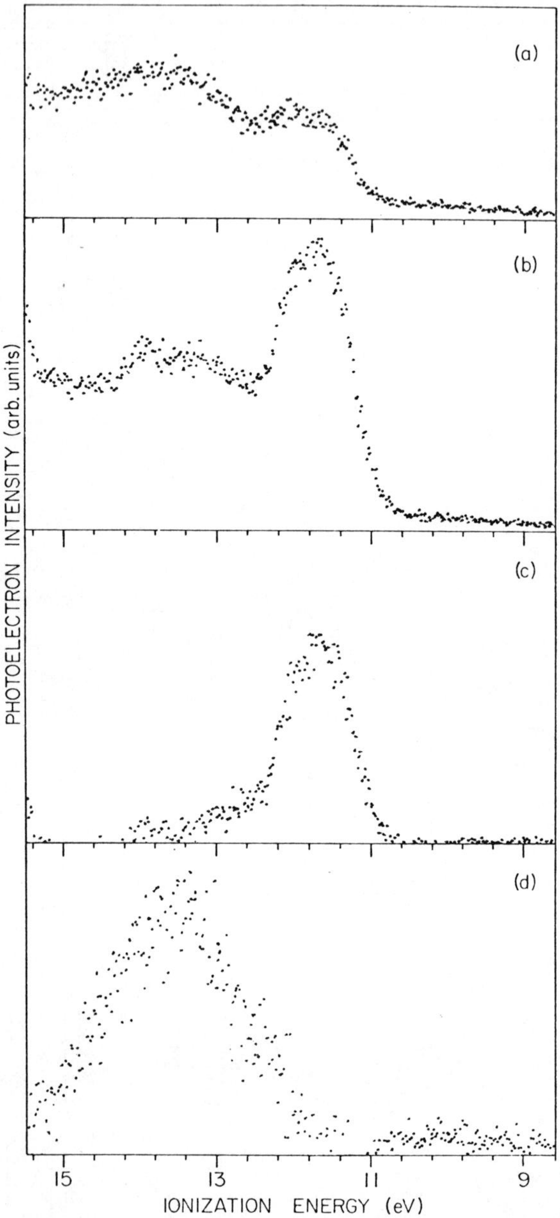

Journal of Chemical Physics

Figure 6. a: He(I) photoelectron spectrum of LiF without superheating, 1400 K. b: He(I) photoelectron spectrum of LiF with superheating, 1300 K. c: "Pure" monomer LiF spectrum, obtained by subtracting a from b, after background subtraction and normalization at dimer band. d: "Pure" dimer (Li_2F_2) spectrum, obtained by subtracting c from a, after normalization at monomer band (18).

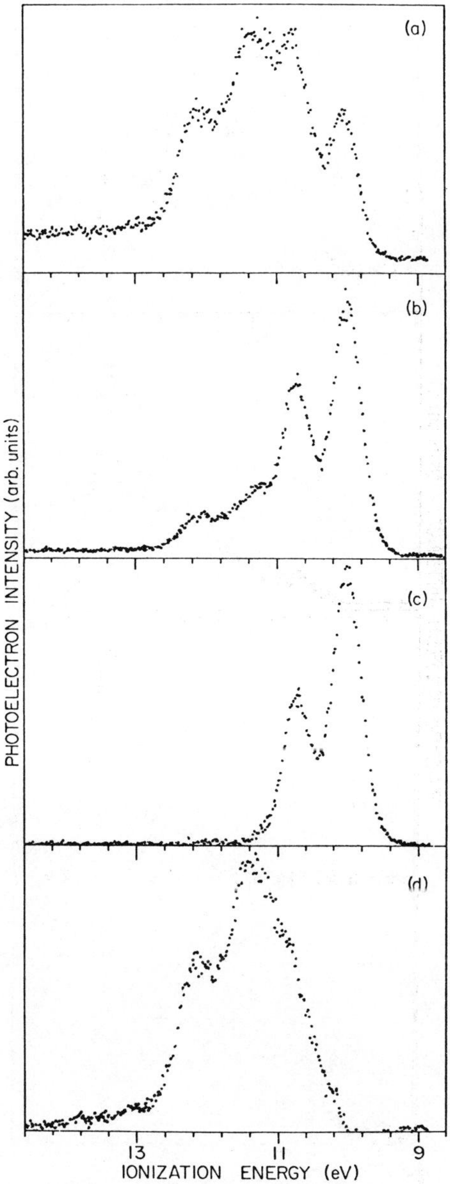

Journal of Chemical Physics

*Figure 7. a: He(I) photoelectron spectrum of LiCl without superheating, 1000 K. b: He(I) photoelectron spectrum of LiCl with superheating, 1000 K. c: Pure monomer LiCl spectrum, obtained by subtracting a from b, after background subtraction and normalization at dimer band. d: Pure dimer (Li_2Cl_2) spectrum, obtained by subtracting c from a, after normalization at monomer band (*18*).*

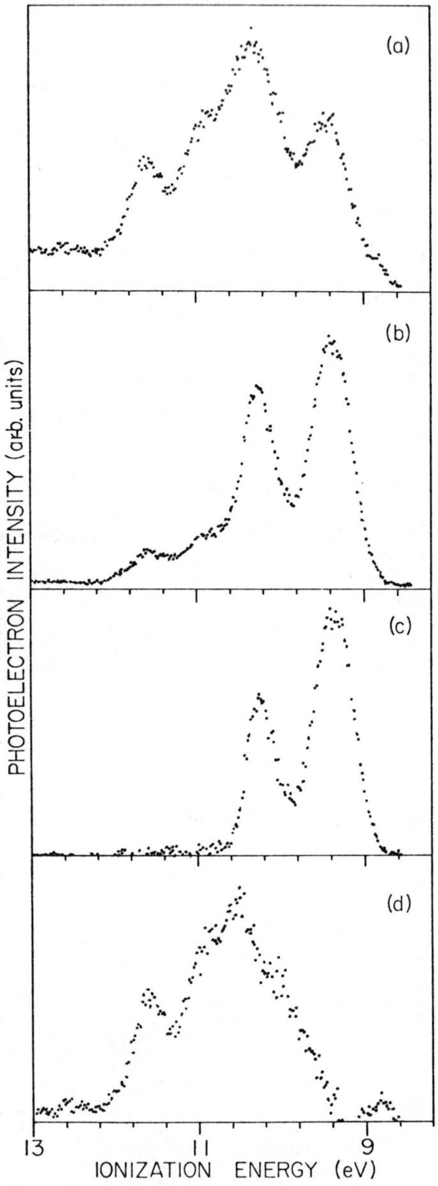

Journal of Chemical Physics

Figure 8. a: He(I) photoelectron spectrum of LiBr without superheating, 850 K. b: He(I) photoelectron spectrum of LiBr with superheating, 850 K. c: Pure monomer LiBr spectrum, obtained by subtracting a from b, after background subtraction and normalization at dimer band. d: Pure dimer (Li_2Br_2) spectrum, obtained by subtracting c from a, after normalization at monomer band (18).

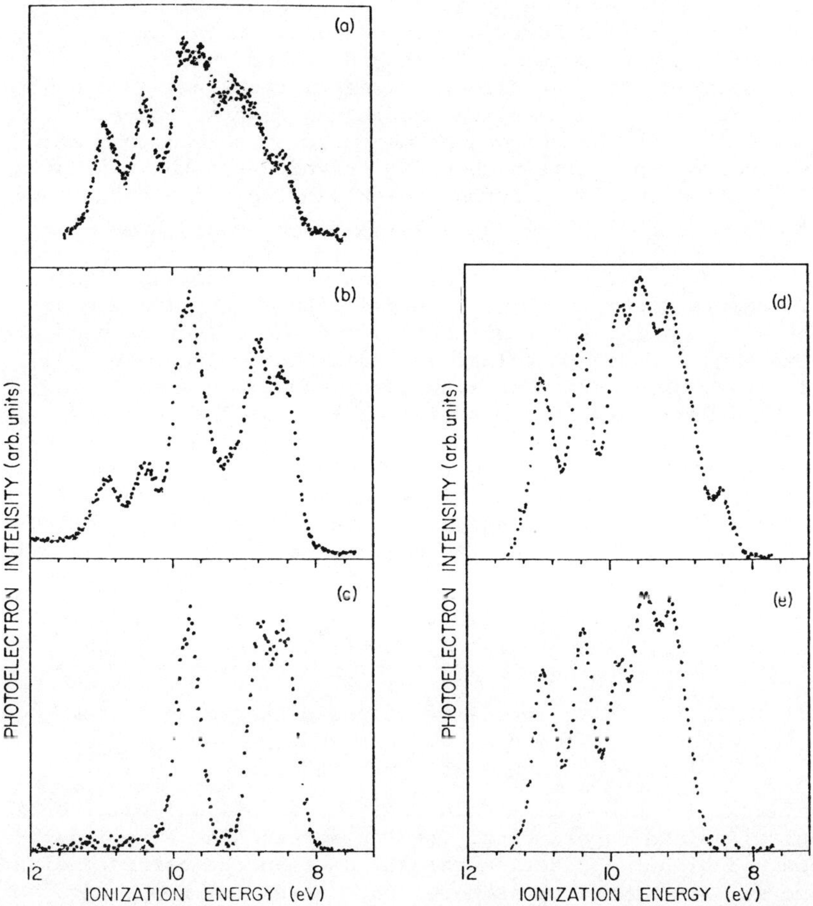

Journal of Chemical Physics

Figure 9. a: He(I) photoelectron spectrum of LiI without superheating, 800 K. b: He(I) photoelectron spectrum of LiI with superheating, 800 K. c: Pure monomer LiI spectrum, obtained by subtracting a from b, after background subtraction and normalization at dimer band. d: Spectrum with 50% of the monomer subtracted, to demonstrate the effect of reduction on the various peaks. e: Pure dimer (Li_2I_2) spectrum, obtained by subtracting c from a, after normalization at monomer band (18).

and (d) in Figures 6—8, (c) and (e) in Figure 9. Both the adiabatic and vertical first ionization potentials are significantly higher in the dimer than in the corresponding monomer. The numerical values for these quantities have been extracted and summarized in Table V. These results did not surprise us, since they were predicted by ionic model calculations (19) as well as one *ab initio* Hartree-Fock calculation for lithium fluoride (20) (a subsequent one is also shown in Table V) which treated both monomer and dimer. However, the trend is opposite to that observed with metal and noble gas dimers, whose I.P.'s are lower than the corresponding monomers. It is simply a consequence of the relative bonding strengths of the two units in the neutral and ionic forms. Alakli halide dimers are more stable as neutrals; metal and noble gas dimers are generally more stable as ions.

We have performed Xα calculations on the various lithium halide dimers, from which we have determined the sequence of molecular orbitals, their symmetries and the energy span of the valence band. These calculations [summarized elsewhere (18)] are in reasonable agreement with experiment, and also indicate that the dimers have a higher first I.P. than the monomers. The geometries chosen for these calculations were the well-established planar-rhombic structures, with angles and distances taken from electron diffraction and infrared matrix isolation results.

The evidence for assigning the higher energy peaks discussed by Goodman, et al. (14) and Timoshenko and Akopyan (16) as dimer peaks can now be based on two documented observations.

(1) The dimers have higher I.P.'s and span a sufficient energy range to extend to higher energy than the monomer valence bands.

(2) The relative intensity of the higher energy peak increases from one sample to another as the known relative abundance of dimer increases.

In summary, ionic model calculations, *ab initio* molecular orbital calculations and experimental results agreed that alkali halide dimers had higher ionization thresholds than the corresponding monomers. The next step is more controversial. Rupp and Ahlrichs (21) extended the Hartree-Fock calculations beyond dimer, to trimer and tetramer for lithium fluoride. They predicted that Li_3F_3 should have a first I.P. 0.43 eV larger then Li_2F_2, with Li_4F_4 approximately the same as Li_3F_3. We (18) had extended our Xα calculations for Li_nCl_n to trimer and tetramer, and found that the first I.P. of trimer was lower than that of dimer and tetramer, indicating an oscillatory behavior. Ionic model calculations indicated a significant increase in I.P. between monomer and dimer, with a tendency (depending upon ionic model and sample) toward a further, but less dramatic increase in I.P. for trimer and tetramer. In our search for an experimental test of these calculations, we found one electron impact

TABLE V

Difference Between Dimer and Monomer First Ionization Potentials, in eV

	Vertical, expt.	Adiabatic, expt.	Ionic model	Hartree-Fock
LiF	(1.3)	0.92 — 1.16	1.50/1.23	
LiCl	0.58_5	0.58 — 0.67	0.62	1.32[a], 1.27[b]
LiBr	0.54_9	0.49 — 0.55	0.86	
LiI	0.69	0.59 — 0.61	0.96	

[a] Baskin et al. (20)

[b] Rupp and Ahlrichs (21)

mass spectrometric study of lithium fluoride (22) with reported I.P.'s of 11.8 ± 0.5 eV, 13.0 ± 0.5 eV and 12.3 ± 0.5 eV for monomer, dimer and trimer, respectively. This single experimental result had the same trends as the Xα predictions. We wanted to apply the more precise photon impact technique. However, even double oven photoelectron spectroscopic experiments would not be able to extract information on trimer, with $\lesssim$10% abundance, in the presence of an overwhelming concentration of monomer and dimer. It was necessary to apply the more species-specific technique of photoionization mass spectrometry.

Photoionization Mass Spectrometry —Its Application to the Alkali Halide Vapors

This technique employs a broad-banded light source and vacuum ultraviolet monochromator to select incremental wavelengths. The photon beam is incident upon a gas or vapor target, and the photoions generated are extracted and passed through a mass spectrometer. The intensity of a particular ion mass (normalized to light intensity) is scanned as a function of wavelength. Ideally, this intensity should decrease to zero (or background) at the adiabatic ionization threshold for a parent ion, and at the thermochemical threshold for a fragment ion. With a high temperature vapor, many vibrational and rotational levels will be populated by the Boltzmann distribution, and some analytical or graphical technique must be used to de-Boltzmannize the ionization yield curve.

We had previously used this technique for studying several alkali halides, including NaI, (23) the cesium halides (24) and the rubidium halides. (25) The apparatus consisted of a one-meter normal incidence monochromator and a magnetic mass spectrometer. In these earlier studies, ionic species attributable to dimer and monomer were studied, but the intensity of trimer was too weak for measurement. With our more recent interest in trimer, we once again turned to the lithium halides. This time, our apparatus consisted of a three-meter normal incidence monochromator and a quadrupole mass spectrometer. (7) For experimental reasons (moderate temperature, moderate mass range, convenient wavelength range) lithium chloride was selected as the sample.

Four ion masses attributable to lithium chloride species were observable (7) in the photoionization mass spectrum: Li^+, $LiCl^+$, Li_2Cl^+ and $Li_3Cl_2^+$. The relative abundance of these ions was 9.2 : 27 : 100 : 5.4 at 1161.1 Å. No measurable intensities of $Li_2Cl_2^+$ and $Li_3Cl_3^+$ were recorded. The ion yield curves of these species are plotted in Figures 10 —13 on semi-logarithmic coordinates in order to "deboltzmannize" the results. At the temperature of these experiments (ca. 1000°K) several vibrational levels of the monomer, and a more complex distribution of excited levels of dimer and trimer, are populated. Ionization

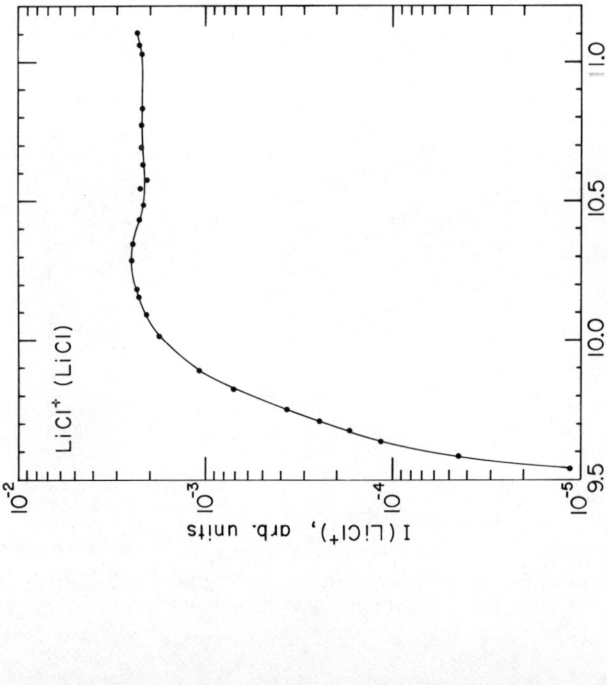

Journal de Chimie Physique

Figure 10. Variation of normalized $LiCl^+$ *intensity with photon energy for the process* $LiCl + h\nu \rightarrow LiCl^+ + e$*, plotted on semilogarithmic coordinates (7).*

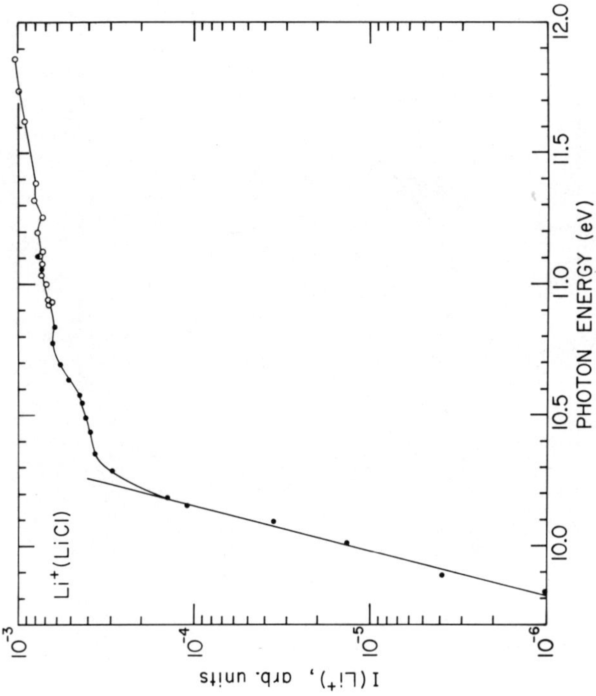

Journal de Chimie Physique

Figure 11. Variation of normalized Li^+ *intensity with photon energy for the process* $LiCl + h\nu \rightarrow Li^+ Cl^0 + e$*, plotted on semilogarithmic coordinates (7).*

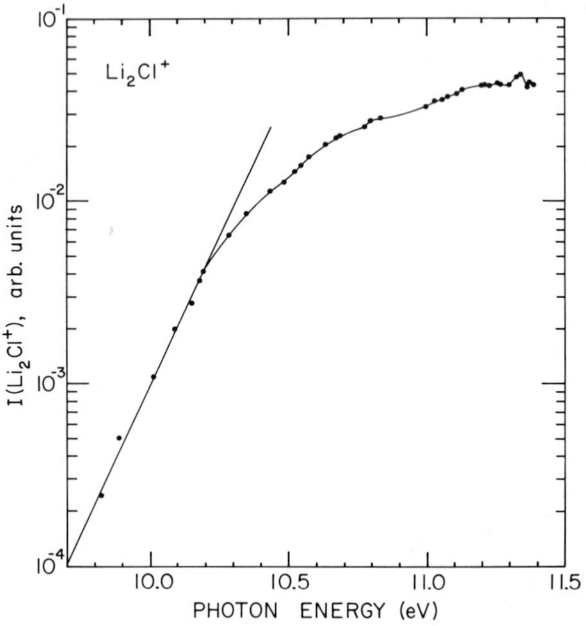

Journal de Chimie Physique

Figure 12. Ion yield of Li_2Cl^+ vs. photon energy for the process $Li_2Cl_2 + h\nu \rightarrow Li_2Cl^+ + Cl^0 + e$, on semilogarithmic plot (7).

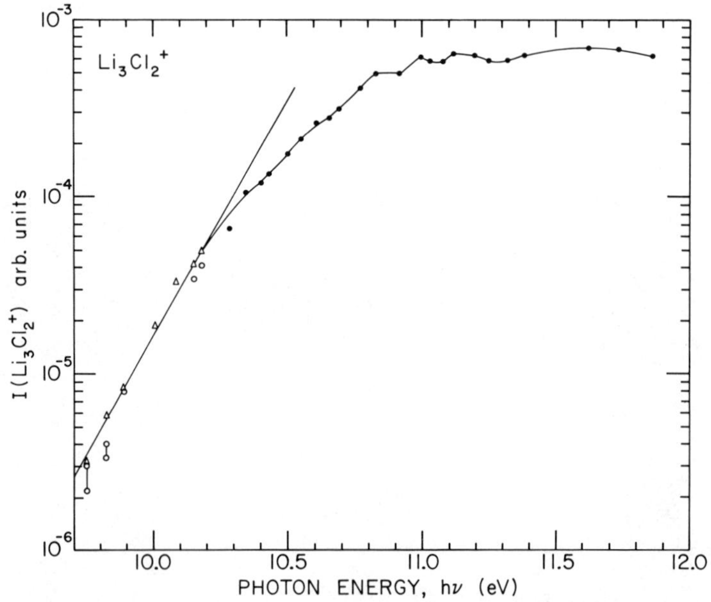

Journal de Chimie Physique

Figure 13. Ion yield of $Li_3Cl_2^+$ vs. photon energy for the process $Li_3Cl_3 + h\nu \rightarrow Li_3Cl_2^+ + Cl^0 + e$, on semilogarithmic plot (7).

will be observed at lowest photon energy for the highest vibrational level that is significantly populated. As the photon energy is increased, successively lower vibrational levels become capable of ionization. This increasing ionization yield near threshold largely reflects the exponential Boltzmann increase in population of vibrational levels, and hence a semi-logarithmic plot should tend to linearize this threshold variation. After the v" = 0 level becomes accessible to ionization, Franck-Condon factors become the controlling factors in the ionization yield curve, and one expects a departure from the exponential threshold behavior.

For $LiCl^+$ (Figure 10) the adiabatic ionization potential (departure from linearity) is barely attained in this experiment, at ca. 9.6 eV. In the earlier photoelectron spectroscopic study, we had obtained 9.55 and 9.59 eV for this value by two different extrapolation methods. This ion yield curve corresponds to transitions between the neutral ground state ($X^1\Sigma$) and the ground state of $LiCl^+$ ($X^2\Pi$), which is a bound state. (26)

The threshold region for Li^+ (Figure 11) has a linear segment over two decades, and departs rather abruptly at ∿10.17 eV. If we take this value as the threshold for the reaction

$$LiCl + h\nu \rightarrow Li^+ + Cl + \bar{e}$$

then subtraction of the ionization energy of Li (5.39177 eV) (27) yields 4.77_8 eV for D_o(LiCl). The best current estimates of this quantity are (28) 4.873 eV and (29) 4.813 ± 0.13 eV.

Li_2Cl^+ (Figure 12) was the most abundant species, and hence the easiest to measure. A reasonably linear segment is apparent over ca. one decade, although this curve is less steep than in the case of Li^+. This may reflect the more complex Boltzmann population of states for this tetratomic molecule, where a pre-exponential factor is necessary. The departure from linearity is also less abrupt, and occurs at ca. 10.20 eV. We take this to be the adiabatic I.P. of Li_2Cl_2. Our earlier photoelectron spectroscopic studies (18) yielded 10.22 and 10.17 eV for this quantity, employing two alternative methods of extrapolation. Hence, the inference seems quite plausible that departure from linearity on the semi-logarithmic plot yields a value very close to the adiabatic ionization potential.

Note, however, that we are dealing with a fragment ion here. The lack of a measurable parent ion, combined with the good agreement between photoelectron spectroscopic (mass independent) and photoionization mass spectrometric adiabatic ionization potentials implies that the parent ion $Li_2Cl_2^+$ is unstable very near to its threshold. We have examined the instability of $M_2X_2^+$ and $M_3X_3^+$, and the high stability of M_2X^+ and $M_3X_2^+$ by various ionic models. (7) The latter two ionic species are sufficiently stable to be observed effusing directly from a sample-containing oven. The heats of formation of a number of the triatomic ions is known, but their structure (linear or bent) is not. The various ionic models differ on this point.

The species of most interest in this investigation was $Li_3Cl_2^+$, and it was the most difficult to measure. It represents only ca. 5% of the Li_2Cl^+ intensity, and is split between two major isotopes. In addition, we are trying to probe the decreasing Boltzmann population in the pre-threshold region. The success of the experiment depended upon optimum tuning, reducing background to a few counts per minute, and stability of the apparatus over a period of hours. The measurement in the Boltzmann region required ∿20 min. per point, with and without the photon beam. The departure from linearity (Figure 13) occurs at ca. 10.17 eV. Within the statistical uncertainty, we can only conclude that Li_3Cl_3 has an adiabatic ionization potential very nearly that of the dimer. Recall, however, that the Xα calculation predicted a substantially lower vertical I.P., whereas the Hartree-Fock calculation for lithium fluoride predicted a vertical I.P. for Li_3F_3 0.43 eV higher than for Li_2F_2. Although the current measurements refer to adiabatic ionization potentials, our photoelectron spectroscopic study displayed the same trends between monomers and dimers whether one was comparing vertical or adiabatic ionization potentials.

Conclusion

This manuscript summarizes the most significant current experimental information regarding the photoionization behavior of alkali halide monomers, dimers and trimers. Some data exist on higher ionization potentials, (30) and ionic model calculations have also been applied to these higher levels. (19, 30) It is tempting to extrapolate to the behavior of the corresponding solids, but the experimental data here have an uncertain energy scale and display time-dependent variations due to charging of the samples, which are insulators. Future work may involve studies of trimers and higher clusters whose abundance has been enhanced by supersonic expansion.

Work supported by the United States Department of Energy, Office of Basic Energy Sciences, under Contract W-31-109-Eng-38.

Literature Cited

1. Ionov, N. I. Dokl. Akad. Nauk SSSR 1948, 59, 467.
2. Friedman, L. J. Chem. Phys. 1955, 23, 477.
3. Miller, R. C.; Kusch, P. J. Chem. Phys. 1956, 25, 860.
4. Klemperer, W.; Rice, S. A. J. Chem. Phys. 1957, 26, 618.
5. Rice, S. A.; Klemperer, W. J. Chem. Phys. 1957, 27, 573.
6. Berkowitz, J.; Chupka, W. A. J. Chem. Phys. 1958, 29, 653.

7. Berkowitz, J.; Batson, C. H.; Goodman, G. L. J. de chim phys. 1980 , 77, 631.
8. Berkowitz, J.; Dehmer, J. L.; Walker, T. E. H. J. Chem. Phys. 1973, 59, 3645.
9. Brewer, L.; Brackett, E. Chem. Rev. 1961, 61, 425.
10. Berry, R. S. in "Alkali Halide Vapors: Structure, Spectra and Reaction Dynamics"; Davidovits, P.; McFadden, D. L., eds. Academic Press, Inc.: N. Y.; 1979; p 78 —126.
11. Potts, A. W.; Williams, T. A.; Price, W. C. Proc. Roy. Soc. Lond. 1974, A341, 147.
12. Koopmans, T. Physica 1934, 1, 104.
13a. Matcha, R. L. J. Chem. Phys. 1967, 47, 4595.
13b. Matcha, R. L. J. Chem. Phys. 1967, 47, 5295.
13c. Matcha, R. L. J. Chem. Phys. 1968, 48, 335.
13d. Matcha, R. L. J. Chem. Phys. 1968, 49, 1264.
13e. Matcha, R. L. J. Chem. Phys. 1970, 53, 485.
13f. Matcha, R. L. J. Chem. Phys. 1970, 53, 4490.
14. Goodman, T. D.; Allen, Jr., J. D.; Cusachs, L. C.; Schweitzer, G. K. J. Electron Spectr. Relat. Phenom. 1974, 3, 289.
15. Berkowitz, J. in "Electron Spectroscopy —Theory, Techniques and Applications"; Brundle, C. R.; Baker, A. D., eds., Academic Press: N. Y., 1977; p 355.
16. Timoshenko, M. M.; Akopyan, M. E. Khim. Vys. Energ. 1974, 8, 211. [Eng. transl. High Energy Chem. (USSR) 1974, 8, 175]
17. Berkowitz, J.; Dehmer, J. L.; Walker, T. E. H. J. Electron Spectr. Relat. Phenom. 1974, 3, 323.
18. Berkowitz, J.; Batson, C. H.; Goodman, G. L. J. Chem. Phys. 1979, 71, 2624.
19. Berkowitz, J. in "Alkali Halide Vapors: Structure, Spectra and Reaction Dynamics"; Davidovits, P.; McFadden, D. L., eds., Academic Press, Inc.: N. Y.; 1979; p 155-188.
20. Baskin, C. P.; Bender, C. F.; Kollman, P. A. J. Am. Chem. Soc. 1973, 95, 5868.
21. Rupp, M.; Ahlrichs, R. Theor. Chim. Acta 1977, 46, 117, and private communication.
22. Grimley, R. T.; Forsman, J. A.; Grindstaff, Q. G. J. Phys. Chem. 1978, 82, 632.
23. Berkowitz, J.; Chupka, W. A. J. Chem. Phys. 1966, 45, 1287.
24. Berkowitz, J. J. Chem. Phys. 1969, 50, 3503.
25. Berkowitz, J. Adv. High Temp. Chem. 1971, 3, 123.
26. See Fig. 5 of ref. 7.
27. Moore, C. E. "Atomic Energy Levels"; Nat. Stand. Ref. Data Ser. (U.S.); NSRDS —NBS 1971, 35.
28. Gilbert, T. L. Second Midwest Chemistry Conf., Madison, Wisc., WIS— TCI-302 1968.
29. "Bond Dissociation Energies in Simple Molecules"; Natl. Bur. Std. NSRDS —NBS 1970, 31.
30. Potts, A. W.; Williams, T. A. J. Chem. Soc. Far. Trans. II 1976, 72, 1892.
31. Berkowitz, J. J. Chem. Phys.1972, 56,2766.

RECEIVED August 26, 1981.

20

Electric Dipole Polarizabilities of Alkali Halide Dimers

R. KREMENS, B. JADUSZLIWER, and B. BEDERSON
New York University, New York, NY 10003

J. A. D. STOCKDALE
Oak Ridge National Laboratory, Oak Ridge, TN 37830

Measurements have been made of the dc electric dipole polarizabilities of several alkali halide dimers, using a molecular beam technique. A combination monomer-dimer beam is passed through an inhomogeneous static electric field; the resulting deflection pattern in the detector plane yields the dimer polarizability. Effects of uncertainties in the determination of apparatus parameters, such as knowledge of the electric field gradient at the beam position, are minimized by normalizing to the known polarizability of atomic potassium. An analysis of the monomer beam deflection pattern is presented. The dimers studied were: $(CsCl)_2$, $(RbCl)_2$, $(KCl)_2$, $(CsF)_2$, and $(KF)_2$. Calculations using a combination of bond and ionic polarizabilities, obtained from measured vibrational constants and calculated ionic polarizabilities are in good agreement with our measurements.

Despite the importance of the induced dipole moment, generated by the interaction of charged particles and laboratory fields with highly polar molecules, no measurement of the molecular parameter governing this quantity, i.e., the average dc electric dipole polarizability, exists in the published literature, to our knowledge. This is partly due to the difficulty associated with unscrambling the induced moment effects from those due to the intrinsic moments of such molecules. Thus, for example, there is considerable uncertainty in the literature concerning the role played by polarization forces in electron-highly polar molecule scattering (1). There are on the other hand several ab initio calculations of these quantities for the alkali halide molecules (2, 3, 4).

The difficulty in distinguishing permanent dipole from induced dipole moment effects in a molecular beam deflection

0097-6156/82/0179-0301$05.00/0

experiment inhibits polarizability measurements on the monomer. This does not obtain for the molecular dimer, since these systems do not possess permanent dipole moments; however, the dimer polarizabilities are attributable partly to modified monomer molecular orbitals, combined of course with the appropriate dimer bond polarizabilities. Thus, the determination of dimer polarizabilities would, in addition to being of intrinsic interest in themselves, constitute a step towards the experimental determination of the monomer polarizabilities.

Previous experimenters have found dimerization in all alkali halides at vapor pressures corresponding to source temperatures appropriate for effusive beams (5, 6, 7). The percent dimerization increases as molecular weight decreases towards lighter alkali halides, with lithium fluoride possessing the largest fractional dimerization.

Deflection Analysis

We now briefly discuss the deflection patterns to be expected when alkali halide monomers and dimers pass between the polefaces of an inhomogeneous electric field (a detailed analysis of the deflection pattern due to molecules possessing strong electric dipole moments will be presented elsewhere). The geometry of the experiment is shown in Figure 1.

The essential elements of the experiment are a) an effusive molecular beam source, b) inhomogeneous deflecting electric polefaces, c) surface ionization detector, capable of translation in order to obtain the deflected beam pattern. ℓ_1, ℓ_2, ℓ_3, are the distance from the source to the front of the polefaces, the length of the polefaces and the distance from the back of the polefaces to the detector, respectively. A general review of deflection methods for determining polarizabilities is given by Miller and Bederson (8).

Consider the monomer, which possesses a permanent dipole moment μ. Then it can be shown (9) that at temperatures ~1000° K or higher, the deflection x' of a molecule of speed v and rotational state J, M_J is

$$x\ (v, J, M_J) = \frac{b^2\mu^2\gamma A E^2}{4mh^2v^2J^2}\left(1-3\,\frac{M_J^{\,2}}{J^2}\right) \qquad (1)$$

where $b^2 = \ell_2\ (2\ell_3 + \ell_2)$, γ is the field gradient constant, equal to the ratio of electric field gradient to electric field in the poleface gap, A is the molecular moment of inertia perpendicular to the internuclear axis, E is the electric field strength, and it is assumed that $\mathcal{E}_\mu \ll kT$ (low field case), where $\mathcal{E}_\mu$ is the Stark energy of the permanent dipole moment. Averaging over M_J and integrating over J we obtain the intensity pattern

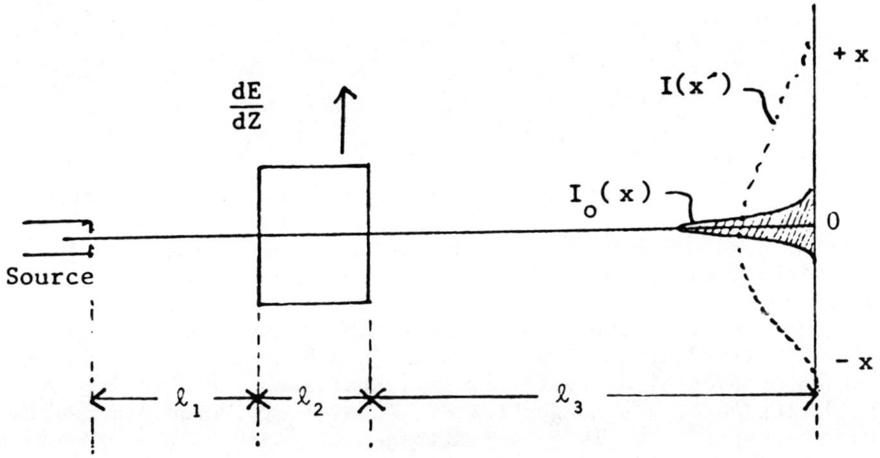

Figure 1. Geometric arrangement. Key: shaded beam, undeflected (δ-function) beam; and I (x′), deflected beam.

$$I\ (v,\ x') = \frac{\sigma^{3/2} v^2 I_o(v)}{\sqrt{3\pi}\ \beta} \int_o^{J_{max}} \frac{J^4 e^{-\sigma J^2}\, dJ}{\sqrt{1 - x' v^2 J^2/\beta}} \qquad (2)$$

where J_{max} depends on whether x> or <0, $\sigma = h^2/2AkT$ and $\beta = b^2\mu^2\gamma AE^2/4mh^2$, and for the time being it is assumed that the undeflected beam is infinitesimally narrow, i.e., $I_o\ (v,0) = I_o(v)\delta(x)$. On the other hand, the polarizability deflection is

$$x'\ (v;pol) = b^2\alpha\gamma E^2/2mv^2 \qquad (3)$$

where α is the (space-averaged) electric dipole polarizability, and the intensity pattern caused by the induced (polarizability) dipole moment for a given beam speed v is

$$I(v,x';pol) = I_o(v)\,\delta\left(x' - \frac{b^2\alpha\gamma E^2}{2mv^2}\right) \qquad (4)$$

that is, the molecular beam is simply displaced in the detector plane by the amount $x' = b^2\alpha\gamma E^2/2mv^2$. The actual intensity pattern, due to the polarizability acting alone, will depend on the velocity distribution, but will always possess a peak in the direction of the stronger field, that is, towards the convex poleface. On the other hand, integrating Equation 2 over the velocity distribution always results in a beam deflection pattern that monotonically decreases as $|x'|$ increases, when moving in the direction of a stronger field. Accordingly, should a peak be observed experimentally on the strong-field side of the deflection pattern, it must be caused by polarizability forces. The deflection caused by the monomer polarizability would be additive, for each monomer molecule, to the permanent dipole deflection, and hence would result in a constant displacement of the entire monomer deflection pattern.

To obtain reliable values for the polarizabilities it is desirable to normalize to a known polarizability. By this means systematic errors that are associated with imperfect knowledge of the electric field and gradient parameters as well as of the precise position of the beam with respect to the polefaces, are minimized. The best "standard" for this purpose is atomic potassium, whose polarizability is well-known from earlier measurements on the same apparatus (10). In figure 2 we show a pair of beam profiles for potassium, with zero electric field and with a field corresponding to potential differences of 3620 V. In Figure 3 we show a typical deflection pattern for one of the alkali halide-dimer combinations (RbCl).

Also shown in Figure 3 is a computer calculation of beam

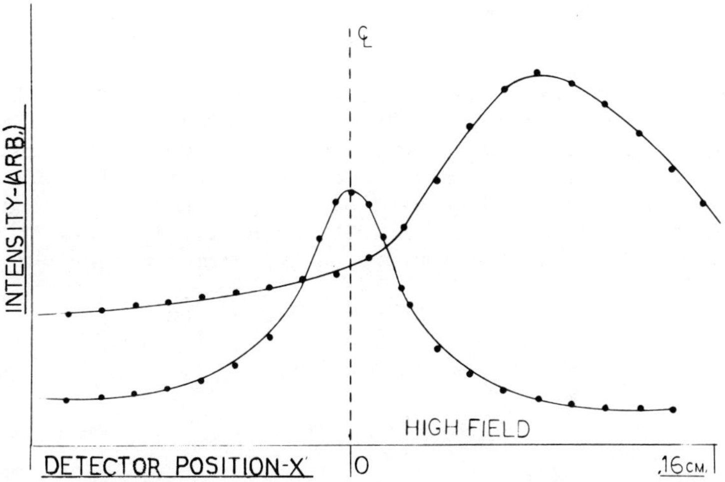

Figure 2. Beam deflection profiles at 503 K for K. Beam intensity vs. detector position, for an applied voltage of 3620 V across inhomogeneous polefaces (upper curve), and 0 V (lower curve). Convex poleface is to right of beam axis.

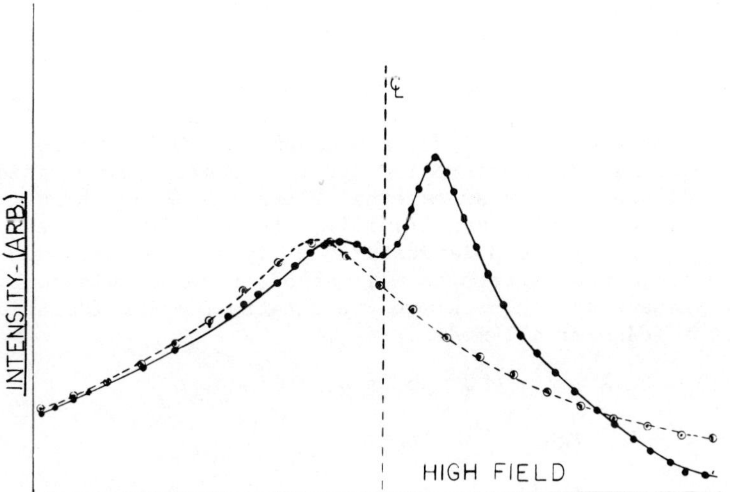

Figure 3. Beam deflection profile at 850 K for RbCl. Beam intensity vs. detector position, for an applied voltage of 3620 V across inhomogeneous polefaces (——). Computed deflection curve for monomer (– – –) assuming an electric dipole moment of 11.6 debye. Key: •, experimental and ⊙, calculated.

deflection versus intensity, using the appropriate beam parameters for RbCl in Equation 2 (the monomer polarizability force is not included). In the calculated curve the undeflected δ-fn beam is replaced by a best-fit Gaussian to the experimental beam shape of Figure 3 (with long tails removed), and Equation 2 is summed over the finite beam width. The calculated deflection pattern is normalized to the experimental curve at the maximum. A best fit is obtained assuming a dipole moment of 11.6 debye, which is approximately 9 percent higher than the vibrational-state-averaged known moment of 10.6 debye (11, 12).

The results of our measurements are summarized below:

$$\bar{\alpha}\,(KCl)_2 = 28.8 \pm 3.5 \times 10^{-24} cm^3$$

$$\bar{\alpha}\,(RbCl)_2 = 41.5 \pm 5.8$$

$$\bar{\alpha}\,(CsCl)_2 = 36.6 \pm 5.0$$

$$\bar{\alpha}\,(KF)_2 = 20.3 \pm 2.5$$

$$\bar{\alpha}\,(CsF)_2 = 14.6 \pm 2.1$$

While quantitative calculations are unavailable for comparison with these results, rough estimates can be made for qualitative comparison, by using bond polarizabilities obtained from calculations of dimer vibrational constants (13). Trugman and Gordon (14) have calculated the normal vibrational frequencies for many of the alkali halide dimers assuming an electron-gas ionic model. Using these frequencies, which agree well with the available experimental values, and assuming simple spring models for the potentials, we can calculate the equilibrium positions of the ionic centers in an external electric field, thence the induced dipole moments, and, finally, the polarizabilities. To these we add the ionic polarizabilities (15). Detailed discussion of the experimental errors as well as of these calculations will also be presented elsewhere; we present the final calculations here, averaged over all modes.

$$\bar{\alpha}\,(KCl)_2 = 32.85 \times 10^{-24} cm^3$$

$$\bar{\alpha}\,(RbCl)_2 = 40.33$$

$$\bar{\alpha}\,(CsCl)_2 = 35.3$$

$$\bar{\alpha}\,(KF)_2 = 19.54$$

$$\bar{\alpha}\,(CsF)_2 = 24.88$$

The agreement with our measurements is seen to be good, possibly excepting $(CsF)_2$. The $(CsF)_2$ calculation was made with older, possibly inaccurate $(CsF)_2$ vibrational frequencies (14). These calculations used the free ionic polarizability of Brumer and Karplus (16).

Acknowledgements

This work was supported by the Department of Energy, Basic Sciences Division.

We thank Mr. Al Tino for his help with the analysis of the data, including the computation of the beam deflection pattern.

Literature Cited

1. Collins, L.A.; Norcross, D.W.; Phys. Rev. A 1978, 18, 467.
2. Yoshioka, Y.; Jordan, K.D.; J. Chem. Phys. 1980, 73, 5899.
3. Bounds, D.G.; Clarke, J.H.R.; Hinchliffe, A.; Chem. Phys. Lett. 1977, 45, 367; 1978, 54, 289.
4. Bounds, D.G.; Chem. Phys. 1979, 42 1979.
5. Berkowitz, J.; Batson, C.H.; Goodman, G.L.; J. Chem. Phys. 1979, 71, 2624.
6. Davidovitz, P.; McFadden, Eds.; "Alkali Halide Vapors", Academic Press, NY, 1979, pp. 25-29.
7. Stull, D.R.; Prophet, H.; "JANAF Thermodynamic Tables" 2nd Ed., NSRDS-NBS No. 37., U.S. Gov't. Printing Office, Washington, D.C., 1971.
8. Miller, T.M.; Bederson, B.; "Advances in Atomic and Molecular Physics", Vol. 13, Academic Press, NY, 1977, pp. 1-55.
9. For a discussion of the Stark energy of polar molecules, see Townes, C.H.; Schawlow, A.L.; "Microwave Spectroscopy", McGraw Hill, NY, 1955, p. 250 tf. For deflection discussion, see Zorn, J.C.; English, R.C.; "Advances in Atomic and Molecular Physic-", Vol. 9, Academic Press, NY, 1973.
10. Molof, R.W.; Schwartz, H.L.; Miller, T.M.; Bederson, B.; Phys. Rev. A 1974, 10, 1131.
11. Story, T.L.; Herbert, A.J.; J. Chem Phys. 1976, 64, 855.
12. Herbert, J.A.; Lovas, F.J.; Melendres, C.A.; Hollowell, C.D.; Story, T.L.; Street, K.; J. Chem. Phys. 1968, 48, 2824.
13. Hirschfelder, J.A.; Curtis, C.F.; Bird, R.B.; "Molecular Theory of Gases and Liquids", J. Wiley, NY, 1954.
14. Trugman, S.; Gordon, R.G.; J. Chem. Phys. 1976, 64, 4625.
15. Brumer, P.; Karplus, M.; J. Chem. Phys. 1973, 58, 3903.
16. Berkowitz, J.; J. Chem. Phys.1960, 32, 1519.

RECEIVED August 26, 1981.

Vibrational Spectroscopy of High Temperature Metal–Halide Vapor Complexes

G. N. PAPATHEODOROU

Argonne National Laboratory, Chemical Engineering Division, Argonne, IL 60439

Raman spectra of vapor-phase compounds $CuFeCl_5$, $HfCl_4 \cdot POCl_3$ and $NH_3 \cdot AlX_3$ (X = Cl, Br) have been measured at ∿500-1000 K. Resonance Raman spectra obtained from an equilibrium vapor mixture containing $CuFeCl_5(g)$ showed two characteristic polarized bands at 441 and 275 cm^{-1}, which were attributed to trigonally coordinated Cu(II) and suggested a C_{2v} symmetry for the vapor complex. Raman spectra of $HfCl_4 \cdot POCl_3$ vapors were characterized by a superposition of (1) strong bands due to $POCl_3(g)$ and $HfCl_4(g)$ and (2) a few new bands (at 512, 1221, and 1263 cm^{-1}) due to the vapor complex. A comparison of the vapor-complex spectra with those of $HfCl_4(g)$ and $POCl_3(g)$ as well as $HfCl_4 \cdot POCl_3$ in the liquid and glass states indicated that the bonding of the complex with a C_{3v} symmetry occurs through an oxygen bridge. Raman spectra of the vapors over liquid NH_4AlX_4 (X = Cl, Br) were measured and compared with the spectra of the liquid NH_4AlX_4 itself as well as NH_3AlCl_3 in the vapor and liquid states. The measurements support the view that dissociative vaporization and further dissociation occur according to the reaction

$$NH_4AlX_4(\ell) \rightleftharpoons NH_3AlX_3(g) + HX(g) \rightleftharpoons NH_3(g) + AlX_3(g) + HX(g)$$

For NH_3AlCl_3, the C_{3v} molecular symmetry of the gaseous state is not preserved in the liquid state whose structure appears to be network-like.

During the past three decades, it has been established that salt vapors at elevated temperatures (typically T>500 K) associate partly to yield dimeric and/or polymeric gaseous molecules. Furthermore, certain vapors react with vapors and/or condensed-phase compounds to form vapor complexes. Vapor complexation appears to

0097-6156/82/0179-0309$05.00/0

be a general phenomenon for many metal-halide salts as well as oxides and other types of salts and is important in determining the rates of vaporization, vapor pressures and apparent volatilities of materials (1, 2, 3). Most of the studies have been focused on the formation of vapor complexes between two metal halides GX_y and RX_k, having a common anion X (F, Cl, Br, or I), which react according to the general scheme:

$$mGX_y(s,\ell,\text{ or }g) + nRX_k(g) \rightarrow G_mR_nX_{(my + nk)}(g) \qquad (1)$$

In most cases RX_k is a more volatile halide and is present only as a vapor whereas GX_y may be in a condensed or vapor phase depending on the temperature and total pressure of the system. Stoichiometric coefficients and thermodynamic quantities for reaction 1 have been measured using a variety of experimental techniques, including mass-spectrometry, chemical transport and spectrophotometry. A large number of halide vapor complexes have been reported, and extensive reviews on their systematics and thermodynamics of formation are already available (4-9).

However, studies regarding the electronic and vibrational properties of vapor complexes are not so extensive. Table I summarizes the metal-halide vapor complexes which have been studied by spectroscopic methods. A major part of the work has been devoted to obtaining electronic absorption spectra of complexes involving transition-metal halides. A limited amount of matrix-infrared (IR) spectroscopy data has also been reported. More recently, high-temperature Raman and fluoresence spectroscopy has been used to investigate the vibrational and electronic states of these vapors. In most cases, the spectra measured served as a means of proposing structural models for the vapor complexes. Conclusions regarding the molecular structure and symmetry of the complexes are also given in Table I.

The present paper is concerned with the characterization of the vibrational properties of certain vapor complexes at elevated temperatures. Raman spectroscopy is used to study the vapor complexes $CuFeCl_5$ and $HfCl_4{\cdot}POCl_3$ as well as the vapor species formed by vaporization of NH_4AlX_4 (X = Cl, Br).

Experimental Methods

The preparation and purification of metal halides for high-temperature spectroscopy present certain difficulties which vary for the different salts. All chemicals used in the present work were "anhydrous" salts purchased from Cerac/Pure Inc. $AlCl_3$, $AlBr_3$, $FeCl_3$, and $HfCl_4$ were purified by repeated slow sublimations in fused-silica tubes under vacuum. Anhydrous $CuCl_2$ was synthesized by reacting $AlCl_3$ with the corresponding oxide at ∿300 C. A final purification of $CuCl_2$ was made by vapor transport according to reaction 1 using $AlCl_3(g)$ as a carrier gas. Two ammonium salts, NH_4Cl and NH_4Br, were purified by recrystallization

TABLE I. Spectroscopic Investigations of Metal-Halide Vapor Complexes

System	Temperature K	Predominant Vapor Complex Species and Molecular Symmetry	Spectroscopic Method of Investigation	Reference
$LiF-BeF_2$	900	$LiBeF_3$; C_{2v}	Matrix IR	(10)
$CsCl-CoCl_2$	1300	$CsCoCl_3$; C_{2v}(?)	VIS-UV	(11)
$AF-AlF_3$ (A=Li to Cs)	800-1100	$AAlF_4$; C_{2v}	Matrix IR	(12,13)
$InCl-AlCl_3$	800-1000	$InAlCl_4$; C_{2v}(?)	Raman	(14)
$InBr-AlBr_3$	900-1100	$InAlBr_4$; C_{2v}(?)	Raman	(14)
$LiI-NdI_3$	1050-1240	$LiNdI_4$; C_{2v}(?)	VIS-UV	(15)
$CsI-NdI_3$	1150-1250	$CsNdI_4$; C_{2v}(?)	VIS-UV	(15)
$TiCl_2-AlCl_3$	460-610	$TiAl_3Cl_{11}$	VIS-UV	(16)
VCl_2-AlCl_3	620-840	VAl_3Cl_{11}	VIS-UV	(17)
$CrCl_2-AlCl_3$	600	$CrAl_2Cl_8$; D_{3d}	VIS-UV	(18)
$CoCl_2-AlCl_3$	600-800	$CoAl_2Cl_8$; D_{3d}; D_{2d}(?)	VIS-UV; Raman	(19-21)
$CoBr_2-AlBr_3$	550-800	$CoAl_2Br_8$; D_{3d}; D_{2d}(?)	VIS-UV	(22,23)
$CoCl_2-GaCl_3$	500-820	$CoGa_2Cl_8$	VIS-UV	(24)
$CoCl_2-InCl_3$	850-1000	$CoIn_2Cl_8$	VIS-UV	(25,26)
		$CoInCl_5$; C_{2v}		
$NiCl_2-AlCl_3$	650-950	$NiAl_2Cl_8$; D_{3d}	VIS-UV	(27)
$NiCl_2-InCl_3$	~1000	$NiIn_2Cl_8$; D_{3d}	VIS-UV	(28)
		$NiInCl_5$; C_{2v}	VIS-UV	
$CuCl_2-AlCl_3$	500-700	$CuAl_2Cl_8$; D_{2h}(?)	VIS-UV; Raman	(29,30)
$CuCl_2-GaCl_3$	500-900	$CuGa_2Cl_8$; D_{2h}(?)	VIS-UV; Raman	(30,31)
		$CuGaCl_5$; C_{2v}		
$CuCl_2-InCl_3$	770-970	$CuInCl_5$; C_{2v}	VIS-UV; Raman	(32,31)
$PdCl_2-AlCl_3$	500-800	$PdAl_2Cl_8$; D_{2h}	VIS-UV; Raman	(27,29)
$PdCl_2-InCl_3$	800-1000	$PdIn_2Cl_8$; D_{2h}	VIS-UV; Raman	(33)
$PdBr_2-AlBr_3$	550-750	$PdAl_2Br_8$; D_{2h}	VIS-UV; Raman	(34)

[a] n = 0.5, 1, 1.5, ...

Continued

Table I. Continued

System	Temperature K	Predominant Vapor Complex Species and Molecular Symmetry	Spectroscopic Method of Investigation	Reference
$PtCl_2$-$AlCl_3$	500-650	$PtAl_2Cl_8$; D_{2h}	VIS-UV	(35)
$SmCl_2$-$AlCl_3$	600-800	$Sm_nAl_{2n}Cl_{(6n+2)}$[a]	VIS-UV	(36)
$EuCl_2$-$AlCl_3$	640-825	$EuAl_3Cl_{11}$ $EuAl_4Cl_{14}$	VIS-UV	(37)
$TiCl_3$-$AlCl_3$	690-850	$TiAlCl_6$; C_{2V}	VIS-UV	(16)
VCl_3-$AlCl_3$	600-790	$VAlCl_6$; C_{2V}	VIS-UV	(17)
		VAl_2Cl_9; D_{3d}		
$CrCl_3$-$AlCl_3$	~800	$CrAl_3Cl_{12}$; D_{3d}	VIS-UV	(17)
$FeCl_3$-$AlCl_3$	420-720	$FeAlCl_6$; C_{2V}	VIS-UV	(38-40)
$NdCl_3$-$AlCl_3$	500-900	$NdAl_3Cl_{12}$	VIS-UV; Fluor.	(41,42)
		$NdAl_4Cl_{15}$		
$NdCl_3$-$GaCl_3$	600-1000	$NdGaCl_6$; C_{2V}	VIS-UV	(15)
		$NdGa_3Cl_{12}$;	VIS-UV	
		$NdGa_4Cl_{15}$	VIS-UV	
$SmCl_3AlCl_3$	600-800	$SmAl_3Cl_{12}$	VIS-UV	(36)
$HoCl_3$-$AlCl_3$	500-800	$HoAl_3Cl_{12}$	VIS-UV	(43)
$ErCl_3$-$AlCl_3$	525-800	$Er_nAl_{2n}Cl_{(6n+2)}$[a]	VIS-UV; Fluor.	(44-46)
$ErCl_3$-$GaCl_3$	750-1000	$Er_nGa_{2n}Cl_{(6n+2)}$[a]	VIS-UV; Fluor.	(44,45)
$ErCl_3$-$InCl_3$	900-1100	$Er_nIn_{2n}Cl_{(6n+2)}$[a]	VIS-UV; Fluor.	(44,45)
$TbCl_3$-$AlCl_3$	800	$TbAl_{2n}Cl_{(6n+2)}$[a]	Fluor.	(47)
UCl_4-$AlCl_3$	1000	UAl_2Cl_{10}; D_{2h}	VIS-UV	(48)
$ThCl_4$-$AlCl_3$	1000	$ThAl_2Cl_{10}$; D_{2h}	Raman	(49)
UCl_5-$AlCl_3$	800	$UAlCl_8$; C_{2V}	VIS-UV; Raman	(48,50)

from aqueous solution and vacuum dehydration. All purified salts were handled in sealed fused silica containers and/or in a helium glove box with a water-vapor content of <20 p.p.m.

Raman spectra of metal-halide vapors were obtained by the "conventional" methods described in detail elsewhere (51, 52). Evacuated and sealed fused-silica cells containing the reacting chemicals (reaction 1) were placed in resistance-heated furnaces (31), and the spectra were excited with CW lasers. The scattered light was dispersed with a double monochromator and detected with a photomultiplier tube with the use of photon counting or modulation techniques. Details for adjusting the partial pressures of the vapor species in the cells have been reported previously (31, 33).

Vibrational Raman Spectroscopy

Raman spectroscopy has been used successfully in the past to investigate the structures of single-component gases at high temperatures (53). For the study of vapor complexes, however, difficulties arise because, according to reaction 1, the complexes exist only in equilibrium with the reacting gas(es). Furthermore, dimerization and/or polymerization of the GX_y and RX_k vapors may lead to a complicated vapor equilibrium situation where the vapor complex is a minor component with relatively low partial pressures. Thus the Raman spectra are expected to be dominated by the bands due to the reacting gas halide(s), and consequently, the bands due to the small amounts of complex may be difficult to recognize.

However, the determination of the vapor-complex vibrational frequencies may be achieved by: (1) careful examination of the spectra of the reacting gases in separate cells and then comparing the spectra with those from cells containing the vapor mixture and (2) enhancement of the Raman band intensities of the complex(es) by the use of resonance Raman techniques (31). The results reported in this paper have been obtained by these methods.

Resonance Raman Spectra of $CuFeCl_5$ Vapors. Copper(II) chloride is known to react at elevated temperatures with $AlCl_3$ (54), $GaCl_3$(55), and $InCl_3$(31, 32) vapors, thereby forming vapor complexes of the type CuL_2Cl_8 (L = Al, Ga) (54) and $CuLCl_5$ (L = Al, Ga, In) (29-30). When iron(III) and copper(II) chlorides are brought together at elevated temperatures, a gas-solid equilibrium, analogous to that for the $CuCl_2$-$InCl_3$ system, is established as follows:

$$CuCl_2(s) + 1/2Fe_2Cl_6(g) \rightleftharpoons CuFeCl_5(g) \qquad (2)$$

Furthermore, at temperatures between 600 and 1000 K, vaporization of $CuCl_2$(s) gives rise to a dimer-monomer equilibrium,

$$Cu_2Cl_4(g) \rightleftharpoons 2CuCl_2(g) \qquad (3)$$

and partial dissociation-decomposition of the $Fe_2Cl_6(g)$ occurs (56). Thus,

$$Fe_2Cl_6(g) \rightleftharpoons 2FeCl_3(g) \quad (4)$$

$$FeCl_3(g) \rightleftharpoons FeCl_2(s) + 1/2Cl_2(g) \quad (5)$$

The copper(II) chloride vapors possess charge transfer (CT) bands (56) in the visible spectral region which overlap with certain laser lines from the Ar and Kr ion lasers. As in the case of $CuLCl_5$ (L = Al, Ga, In) (29, 30), $CuFeCl_5$ is also expected to have a CT band near 20000 cm^{-1}. The maxima of the CT bands for $FeCl_3(g)$ and $Fe_2Cl_6(g)$ lie above 30000 cm^{-1} (56), and only the band tails overlap with the laser lines used in this study. Thus in measuring the Raman spectra of vapors in the $CuCl_2$-$FeCl_3$ system, one expects resonance enhancement of certain bands due to $CuCl_2(g)$, $Cu_2Cl_4(g)$ and $CuFeCl_5(g)$ and pre-resonance enhancement of the $FeCl_3(g)$ and $Fe_2Cl_6(g)$ bands.

Figure 1 gives representative Raman spectra (λ = 501.7 nm) obtained from cells containing the reaction mixture and the vapor components of equilibria 2-5. In order to avoid the decomposition reaction 5 small amounts of $Cl_2(g)$ were introduced into the cell. Measurements of spectra obtained at constant temperature showed that the intensity of the $CuFeCl_5(g)$ bands, relative to that of the iron(III) chloride bands, increases with increasing laser-excitation frequency. These trends and the appearance of strong combination and overtone bands are characteristic of the resonance nature of the spectra (29-30). Temperature-dependence studies have shown that the vapor complex bands decrease drastically with increasing temperature and/or power density of the excitation line. At temperatures T>800 K and laser powers I_o>2W, the spectra were mainly characterized by the $CuCl_2(g)$ and $FeCl_3(g)$ bands, indicating that the vapor complex dissociates as follows:

$$CuFeCl_5(g) \rightleftharpoons CuCl_2(g) + FeCl_3(g) \quad (6)$$

Similar laser-induced dissociation of "colored" vapors and vapor complexes has been recently reported for a number of metal-halide systems (22, 31, 50).

In Table II, the frequencies of the resonance Raman fundamentals of the $CuFeCl_5$ measured in the present study are compared with those of other $CuLCl_5$ complexes and $Cu_2Cl_4(g)$. The frequency of the strong band at $\sim$440 cm^{-1} is near to that expected for Cu(II) in a trigonal coordination (29, 30); this finding suggests the following bidentate structure (C_{2v} symmetry) for the $CuLCl_5(g)$ (L = Al, Ga, In, Fe) molecules:

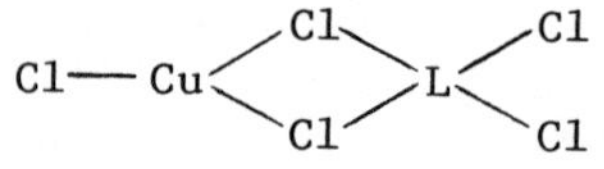

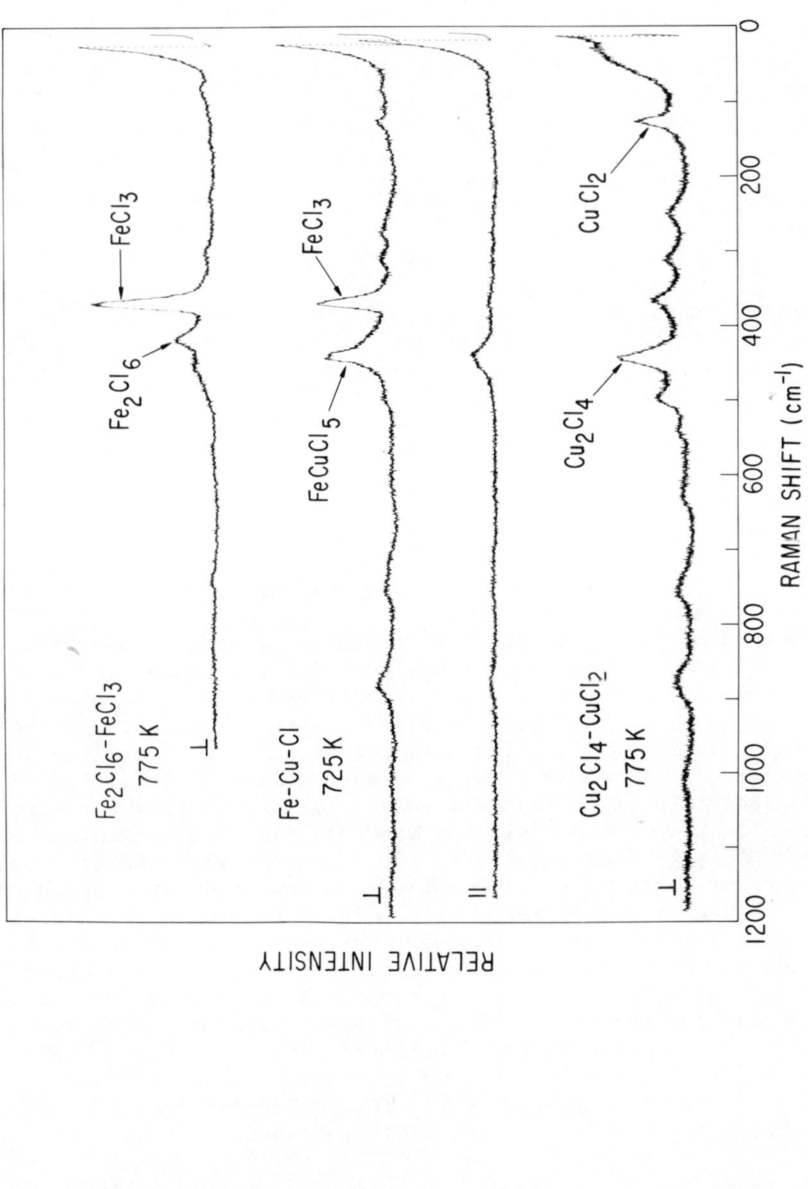

Figure 1. Resonance Raman spectra of $CuFeCl_5(g)$ and the component vapors $CuCl_2$–Cu_2Cl_4 and $FeCl_3$–Fe_2Cl_6. $\lambda_0 = 501.7$ nm; power, 600 mW; slit width, ~6 cm^{-1}; time constant, 2 s; and scan speed, 10 cm/min.

Table II. Resonance Raman Bands Attributed to $CuLCl_3$ Vapor Complexes (Frequency in cm^{-1})

$CuFeCl_5$	$CuAlCl_5$ (22)	$CuGaCl_5$ (30)	$CuInCl_5$ (30,31)	Cu_2Cl_4 (31,57)	Assignment[a]
275	291	278	275(300)	314	ν_{Cu-Cl_b}
441	448	443	442	442	ν_{Cu-Cl_t}

(a) Assignments to bridged (b) and terminal (t) Cu-to-Cl vibrations.

Raman Spectra of $HfCl_4 \cdot POCl_3$ Vapors. Phosphoryl chloride reacts with a number of metal halides to form stable liquids with low melting points (58). A large number of these liquids volatilize dissociatively. There are, however, cases (e.g., $POCl_3 \cdot FeCl_3$, $POCl_3 \cdot AlCl_3$) where the predominant vapor species is a stable vapor complex (59). The reaction of $POCl_3$ with $HfCl_4$ gives a glass-forming liquid whose vapors at elevated temperatures contain small amounts of $HfCl_4 \cdot POCl_3$ complex (59) in equilibrium with the component gases. Thus,

$$HfCl_4(g) + POCl_3(g) \rightleftharpoons HfCl_4 \cdot POCl_3(g) \qquad (7)$$

The present section is concerned with the Raman spectra of these vapors. Cells which contained $HfCl_4 \cdot POCl_3(\ell)$ were made, and spectra were measured for both the condensed phase(s) (liquid-glass) and the vapors over the liquid compound. These Raman spectra are given in Figure 2. As expected from reaction 7, the overall spectrum in Figure 2 is mainly a superposition of the $HfCl_4(g)$ and $POCl_3(g)$ spectra plus certain new bands which arise from the vapor complex. The Raman band positions and relative intensities measured for $HfCl_4(g)$ and $POCl_3(g)$ are in agreement with the data and assignments of Clark et al. (60, 61). The new bands attributed to the complex are listed in Table III. Higher temperatures changed the relative intensities of the upper spectrum shown in Figure 2; at $T > 800$ K, the vapor complex was completely dissociated (reaction 7) and only Raman bands due to the component vapors could be recognized. Raman spectra of liquid and glassy $HfCl_4 \cdot POCl_3$ were also obtained, and the measured frequencies are listed in Table III.

The bonding in the "adduct" vapor complex presumably occurs (58) by bridging the oxygen of $POCl_3$ (C_{3v} molecular symmetry) to the Hf throughout the face of the $HfCl_4$ tetrahedron. Thus,

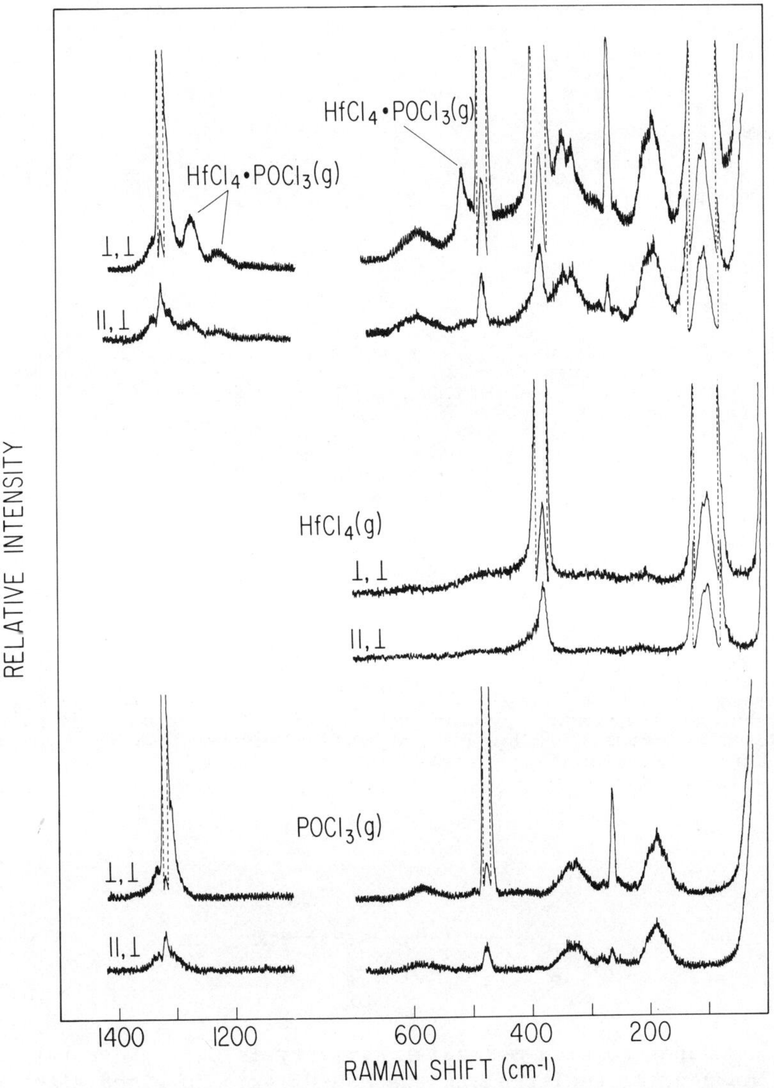

Figure 2. Raman spectra of $HfCl_4 \cdot POCl_3$, $HfCl_4$ and $POCl_3$ vapors at 675 K. $\lambda_0 = 514.5$ µm; power, 1 W; slit width, ~4 cm^{-1}; time constant, 1 s; and scan speed, 20 cm^{-1}/min.

TABLE III. Vibrational Frequencies (cm^{-1}) of $POCl_3$, $HfCl_4$ and $HfCl_4 \cdot POCl_3$ Molecules

$POCl_3$	$HfCl_4$	$HfCl_4 \cdot POCl_3$			
Vapor (675 K)	Vapor (675 K)	Glass (298 K)	Liquid (525 K)	Vapor (675 K)[a]	Assignment
	102.2s, $\nu_2(E)$	105m	101m		E
	109.5s, $\nu_4(T_2)$	133s	123s		E
186.5m, $\nu_6(E)$					
		207m	205m		E
265.5s(p), $\nu_3(A_1)$					
		307m(p)	305m(p)		A_1
325.7w, 341.6w, } $\nu_5(E)$[b]		358sh(p)			A_1
	380.7s(p), $\nu_1(A_1)$	371m(p)	367s(p)		A_1; ν_{Hf-Cl}
	392(1), $\nu_3(T_2)$				
478.5s(p), $\nu_2(A_1)$		465w(p)	453w(p)		A_1
		536s(p)	534s(p)	512m(p)	A_1; ν_{P-Cl}
587.2w, $\nu_4(E)$			640w		E
		~1200vw	1200w	1221w	E
			1220w(p?)		
				1263w(p)	A_1; ν_{P-O}
1319.4s(p), $\nu_1(A_1)$					

[a]Due to the high partial pressures of $HfCl_4$ and $POCl_3$, the Raman bands of the vapor complex are overcome by the strong bands of the component gases.
[b]Vibrational-rotational contour.

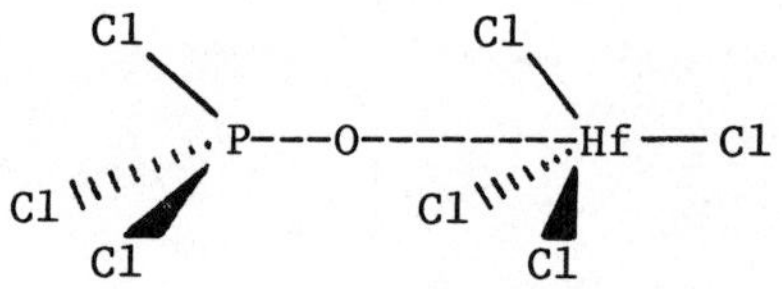

The vapor-complex molecular symmetry is C_{3v}, and rotation of the component molecules along the P-O-Hf axis does not alter this symmetry. The distribution of the 16 vibrational modes of the complex can be classified by group theory as follows:

$$\Gamma_{complex} = 7A_1(R,IR) + 8E(R,IR) + A_2(IR)$$

An assignment of the observed bands is given in Table III.

It is noteworthy that the ν_{P-Cl} stretching frequency in the vapor complex is blue shifted relative to the corresponding frequency of the free $POCl_3(g)$. On the other hand, the ν_{P-O} stretching frequency is red shifted for the complex. Judging from the glass-liquid spectra, a similar red shift occurs for the ν_{Hf-Cl} stretching frequency. All these shifts are compatible with the above-proposed structure, indicating that the formation of the P-O-Hf bridging bond weakens the P-O and Hf-Cl bond yet strengthens the P-Cl bond.

Raman Spectra of Vapors over Liquid NH_4AlX_4 (X = Cl, Br). The reaction of AlX_3 (X = Cl, Br) with NH_4X at ~500 K leads to the formation of low-melting NH_4AlX_4 salts which are analogous to alkali metal-tetrachloroaluminate melts (14). In earlier studies, it was argued that the NH_4AlX_4 volatilizes congruently; evidence, however, was presented recently indicating a dissociative vaporization (62):

$$NH_4AlX_4(s,\ell) \rightleftharpoons NH_3AlX_3(g) + HX(g) \qquad (8)$$

Increasing temperatures, above the boiling points, lead to a further dissociation,

$$NH_3AlX_3(g) \rightleftharpoons NH_3(g) + AlX_3(g) \qquad (9)$$

which is of particular importance for the bromide salt. The $NH_3AlCl_3(g)$ appears to be very stable up to ~1000 K, where decomposition occurs and solid AlN is formed.

This section presents results of Raman spectroscopy used to (1) identify the vapors formed over NH_4AlX_4 and NH_3AlCl_3 melts and (2) characterize the vibrational modes of these melts and vapors.

Figure 3 presents the spectra of the liquid compounds obtained at temperatures slightly above their respective melting points. All liquid spectra were superimposed on a strong fluorescence background but the Raman bands were easily recognized from their high intensity. For the liquid NH_3AlCl_3, two additional bands were measured at ~3170 and 3240 cm^{-1}.

The number, position, and polarization characteristics of the bands measured below 500 cm^{-1} for the $NH_4AlCl_4(\ell)$ are attributed to the existence of the well-established AlX_4 tetrahedral anions in these melts. The NH_4 Raman bands are presumably weak and only two broad and weak bands could be measured at ~3180 and ~3160 cm^{-1}.

The Raman spectra of vapors over $NH_3AlCl_3(\ell)$ (Figure 4) were found to be different from those of the liquid itself. The difference can be clearly seen by comparing the spectra in Figures 3 and 4. To a first approximation, the strong bands observed for the $NH_3AlCl_3(g)$ spectrum can be considered as arising from a combination of bands due to the two component molecules (i.e., NH_3 and $AlCl_3$), which have geometries that deviate from those of the corresponding free gases (i.e., C_{3v} for NH_3 and D_{3h} for $AlCl_3$). Table IV lists the vibrational frequencies of the NH_3AlX_3 and

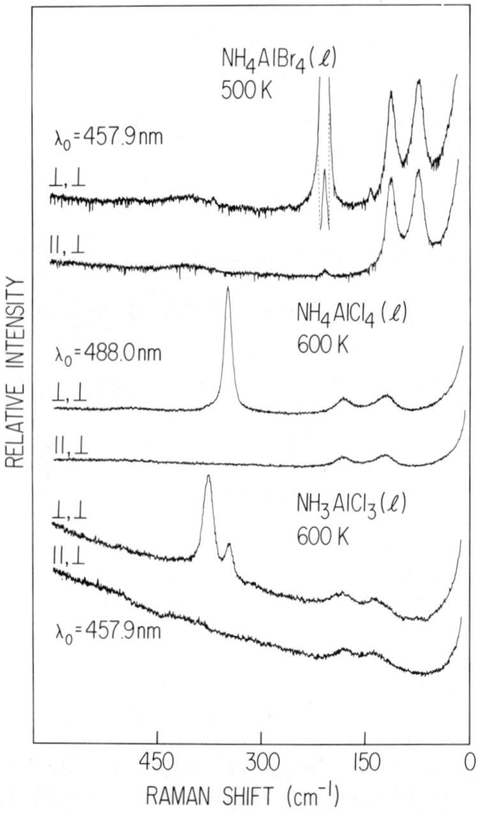

Figure 3. Raman spectra of liquid NH_3AlCl_3 and NH_4AlX_4 (X = Cl, Br). Key: power ~100 mW; slit width, ~4 cm^{-1} time constant, 0.2 s; and scan speed, 50 cm^{-1}/min.

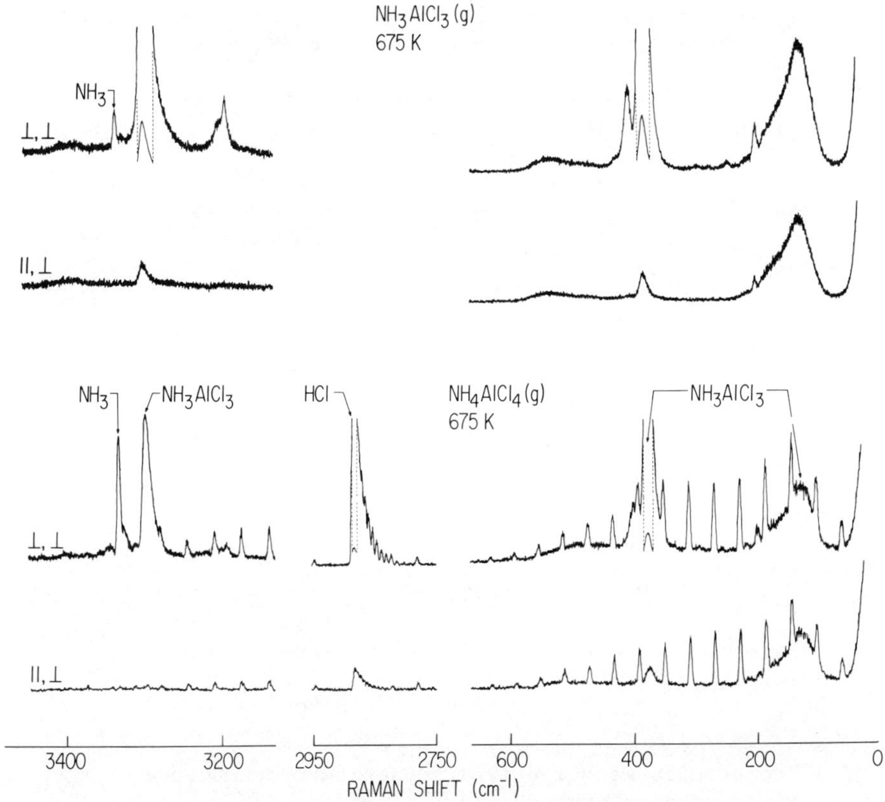

Figure 4. Raman spectra of vapors over NH_4AlCl_4(l) and of NH_3AlCl_3(g). Key: λ_0 = 514.5 μm; power, 2W; slit width, ~6 cm^{-1}; time constant, 0.2 s; and scan speed, 10 cm^{-1}/min.

compares them with those of gaseous NH_3 and AlX_3. It was suggested previously (63) that a C_{3v} geometry is obtained by the $AlCl_3$ in the adduct molecules. However, the frequencies used for the normal coordinate analysis (63) and assignments were those of the liquid NH_3AlCl_3 which, as stated above, are different from those measured for the gas. Presently, we are undertaking the vibrational analysis of $NH_3AlX_3(g)$ spectra using our new gaseous Raman data.

Table IV. Comparison of the Vibrational Frequencies (cm^{-1}) of the NH_3AlX_3 and AlX_3, and NH_3 Vapors

$NH_3AlCl_3(g)$	$AlCl_3(g)$	$NH_3(g)$	$NH_3AlBr_3(g)$ [a]	$AlBr_3(g)$
			85	94
129s	135			
			149s(p)	155
173sh				
	183			
200m(p)				
243wn(?)				
			255s(p)	230s(p)
378s(p)	375s(p)			
402m(p)				
			450m	510
529m				
	610			
		920w(p)		
		1627vw		
			3189w	
3197w(p)			3229w	
3301s(p)		3377s(p)	3294s(p)	
3388w		3444w	3390w	

(a) Frequencies measured from temperature dependence vapor spectra over NH_4AlBr_4 liquid.

It should be noted that the frequencies and polarization of certain low-frequency (<500 cm^{-1}) bands of liquid NH_3AlCl_3 are similar to those of the $AlCl_4$ anion of $NH_4AlCl_4(\ell)$. This has been considered to indicate that the $NH_3AlCl_3(g)$ molecules associate in the liquid (64) according to the scheme:

$$xNH_3AlCl_3 \rightarrow [AlCl_{4-x}(NH_3)_x]^{+(x-1)} \cdot (x-1)[AlCl_4]^- \quad (x = 2, 3, 4)$$

which gives rise to the formation of an $AlCl_4^-$ tetrahedron.

Spectra of vapors over $NH_4AlX_4(\ell)$ were measured at temperatures between 600 and 900 K and were compared with spectra of $NH_3(g)$, $HX(g)$, and $NH_3AlX_3(g)$ obtained in the same temperature range. The observed bands in the spectra for the vapors over $NH_4AlCl_4(\ell)$(e.g., Figure 4) were attributed to rotational bands due to HX(g) and $NH_3(g)$ and to vibrational contours due to HX(g), $NH_3(g)$, and $NH_3AlX_3(g)$. With increasing temperature, an enhancement of the $NH_3(g)$ bands relative to the $NH_3AlX_3(g)$ bands was noted. This temperature effect was more pronounced for the bromide system. All these observations confirm the dissociative vaporization of the NH_4AlX_4 liquids (reaction 8) and the partial dissociation of NH_3AlX_3 according to reaction 9.

Acknowledgements

This work has been supported by the Division of Basic Energy Sciences, U.S. Department of Energy. Many thanks to C. Anderson, J. O'Keefe, and P. Radloff for their assistance.

Literature Cited

1. Schafer, H.; "Chemical Transport Reactions" (Academic Press, NY, 1964).
2. Brewer, L.; "Chemistry and Metallurgy of Miscellaneous Materials: Thermodynamics" ed. L. L. Quill (McGraw-Hill, New York, 1950) p. 193.
3. Brewer, L.; and Kane, J. S.; J. Phys. Chem. 1955, 59, 105.
4. Novikov, G. I.; and Gavrychenkov, F. G.; Russ. Chem. Rev. 1967, 36, 399.
5. Emmenegger, F. P.; J. Crystal Growth 1972, 17, 31.
6. Hastie, J. W.; "High Temperature Vapors" (Academic Press, NY, 1975).
7. Schafer, H.; Angew. Chem. 1976, 88, 775.
8. Hildebrand, D. L.; and Cubicioti, D. D.; eds. Proc. of the Symp. on "High Temperature Metal Halide Chemistry" (The Electrochemical Soc. Inc. Proc. 78-1, Princeton, NJ, 1978).
9. Hastie, J. W.; ed. Proc. of the 10th Materials Research Symp. on "Characterization of High Temperature Vapors and Gases" (National Bureau of Standards Special Publication 561, Washington, D.C., 1979).
10. Snelson, A.; Cyvin, B. N.; and Cyvin, S. J.; J. Mol. Struct. 1975, 24, 165.
11. Gruen, D. M.; and DeKock, C. W.; J. Chem. Phys. 1965, 43, 3395.
12. Cyvin, S. J.; Cyvin, B. N.; and Snelson, A.; J. Phys. Chem. 1971, 75, 2609.
13. Huglen, R.; Cyvin, S. J.; Øye, H. A.; Z. Naturforsch. 1979, 34a, 1118.
14. Radloff, P. L.; and Papatheodorou, G. N.; J. Chem. Phys. 1980, 72, 992.

15. Foosnaes, T. "Gas Complexation of Neodymium Halides," Thesis No. 37, Institute of Inorganic Chemistry, University of Trondheim (1979).
16. Sørlie, M.; and Øye, H. A.; Inorg. Chem. 1978, 17, 2473.
17. Anundskas, A.; and Øye, H. A.; J. Inorg. Nucl. Chem. 1975, 37, 1609.
18. Aits, M.; and Schafer, H.; Z. Anorg. Allg. Chem. 1974, 408, 37.
19. Papatheodorou, G. N.; in Ref. (9), p. 660.
20. Papatheodorou, G. N.; Z. Anorg. Allg. Chem. 1975, 411, 153.
21. Dell'Anna, A.; and Emmenegger, F. P.; Helv. Chim. Acta 1975, 58, 1145.
22. Papatheodorou, G. N.; and Kucera, G. H.; Inorg. Chem. 1977, 16, 1006.
23. Emmenegger, F. P.; Inorg. Chem. 1977, 16, 343.
24. Anundskas, A.; Mahgoub, A. E.; and Øye, H. A.; Acta. Chem. Scand. 1976, A30, 193.
25. Dienstbach, F.; and Emmenegger, F. P.; Helv. Chim. Acta. 1977, 60, 1966.
26. Kucera, G. H.; and Papatheodorou, G. N.; J. Phys. Chem. 1979, 83, 3213.
27. Papatheodorou, G. N.; J. Phys. Chem. 1973, 77, 472.
28. Dienstbach, F.; and Emmenegger, F. P.; Inorg. Chem. 1977, 16, 2957.
29. Papatheodorou, G. N.; and Capote, M. A.; J. Chem. Phys. 1978, 69, 2067.
30. Schlapfer, C. W.; and Rohrbasser, C.; Inorg. Chem. 1978, 17, 1623.
31. Papatheodorou, G. N.; in Ref. (9), p. 647.
32. Dienstbach, F.; Emmenegger, F. P.; J. Inog. Nucl. Chem. 1978, 40, 1299.
33. Papatheodorou, G. N.; and Capote, M. A.; in Ref. (8), p. 334.
34. Capote, M. A.; and Papatheodorou, G. N.; Inorg. Chem. 1978, 17, 3414.
35. Papatheodorou, G. N.; Inorg. Chem. 1973, 12, 1899.
36. Papatheodorou, G. N.; and Kucera, G. H.; Inorg. Chem. 1979, 18, 385.
37. Sørlie, M.; and Øye, J. A.; J. Inorg. Nucl. Chem. 1978, 40, 493.
38. Aits, M.; and Schafer, J.; J. Less-Comm. Metal. 1978, 57, 219.
39. Shieh, C. F.; and Gregory, N. W.; J. Phys. Chem. 1975, 79, 828.
40. Gregory, N. W.; and Laughlin, W. C.; J. Phys. Chem. 1977, 81, 2228.
41. Øye, H. A.; and Gruen, D. M.; J. Am. Chem. Soc., 1969, 91, 2229.
42. Jacobs, R. R.; Krupke, W. F.; Hessler, J. P.; and Carnall, C. W.; Opt. Commun. 1977, 21, 395.
43. Hoekstra, H. R.; Hessler, J. P.; Williams, C. W.; and Carnall, W. T.; in Ref. (8), p. 123.

44. Berg, R. W.; and Papatheodorou, G. N.; Inorg. Chem. Acta. 1980, 45, L211.
45. Papatheodorou, G. N.; and Berg, R. W.; Chem. Phy. Let. 1980, 75, 483.
46. Carnall, W. T.; Hessler, J. P.; Hoekstra, H. R.; and Williams, C. W., Jr.,; Chem. Phys. 1977, 68, 4304.
47. Hessler, J. P.; Wagner, F.; Williams, C. W.; Carnall, W. T.; J. Appl. Phys. 1977, 48, 3260.
48. Gruen, D. M.; and McBeth, R. L.; Inorg. Chem. 1969, 8, 2625.
49. Taylor, D. R.; and Larsen, E. M.; J. Inorg. Nucl. Chem. 1979, 41, 481.
50. Papatheodorou, G. N.; and Buttry, D. A.,; Inorg. Nucl. Chem. Lett. 1979, 15, 51.
51. Barraclough, C.; Beattie, I. R.; and Everett, D.; in "Vibrational Spectra and Structure, Vol. 5" ed. J. R. During (Elsevier, NY, 1976) p. 1.
52. Devlin, J. P.; in "Advances in Infrared and Raman Spectroscopy" eds. R. J. H. Clark and R. E. Hester (Heyden, NY, 1976).
53. Drake, M. C.; and Rosenblatt, G. M.; in Ref. (9), p. 609.
54. Sutakshuto, S.; "Copper(II) Chloride Gas Complexation with Acidic Halides", Thesis No. 31, Institute of Inorganic Chemistry, University of Trondheim, 1976.
55. Dienstbach, F.; and Emmenegger, F. P.; Z. Anorg. Allg. Chem. 1977, 436, 127.
56. Shieh, C. F.; and Gregory, N. W.; J. Phys. Chem. 1975, 79, 828 and references therein.
57. Dienstbach, F.; Emmenegger, F. P.; and Schlapfer, C. W., Helv. Chim. Acta. 1977, 60, 2460.
58. Greenwood, N. N.; and Wade, K.; in "A Comprehensive Treatise on Inorganic Chemistry", J. N. Mellor ed., Longman Publications, London (1971) Vol. III, p. 499.
59. Suvorov, A. V.; Zh. Koord. Khim. 1977, 3, 1146.
60. Clark, R. J. H.; and Rippon, D. M.; Molec. Phys. 1974, 28, 305.
61. Clark, R. J. H.; Hunter, B. K.; and Rippon, D. M.; Inorg. Chem. 1972, 11, 59.
62. Laughlin, W. C.; and Gregory, N. W.; (a) Inorg. Chem. 1975, 14, 1263; (b) J. Chem. Eng. Data 1975, 20, 137 and references therein.
63. Cyvin, S. J.; and Hargittai, I.; Spectrosc. Lett. 1976, 9, 131 and references therein.
64. Papatheodorou, G. N.; work in progress.

RECEIVED August 26, 1981.

22

Matrix Isolation Studies of Alkali Halide Salt Molecules with Lewis Acids and Bases

BRUCE S. AULT

University of Cincinnati, Department of Chemistry, Cincinnati, OH 45221

The matrix isolation technique has been applied, in conjunction with the salt/molecule reaction technique, to model the high temperature gas phase reactions of alkali halide salt molecules. The reactions with Lewis acids such as SiF_4, HF and CO_2 yielded ion pair products which were quenched into inert matrices for spectroscopic study. Difficulties arising from lattice energy considerations in alkali halide salt reactions are minimized by the initial vaporization of the salt reactant. The reaction of such salt molecules with Lewis bases, including H_2O and NH_3, yielded complexes similar in nature to transition metal coordination complexes, with binding through the alkali metal cation to the base lone pair.

Interest in the chemistry of materials at high temperatures has increased rapidly in the past few years. The interest in the field of high temperature chemistry is well demonstrated by the diversity of techniques which have been applied, ranging from chemiluminescent reactive collisions to low temperature quenching into inert gas matrices. Of considerable importance also is the fact that high temperature chemical reactions occur under solvent-free conditions, and chemical behavior may be altered under these conditions. The alkali halide salts, for example, have been investigated thoroughly in a variety of polar solvents, where they dissolve readily into separated ions. However, the chemistry of the alkali halide salts in the absence of solvent in the gas phase has not been nearly as well characterized. Recent developments now permit the exploration of the chemistry of the alkali halide salts, and others, under high temperature conditions.

One area of interest in high temperature chemistry, and certainly in all branches of chemistry, is the determination of the nature of the initial complex formed through the single collision reaction of two partners. For example, the initial complex

0097-6156/82/0179-0327$05.00/0

might be a highly reactive complex, a weakly bound transition state complex, or a relatively stable ion pair. Knowledge of the final products of the system only permits inferences as to the nature of the initial reaction product or complex. The matrix isolation technique was specifically developed to assist in the investigation of initial reaction products and has often been applied to the study of both neutral and charged chemical species. The initial thrust of work in the matrix isolation field was toward free radicals (1-4), which are frequently postulated in reaction mechanisms, but are sufficiently reactive that they often escape detection and characterization by traditional techniques. In the mid-1960's, the technology of high temperature evaporative ovens was applied to matrix isolation, and increased the range of species available for study. Linevsky (5) and others (6-9) demonstrated the feasibility of vaporization of ionic salts, while several groups developed techniques for the vaporization of metal atoms for use as reactants in matrix isolation studies (10,11,12). The late 1960's and the 1970's have seen the development of high energy sources, such as vacuum ultraviolet photolysis (13-16), microwave discharge (17,18) and proton beam radiolysis (19,20) for the formation and trapping of ionic species in inert gas matrices.

In many matrix isolation studies, including those which employ high temperature evaporative ovens, two reactants are allowed to mix, in a large excess of inert gas, immediately in front of the cold deposition surface. The gas phase mixture is rapidly quenched and condensed onto the cold surface, usually maintained between 10 and 20 K, so that very few collisions can occur between reactive partners, and the initial reaction products of the system are isolated. When photolysis is employed in the formation of intermediate species, irradiation can be carried out either during deposition of the reaction mixture, or after the inert gas matrix is rigidly frozen in place.

Once deposition is complete and the initial reaction product is trapped in an inert gas matrix, characterization is carried out spectroscopically. Several spectroscopic techniques have been used; the most common is infrared spectroscopy, either dispersive or Fourier transform. Raman spectroscopic studies have been carried out as well, but low signal levels have made this approach difficult. When the trapped intermediate is a free radical, electron spin resonance techniques are valuable as well. Finally, a number of researchers are employing electronic spectroscopy, when the species of interest has an absorption in the visible or ultraviolet region.

The studies which will be described below were carried out using a conventional matrix isolation apparatus, including a CTi model 21 closed cycle refrigerator as a low temperature source. Such refrigerators can cool to about 10 K with little or no heat load, and to 14-16 K with the typical heat load generated by a high temperature oven operating at 500°C. A schematic diagram of the vacuum vessel used in these experiments, for single or double oven experiments, is shown in Figure 1. The ovens used in this work were small, resistively heated quartz ovens, with Chromel A heating wire, and were generally capable of 600°C. A small stainless steel Knudsen cell with 1-2 mm orifice was loaded with the salt of interest, placed in the oven, and the system evacuated. Since the alkali halide salts are relatively hygroscopic, the oven was heated to slightly below the vaporization temperature under vacuum to dry the salt. Small residual amounts of water were always detected, and interference was generally minimal; however, as discussed below, the complex of the alkali halide salt with H_2O was often weakly observed. Argon was used as the matrix material in all of the experiments described here, and was used without further purification. Infrared spectroscopy was the primary means of characterization of the reaction products although Raman spectra were also recorded for certain systems. When appropriate, normal coordinate calculations were run to confirm the band assignments and to deduce force fields for the trapped species.

Alkali Halide Salt Reactions

Since the adaptation of high temperature oven technology to matrix isolation, the vaporization of halide and oxide salts has become not only feasible but readily accomplished. Many early studies were devoted to investigating the salts themselves, in an environment free from lattice interactions and perturbations. These led to matrix studies of the structure of ion pairs and triple ions, such as the thorough studies by Devlin and coworkers on matrix isolated alkali nitrate (21), chlorate (22) and perchlorate ion pairs (23). For relatively simple salts, such as the alkali halides, investigations were conducted into the structure of the dimeric salt species (6,7,8), which is present in a gas phase equilibrium with the monomeric salt species. These dimers have been found to be very strongly bound in a cyclic structure.

Only occasionally have salt molecules been vaporized for use as a reactant toward another species in matrix isolation studies. Devlin (24,25,26) conducted extensive experiments in which salt molecules were vaporized and condensed into argon matrices containing from 1% to 90% H_2O or NH_3, to study the effects of stepwise solvation of the salt molecule, as a model for solution studies. Margrave (27) and Snelson (28) each have used salt molecules as reactants, but most commonly toward another salt molecule to form a mixed salt dimer. The work described below, which was initiated at the University of Virginia and has been continued at the University of Cincinnati, employs alkali halide salt molecules as reactants toward a variety of species, including both Lewis acids and Lewis bases. The initial intent was to react a salt molecule such as NaCl with HCl in an excess of argon to bring

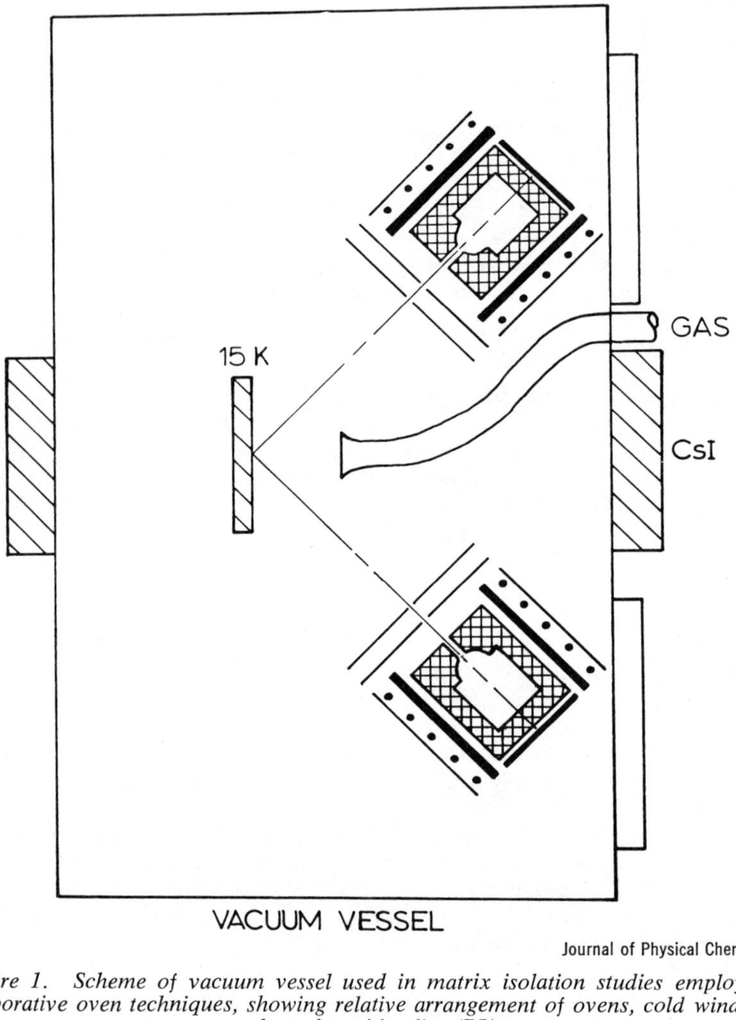

Journal of Physical Chemistry

Figure 1. Scheme of vacuum vessel used in matrix isolation studies employing evaporative oven techniques, showing relative arrangement of ovens, cold window, and gas deposition line (75).

about halide anion transfer from NaCl to HCl, forming the $Na^+HCl_2^-$ ion pair. While the reaction $NaCl + HCl \rightarrow Na^+ + HCl_2^-$ is endothermic by 122 Kcal/mole, energy is regained by the electrostatic attraction of the cation and anion (ion pairing energy), and the reaction to form the ion pair $Na^+HCl_2^-$ becomes exothermic for charge separations less than 2.72 Å. For larger alkali metals these reactions become more exothermic as a consequence of the lower ionization potentials of the larger alkali metals. These simplistic thermodynamic arguments provided at least some basis for the anticipation that ion pair formation could occur under these conditions, and the formation of new and unusual anionic species might be feasible. These new species, to the degree that they can be viewed as distinct ion pairs, are themselves salts free from lattice interactions and information can be gained both about the structure and stability of the anion, and also about the nature of ion pairing. As indicated above, when alkali halide salts are vaporized both the monomer and dimer species are present, as well as higher polymers in lesser amounts. However, the dimer is sufficiently strongly bound, by 50-80 Kcal/mole, that it does not enter into the chemistry of these systems; instead, the reactions of the monomer only are observed.

The first application of this salt/molecule technique was to the study of the hydrogen bihalide anions, HX_2^- and HXY^-, where X and Y represent halogens. The HF_2^- anion is well known and is a very strongly hydrogen bonded, centrosymmetric anion in a variety of crystals (29). Studies of the HCl_2^- anion provided evidence for two different forms of the anion, a type I asymmetric anion and a type II centrosymmetric anion (30). Tetraalkylammonium cations were employed in this original work, and the cation was thought to determine whether the type I or the type II anion was formed. While these two anions, HF_2^- and HCl_2^- were well investigated, little was known about either HBr_2^- or HI_2^-, as well as the mixed halogen anions.

The matrix codeposition of CsF with HF (31,32) in an excess of argon, up to Ar/HF ratios of 3000, yielded intense bands at 1364 and 1218 cm^{-1}, within a few wavenumbers of the HF_2^- band positions in ionic lattices (29). The deuterium shift, upon formation of DF_2^-, was slightly above the harmonic value of 1.41, indicating that the anion maintained a center of symmetry. Similar results were obtained with other alkali fluoride salts, but the product yield decreased as the radius of the alkali metal cation decreased. These results provided immediate confirmation of the salt/molecule technique, demonstrating that for a known system ion pair formation occurred and that the spectrum of the product anion resembled closely that of the anion in known environments.

The reaction of Ar/HCl mixtures with alkali chloride salt molecules (33), and in particular CsCl gave rise to a very intense infrared absorption near 723 cm^{-1}. Deuteration gave evidence for a center of symmetry in the product HCl_2^- anion, consistent with the formation of a type II species. No evidence was obtained for formation of a type I HCl_2^- anion under these conditions, suggesting that type I anion formation in solution resulted from distortion of the more stable type II anion by certain bulky cations. Similar studies were conducted on both (34,35) the HBr_2^- and HI_2^- anions; HBr_2^- was characterized by a single, very intense infrared fundamental near 727 cm^{-1}, while HI_2^- showed a strong absorption near 673 cm^{-1}. Both anions showed distinct evidence for a center of symmetry, and were assigned as type II anions, while no evidence was obtained for a type I anion. Pimentel and coworkers observed similar, although sharper, bands when they passed a mixture of $Ar/HX/X_2$ through a microwave discharge prior to matrix condensation (36,37,38). They assigned these bands to the HX_2 radical, assignments which were disputed shortly thereafter by Milligan and Jacox (39,40). The close agreement of the band positions of the $M^+HX_2^-$ ion pairs to the bands observed by these researchers suggests that these features be reassigned to the isolated HX_2^- anions.

The formation of the mixed halogen anions is of interest as well; a center of symmetry is not possible under these conditions, and the potential exists for both types I and II anions. All of the combinations of halides were investigated, from $ClHF^-$ to $BrHI^-$, with alkali metal cations. The results were consistent throughout the entire series, and the spectra were considerably more complex than those of the HX_2^- anions. Moreover, each anion could be formed in either of two ways; the $BrHCl^-$ anion might be formed through either the reaction of HBr with CsCl or HCl with CsBr. Identical spectra were obtained in each case, suggesting that the reactions proceeded through identical transition states. Collectively, the spectra indicated formation of both type I and type II anions simultaneously, and the yield of each type was determined by the location of the cation in the ion pair. The type I/type II ratio varied with the cation; the smaller alkali metal cations favored the type II anion, while the larger cations favored the type I anion. Table I lists the position of the hydrogen stretching frequency ν_3 for each of the hydrogen bihalide anions studied.

In a related study (41), the role of the cyanide anion CN^- as a pseudohalogen was investigated; the reaction of either CsF with HCN or CsCN with HF yielded the $FHCN^-$ anion, ion paired with the Cs^+ cation. This anion appeared to have a very strong hydrogen bond, as anticipated, and the hydrogen stretching frequency occurred at 1800 cm^{-1}, indicating a type I anion, not a centrosymmetric analog of the HX_2^- anions.

Table I

Hydrogen Stretching Band Positions for the Hydrogen Bihalide Anions

Anion	Type	ν_3	Ref.	Anion	Type	ν_3	Ref.
HF_2^-	II	1364 cm^{-1}	32	HIF^-	I	2960 cm^{-1}	32
HCl_2^-	II	723	33	HIF^-	II	763	32
HBr_2^-	II	727	34	$HBrCl^-$	I	1110	34
HI_2^-	II	673	35	$HBrCl^-$	II	742	34
$HClF^-$	I	2500	32	$HICl^-$	I	1560	35
$HClF^-$	II	933	32	$HICl^-$	II	644	35
$HBrF^-$	I	2803	32	$HIBr^-$	I	920	35
$HBrF^-$	II	849	32	$HIBr^-$	II	666	35

Cesium Fluoride Catalysis Studies

A major application of the fluoride ion in solution, as well as solid CsF, is the catalysis of a wide range of reactions (42-45). The mechanism of this catalysis is not entirely clear; both addition and abstraction reactions have been postulated as the first step. The vaporization of individual alkali fluoride salt molecules and reaction with suitable partners before matrix condensation might serve as an appropriate model for this solution phase catalysis. One system in which CsF has been employed as a catalyst (46) is in oxidative fluorination reactions of F_2, which might proceed through the trifluoride anion, F_3^-, analogous to the known Cl_3^-, Br_3^- and I_3^- anions. The salt/molecule reaction was employed to study the reaction (47) of CsF with F_2 in argon matrices; a single intense infrared absorption was detected at 550 cm^{-1}, while the Raman spectrum of a similar sample showed an intense peak at 461 cm^{-1}. Similar results were obtained with RbF, the product bands showing virtually no cation shift, while with KF and NaF no distinct product bands were observed. These two bands have been assigned to the F_3^- anion in the $Cs^+F_3^-$ ion pair, and suggest that oxidative fluorination reactions may well proceed via this intermediate species. While a valence bond structural model cannot readily account for the F_3^- anion, simple triatomic molecular orbital theory predicts a spectrum similar to the isoelectronic XeF_2 and KrF_2 (22 valence electrons). The agreement between the spectra of XeF_2 and F_3^- is remarkably good, and supports this bonding model for the F_3^- anion.

These efforts were extended to the Cl_3^- and Br_3^- anions, as well as to the mixed trihalide anions. The reaction of alkali chloride salts with Cl_2 in argon gave rise to distinctive bands between 200 and 400 cm^{-1}, in both the infrared and Raman spectra (48). These have been assigned to Cl_3^-, which showed similar band positions in solution with bulky tetraalkylammonium cations (49). A shift in band position was observed with a change in cation, and the infrared and Raman spectra demonstrated that mutual exclusion did not hold. Since the Cl_3^- is expected to be centrosymmetric in its isolated form, by analogy to I_3^-, it is likely that the alkali metal cation is causing some perturbation, and lowering the spectroscopic symmetry of the Cl_3^- anion. It is noteworthy that the F_3^- was not detectably perturbed by the Cs^+ cation, and mutual exclusion did hold within the limits of detectability. The mixed anions have been synthesized as well, for all combinations of F, Cl and Br (50,51). In the cases in which two product anions were possible, such as the reaction of CsF with ClF to form either the $FClF^-$ or $ClFF^-$ anions, both reactions were observed to occur. Recently, Andrews (52) and coworkers completed this series of studies by investigating the matrix isolated I_3^- anion. The vibrational frequencies of this anion are too low to be detected in an infrared spectrum, and only Raman spectra were recorded. I_3^- has an electronic absorption in the visible portion of the spectrum, so that resonance enhancement of the Raman spectrum was observed, as well as an overtone progression.

A second class of chemical compounds whose reactions are catalyzed by the fluoride ion, and by CsF are organosilanes (42). The diversity of chemistry exhibited by these compounds is considerable, and with the potential for expanded valence in these compounds to five and six coordinate intermediate anions, the application of the salt/molecule technique was suggested. The gas phase ion/molecule reactions of several silane systems have been investigated through ion cyclotron resonance techniques (53,54), providing thermochemical information about the product ions in these systems. Perhaps the simplest reaction is that of F^- with SiF_4; the intermediate anion SiF_5^- has been stabilized under carefully controlled circumstances (55), while the 2:1 adduct $SiF_6^=$ is well known (56). Less is known about the effects of alkylation on the stability of the product ions.

Salt/molecule studies (57) of the reaction of either CsF or KF with SiF_4 diluted in argon gave rise to a series of product bands over concentration ranges from 200/1 to 1000/1. Bands were observed at 812, 855 and 932 cm^{-1} with CsF, as well as weaker bands between 400 and 500 cm^{-1} as can be seen in Figure 2. These have all been assigned to the SiF_5^- anion in reasonably good agreement with the crystalline spectrum of the anion. No evidence was detected for the $SiF_6^=$ anion until the sample was allowed to diffuse at somewhat elevated temperatures (45 K). Under these conditions, the bands assigned to SiF_5^- disappeared, and intense bands grew in near 730 and 480 cm^{-1}, which were assigned to the $SiF_6^=$ anion. The observation of three bands above 800 cm^{-1}, the Si-F stretching region, demonstrated that the symmetry of the anion in the ion pair is not strictly the D_{3h} trigonal bipyramid structure anticipated from valence bond considerations. Rather, the local symmetry is lowered by the alkali metal cation, either activating an otherwise forbidden

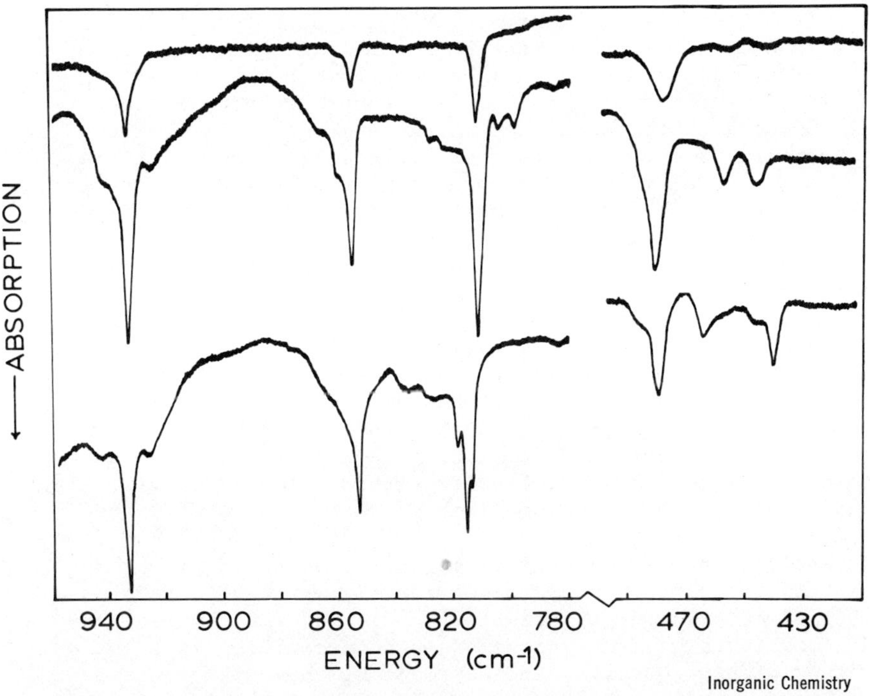

Inorganic Chemistry

Figure 2. IR spectra of reaction products of CsF and KF with samples of Ar/SiF$_4$ at varying dilution, over the spectral regions of interest (57). Key: top, reaction products of CsF with a sample of Ar/SiF$_4$ = 1000; middle, comparable reaction at a dilution of 400:1; and bottom, reaction products of KF with a sample of Ar/SiF$_4$ = 150.

mode of the anion, or removing the degeneracy of the E' equatorial fluorine stretching vibration. The preferred model is an axial interaction of the alkali metal cation with one axial fluorine, activating the symmetric axial Si-F stretch but maintaining the degeneracy of the antisymmetric equatorial stretch. However, the C_{2v} and C_{4v} symmetries could not be conclusively ruled out.

The synthesis of the pentavalent mixed halogen silicate anion was achieved in part as well. The reaction of CsCl with SiF_4 yielded the $SiClF_4^-$ anion, for example, and the series was extended up to and including the $SiCl_4F^-$ anion. However, attempts to synthesize the $SiCl_5^-$ anion with either the Cs^+ or K^+ cation were unsuccessful. This may indicate that either this anion is not stable, contrary to one published report, or that there is some inherent limitation to the salt/molecule technique. This point will be discussed in more detail below.

The dependence of the stability and reactivity of the pentacoordinate silicon anions on the number of alkyl groups is of considerable interest, as most organosilicon chemistry is carried out on alkylsilane compounds, not SiF_4. Much less is known about the potential product anions in these studies; an ICR study (54) did indicate that the fluoride ion affinity of the parent silane decreased dramatically with an increasing number of methyl groups. The reaction (58) of CsF with (CH_3)-SiF_4 in argon gave rise to four product bands, at 742, 825, 837 and 862 cm^{-1}, which might be anticipated for a C_{2v} product anion. These bands were formed with relative ease, suggesting that the product anion is quite stable. However, efforts to synthesize the $(CH_3)_2SiF_3^-$ anion were marginally successful. Weak bands were observed in the Si-F stretching region, but their breadth and lack of intensity indicated that this anion is only weakly bound at best under these conditions. Attempts to form the remaining members of the series, $(CH_3)_3SiF_2^-$ and $(CH_3)_4SiF^-$, were completely unsuccessful, supporting the conclusion from the ICR data that the stability decreases rapidly with an increasing number of methyl groups. However, these results suggest that pentacoordinate silicon anions are stable entities in solution, although highly reactive, and may well be the active intermediates in organosilane reactions catalyzed by CsF.

Another example (59) of the catalytic role of the fluoride ion is in the chemistry of COF_2, including the reaction of COF_2 with F_2 in the presence of CsF. This reaction proceeds readily at -78°C and is thought to involve attack of the F^- on COF_2 to form the COF_3^- anion. This anion was first detected in 1965, but only recently characterized spectroscopically (60). Even here, a very broad spectrum was obtained, as a consequence of the crystalline environment. The matrix reaction of CsF with COF_2 was investigated to determine the solvent-free chemistry of this system, as well as to possibly sharpen up spectral features (61). At Ar/COF_2 dilutions as high as 3000/1, intense product bands were observed at 555, 808, 919, 1039 and 1514 cm^{-1}, which were assigned to the COF_3^- anion. These occur near the absorptions observed in the crystalline spectrum, but are much sharper and better resolved. However, the anion shows less than C_{3v} symmetry, which was determined for the crystalline anion, and this too can be attributed to the presence of the alkali metal cation. The doubly degenerate mode is split into components at 919 and 1039 cm^{-1}, from a single E mode at 960 cm^{-1} in the crystal. This splitting, which is typical of ion pairing effects on degenerate bands, will be discussed in more detail below. The band at 1514 cm^{-1} was assigned to the carbonyl stretching vibration and shows character intermediate between a single and double bond. While this reaction proceeded rapidly at high dilutions to yield intense product bands, the reactions of CsCl with COF_2 and CsF with COClF, either of which might have yielded the $COClF_2^-$ anion, did not occur. This highlights the considerable difference in chemistry between the chloride and fluoride ions, not only in solution but also at relatively high temperatures in the gas phase.

One simple anion which has been of considerable interest to inorganic chemists is the fluoroformate anion, CO_2F^-. Several attempts (62,63) have been made at the solution or solid phase synthesis of this anion, and all have been unsuccessful, even though this anion is isoelectronic with the very stable carbonate anion, $CO_3^=$. However, gas phase ICR (64) results indicate that the fluoride ion affinity of CO_2 to form the CO_2F^- anion is roughly 32 Kcal/mole. Typical condensed phase synthetic attempts involved subjecting CsF powder to a high pressure CO_2 atmosphere at elevated temperatures. It has been postulated (62) that the lack of reaction, in view of the predicted stability of the anion, is due to the high lattice energy of CsF, which must be overcome during the reaction. However, the high temperature salt/molecule technique overcomes the lattice energy requirement by vaporization of the salt prior to reaction. Hence, application of this technique could well lead to the synthesis of the CO_2F^- anion.

The matrix codeposition of CsF molecules (65) with samples of Ar/CO_2 at dilutions between 250/1 to 1000/1 gave rise to a set of sharp, relatively intense bands at 870, 1310 and 1750 cm^{-1}, along with weak, broad bands near 810, 1290 and 1800 cm^{-1}. When such a sample was annealed to 45 K and recooled, the sharp bands diminished in intensity and the broad bands grew in. Deposition of CsF into a pure CO_2 matrix gave rise to these broad product bands as well. The codeposition of CsF with a sample of $Ar/^{13}CO_2$ gave rise to a shifted set of sharp bands, indicative of vibrations involving the central carbon atom. These bands have been assigned to the C-F and the two C-O stretching vibrations in the CO_2F^- anion, ion paired with Cs^+. This anion should have two C-O stretching vibrations near the E mode of the $CO_3^=$ anion, as these vibrations are derived from the E mode upon reducing the symmetry from D_{3h} to C_{2h}. The E mode of the carbonate anion occurs at 1495 cm^{-1}, nearly halfway between the 1310 and the 1750 cm^{-1} bands. The 870 cm^{-1} band is in the C-F stretching region for the similar COF_3^- anion, and is easily assigned to the C-F stretch of the CO_2F^- anion. While characterization of the fluoroformate anion is still in progress, this system provides another example of the utility of the salt/molecule technique, allowing for the synthesis of an unusual anionic species without regard to the lattice energies of the reactants or the products.

Ion Pairing Effects

Salt/molecule reactions, as used here in conjunction with the matrix isolation technique, have been employed to mimic high temperature gas phase ion-molecule reactions. However, this is not entirely correct; there are distinct, observable effects due to the presence of the alkali metal cation. It might well be questioned whether or not the product species represent true ion pairs. However, the observation of only small shifts in band positions with a change of cation, and the good agreement of matrix band positions with crystal data when available suggest that the species are very ionic if not true ion pairs. However, the alkali metal cation does manifest itself in the splitting of degenerate vibrational modes for a number of anions, and the magnitude of splitting is cation-dependent. Also, this technique has proven unsuccessful at synthesizing a number of species which might reasonably have been expected to be stable, such as $COCl_3^-$ and $SiCl_5^-$. This may be due to either an inherent lack of stability of the anions or to a fundamental limitation of the technique.

While the species formed are ion pairs rather than the isolated anion of interest, information can be gained about the nature of the cation-anion interaction. These species can be considered contact ion pairs, in view of the nonpolar nature of the argon "solvent", and analyzed in a manner similar to that used for contact ion pairs in solution (66). There has been an upsurge of interest (67) in ion pairing effects in solution, and the differences in chemical reactivity of ions in different types of ion pairing arrangements, such as contact ion pairs and solvent-separated ion pairs. The matrix isolated ion pair can be used as a model to gain information about the contact ion pair in nonpolar solutions. Devlin and coworkers have exploited this approach in a number of studies, investigating the alkali nitrate, chlorate and perchlorate ion pairs in inert matrices (21,22,23) after direct vaporization of the salt. Generally, the alkali metal cation was found to be in a bidentate arrangement with two oxygens of the anion under investigation, causing a splitting of the degenerate stretching mode of the anion. Normal coordinate calculations were used to model this splitting by lowering of the N-O or Cl-O force constant of the coordinated oxygens relative to the noncoordinated oxygens. These workers also included the possibility of a nonzero force constant between the metal cation and the coordinated oxygens, and were able to reproduce the observed spectra quite well for a variety of cations. The difference, ΔK, between the force constants of the coordinated and free oxygens was found to be substantial, on the order of 3 mdyne/Å, suggesting that ion pairing effects are very important in these matrix isolated systems. The interaction of the cation with the anion was described in terms of polarization of the anion by the cation, and force constant trends analyzed relative to the polarizing power of the cations. This provided an adequate but not perfect description, so that a covalent interaction term was introduced as well, following the lead of earlier workers (68). A similar study (69) of the $M^+BF_4^-$ ion pair in this laboratory led to comparable results, including a splitting of the F_2 stretching mode of the anion, and a ΔK value of 4.17 mdyne/Å. Figure 3 shows a plot of the B-F stretching vibrations as a function of ΔK in a monodentate, C_{3v}, arrangement.

An alternative view of the interaction of an alkali metal cation with a fluoride-containing anion is one of Lewis acid/base competition. The reactions discussed in the preceding section involved the reaction of an alkali fluoride salt with a Lewis acid with subsequent fluoride ion transfer to the Lewis acid. However, the alkali metal cation is a Lewis acid as well, and the degree of perturbation of the anion by the cation may be dependent on the differences in fluoride ion affinity of the Lewis acid and the alkali metal cation. The fluoride ion affinities for a variety of Lewis acids are well known from ICR (53,54,64) studies, while the fluoride ion affinities for alkali metal cations are the heterolytic bond dissociation energies of the gas phase alkali fluoride molecules

$$MF \rightarrow M^+ + F^-$$

These can be evaluated through a thermodynamic cycle (70), and the results are given in Table II for both Lewis acids and alkali metal cations. Clearly, the alkali metal cations are stronger Lewis acids than the neutral species listed in Table II;

Table II

Lewis Acid	F^- Affinity	M^+	F^- Affinity
HF	50 Kcal/mole	Li^+	182 Kcal/mole
SiF_4	68	Na^+	153
BF_3	71	K^+	139
COF_2	34	Rb^+	133
CO_2	32	Cs^+	130
PF_5	>71	Tl^+	162

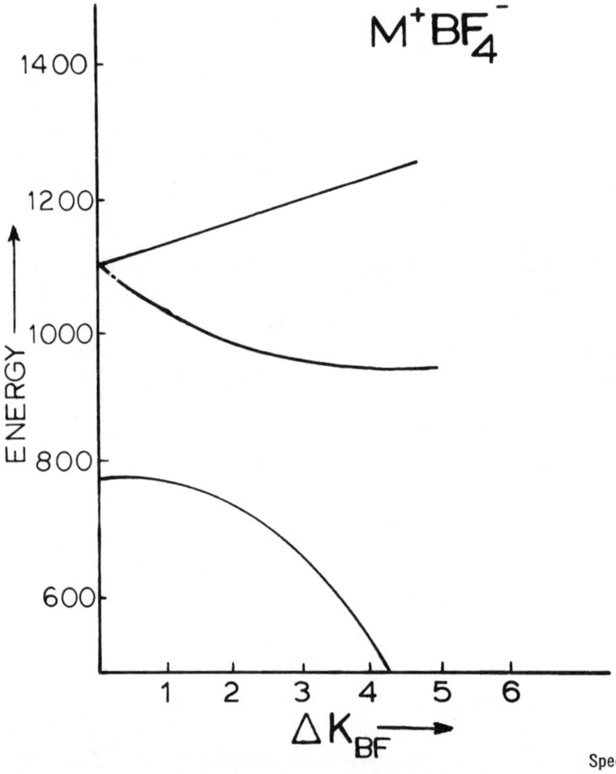

Spectrochimica Acta

Figure 3. A plot of the B–F stretching frequencies of the BF_4^- anion ion paired in a monodentate arrangement with an alkali metal cation, as a function of ΔK, the difference in force constant between coordinated and noncoordinated fluorines (69).

however, energy is gained back through the electrostatic attraction in an ion pair, and ion pair formation is feasible. Nonetheless, the difference in fluoride ion affinity between the Lewis acid and metal cation may determine the degree of perturbation and anion distortion. Certainly, Table II demonstrates why Cs^+ is generally the cation of choice for both the best yield of product ion, and the least perturbation of the anion. These values also suggest that when the metal cation has a sufficiently high fluoride ion affinity relative to a given Lewis acid, product formation will not occur. This was observed for the F_3^- system; reaction of CsF or RbF with F_2 gave large product yields, while the reaction of KF or NaF with F_2 did not yield any distinct product species. These results indicate that there is a limit to the effectiveness of the salt/molecule technique, and the lack of formation of such species as $COCl_3^-$ may be due to this limit rather than the inherent stability of the anion.

Clearly, the effects of ion pair formation in matrix isolation spectra are apparent, particularly in the removal of degeneracy of vibrational modes. However, these can be fit quite well with normal coordinate calculations which show that ion pairing effects cause significant force constant changes. Moreover, it is apparent that the energetics of ion pair formation do provide a limit as to the species which can be studied in this fashion. However, for the species which can be stabilized in this manner, the technique is still effective and provides a model for both ion-molecule reactions and CsF catalysis.

Alkali Halide Salt Reactions with Lewis Bases

The application of the salt/molecule reaction technique to the study of reactions with Lewis bases such as H_2O and NH_3 presents the possibility for a different type of interaction which may find some analogy in transition metal coordination chemistry. The structure of small complexes such as $MX \cdot H_2O$ are of considerable interest both experimentally and theoretically. These studies were initiated as a result of the observation of several bands in the spectrum of alkali halide salts in argon which could not readily be assigned to the isolated salt species. Rather, it was shown that these bands were due to reaction of the salt with impurity H_2O, which was always present in these experiments to some degree. A study was then initiated to investigate these bands, and the nature of the reaction complex.

Numerous experiments (71) were conducted using a wide range of alkali halide salts, including all of the halogens, and all alkali metals except Li. In addition, several cyanide salts were also employed. The resulting band positions were relatively consistent, except when CsF was employed, where very different spectra were obtained. This was rationalized in terms of the unique reactivity of the fluoride ion, and the focus of the study was directed toward the remaining salts. In most cases, four product bands were observed, two in the region 3100-3250 cm^{-1}, and two between 400 and 700 cm^{-1}. Complete isotopic studies, using both deuterium and ^{18}O labeling, indicated that all four vibrations were hydrogenic in nature and that the two hydrogens were equivalent in the complex. A number of structures for the complex were considered; a definite dependence of the band positions on M^+ was noted, as well as a slight dependence on X^-. In addition, the positions of the two upper bands, which are assigned to O-H stretching modes, indicated some hydrogen bonding was occurring. The remaining features in the spectrum were reminiscent of coordinated aquo complexes suggesting a definite M^+ interaction, which was consistent with the M^+ dependence of the band positions. The favored structure of the complex was a C_s symmetry trigonal pyramid with the metal cation bound directly to the oxygen of the H_2O and the anion sitting at the base of the pyramid, interacting electrostatically with the alkali metal cation and hydrogen bonding to the two hydrogens. Other structures might be envisioned for the complex which fit the observed data but there was insufficient data to reach a more definitive conclusion.

Devlin and coworkers (24,25,26) studied the interaction between metal nitrate and chlorate ion pairs and water, and concluded that the interaction is between the metal cation and the oxygen on the coordinated H_2O. In addition, they observed some evidence for hydrogen bonding in the second coordination sphere similar to the results here. These researchers worked in a quite different concentration regime and monitored the complex by observing perturbed bands of the anion rather than the water molecule, but the results were nonetheless consistent.

The analogous complex between a salt molecule and NH_3 was studied as well to investigate similarities to ammine coordination chemistry. The deposition of a variety of salt molecules with mixtures of argon and NH_3 gave rise to several product bands which were indicative of a strongly bound complex (72). The most prominent feature was an intense band near 1100 cm^{-1}, which showed a distinct metal cation dependence and very little anion dependence. This band has been assigned to the NH_3 deformation mode in the complex which is generally the most intense in coordinated ammine complexes (73), shifts up from free NH_3 near 970 cm^{-1}, and is very sensitive to the strength of interaction and hence the cation. Other modes detected included two N-H stretching modes near 3400 cm^{-1} and an NH_3 rocking mode at low frequencies. The N-H stretching modes were not shifted by a significant amount from free NH_3, suggesting that no hydrogen bonding was taking place, and the interaction was directly through the metal cation and the nitrogen lone pair as in traditional coordination chemistry. Similar results were obtained by Devlin (24,25,26) et al. for the complexes formed between the metal nitrate salts and NH_3 in argon matrices, although the conclusions were again based on the shifts of the NO_3^- bands, not NH_3 complex bands.

These studies were extended briefly to a number of other Lewis bases, including dimethyl ether to look for trends relating to base strength or proton affinity (74). The major spectral features of the complexes were shifted vibrational modes of the base, but these shifts were much less substantial than for H_2O or NH_3, and very little information about the nature or structure of the complex could be extracted.

Conclusions

The reactions of alkali halide and other salt molecules in the gas phase are of considerable interest to high temperature chemists; reactions of CsF in either the gas phase or condensed phases are also of interest to catalytic chemists. The matrix isolation technique has proven itself valuable in the area of high temperature chemistry; while inert matrices are condensed at 15 K, the technique allows a high temperature reaction to be initiated in front of the cold surface and then rapidly quenched to trap the initial products of the high temperature reaction.

Salt/molecule reactions with fluoride acceptors or Lewis acids appear, in all cases studied to date, to form ion pair complexes as the initial reaction product for such Lewis acids as HF, SiF_4, BF_3 and COF_2. There has been no evidence of abstraction as the initial step, although some systems may exist where this step will be preferred. Some spectroscopic effects of ion pairing are observed in the spectra of the product anions, but these still allow for the study of unusual anions in a relatively unperturbed environment.

The reaction of a salt molecule with Lewis bases has been effective as well in forming intermediate complexes. In the systems studied, the complexes appear bound through the metal cation of the salt molecule to the lone pair on the Lewis base. However, in the case of H_2O, there appears to be some hydrogen bonding interaction as well. These products generally mimic transition metal coordination complexes, although the strength of the interaction is considerably weaker as might well be expected.

The salt molecule technique, in conjunction with matrix isolation, appears to be a promising technique for the study of high temperature reactions, and a further development of this approach should be beneficial. Extension to additional ion pair adducts would appear feasible, allowing for the study of unusual fluoride-containing anions. Extension may also be possible to oxide salt vaporization and reaction to form oxyanions with a -2 charge in matrix isolated triple anions, but this avenue has only been briefly explored. Perhaps the results obtained to date and described here as well as the numerous studies which could not be mentioned, will stimulate further study, both experimental and theoretical, in this area of high temperature chemistry.

Acknowledgments

The author gratefully acknowledges support of much of the work described above by the Research Corporation under grant #8305, the National Science Foundation under grant CHE78-27643, and the Henry and Camille Dreyfus Foundation. The considerable assistance of Dr. Lester Andrews, particularly in the early phases of this research, is also gratefully acknowledged.

Literature Cited

1. Andrews, L. Ann. Rev. Phys. Chem. 1971, 22, 109.
2. Pimentel, G. C.; "Formation and Trapping of Free Radicals" A. M. Bass and H. P. Broida, ed. Chapt. 4, Academic Press, New York, 1960.
3. Milligan, D. E.; Jacox, M. E. in "Physical Chemistry, An Advance Treatise, 4" Chapter 4, Academic Press, New York, 1969.
4. Craddock, S.; Hinchcliffe, A. J. "Matrix Isolation" Cambridge University Press, New York, 1975.
5. Linevsky, M. J. J. Chem. Phys. 1961, 34, 587.
6. Snelson, A.; Pitzer, K. S. J. Phys. Chem. 1963, 67, 882.
7. Schick, S.; Schnepp, O. J. Chem. Phys. 1964, 41, 463.
8. Ismail, Z. K.; Hauge, R. H.; Margrave, J. L. J. Mol. Spectrosc. 1975, 54, 402.
9. Mann, D. E.; Calder, G. V.; Seshadri, K. S.; White, D.; Linevsky, M. J. J. Chem. Phys. 1967, 46, 1138.
10. Hastie, J. W.; Hauge, R. H.; Margrave, J. L. Ann. Rev. Phys. Chem. 1970, 21, 475.
11. Andrews, W. L. S.; Pimentel, G. C. J. Chem. Phys. 1966, 44 2361.
12. Andrews, L. J. Chem. Phys. 1969, 50, 4288.
13. Milligan, D. E.; Jacox, M. J. J. Chem. Phys. 1969, 51, 2150.
14. Brus, L. E.; Bondybey, V. E. J. Chem. Phys. 1975, 63, 3123.
15. Prochaska, F. T.; Andrew, L. J. Chem. Phys. 1978, 68, 5577.
16. Wight, C. A.; Ault, B. S.; Andrews, L. J. Chem. Phys. 1976, 65, 1244.
17. Jacox, M. E. Chem. Phys. 1976, 12, 51.
18. Andrews, L.; Tevault, D.; Smardzewski, R. R. Appl. Spectrosc. 1978, 32, 157.
19. Andrews, L.; Grzybowski, J. M.; Allen, R. O. J. Phys. Chem. 1975, 79, 904.
20. Prochaska, F. T.; Andrews, L. J. Chem. Phys. 1977, 67, 1091.
21. Smith, D.; James, D. W.; Devlin, J. P. J. Chem. Phys. 1971, 54, 4437.
22. Smyrl, N.; Devlin, J. P. J. Chem. Phys. 1974, 60, 2540.
23. Ritzhaupt, G.; Devlin, J. P. J. Chem. Phys. 1975, 62, 1982.
24. Ritzhaupt, G.; Devlin, J. P. J. Phys. Chem. 1975, 79, 2265.
25. Smyrl, N.; Devlin, J. P. J. Chem. Phys. 1974, 61, 1596.
26. Ritzhaupt, G.; Devlin, J. P. Inorg. Chem. 1977, 16, 486.

27. Kana'an, A. S.; Hauge, R. H.; Margrave, J. L. J. Chem. Soc. Faraday Trans. II 1976, 1991.
28. Snelson, A.; Cyvin, B. N.; Cyvin, S. J. J. Mol. Struc. 1975, 24, 165.
29. Jones, L. H.; Penneman, R. A. J. Chem. Phys. 1954, 22, 781.
30. Evans, J. C.; Lo, G. Y-S. J. Phys. Chem. 1966, 70, 11.
31. Ault, B. S. J. Phys. Chem. 1978, 82, 844.
32. Ault, B. S. J. Phys. Chem. 1979, 83, 837.
33. Ault, B. S.; Andrews, L. J. Chem. Phys. 1975, 63, 2466.
34. Ault, B. S.; Andrews, L. J. Chem. Phys. 1976, 64, 1986.
35. Ellison, C. M.; Ault, B. S. J. Phys. Chem. 1979, 83, 832.
36. Noble, P. N.; Pimentel, G. C. J. Chem. Phys. 1968, 49, 3165.
37. Bondybey, V.; Pimentel, G. C.; Noble, P. N. J. Chem. Phys. 1971, 55, 540.
38. Noble, P. N. J. Chem. Phys. 1972, 56, 2088.
39. Milligan, D. E.; Jacox, M. J. J. Chem. Phys. 1970, 53, 2034.
40. Milligan, D. E.; Jacox, M. J. J. Chem. Phys. 1971, 55, 2550.
41. Ault, B. S. J. Phys. Chem. 1979, 83, 2634.
42. Clark, J. H. Chem. Rev. 1980, 80, 429.
43. Corey, E. J.; Venkateswarill, A. J. Amer. Chem. Soc. 1974, 94, 6190.
44. Kraihanzel, C. S.; Poist, J. E. J. Organomet. Chem. 1967, 8, 239.
45. Chan, T. H.; Mychajlowski, W. Tetrahedron Lett. 1974, 3479.
46. Hohorst, F. A.; Shreeve, J. M. J. Amer. Chem. Soc. 1967, 89, 1809.
47. Ault, B. S.; Andrews, L. J. Amer. Chem. Soc. 1976, 98, 1591.
48. Ault, B. S.; Andrews, L. J. Chem. Phys. 1976, 64, 4853.
49. Evans, J. C.; Lo, G. Y-S. J. Chem. Phys. 1966, 44, 3638.
50. Ault, B. S.; Andrews, L. Inorg. Chem. 1977, 16, 2024.
51. Miller, J. H.; Andrews, L. Inorg. Chem. 1979, 18, 988.
52. Andrews, L.; Prochaska, E. S.; Loewenschuss, A. Inorg. Chem. 1980, 19, 463.
53. Haartz, J. C.; McDaniel, D. H. J. Amer. Chem. Soc. 1973, 95, 8562.
54. Murphy, M. K.; Beauchamp, J. L. J. Amer. Chem. Soc. 1977, 99, 4992.
55. Clark, H. C.; Dixon, K. R.; Nicholson, J. G. Inorg. Chem. 1969, 8, 450.
56. Begun, G. M.; Rutenberg, A. C. Inorg. Chem. 1967, 6, 2212.
57. Ault, B. S. Inorg. Chem. 1979, 18, 3339.
58. Ault, B. S.; Tandoc, U. Inorg. Chem. in press.
59. Lustig, M.; Pitochelli, A. R.; Ruff, J. K. J. Amer. Chem. Soc. 1967, 89, 2841.
60. Christie, K. O.; Curtis, E. C.; Schack, C. J. Spectrochim. Acta A 1975, 31, 1035.
61. Ault, B. S. J. Phys. Chem. 1980, 84, 3448.
62. Lawlor, L.; Passmore, J. Inorg. Chem. 1979, 18, 2923.
63. Martineau, E.; Milne, J. B. Chem. Comm. 1971, 1327.
64. McMahon, T. C.; Northcott, C. J. Can. J. Chem. 1978, 56, 1069.

65. Ault, B. S. to be published.
66. Edgell, W.; Lyford, J. J. Amer. Chem. Soc. 1971, 93, 6407.
67. Hogen-Esch, T. E.; Smid, J. J. Amer. Chem. Soc. 1966, 88, 307.
68. Hester, R. E.; Scaife, C. W. J. J. Chem. Phys. 1967, 47, 5253.
69. Hunt, R. L.; Ault, B. S. Spectrochim. Acta Part A, 1981, 37, 63.
70. Handbook of Chemistry and Physics, Chemical Rubber Co., 50th ed.
71. Ault, B. S. J. Amer. Chem. Soc. 1978, 100, 2426.
72. Ault, B. S. J. Amer. Chem. Soc. 1978, 100, 5773.
73. Nakamoto, K. "Infrared Spectra of Inorganic and Coordination Compounds" 3rd ed., Wiley-Interscience, New York, 1978.
74. Ault, B. S. unpublished results.
75. Kallendorf, C. J.; Ault, B. S. J. Phys. Chem. 1981, 85, 608.

RECEIVED August 26, 1981.

23

Electronic Absorption Spectroscopy of Aluminum Metal Atom Hydration Products Isolated in Rare Gas Matrices at 15 K

M. A. DOUGLAS, R. H. HAUGE, and J. L. MARGRAVE

Rice University, Rice Quantum Institute and Department of Chemistry, Houston, TX 77001

Assignments are made for the electronic transitions corresponding to the Al metal atom, its dimer, the insertion product of Al and H_2O (HAlOH), and the HAlOH photolysis products: AlO and AlOH. Al_2 shows a (0,0) transition at 4049 Å with an average $\Delta G'_{v+\frac{1}{2}}$ of 252 cm^{-1}. HAlOH displays a very broad absorption with a short wavelength onset at 4000 Å tailing out to about 6300 Å. AlOH has a broad absorption with a maxima at 2490 Å.

Much recent interest has attended the characterization of the oxidative Group IIIA metal atom hydration reaction intermediates and products. Margrave and co-workers (1) have performed infrared matrix isolation studies of the interactions of Group IIIA metal atoms with H_2O demonstrating that Al atoms spontaneously undergo oxidative insertion to form the divalent HMOH species, whereas Ga, In, and Tl atoms form the M...OH_2 adduct as evidenced by a decrease in the ν_2 bending mode of H_2O. Upon photolysis, each of these adducts except for Tl...OH_2 form the HMOH molecule and extended photolysis of HMOH generates the monovalent MOH molecule.

Theoretical calculations (2,3) have been performed on the M...OH_2 and HMOH molecules to evaluate their stabilities, geometries and reaction pathway potential energy barriers. Preliminary results from Kurtz and Jordan (2) indicate the HAlOH molecule is nonlinear, experiences no potential energy barrier upon formation from Al + H_2O, and is stable by at least 38 kcals/mole relative to Al + H_2O. These theoretical predictions are corroborated by experimental results. Margrave and co-workers (1) observed spontaneous insertion of the Al atom into the OH bond of H_2O to form HAlOH species of C_s symmetry. Oblath and Gole (4) proposed that the excited HAlOH molecule was a plausible candidate to explain the white-blue chemiluminescent emission continuum emanating from

0097-6156/82/0179-0347$05.00/0

the oxidative interaction of Al and H_2O vapors under single and multiple collision conditions.

Finally, there is some controversy surrounding the characterization of the visible chemiluminescent continuum emitter observed upon gas-phase oxidation of aluminum or aluminum-containing compounds. In particular, upper atmospheric (90-220 km) seeding studies of winds, diffusion, temperature and other atmospheric properties involve the release of trimethyl aluminum (TMA) or the explosion of aluminum grenades in the upper atmosphere which at nighttime generates a chemiluminescent glow characterized by an unresolvable emission continuum. This chemiluminescent continuum has been reproduced in a number of laboratories, and a variety of suggestions with regard to the nature of the continuum emitter have been forwarded.

Oblath and Gole (4) have reviewed these aluminum oxidation studies whose combined work encompasses a wide variety of potential upper atmospheric oxidizers. However, only Oblath and Gole (4) have investigated the oxidative gas-phase hydration of aluminum vapors, finding that the interaction, under single and multiple collision conditions, generates a chemiluminescent flame having a spectrum very reminiscent of the upper atmospheric glows. Further investigation by this group of other oxidizers (O_2, SO_2, CO_2 and H_2O_2) interacting with aluminum vapors yielded no observable chemiluminescence. In contrast, previous studies (5) of the reaction of aluminum vapors with N_2O, O_2, and CO_2 reported the observation of chemiluminescence similar to the upper atmospheric glows. Oblath and Gole note that intrinsic to the detection of fluorescence is an enhanced sensitivity to contamination and suggest that previous investigations (5) may have been plagued by water contamination. They also point out that most of the water vapor present in the upper atmosphere is deposited by the payload rockets dispersing the TMA and aluminum grenades. Based on the limited possibilities for the emitter from single-collision results and after a consideration of reaction and excited state energetics, Oblath and Gole (4), in conjunction with matrix-isolation studies, suggest that the continuum emitter is a complex involving aluminum and several water molecules or a 1:1 aluminum-water adduct that rearranges to form aluminum hydroxy-hydride in an excited state, HAlOH*.

In view of the matrix-isolation infrared studies, the theoretical calculations, and those studies of aluminum beam-water reactions, it is of considerable interest to probe the electronic structure of aluminum metal atom hydration reaction intermediates and products isolated in rare-gas matrices. Such a study will form a useful adjunct to the infrared research. A mapping of the electronic structure of these species will provide a data set to which further theoretical calculations may compare. A more focused elucidation of the nature of the gas-phase aluminum hydration reaction's chemiluminescent continuum emitter may be provided since the HAlOH molecular species may be isolated and spectroscopically analyzed in a rare gas mixture, and its stability may be probed by selective photolysis. In addition, this study will report the spectrum of the AlOH molecule, which is a photolytic product of the HAlOH species. There is interest in the electronic structure of the AlOH molecule since it is a member of the HBA class of compounds where B is more electronegative than A (6). Quantum-chemical calculations predict the first electronically excited states of AlOH are at least 30,000 cm^{-1} above the $^1\Sigma^+$ ground state (6).

Experimental

The matrix-isolation apparatus employed in this study is described elsewhere (7). Briefly, cryogenic temperatures are achieved utilizing an Air Products Displex Closed Cycle Helium Refrigerator, Model CS-202. Matrices are formed on aluminum coated, second surface, UV grade sapphire mirrors in good thermal contact with an aluminum block which, in turn, is thermally bound to the second stage of the refrigerator. UV-visible absorption spectra are taken utilizing a 1/4 meter Perkin-Elmer 210 grating monochromator (calibrated with mercury emission lines), a Hamamatsu R-928 photomultiplier tube, and a pair of PAR phase sensitive amplifiers (Models 128A and 186). The optical design is such as to allow referencing of the host matrix to circumvent severe light scattering problems that normally plague the UV spectra. A quartz crystal microbalance mounted on the aluminum cold block allows measurement of the background contaminant, the host, and the guest species deposition rates thereby permitting highly accurate calculation of the M/R ratios. Thermocouple gauges positioned upstream from the gas ports are used to monitor the faithfulness of the gas flow rates after deposition rates have been established with the microbalance.

The matrices were photolyzed subsequent to trapping with a 100-watt short-arc mercury lamp focused down to a 1-inch diameter spot. Photolysis times and filters utilized in wavelength selective studies will be reported for each experiment in the next section.

Aluminum was vaporized from a molybdenum cell with carbon spacers to support the cell off a resistively heated, 1-mil Ta furnace. The furnace-cell assembly was outgassed for two hours previous to deposition at a temperature of about 950°C. System pressure after cooldown and oven outgassing was less than 1×10^{-7} torr. M/R ratios were measured and will be reported for each experiment. Typical Kr/Al ratios range from 750/1 to 1500/1 which corresponds to deposition rates of 2Hz/minute for Al and 7000 Hz/minute for Kr on the microbalance. The background drift rate was less than 0.5 Hz/minute. The depositions took 15 minutes, and the aluminum cold block temperature was 15K. The aluminum cold block temperature was continually monitored by measuring the resistance of a 68 Ω (298K) Allan-Bradley carbon resistor sunk into the cold block and calibrated with a chromel vs. gold -0·07% at Fe thermocouple. H_2O and D_2O were vacuum distilled going through at least five freeze-thaw cycles. High purity krypton was used (Matheson Research Grade, 99.995%) with a liquid nitrogen cold trap.

Results

Co-condensation of aluminum atoms with H_2O molecules at 15K in an excess of Kr gives a transparent, colorless matrix.

Figure 1 shows the electronic absorption spectra corresponding to four experiments: Al/Kr; Al/H_2O/Kr; Al/H_2O/Kr + hν; and Al/D_2O/Kr. Table 1 lists the assignments corresponding to band maxima for each of these spectra. Figure 1(a) shows the strongly allowed electronic resonance transitions of the Al atom. Comparison with gas phase transition energies (8,9) allows assignment of the peaks denoted by "a" to the $4s(^2S_{1/2}) \leftarrow 3p(^2P_{1/2})$ resonance transition and "b" to the $3d(^2D_{3/2}) \leftarrow 3p(^2P_{1/2})$ transition. The major "a" and "b" peaks are blue-shifted by 1900 cm^{-1} and 500 cm^{-1}, respectively, relative to their respective gas phase transitions. These expected matrix blue-shifts are attributable to a strongly repulsive matrix interaction in the atom's excited state (10). Subordinate "a" and "b" bands may also be attributed to the Al atom in multiple and/or low symmetry trapping sites since a series of dilution and contaminant dopant (O_2, CO, H_2O) experiments show no relative change in these peak intensities. ESR results (11,13) indicate there are three sites for Al/Kr. These assignments basically concur with those made by Ammeter and Schlosnagle (11) for atomic aluminum. However, at higher dilutions (Kr:Al > 1000) the "c" bands disappear yet the "d" bands persist at the same intensity relative to the other atomic bands. Also, upon annealing, the 390.2 nm "a" band disappears, revealing a previously buried, weak third "c" band at 396.8 nm forming a Franck-Condon envelope with an average $G'_{v+\frac{1}{2}}$ of 252 cm^{-1}. The "c" bands may possibly be correlated to the $A\ (^3\Sigma_u^-) \leftarrow X\ (^3\Sigma_g^-)$ transition for Al_2 in view of their concentration dependence and the gas-phase $\Delta G_{\frac{1}{2}}$ value of 279 cm^{-1} for Al_2 (12). Assuming the long wavelength "c" band is the (0,0) transition, this identification corresponds to a 7460 cm^{-1} matrix blue-shift from the gas phase transition. This is a very large shift, and leads one to believe the lower state observed in the gas-phase emission of Al_2 may not correspond to the ground state of Al_2. Emission studies are presently being pursued to probe the nature of the dimer's ground state. In view of their triplet structure, which is similar to the other Al resonance transitions, and their concentration dependence, the "d" bands are tentatively assigned to the $5s(^2P_{1/2})$ and $4d(^2D_{3/2}) \leftarrow 3p(^2P_{1/2})$ resonance transitions. The large line width is probably partially due to the overlapping of the two transitions.

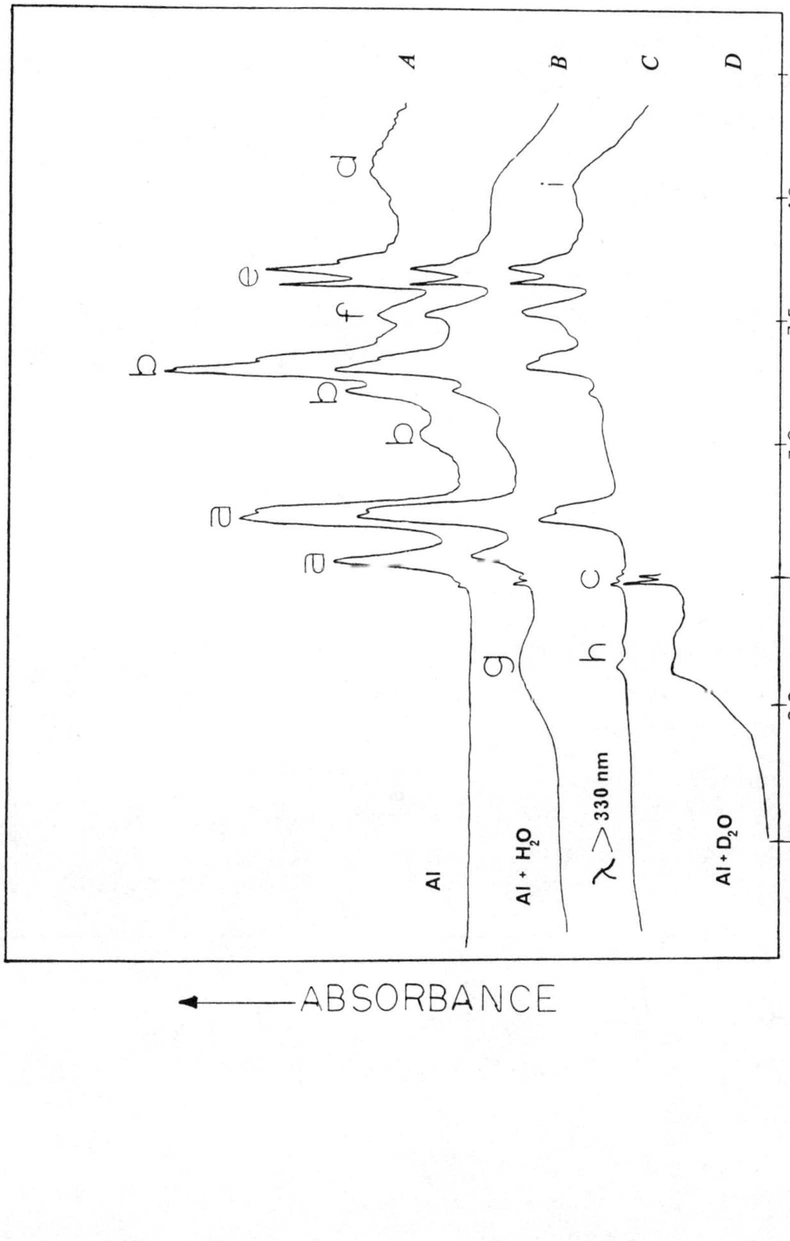

Figure 1. Kr matrix. A, Al:Kr = 1:500; B, Al:H_2O:Kr = 1:10:675; C, same as B but photolyzed for 5 min with light of $\lambda > 3300$ Å, and D, Al:D_2O:Kr = 1:3.5:350.

Table I.
Optical Absorption Spectra of the Al/Kr, $Al/H_2O/Kr$, $Al/H_2O/Kr + h\nu$, and $Al/D_2O/Kr$ Systems.

Peak	λ(Å)	$\bar{\nu}$[c] (cm^{-1})	Species	Assignment	Gas Phase[b] (cm^{-1})
A	3902	25 627	Al	$4s(^2S_{1/2}) \leftarrow 3p(^2P_{1/2})$	25 348
A	3695	27 062	Al	"	"
A	3651	27 389	Al	"	"
B	3292	30 375	Al	$3d(^2D_{3/2}) \leftarrow 3p(^2P_{1/2})$	32 435
B	3122	32 027	Al	"	"
B	3040	32 891	Al	"	"
B	3002	33 314	Al	"	"
C	4049	24 695	Al_2	$A \xleftarrow{(0,0)^a} X$	17 234
C	4007	24 956	Al_2	$A \xleftarrow{(1,0)} X$	
C	3968	25 199	Al_2	$A \xleftarrow{(2,0)} X$	
D	2624	38 115	Al	$5s(^2S) \leftarrow 3p(^2P_{1/2})$	37 689
D	2401	41 649	Al	$4d(^2D_{1/2}) \leftarrow 3p(^2P_{1/2})$	38 929
E	2744	36 444	Al_2O	$A(0,0,0) \leftarrow X(0,0,0)^d$	
E	2696	37 085	Al_2O	$A(1,1,0) \leftarrow X(0,0,0)$	
E	2677	37 352	Al_2O	$A(2,0,0) \leftarrow X(0,0,0)$	
F	2849	35 100	IMPURITY		
G	4524	22 103	HAlOH	$A \longleftarrow X$	
H	4576	21 851	AlO	$B(^2\Sigma^+) \xleftarrow{(0,0)} X(^2\Sigma^+)$	20 635
H	4400	22 723	AlO	$B(^2\Sigma^+) \xleftarrow{(1,0)} X(^2\Sigma^+)$	
I	2524	39 613	AlOH	$A \longleftarrow X(^1\Sigma^+)$	
I	2486	40 227	AlOH	$A \longleftarrow X(^1\Sigma^+)$	
I	2450	40 823	AlOH	$A \longleftarrow X(^1\Sigma^+)$	

[a] (ν', ν''). [b] From Refs. 7, 8, and 11. [c] $\pm$ 15 cm^{-1}.
[d] (ν_1, ν_2, ν_3).

The "e" bands correspond to vibronic transitions for the Al_2O molecule (7) which is generated by aluminum reducing the aluminum sesquioxide residing in the metal sample. Extended operation of the furnace discriminated against these bands. From dilution and contaminant dopant experiments, the "f" band could not be associated with any of the other bands; however, it did decrease in intensity as the oven temperature was lowered. This band has also been observed in other metal atom isolation experiments in this laboratory (Fe and Ti) where alumina spacers were utilized at high temperatures. This band is presently left unexplained.

In Figure 1(b), the band denoted by "g" may be assigned to the reaction product of aluminum and water which, in collaboration with previous infrared studies (1), may be identified with the HAlOH molecule. This very broad absorption has a short wavelength onset at ~4000 Å, maximizes at ~4500 Å, and tails out to ~6300 Å. This absorption correlates well with the densely structured chemiluminescent continuous emission spectra observed in beam-gas experiments involving the hydration of aluminum which typically ranged from 3500 Å to 8500 Å. Due to the enhanced sensitivity to weak features intrinsic to fluorescence detection, it is easy to understand a discrepancy in the red cutoff compared to the absorption technique. The Kr:H_2O ratio was varied from 300:1 to 65:1 while maintaining the Kr:Al ratio at 650:1. Over this concentration range, only the "g" band was observed whose intensity was maximized at a ratio of 110:1. At higher concentrations the band diminished in intensity, perhaps indicative of a higher order water-metal interaction. Neither the water concentration nor annealing studies noticeably changed the structure of the "g" band. Photolysis of the "g" band with light having a 4500 Å short wavelength cutoff (Corning 3-72 + water-pyrex filters) for ten minutes caused no change in the band structure. Figure 1(c) shows that a five minute photolysis with light having a 3300 Å short wavelength cutoff (water-pyrex filter) destroys the "g" band generating the "h" and "i" bands. The "h" bands are assigned to the AlO species since the peak positions correlate well with previous matrix isolation studies of the $B(^2\Sigma^+) \leftarrow X(^2\Sigma^+)$ transition of AlO (13). The "i" bands are assigned to transitions of the AlOH molecule since this is the major photolysis product found in infrared matrix studies (1). In addition, the feature is energetically very similar to the analogous gas phase AlF $A(^1\pi) \leftarrow X(^1\Sigma^+)$ transition (T_e = 43,949 cm^{-1}) (12), and quantum-chemical calculations indicate the lowest-lying excited states of AlOH lie at energies in excess of 30,000 cm^{-1} above the ground $^1\Sigma^+$ state (4,6). The "g" band in Figure 1(d) is assigned to the DAlOD molecule which is identical in structure to the HAlOH molecules except for superimposed AlO $B^2\Sigma \leftarrow X^2\Sigma$ "h" bands. Further study is needed to understand this phenomena.

Discussion

In view of the very close agreement between the matrix isolation visible absorption spectra of the HAlOH molecule and the chemiluminescence features observed in the gas-phase oxidation of aluminum in the upper atmosphere (14) and in the laboratory (4), this study further substantiates the plausibility that the continuum emitter is the divalent oxidative insertion product of a 1:1 aluminum hydration reaction, HAlOH. The insertion reaction is probably facile; however, it is possible that radiation from the furnaces may have photolyzed the $Al...OH_2$ adduct to the insertion product during co-condensation. Preliminary theoretical studies indicate there is no potential energy barrier in going directly to the insertion product from $Al + H_2O$ (2).

AlO is another product from photolysis of HAlOH which is not observed in the infrared studies (1) since the AlO 917 cm^{-1} band (13) is obscured by the presence of other product bands. In multiple-collision studies, Gole and Oblath observe emission from the $B(^2\Sigma^+)$ state of AlO postulating that the state is collisionally activated by HAlOH* since single-collision studies display no AlO emission.

Acknowledgments

This work is supported by funds from the National Science Foundation, the U.S. Army Research Office (Durham), and The Robert A. Welch Foundation.

Literature Cited

1. Hauge, R.H.; Kauffman, J.W.; Margrave, J.L. J. Am. Chem. Soc. 1980, 102, 6005.
2. Kurtz, H.A.; Jordan, K.D. J. Am. Chem. Soc. 1980, 102, 1177.
3. Trenary, M.; Schaefer, J.F. III; Kellman, P.A. J. Am. Chem. Soc. 1978, 68, 4047.
4. Oblath, S.B.; Gole, J.L. J. Chem. Phys. 1979, 70, 581; Comb. and Flame 1980, 37, 293.
5. Rosenwaks, S.; Steele, R.E.; Broida, H.P. J. Chem. Phys. 1975, 63, 1963.
6. Gole, J.L. (in preparation).
7. Douglas, M.A.; Hauge, R.H.; Margrave, J.L. (in preparation).
8. Moore, C.E. Circ. U.S. Nat'l. Bur. Stand. 1949, 467.
9. Corliss, C.H.; Bozman, W.R., Managr. U.S. Nat'l. Bur. Stand. 1962, 53, xxx.
10. Meyer, B. Low Temperature Spectroscopy (American Elsevier, New York, 1971).
11. Ammeter, J.H.; Schlosnagle, D.C. J. Chem. Phys. 1973, 59, 4784.
12. Huber, K.P.; Herzberg, G. Constants of Diatomic Molecules (Van Nostrand, New York, 1979).
13. Knight, L.B. Jr.; Weltner, W. Jr. J. Chem. Phys. 1971, 55, 5066.
14. Rosenberg, N.W. Science 1966, 152, 1017.

RECEIVED August 26, 1981.

24

IR Matrix Isolation Studies of Group IVA Metals With Water

J. W. KAUFFMAN, R. H. HAUGE, and J. L. MARGRAVE

Rice University, Chemistry Department, Houston, TX 77001

Silicon, germanium, tin and lead metal atoms and dimers react with water molecules to form an adduct on cocondensation in argon, which can then undergo a photoinsertion reaction of the metal into the O-H bond. The water molecule ν_2 bending mode frequency is red shifted in the initially formed $M_1 \cdots OH_2$ and $M_2 \cdots OH_2$ adducts by 18.6, 15.9, 15.3 and 8.0 cm^{-1} for Si, Ge, Sn and Pb and 32.1, 19.9 and 16.0 cm^{-1} for Ge_2, Sn_2 and Pb_2 respectively. Data suggest that the HMOH and MO species are formed on further reaction of the Si, Ge and Sn metal atom-water molecule adducts. The Ge_2 and Sn_2 adduct reacts to form a dimetal species containing a bridging hydrogen atom. Lead does not react beyond adduct formation.

This paper is a continuation in the series of reports dealing with the interaction of water molecules with metal atoms utilizing infrared matrix isolation spectroscopy. Previous experimental (1,2,4) and theoretical (3,5) works have described the initial reaction of the metal atom with a water molecule to form a C_{2v} adduct, with the metal atom lying in the plane of the water molecule and bonding through the oxygen atom. The formation of the adduct involves a transfer of electron density from the $3a_1$ orbital of water to the metal atom. This results in a decrease in the ν_2 vibrational mode, thus producing a band that is shifted down from the uncomplexed water bending mode at 1593 cm^{-1}. The adduct is then photolyzed to induce further reaction. In general, reaction products can best be described as resulting from an insertion reaction of the metal into the O-H bond of water.

The purpose of this study is to investigate the reaction pathways involved in the Si, Ge, Sn and Pb metal atom reactions

0097-6156/82/0179-0355$05.00/0

with water and to identify photolysis products in light of the previous findings. Only the preliminary results involving $H_2{}^{16}O$ are reported here. These studies might serve as a prototype model for understanding the interactions and reactions of water with Si, Ge, Sn and Pb bulk metal surfaces.

Experimental

The experimental set-up has been previously described (4). Silicon and germanium metal were vaporized from a vitreous carbon crucible over the temperature range of 1475-1675°C and 1325-1500°C respectively. Tin and lead were vaporized from a molybdenum crucible over a temperature range of 1125-1240°C and 644-742°C respectively. This corresponds to an approximate metal vapor pressure range of $1x10^{-3}$-$1x10^{-2}$ torr for each metal.

In each case, a relative measure of the metal concentration in the matrix was determined by the color of the matrix. The silicon matrix varied from a pale orange to dark orange color as the metal concentration increased. Germanium and tin varied from a pale orange to an orange-green color. Lead gave a light green colored matrix at low metal concentrations, which changed to a very dark green color with the heaviest metal concentration. The water-argon ratios used were 6.5:1000, 13:1000 and 36:1000.

Water and metal concentration studies were carried out in which each substrate was incrementally increased in successive experiments while the other substrate remained constant. This was done in order to identify reactions containing more than one metal atom and/or water molecule. The matrix was photolyzed in each experiment with a 100 watt medium pressure Hg lamp to induce further reaction. Isotopically substituted water, $H_2{}^{16}O$, $H_2{}^{18}O$ and D_2O were used to aid in product identification.

Results

Silicon. Silicon metal atoms reacted with water on cocondensation in excess argon at 15°K, to form the metal atom-water molecule adduct. The absorption band assigned to the ν_2 water bending mode in the adduct is shown in Figure 1. The adduct reacted slowly under no light conditions over a five hour period; however, it reacted much faster when photolyzed with $\lambda > 360$ light. The adduct reacted to form the HSiOH product, in a reaction similar to that observed in previously studied Group IIIA metal atom-water reactions (4). In addition, at relatively high water concentrations, a reaction product was formed on cocondensation which can best be described as a hydrated silicon hydroxyhydride molecule, $HSiOH(H_2O)$.

Prolonged photolysis of the matrix with $\lambda > 360$ nm light caused the bands assigned to the HSiOH molecule to disappear with the appearance of a single band, assignable to silicon monoxide,

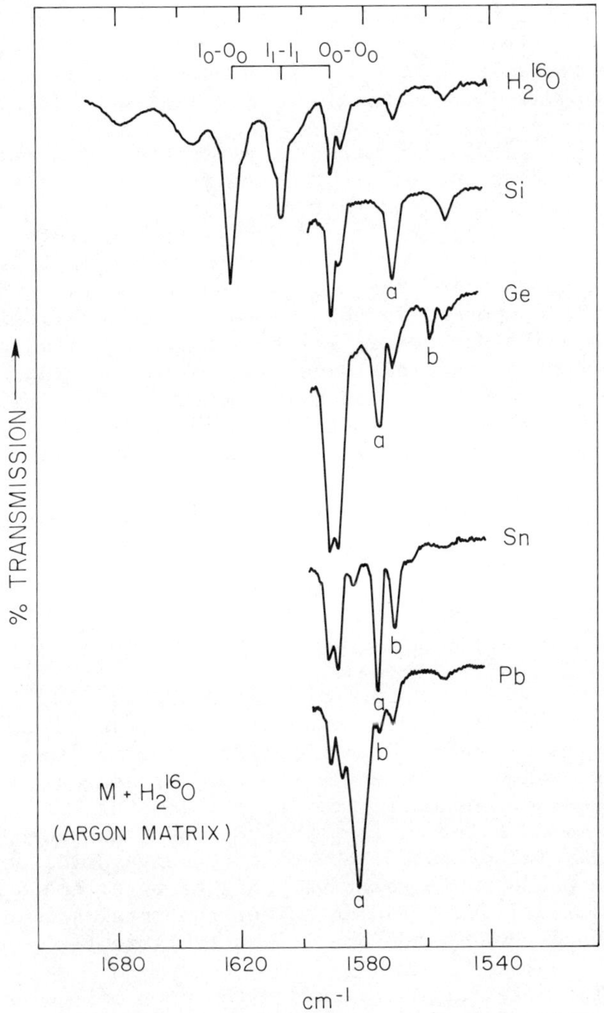

Figure 1. IR spectra of water bending mode region immediately after the cocondensation of metal and water in Ar. Key: a, metal atom; and b, metal dimer–water molecule adduct bands.

SiO, which has been previously studied by Ogden (8). This suggested that the hydrogen bonds in HSiOH are cleaved on photolysis in a concerted reaction. Photolysis also caused the bands assigned to the HSiOH(H_2O) product to disappear with the appearance of bands assigned to the silicon dihydroxy dihydride, $H_2Si(OH)_2$, in analogy to H_2SiF_2, which has been studied by Milligan (9). This suggests that the metal center in HSiOH(H_2O), inserts into the coordinated water molecule on photolysis. The adduct band frequency is listed in Table I. The product band frequencies for HSiOH, HSiOH(H_2O), SiO, and $H_2Si(OH)_2$ are listed in Table II.

Germanium. Germanium metal atoms reacted with water on cocondensation to form the metal atom and metal dimer-water molecule adducts. The bands assigned to the ν_2 water bending mode in the adducts are shown in Figure 1, and are listed in Table I. The metal dimer-water adduct reacted on photolysis with 400-340nm light to produce a set of three bands that show only a deuterium dependence. This species is thought to contain a bridging hydrogen atom. The absence of a metal-oxygen stretching mode band also suggested that this species contains a bridging hydroxyl group, which is expected to lie below $400cm^{-1}$, the limit of the spectrometer. This species is tentatively assigned to structure (I), resulting from an insertion reaction of the metal dimer into the O-H bond of water.

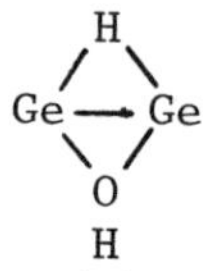

(I)

Photolysis of the metal atom-water molecule adduct with 340-300nm light caused a metal atom insertion reaction to give the germanium hydroxyhydride molecule, HGeOH. Prolonged photolysis (ca. 60 minutes) of the matrix caused the loss of both hydrogen atoms in the HGeOH molecule to form GeO, which has been previously assigned (10). The hydrated germanium hydroxyhydride molecule, HGeOH(H_2O) is produced at relatively higher water concentrations on cocondensation. The frequencies for the metal atom and metal dimer reaction product bands are listed in Table II.

Tin. Tin metal atoms reacted with water molecules on cocondensation to form the metal atom and metal dimer-water molecule adducts. The adduct bands are shown in Figure 1, and the frequencies are listed in Table I. The metal dimer-water molecule adduct reacted on photolysis of the matrix with 400-340nm light. The resulting product bands suggest that a species containing a bridging hydrogen atom and hydroxyl group is formed. These bands

Table I

Metal-Water Adduct Frequencies

and Frequency Shifts $\Delta\nu$ for the $H_2{}^{16}O$ ν_2

Bending Mode in the Metal-Water Adducts

(cm^{-1})

	ν_2	$\Delta\nu$		ν_2	$\Delta\nu$
H_2O	1593.3			1593.3	
$Si\cdots OH_2$	1574.7	18.6	$Si_2\cdots OH_2$	--	
$Ge\cdots OH_2$	1577.4	15.9	$Ge_2\cdots OH_2$	1561.2	32.1
$Sn\cdots OH_2$	1578.0	15.3	$Sn_2\cdots OH_2$	1573.4	19.9
$Pb\cdots OH_2$	1583.3	8.0	$Pb_2\cdots OH_2$	1577.3	16.0

Table II

Measured Frequencies (cm^{-1}) For

Si, Ge and Sn Reaction Products

HMOH

M	M-H Stretch	H-M-O bend	M-O stretch	H-M-O-H Torsion
Si	1885.2	936.9	850.5	595.8
	1849.1			
Ge	1741.3		661.3	566.2
Sn	1608.0	782.6	569.3	475.5
	1597.7			

MO

M	M-O Stretch
Si	1222.2
Ge	975.3, 972.9, 970.8
Sn	810.5

$HMOH(H_2O)$

M	M-H Stretch	M-O Stretch
Si	1929.4	778.5
Ge	1763.1	609.1

HM_2OH

M				
Ge		784.8	710.8	683.0
Sn	1100	748.9	690.0	620.3

are tentatively assigned to structure (II). This species apparently results from insertion of the metal dimer into the O-H bond of water.

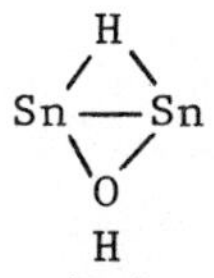

(II)

Photolysis of the matrix with 340-300nm light caused the metal atom-water molecule adduct to react to form the tin hydroxyhydride molecule, HSnOH. Prolonged photolysis (ca. 60 minutes) of the HSnOH molecule, caused cleavage of the hydrogen bonds to form SnO, which has previously been assigned (11). The reaction product frequencies are listed in Table II.

Lead. Lead formed only the metal atom and metal dimer-water molecule adduct on cocondensation; however no further reaction was observed on photolysis. The adduct bands are shown in Figure 1 and the frequencies are listed in Table I.

Discussion

Si, Ge, Sn and Pb metal atoms initially reacted with water on cocondensation to form the metal atom-water molecule adduct as has been generally observed for metal atom-water reactions (4). The shift in the water bending mode frequency on adduct formation decreases with increasing atomic number as illustrated in Table I. Using the results obtained from theoretical studies (3,5), it is believed that the water bending mode shift might serve as a relative measure of the extent of interaction between the water molecule and the metal atom, for metals within a particular group.

The metal dimer apparently interacts with water in a similar manner bonding through the $3a_1$ orbital of water, but to a different degree. An interesting effect is observed in comparing the shift values of the Ge, Sn and Pb-metal dimer-water adducts with the corresponding metal atom-water adduct. On a relative basis the metal dimer appears to interact more strongly with water than the metal atom, as suggested by a larger shift in the water bending mode frequency.

The reactivity of the metal is seen to decrease down the group with silicon metal atoms reacting with water on cocondensation to form the HMOH molecule, whereas lead does not react even on photolysis. Ge and Sn are intermediate cases in which reaction occurs on photolysis. Both the metal atom and metal dimer are observed to undergo a similar insertion reaction of the metal into the H-O bond of water as previously found for the Group IIIA metals (4). It is found that the formation of the metal hydroxy-hydride molecule can most likely be predicted by using simple heats of formation of the metal hydride and metal hydroxides (6, 7). The HMOH molecule is observed for Si, Ge, and Sn but not Pb in agreement with the following values for standard heats of formation: HSiOH (-67 kcal/mole); HGeOH (-49 kcal/mole); HSnOH (-32 kcal/mole); HPbOH (+3.7 kcal/mole).

Literature Cited

1. Hauge, R.H.; Meier, P.H.; Margrave, J.L. Ber. Bunsenges. Phys. Chem., 1978, 82, 102.
2. Meier, P.H.; Hauge, R.H.; Margrave, J.L. J. Am. Chem. Soc., 1978, 100, 2108.
3. Trenary, M.; Schaefer III, J.F.; Kollman, P.A. J. Chem. Phys. 1978, 68, 4047; J. Am. Chem. Soc., 1977, 99, 3885.
4. Hauge, R.H.; Kauffman, J.W.; Margrave, J.L. J. Am. Chem. Soc., 1980, 102, 6005.
5. Curtiss, L.A.; Frurip, D.J. Chemical Physics Letters, submitted.
6. Gaydon, A.G. "Dissociation Energies and Spectra of Diatomic Molecules", Chapman and Nall, London, 1968.
7. Jackson, D.D. "Thermodynamics of the Gaseous Hydroxides", UCRL-51137, Lawrence Livermore Laboratory, Livermore, Calif., 94550.
8. Anderson, J.S.; Ogden, J.S. J. Chem. Phys., 1969, 51, 4189.
9. Milligan, D.E.; Jacox, M.E. J. Chem. Phys., 1968, 49, 4269.
10. Ogden, J.S.; Ricks, M.J. J. Chem. Phys., 1970, 52, 352.
11. Ogden, J.S.; Ricks, M.J. J. Chem. Phys., 1970, 53, 896.

RECEIVED August 26, 1981.

25

Matrix Isolation IR Studies of Metal Atom–Water Reactions

R. H. HAUGE, J. W. KAUFFMAN, L. FREDIN, and J. L. MARGRAVE

Rice University, Chemistry Department, Houston, TX 77001

Metal atom and small metal molecules of Li, Na, K, Cs, Mg, Ca, Sr, Ba, Al, Ga, In, Tl, Si, Ge, Sn, Pb, and all of the first row transition metals have been cocondensed with H_2O in an argon matrix. An initial $M_x...OH_2$ adduct is formed for most metals where x varies from one to three. Certain metals, Al, Sc, Ti and V, reacted spontaneously with H_2O during co-condensation to form the HMOH species. The $M_x...OH_2$ adducts are characterized by a $\Delta\nu_2$ decrease in the bending mode of H_2O. Many of the $M_x...OH_2$ adducts react further when photoexcited to form the HMOH species. The photo reactions observed for any one metal appear to be governed by the exothermicity of the reaction and the preference for particular valence states.

Our studies have shown that almost all metal atoms will form an adduct with water and in many instances undergo further reaction. Theory and experiment (1-7) support the concept of water acting as a Lewis base which donates electron density to the metal. Theory has also shown for Be and Mg that donation occurs extensively from the $3a_1$ (σ lone pair) orbital of water to the metal. This is of interest since this orbital is largely responsible for the nonlinearity of the water molecule. It is known, for instance, that ionization from this orbital causes water to become linear and results in a decrease in its bending frequency by approximately 700 cm^{-1}. This is consistent with our finding that only the ν_2 bending mode of water decreases measurably upon adduct formation. The observed $\Delta\nu_2$ changes for metal-water adducts are compared and discussed below.

Our studies have also shown that water will react spontaneously with some metals or can be caused to react further by photoexciting the adduct. Reaction in all cases leads initially to metal insertion into the hydrogen-oxygen bond. This metal hydroxy-hydride will, for certain metals, undergo photolysis to

0097-6156/82/0179-0363$05.00/0

a metal hydroxide or metal oxide. The observed metal hydride or hydroxide stretching frequencies are given in Table IV and discussed with respect to bonding trends within groups and across the first row transition metal series.

In some cases reactions involving two metal atoms with a single water or two water molecules with one metal atom are observed. It is particularly interesting that a dimetal atom-water adduct can in general be photolyzed by longer wavelength light than that required for the metal atom adduct. Thus its photoproducts can be studied independently. A number of examples are given.

Experiment

The apparatus consists of a closed cycle helium refrigerator which cools a square copper block with three polished surfaces and a 6 MC quartz crystal mounted on the fourth surface (8). Infrared spectra are obtained by reflection off a polished surface after cocondensing an argon-reactant gas mixture with atomic metal for periods of approximately one hour. The ratios of metal/reactant gas/argon are measured directly with the cooled quartz crystal microbalance. Typical ratios were 1/1/500 to 1/1/100 metal/reactant gas/argon. Metals were generally vaporized from 0.25" high-purity alumina or graphite cells which were heated by cylindrical Ta foil furnaces. Infrared spectra were obtained with a Beckman IR-9 spectrometer. Matrices were photolyzed by a short arc 100W mercury lamp through selected Corning long pass filters either during or following deposition.

Metal-Water Adducts

Figures 1-3 illustrate the growth of new peaks in the bending mode ν_2 region of water when metal atoms are cocondensed with water into an argon matrix. The peak labeled "a" is the first to appear at low metal concentrations. A second peak "b" appears at higher metal concentrations and for Mg a third peak also appears at the highest metal concentrations. The peak positions are measured relative to the 0_0-0_0 peak located at 1593.3 cm^{-1} which is labeled with an "*". One also notices a peak slightly below the 0_0-0_0 peak which also seems to grow with metal concentration. This is thought to be due to water itself, most likely in a different matrix site in which it does not rotate. The same peak has also been observed by M. E. Jacox in photoionization studies and assigned to a nonrotating site of water (9). Its position is independent of the nature of the cocondensed metal. The a, b, and c peaks are assigned to $M...OH_2$, $M_2...OH_2$ and $M_3...OH_2$ adducts respectively.

Table I lists the measured frequency changes for all of the metals we have studied. It is interesting to compare the trends in $\Delta\nu_2$ within groups. The alkali metal interaction clearly

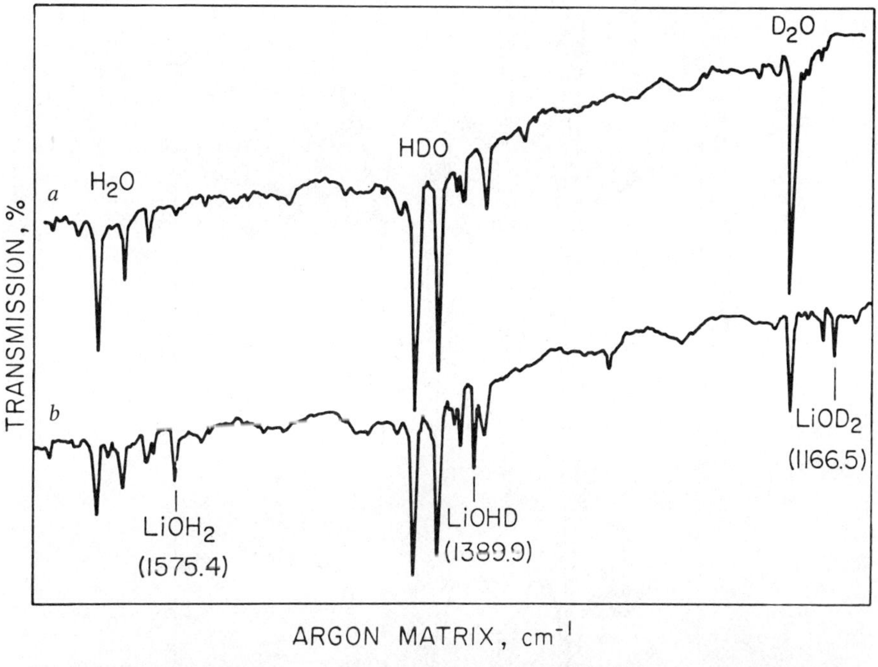

Figure 1. a: The ν_2 bending mode region of water for a mixture of H_2O, HDO, and D_2O in an argon matrix. b: Same as a but with Li cocondensed with water into an Ar matrix.

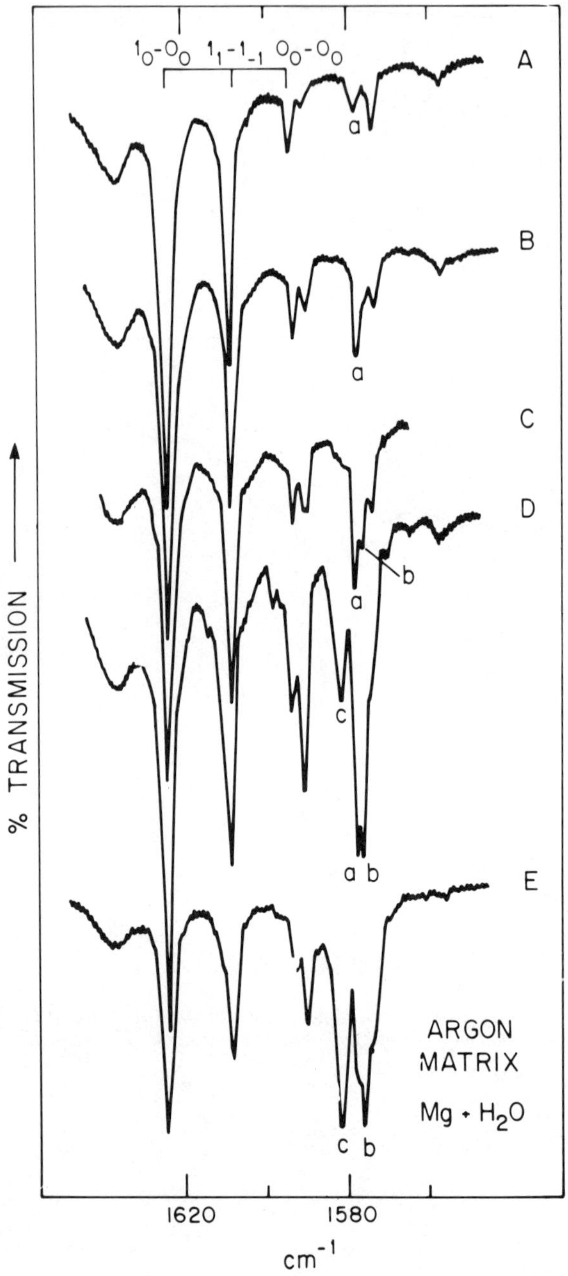

Figure 2. The ν_2 bending mode region of water where water and Mg are cocondensed in an excess of Ar. A–E illustrate the effect of increasing Mg concentration. Key: A, Mg (350°C); B, Mg (360°C); C, Mg (370°C); D, (380°C); and E, Mg (390°C).

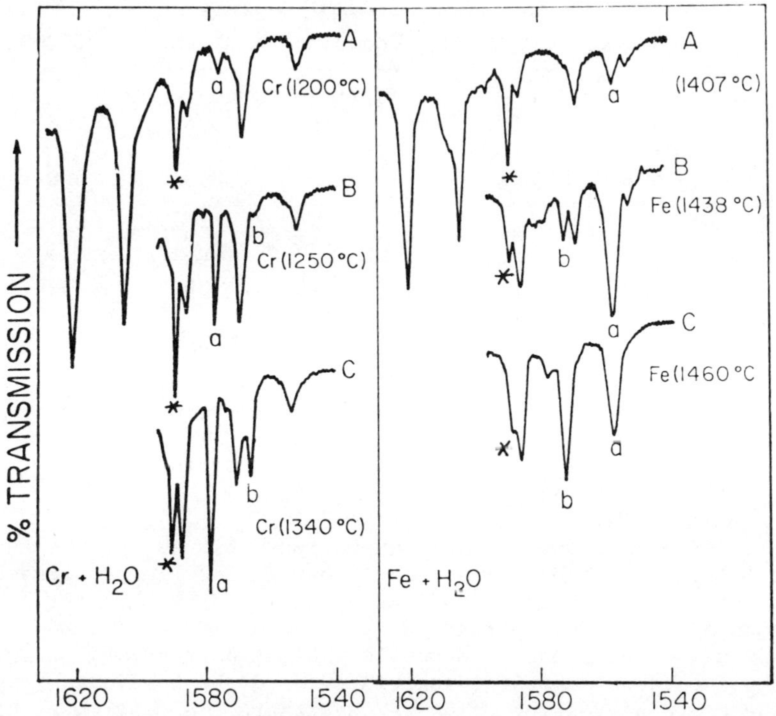

Figure 3. The ν_2 bending mode region of water where Cr and Fe have been cocondensed with water. A–C illustrate the effect of increasing metal atom concentration. Left: A, Cr (1200°C); B, Cr (1250°C); C, Cr (1340°C). Right: A, Fe (1407°C); B, Fe (1438°C); C, Fe (1460°C).

Table I.
Measured Frequency Changes (cm^{-1}) for the H_2O (ν_2) Bending Mode in Main Group Metal-Water Adducts.

Metal	$M_1...OH_2$	Metal	$M_1...OH_2$	$M_2...OH_2$	$M_3...OH_2$
Li(7)	17.5	Be	-		
Na	7.4	Mg(13)	15.7	17.3	12.0
K	6.6	Ca	30.8		
Rb	-	Sr	27.7		
Cs	2.7	Ba	34.0		

Metal	$M_1...OH_2$	$M_2...OH_2$	Metal	$M_1...OH_2$	$M_2...OH_2$
Al(10)	*	-	Si(16)	19.5	-
Ga	21.4	14.4	Ge	15.4	31.6
In	16.5	10.6	Sn	15.8	20.8
Tl	9.6	-	Pb	8.0	16.0

Metal[a,b]	Cr(15)	Mn	Fe	Co	Cu	Zn
$M_1...OH_2$	12.2	16.0	31.0	28.5	20.5	5.0
$M_2...OH_2$	24.3	19.2	16.2			
$M_3...OH_2$		23.5				

*Reacted during cocondensation
[a]Sc(14), Ti, and V reacted during cocondensation
[b]Ni does not form an adduct

decreases with increasing atomic number but this is not the case for the alkaline earth series. The Group IIIA series decreases but not as rapidly as the alkali metals. The Group IVA series undergoes a less regular trend but Pb does appear to interact the least of all. The transition metals exhibit a trend which suggests an important d-orbital contribution to the bonding. Perhaps the most interesting point is that there are many metal atoms which undergo significant interaction with water.

It is also very interesting to note the large differences in $\Delta\nu_2$ for the metal atom and diatomic species. Chromium and iron atoms differ by a factor of two from the diatom but in opposite directions. Finally, Al, Sc, Ti and V reacted rather than form a stable metal-water adduct while Ni did not form an adduct or react.

The structure of the metal-water adduct has been determined experimentally for Li(6,7) and theoretically for Li, Na(1), Be, Mg(3,4), Ca(5), and Al(2). In all cases water is acting as a Lewis base and donating electron density to the metal through the oxygen. For all metals but Be the minimum energy geometry is planar. For Be, the Be-O bond is calculated to be symmetrical and at 73° with respect to the water plane. However, it can be very easily distorted to a planar configuration.

Table II compares the measured $\Delta\nu_2$ values to the calculated binding energy. It is seen that, where both are available,

Table II.
Comparison of Frequency Changes (cm^{-1}) of H_2O ($\Delta\nu_2$) to Calculated Binding Energies and Measured Changes in the Nuclear Hyperfine Interaction of the Alkali Metals.

	ΔE	$\Delta\nu_2$	%ΔW[6]		ΔE	$\Delta\nu_2$		ΔE	$\Delta\nu_2$
Li(1)	11.7	17.5	34.4	Be(3)	0.62				
Na	5.2	7.4	17.0	Mg	2.34	15.7	Al(1,2)	4.4	-
K		6.6	13.8	Ca(5)	7.18	30.8			
Rb			9.5						
Cs		2.7	0						

they parallel one another in their behavior within a group, however, it is evident that similar $\Delta\nu_2$ values do not imply the same binding energy between groups of the periodic table. Table II also shows the percent change in nuclear hyperfine splitting of the alkali metals upon formation of a water adduct. This measure of the perturbation of the alkali metal clearly parallels the $\Delta\nu_2$ values as expected.

Table III compares the measured $\Delta\nu_2$ values to the shifts in the carbonyl frequency of MCO and M_2CO. It is seen that the de-

Table III.
Comparison of Frequency changes (cm^{-1}) for H_2O ($\Delta\nu_2$) and CO in First Row Transition Metal Adducts.

	CO	H_2O		CO	H_2O
V	139	-			
Cr	-	12.2	Cr_2		24.3
Mn(11)	291	16.0	Mn_2	177(t)452(b)	19.2
Fe(12)	243	31.0	Fe_2		16.2
Co	187	28.5			
Ni	145	-	Ni_2(12)	168(t)203(b)	-
Cu	131	20.5			
Zn	-	5.0			

t = terminal; b = bridging

crease in the CO frequency follows that for H_2O except for Mn. The decrease for CO is thought to be due to π backbonding from the metal atom. The extent to which this occurs is dependent on the amount of σ donation to the metal from CO. The $\Delta\nu_2$ value of water primarily represents σ donation from water to the metal and thus provides a measure of the sigma bonding ability of a metal without appreciable π backbonding.

Metal Hydroxy-Hydrides

Figures 4 and 5 illustrate the chemistry of iron and scandium atoms reacting with water. As noted earlier, both Fe and Fe_2 water adducts are formed and labeled "a" and "b", respectively, in Figure 4. It is interesting that $Fe_2...OH_2$ can be photolyzed without photolyzing the $Fe...OH_2$ adduct. The "d" peaks which result from photolysis of $Fe_2...OH_2$ indicate that a species with terminally bonded H and OH groups is formed. The photolysis of $Fe_2...OH_2$ would appear to lead to formation of the HFeFeOH species.

Irradiation with shorter wavelength light causes photolysis of $Fe...OH_2$ with formation of the HFeOH "c" peaks. Figure 5 shows that Sc does not form an adduct but rather spontaneously reacts with water to form the HScOH insertion product "a" which itself can be photo-decomposed to ScO "d" and H_2.

The metal hydroxy-hydrides which have been observed are listed in Table IV along with their measured metal hydride and

Table IV.
Observed Metal-Hydroxy Hydride Vibrational Stretching Mode Frequencies (cm^{-1}), Argon Matrix, and Corresponding Diatomic Metal Fluoride Frequencies.

	HMOH		MF		HMOH		MF
	M-H	M-OH	M-F		M-H	M-OH	M-F
Mg(13)	1432.2	742.3	701.8	Al(10)	1743.3	817.9	792.7
Ca	1260.0	574.6	575.6	Ga	1669.8	649.4	615.8
Sr	1172.2	498.4	497.9	In	1486.3	548.0	530.1
Ba	1139.3	458.8	465.3				
Si(16)	1885.2	850.5	897.7	Sc(14)	1482.1	712.8	728.0
Ge	1741.3	660.6	659.4	Ti	1538.9	699.7	-
Sn	1597.9	569.1	572.3	V	1583.9	702.0	-
				Cr(15)	1639.9	669.8	-
				Mn	1663.4	648.9	612.8
				Fe	1731.5	680.2	664
				Co	1784.9	633.2	-

hydroxide stretching frequencies. The close agreement of the M-OH stretching frequency with the M-F frequency indicates the bonding is very similar for F and OH near the minimum of their potential wells. The lower measured bond energies of M-OH bonds in comparison to M-F bonds probably results to a great extent from the lower electron affinity of OH. Figure 6 illustrates the variation of the metal hydride stretching frequency with atomic number from Sc to Co. This probably reflects a linear change of the metal ionic radius with atomic number and also suggests a linear increase in metal hydride bond energy from Sc to Co. The irregular variation in the metal hydroxide frequency with atomic number probably reflects promotional energy and ionization energy trends across the series.

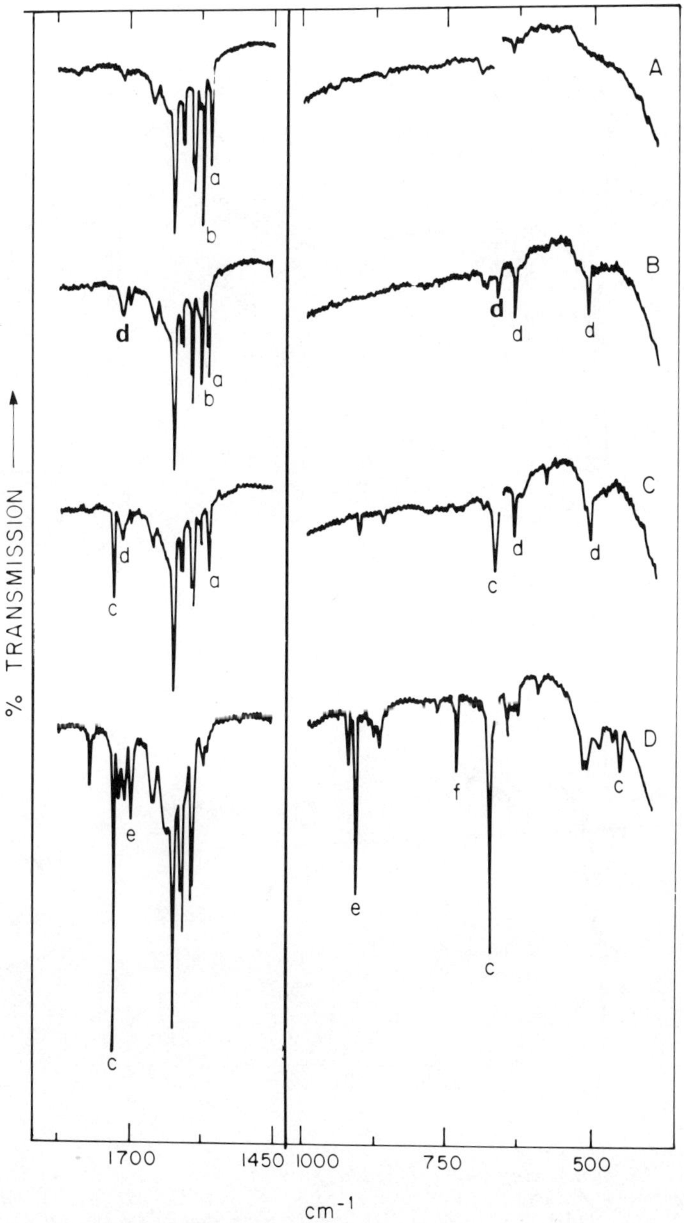

Figure 4. IR Ar matrix isolation products from the reaction of Fe atoms with water. Key: A, Fe (1460°C) and H_2O (110μ), 1 hour; B, matrix A irradiated with (λ > 620 nm) for 30 min; C, matrix C irradiated with (λ > 330 nm); and D, Fe (1440°C) and H_2O (150μ) irradiated during trapping (λ > 330 nm).

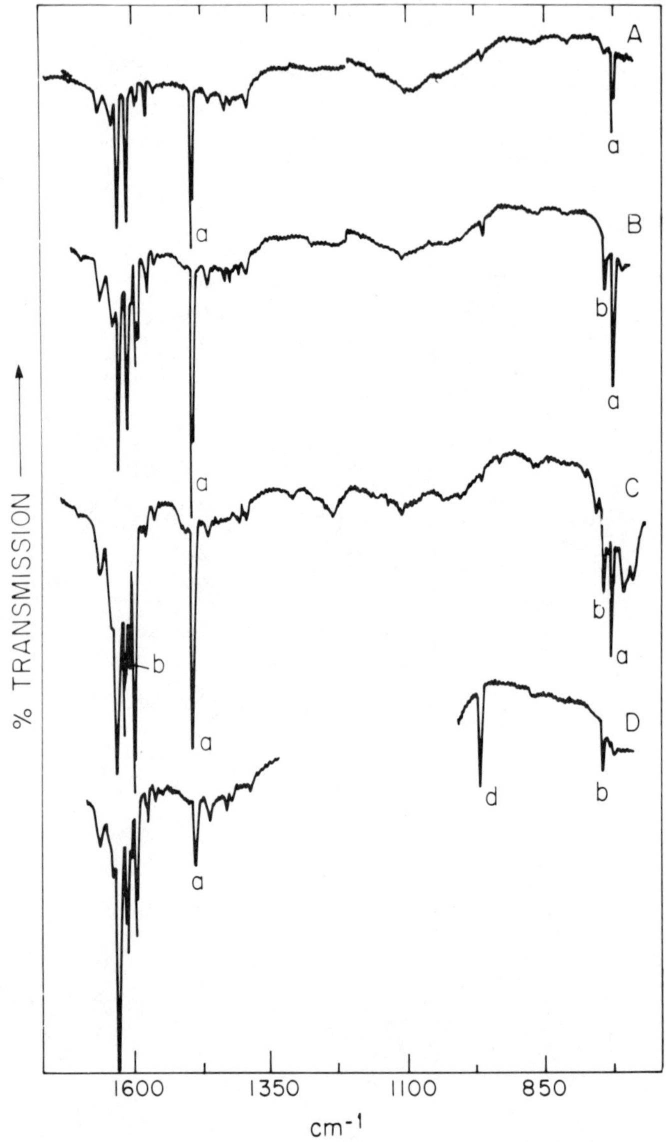

Figure 5. IR Ar matrix isolation spectra of products from the reaction of Sc atoms with water. Key: A, Sc (1255°C) and H_2O (57μ); B, Sc (1255°C) and H_2O (120μ); C, Sc (1255°C) and H_2O(220μ); and D, matrix B irradiated for 10 min with λ > 330 nm.

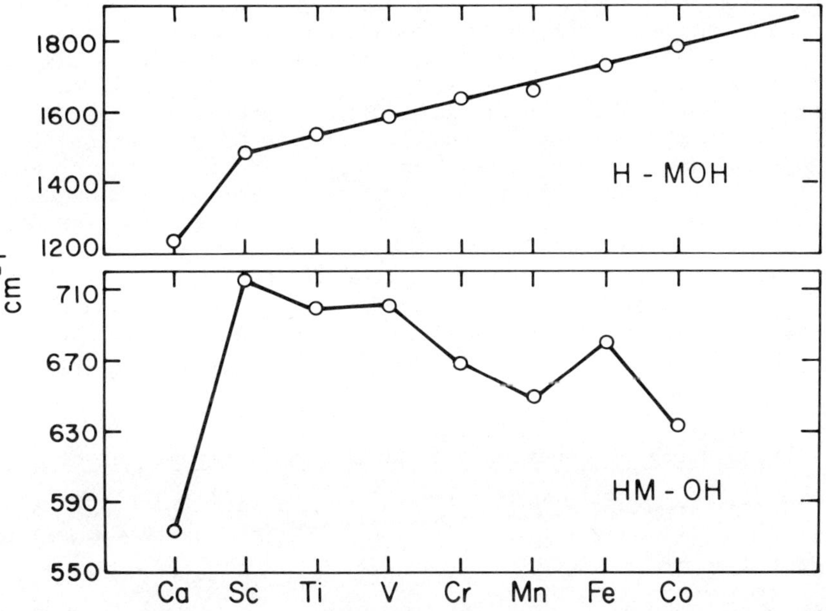

Figure 6. Variation of the stretching frequencies for HMOH vs. atomic number.

Dimetal-Water and Metal-Diwater Products

Figure 4D shows additional peaks labeled "e" and "f" which have been found to result from HFeOFeH and $Fe(OH)_2$, respectively. The product HFeOFeH appears to be favored if photolysis is carried out during the trapping process which suggests it is forming by sequential insertion of iron atoms. It is interesting that sequential reaction of two iron atoms differs from the reaction of the iron dimer.

Metal diwater reactions must occur if the metal dihydroxide is formed. This reaction may proceed by initial formation of a symmetric diwater adduct as was observed in lithium EPR studies (6) or by formation of a metal water-dimer adduct as was observed in lithium IR studies. Sufficient work has not been done to indicate whether one path is more effective than the other in producing the dihydroxide. For metals which can achieve a higher valence state, formation of $HM(OH)_2$ and $H_2M(OH)_2$ is possible. Formation of $HSc(OH)_2$ is observed at higher water concentrations, and is responsible for the "b" peaks in Figure 5. Formation of $HAl(OH)_2$, $H_2Ti(OH)_2$ and $H_2Si(OH)_2$ has also been observed. A schematic illustrating possible reaction pathways for metal atom-water reactions is presented in Figure 7. The path which a particular metal follows can to some extent be predicted from bond energy considerations as presented in Table V. This table lists a number of estimated reaction energies. In all cases, the reactions observed have been only those which have a net energy release. For example, a zinc-water adduct will form but does not undergo

Table V.
Estimated Heats of Reaction for Metal Atom-Water Reactions.

Reaction	H(kcal/mole)
$Li + H_2O \xrightarrow{h\nu} H + LiOH$	+15
$Mg + H_2O \longrightarrow HMgOH$	-20
$Al + H_2O \longrightarrow HAlOH$	-60
$\xrightarrow{h\nu} H + AlOH$	-10
$\longrightarrow H_2 + AlO$	0
$Sc + H_2O \longrightarrow HScOH$	-40
$\longrightarrow H + ScOH$	+ 5
$\xrightarrow{h\nu} H_2 + ScO$	-31
$Fe + H_2O \longrightarrow HFeOH$	- 0
$Ni + H_2O \longrightarrow HNiOH$	-19
$Zn + H_2O \longrightarrow HZnOH$	+25
$Si + H_2O \longrightarrow HSiOH$	-57
$\longrightarrow H_2 + SiO$	-65
$\longrightarrow H_2SiO$	-54

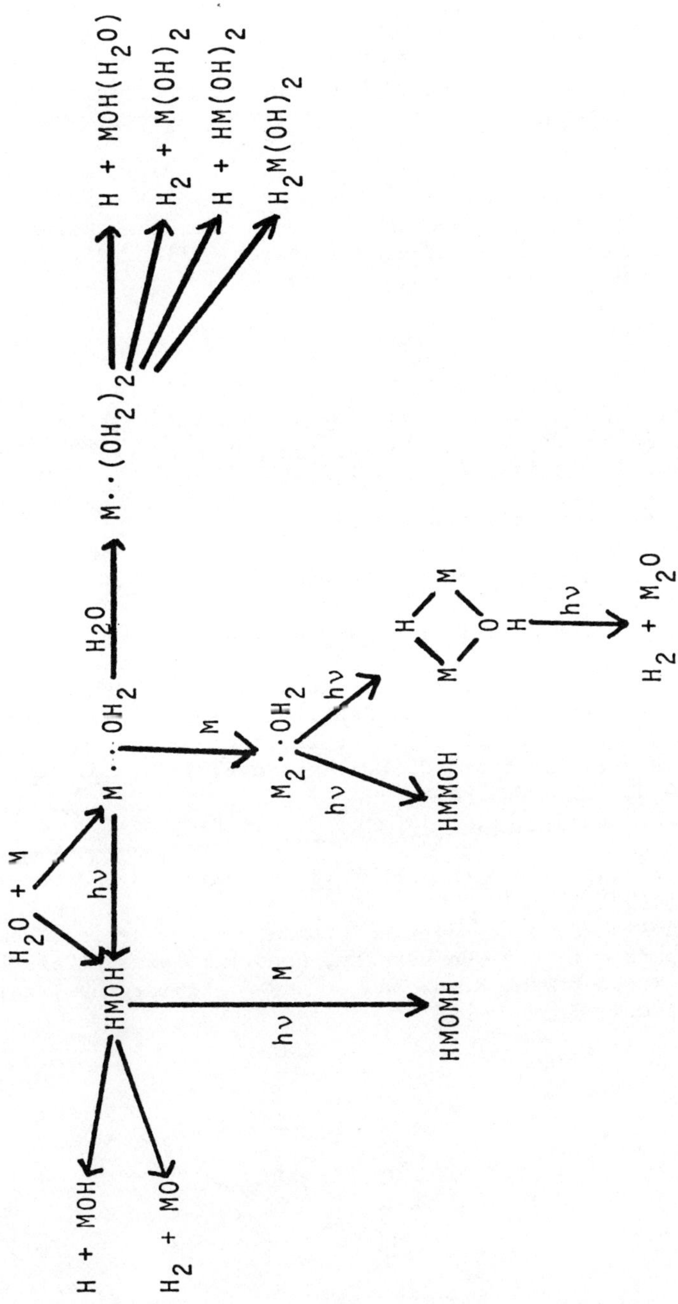

Figure 7. Scheme of reaction pathways for metal atom–water reactions.

photo-insertion. Perhaps the most interesting result is the lack of reactivity for nickel atoms. Bond energy considerations clearly indicate that formation of HNiOH is energetically favorable. We attribute the lack of reaction to the inability of nickel to form a water adduct and suggest that proper adduct formation is an essential first step in matrix photochemical reactions.

Acknowledgment

This work has been supported by funds from the National Science Foundation, the U. S. Army Research Office, Durham, and The Robert A. Welch Foundation.

Literature Cited

1. Trenary, M.; Schaefer, H. F. III; Kollman, P. A. J. Chem. Phys. 1978, 68, 4047; J. Am. Chem. Soc. 1977, 99, 3885.
2. Kurtz, H. A.; Jordan, K. D. J. Am. Chem. Soc., in press.
3. Curtiss, L. A.; Frurip, D. J. Chem. Phys. Ltrs. 1980, 75, 69.
4. See K. Jordan, this symposium.
5. Curtiss, L. A.; Frurip, D. J. Private communication.
6. Meier, P. F.; Hauge, R. H.; Margrave, J. L. J. Am. Chem. Soc. 1978, 100, 2108.
7. Hauge, R. H.; Meier, P. F.; Margrave, J. L. Ber. Buns. Phys. Chem. 1978, 82, 102.
8. Hauge, R. H.; Margrave, J. L. 10th Materials Research Symposium, NBS Pub. 561, U. S. Dept. of Commerce, NBS, 1978.
9. Jacox, M. E. Chem. Phys. 1976, 12, 51.
10. Hauge, R. H.; Kauffman, J. W.; Margrave, J. L. J. Am. Chem. Soc. 1980, 102, 6005.
11. Huber, H.; Kundig, E. P.; Ozin, G. A.; Poe, A. J. J. Am. Chem. Soc. 1975, 97, 308.
12. Moskovits, M. "Cryochemistry", G. A. Ozin, ed., John Wiley & Sons, 1976.
13-15. Unpublished work. J. W. Kauffman, Ph.D. Thesis, Chemistry Department, Rice University, Houston, Texas, 1981.
16. See J. W. Kauffman, R. H. Hauge, and J. L. Margrave, this symposium.

RECEIVED August 26, 1981.

26

Theory of Metal Atom–Water Interactions and Alkali Halide Dimers

KENNETH D. JORDAN and HENRY A. KURTZ[1]

University of Pittsburgh, Department of Chemistry, Pittsburgh, PA 15260

The results of theoretical calculations of the structures and energetics of the adducts and insertion products formed in the reactions of Be, B, Mg, and Al are presented. In addition, preliminary results for the reaction of Al with HF, NH_3, H_2S, and CH_3OH are described.

In the second portion of the paper, recent ab initio theoretical work on the properties of alkali halides and their aggregates is briefly reviewed. Special attention is placed on the plarizabilities of LiF and $(LiF)_2$.

Metal-Atom Reactions with Water and its Isoelectronic Analogs

The first part of this paper deals with theoretical studies of the interactions of metal atoms with water and some of its isoelectronic analogs. Much of the interest in these reactions has been stimulated by the matrix isolation studies (1,2,3) of Hauge, Margrave and co-workers at Rice University and by the gas phase studies of Liu and Parson (4) and Gole and coworkers (5,6). Since part of the experimental work has been presented at this Symposium by the Rice group, only a brief review will be given here.

Li, Na, K, Cs, Mg, Ca, Sr, Ba, Ga, In, Tl, Si, Ge, Sn, Pb, Cr, Mn, Fe, Co, Cu, and and Zn have all been observed to form weakly bound complexes with water. In all cases it has been found that the H_2O bending frequency is decreased upon formation of the adduct, consistent with the observations that H_2O^+ has a lower bending frequency than H_2O and that the ESR (2) and theoretical (7-9) studies indicate that charge is transferred from

[1] Current address: Research Institute of Physics, Stockholm, Sweden, and Quantum Chemistry Group, University of Uppsala, Sweden.

0097-6156/82/0179-0377$05.00/0

the H_2O group to the metal atom yielding a $M^{\delta-}...OH_2^{\delta+}$ charge distribution.

Photoexcitation of the adduct generally gives rise to the hydroxyhydride, HMOH, which may then undergo subsequent reactions. Adducts were not observed with Al, Sc, Ti, and V; rather the reaction to form the hydroxyhydrides proceeded spontaneously in these cases (1). The Ni atom formed neither the adduct nor the hydroxyhydride. Recently the Rice group has observed that metal atoms, for example, Fe, also form adducts and insertion products with HF, NH_3, and CH_4, which are isoelectronic with H_2O.

Although some experimental data are now available for reactions of metal dimers and trimers with H_2O, the results to be presented here will be limited to reactions of a single metal atom. The calculations to be discussed have been performed using the LCAO-MO-SCF Hartree-Fock method using double-zeta or higher quality basis sets (7-10). The importance of electron correlation effects for these reactions is not yet known. However, very recent calculations (10) on HAlOH give vibrational frequencies in good agreement with experiment supporting use of SCF calculations for the study of such reactions. The SCF results should be particularly valuable for predicting charge densities and geometries and should qualitatively represent the trends in heats of reaction.

The geometry optimizations in the present study were performed by carrying out a series of pointwise calculations with one parameter varied at a time. Until recently all ab initio geometry optimizations were performed with such a pointwise procedure. The difficulty with pointwise optimizations on polyatomics is two-fold: (1) they are very expensive, (2) it is easy to miss the true minima.

Our results for the Be, B, Mg, and Al + H_2O systems are summarized in Table I together with the theoretical results of Curtis and Frurip (8) for Be and Mg, and Trenary et al (7) for Li, Na, and Al. For Al and Mg, there is good agreement between the calculated and measured heats of reaction for the insertion process $M + H_2O \rightarrow HMOH$. Experimental values are not yet available for the B or Be reactions.

The discrepency between the two theoretical values for the stability of the $Be..OH_2$ adduct appears to be due to the neglect of d functions in our basis sets. Curtis and Frurip (8) have found that the addition of d functions (on Be and O) results in about a 3 kcal/mole decrease in stability of $Be..OH_2$ compared to that predicted using an sp basis set. On the other hand, d functions had very little effect on the stability of $Mg..OH_2$. We expect that beryllium is an extreme case and that the corrections due to d functions will in general be smaller for adducts of B, Mg, and Al with HF, H_2O, NH_3, and CH_4. More detailed studies including both d functions and correlation effects are required to develop a good understanding of the trends in the energies and force constants of the adducts.

Table I. Comparison of Experimental and Theoretical Heats of Reaction (kcal/mole)[a]

M	M...OH_2	HMOH	Study
Li	11.7	-	Th, TSK (Ref. 7)
Be	6.2	81.4	Th, KJ (Present Study)
	0.6	78.1	Th, CF (Ref. 8)
B	3.9	68.1	Th, KJ (Present Study)
Na	5.2	-	Th, TSK (Ref. 7)
Mg	-	20	Exp, HKFM (Ref. 1)
	2.4	24.4	Th, (Present Study)
	2.3	18.1	Th, CF (Ref. 8)
Al	-	60	Exp, HKFM (Ref. 1)
	8.5	42.8[b]	Th, (Present Study)
	4.4	-	Th, TSK (Ref. 7)

a) The theoretical values in this table were obtained from all-electron SCF calculations.

b) This value is somewhat larger than that published previously in Reference 9. The new value was obtained using basis sets with d functions on both the Al and O atoms.

In our original calculations (9) we found the AlOH bond of HAlOH to be linear but with a relatively small force constant for bending. In the calculations on the other hydroxyhydrides the MOH group was restricted to be linear. In very recent work, (10), carried out after this paper was nearly completed, it was found using the gradient method that the AlOH bond is actually bent with an angle of 135°. Apparently, this is another case where one can converge on a false minimum by treating the various geometry parameters separately. We will turn to this point again later in the paper. In any case, the treatment of the MOH bonds as linear should not introduce large errors in the heats of reaction, the bond lengths, or the HMO angles.

The geometries of the insertion products of the Al + HF, Al + H_2O, Al + NH_3, Al + H_2S, Be + H_2O, and Mg + H_2O reactions are given in Figure 1. The resulting heats of reaction, together with that for Al + CH_3OH $\rightarrow$ $HAlOCH_3$ are given in Table II. It may be seen from these results that the heats of reaction correlate with the polar character of the R-H bond into which insertion occurs, suggesting that the ionic character of the new bonds is an important driving force in these reactions.

For the Al + H_2O insertion reaction three different calculations of increasing sophistication were performed. In all cases Dunning's double-zeta s and p basis sets were utilized for hydrogen and oxygen. The first calculation employed a core potential (11) for Al, effectively reducing the treatment of aluminum to a three electron problem. In this case the 3s4p basis set of Topiol et al (11) was employed for Al. The calculations were then repeated treating all electrons explicitly and employing the 6s4p basis set of Trenary et al (7) for Al. Finally, a second set of all-electron calculations was performed but with an enlarged basis set with d functions on the aluminum (α_d = 0.8) and oxygen (α_d = 0.4) atoms.

The geometries obtained from the three sets of calculations are in fairly good agreement in spite of the large discrepency between the heats of reaction obtained from the all-electron (-38.1 kcal/mole) and effective potential (-54.3 kcal/mole) methods. While the effective potential value is in good agreement with the experimental estimate, this could be fortuitous as it is possible that there is a partial cancellation of errors resulting from the use of the core potential and the neglect of correlation effects. d functions appear to be relatively unimportant for the heat of reaction. However, the role of d functions could be more significant when correlation effects are included.

Earlier it was mentioned that gradient calculations reveal that the AlOH portion of the molecule is actually bent in contrast to the conclusion reached using a pointwise optimization. Furthermore, these calculations indicate that both the cis and trans isomers are stable, with the trans form more stable by 0.3 kcal/mole. While the basis employed in the gradient study employed fewer primitive functions than that employed in the

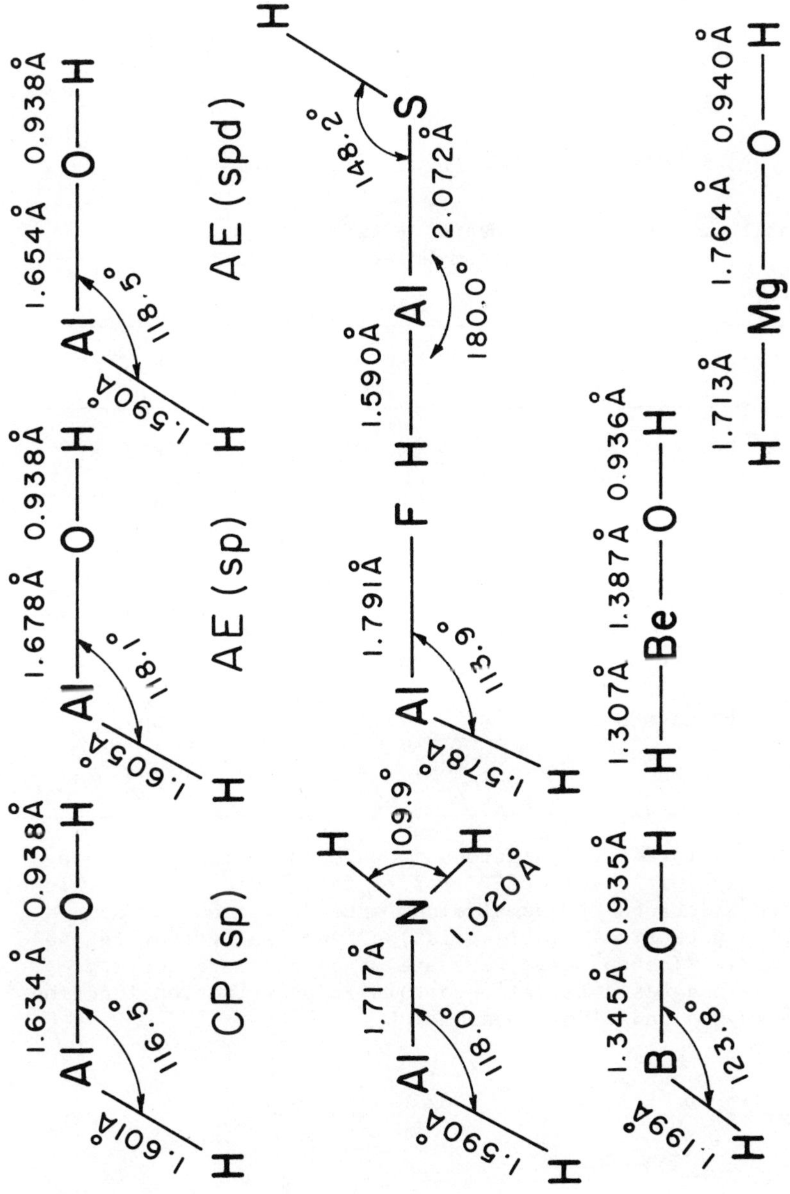

Figure 1. Calculated geometries of HAlOH, HAlNH$_2$, HAlF, HAlSH, HBOH, HBeOH, and HMgOH. The three geometries given for HAlOH are obtained using an effective core potential on Al[CP(sp)], an all-electron calculation employing a double-ζ sp basis set [AE(sp)], and an all-electron calculation employing d polarization functions on the Al and O atoms in addition to the double-ζ sp basis [AE(spd)].

Table II. Calculated Heats of Reaction (kcal/mole) for Various Aluminum-Molecule Reactions

Reaction	ΔH_f	Method[a]
$Al + HF \rightarrow HAlF$	-48.0	AE(sp)
$Al + H_2O \rightarrow HAlOH$	-38.1	AE(sp)
	-42.8	AE(spd)
	-54.3	CP(sp)
$Al + NH_3 \rightarrow HAlNH_2$	-37.4	CP(sp)
$Al + CH_3OH \rightarrow HAlOCH_3$	-27.4	CP(sp)
$Al + H_2S \rightarrow HAlSH$	-28.2	CP(sp)

a) "AE" refers to SCF calculations in which all electrons were treated explicitly while "CP" refers to calculations in which the core electrons of the aluminum were "replaced" by an effective potential. The labels "sp" and "spd" refer to the basis sets. The sp basis sets are of double-zeta quality while the spd basis set also contains d polarization functions on the oxygen and aluminum atoms.

pointwise optimizations, the geometries optimized using the gradient procedure should be more reliable.

The bonding is quite ionic in these compounds. The charge distribution obtained via a Mulliken population analysis on HAlOH is

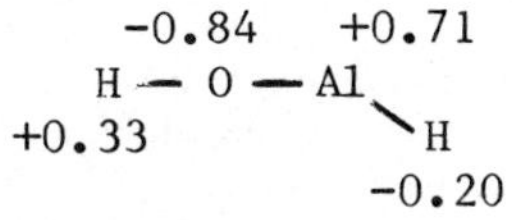

This is consistent with a low frequency for bending the HOAl bond. In fact, many related ionic molecules, for example, LiOH and Li_2O, are known to have linear geometries.

Whereas HBeOH and HMgOH are found to be linear, HAlOH and HBOH have HMO angles close to 120°. This behavior follows from simple molecular orbital considerations, or from analogy with the behavior for the corresponding metal hydrides: BeH_2, MgH_2, BH_2, and AlH_2. Curtis and Frurip (8) have noted that the HMO bending frequency is quite low in HMgOH and HBeOH and that upon the inclusion of d functions HBeOH actually assumes a bent geometry.

The bond lengths of HAlOH, HBOH, HBeOH and HMgOH are compared in Table III with those of relevant fragment molecules. In all cases the HM bond of the hydroxyhydride is seen to be slightly shorter than that of the corresponding diatomic MH molecules, consistent with greater ionic character in the hydroxyhydrides.

The above discussion has focused on the structures of the insertion products. The remainder of this section will briefly discuss the equally important problem of the dynamics of metal atom-water interactions. Of special interest is the possible role of the adducts in the route to the insertion products. The matrix isolation experiments of Hauge and co-workers (1-3) suggest that for most metal atoms the first step of the reaction involves formation of the adduct. Upon irradiation the adduct can interconvert to the insertion product, which, in turn, can give rise to still other products. Two exceptions to this are Sc and Al which in the matrix isolation experiments react spontaneously to give the hydroxyhydrides. Upon irradiation these decompose to yield metal oxides and hydroxides.

Under single collision gas phase conditions Liu and Parson (4) have found that Sc + H_2O gives rise to ScO + H_2 while Gole and Pace (5) have observed $BO(A^2\Pi) \rightarrow BO(X^2\Sigma)$ emission resulting from the B + H_2O reaction under similar conditions. On the other hand Oblath and Gole (6) have observed emission attributed to HAlOH* in the gas phase Al + H_2O reaction. Apparently different reaction pathways are followed in the gas phase than in the matrices. Detailed knowledge of the ground and electronically excited states originating from ^2P B, ^2P Al and ^2D Sc are

Table III. Comparison of Calculated and Experimental Bond Lengths (Å)[a]

Molecule	R(M-H)	R(M-O)	R(O-H)
HAlOH	1.605	1.654	0.938
AlH	1.657(1.648)	-	-
AlH_2	(1.59)	-	-
AlOH	-	1.698(1.678)	0.943(0.939)
AlO	-	1.740(1.618)	-
OH	-	-	0.974(0.970)
H_2O	-	-	0.955(0.956)
HBOH	1.199	1.343	0.935
BH	(1.215)	-	-
BH_2	(1.18)	-	-
BO	-	(1.205)	-
MHgOH	1.713	1.764	0.940
MgH	(1.730)	-	-
MgH_2[b]	1.733[b]	-	-
MgO	-	(1.749)	-
HBeOH	1.307	1.387	0.936
BeH	(1.343)	-	-
BeH_2[b]	1.333	-	-
BeO	-	(1.331)	-

a) The experimental bond lengths are given in parenthesis. The experimenal values for the diatomic molecules are from K.P. Huber and G. Herzberg, "Constants of Diatomic Molecules", (Van Nostrand Reinhold Co., 1979), and those for the triatomics from G. Herzberg, "Molecular Spectra and Molecular Structure III"; (Van Nostrand Reinhold Co., 1960).

b) The bond lengths for BeH_2 and MgH_2 are from R.P. Hosteny and S.A. Hagstrom., J. Chem. Phys. 58, 4396 (1972) and R. Alrichs, F. Keil, H. Lischka, W. Kutzelnigg, V. Staemler, J. Chem. Phys. 63, 455 (1975), respectively.

required to understand the reaction mechanisms. Our knowledge of the surfaces is far from complete at the present time. Even the most basic question concerning the existence and location of barriers is unanswered at present. Only for the Al + H_2O → HAlOH has it been established (9) that there is a pathway on the surface which does not encounter a barrier.

Alkali Halides and Their Dimers

In the second part of this paper we turn our attention to the alkali halides and their dimers. While these species have been the subject of numerous theoretical treatments, here we restrict our attention to recent ab initio work. We also briefly review the status of experimental data on these systems. This survey is not meant to complete but rather is intended to point out problems deserving theoretical attention.

The alkali halides, have long been known to form clusters in the gas phase. With the rapid growth of the interest in molecular dimers and larger aggregates in general, several groups have again turned their attention to the chemistry of alkali halide vapors (12-17).

The heats of formation and geometries are known for most of the dimers, $(MX)_2$. In addition, ionization potentials (IP) and vibrational frequencies are also known for some of the dimers (13). Very little additional information is available for the dimers or larger aggregates. The energies and other properties of the electronically excited states of the dimers are unknown. In fact, the situation with regards to the excited states of the monomers is not much better. The first values for polarzabilities of some of the alkali halide dimers were reported at this Symposium by Bederson and co-workers (18). Experimental values of the polarizabilities of the monomers are not available and would be difficult to determine by deflection studies due to contributions resulting from the permanent dipole moments.

High quality theoretical data for these systems are also sparse. Near Hartree-Fock quality SCF calculations have been performed on the ground states of most of the monomers and on $(LiF)_2$, $(LiF)_3$, $(LiF)_4$ and $(NaF)_2$. The ground and low-lying electronically excited states of LiF have been studied using configuration interaction (19) and valence bond (20) techniques. No theoretical studies of the excited states of the dimers have been carried out, and little work is available for the other monomers. Only for LiF, LiCl, and KF have SCF calculations (21,22) been performed for the cation states. For the other monomers and for $(LiF)_2$, $(LiF)_3$, $(LiF)_4$, and $(NaF)_2$ Koopmans' Theorem estimates of the IP's are available.

The negative ions of the monomers have attracted considerable attention over the past few years (23-31). Photodetachment studies

$$MX^- \xrightarrow{h\nu} MX + e^-$$

by Lineberger and co-workers (23,24) have been carried out for LiCl, NaCl, NaBr, and NaI. Both the photodetachment studies and the X^- + M scattering work of DeVreugd et al (25) have provided molecular electron affinities and some information on the shapes of the anion potential energy curves. It has been established (26,27) that the Hartree-Fock approximation yields fairly accurate electron affinities for the alkali halides. For the monomers the "extra" electron binds sufficiently far back in the dipole field that its overlap with the core electrons is small. As a result correlation corrections to the EA's of these species are relatively small. Both the theoretical calculations and the experimental studies suggest that the EA's of the alkali halides increase nearly linearly with the dipole moments.

It appears unlikely that the cyclic dimers will have stable anions. Calculations (32) on cyclic $(LiH)_2$ failed to give a stable anion. It is likely that the quadrupole fields of the cyclic alkali halide dimers are also too small to bind an electron. $(BeO)_2$, which has a much larger quadrupole field, has been shown (33) to have a stable anion. The dissociative attachment process:

$$e^- + (MX)_2 \rightarrow (MX)_2^- \rightarrow M + X + MX^-$$

has been employed by Von Ebinghaus (28) to prepare the MX^- ions. However, the symmetry of the anion state responsible for this dissociative attachment process is unknown. In addition, it is possible (32) that the MX^- produced in these experiments was the result of electron attachment to the linear, less stable forms of the dimers.

The remainder of this section will be devoted largely to a summary of recent results from our group on the polarizabilities of LiF and $(LiF)_2$. These represent, respectively, the first calculation of an alkali halide polarizability in which electron correlation effects have been included and the first calculation of a polarizability of an alkali halide dimer. The section concludes with a summary of recent theoretical results for the ionization potentials for $(LiF)_n$, n = 1-4.

We have recently completed a detailed study (27) of the dipole moment, electron affinity, and polarizability of LiF and the isoelectronic BeO molecule. A wide range of basis sets were investigated to assure convergence of each of the properties and to serve as a guide for basis set selection for the heavier alkali halides. The results at the equilibrium geometries of LiF and BeO are compared in Table 1 together with those for HF, for which experimental polarizabilities are known. The columns

Table IV. Effect of Correlation on Predicted Dipole Moment, Polarizability, and Electron Affinity of HF(R = 1.7328 bohr), LiF(R = 2.9549 bohr), and BeO(R = 2.5150 bohr).

	HF[a,b]			LiF			BeO		
	SCF[c]	MBPT	Exptl	SCF[c,d]	MBPT	Exptl	SCF[c,d]	MBPT	Exptl
μ (D)	1.93	1.78	1.80	6.51 (6.50)	6.34	6.33	7.46 (7.53)	6.40	...
α_{11} (a_0^3)	5.76	6.45	6.40	7.38 (7.47)	9.73	...	20.37 (20.56)	29.51	...
$\alpha_{\perp}$ (a_0^3)	4.48	5.31	5.08	7.33 (7.48)	9.73	...	21.06 (21.26)	29.73	...
E.A. (eV)	...	...	...	0.29/0.31	0.33	...	1.73/2.11	1.98	...
				(0.29/0.31)			(1.76/2.12)		

[a]From Ref. 35.

[b]The experimental values for α_{11} and $\alpha_{\perp}$ for HF are obtained from $\bar{\alpha} = \frac{1}{3}(\alpha_{xx} + \alpha_{yy} + \alpha_{zz})$ and $(\alpha_{11} - \alpha_{\perp})$, the quantities actually measured.

[c]The dipole moments and polarizabilities listed under SCF are obtained from finite-field SCF calculations. The first and second values of the electron affinities listed under the SCF column are obtained using the Koopmans' theorem and ΔSCF approximation, respectively.

[d]The values given in parentheses are obtained using the following basis sets: Li[12s6p/6s4p], F[11s6p3d/5s4p3d], Be[11s5p1d/5s4p1d] and O[11s6p3d/5s4p3d]. The other SCF and MBPT results are obtained using smaller basis sets with only two d functions (α_d = 0.80, 0.16) on flourine, three d functions (α_d = 0.80, 0.18, 0.06) on oxygen; and [6s4p/4s2p] basis sets on Li and Be.

headed SCF and MBPT give the results obtained from finite-field SCF and second-order many-body perturbation theory (MBPT) calculations, respectively. As can be seen from the Table, the MBPT results are in excellent agreement with experiment, where the values are known. The correlation effects increase both components of the polarizability of LiF by 33%, indicating that it will be necessary to go beyond the Hartree-Fock approximation for the heavier alkali halides and for the dimers. For comparison we note that a similar treatment of correlation results in an increase of the polarizability of a free F^- ion from 10.6 a_o^3 to 16.5 a_o^3 or by 66%. Since the polarizability of F^- is nearly two orders of magnitude larger than that of Li^+, we can view the molecular polarizability components in terms of those of F^- in the field of Li^+ ions. The presence of the Li^+ ion decreases the SCF and correlated polarizabilities of F^- by 30% and 41%, respectively. We expect the importance of correlation for the polarizability of Li_2F_2 to be roughly the same as in LiF; although there are now two F^- ions, each is in the field of two Li^+ ions which should decrease the polarization (and the importance of correlation to it) of each F^- relative to that of F^- in LiF.

Correlation effects cause an even larger increase in the polarizability in BeO. In both molecules the inclusion of correlation results in a decrease in the dipole moment with the change again being greater for BeO. Finally it should be noted that the SCF and correlated EA's of LiF agree within 0.01 eV, supporting the earlier statement that the Hartree-Fock model is quite good for the EA's of alkali halides.

The results of SCF calculations of the polarizabilities of $(LiF)_2$ and Li_2F^+, restricted to the geometry of the LiFLi portion of $(LiF)_2$ are summarized in Figure 2. The polarizability of the dimer is about 1.7 times that of the monomer. Moreover if the polarizabilities of the Li_2F^+ fragment are doubled they agree to within 7% of those from the complete calculation on Li_2F_2. This supports the view that the molecular polarizabilities can be interpreted as a sum of F^- polarizabilities, where the F^- ions are "constrained" by the Li^+ ions. This approximate "additivity" of halogen polarizabilities should work best for the lithium halides and sodium halides, but will be less useful for the halides of the heavier alkali's for which the cation polarizabilities are significant.

In concluding, we briefly consider the IP's of the LiF aggregates. The motivation for this is the observation that the two published theoretical values (16,17) for Li_2F_2 differ by 1 eV making comparison with our results of interest. The contracted Gaussian basis set, Li[10s4p/4s2p] and F[11s6p2d/5s4p2d], used in our calculations yields polarizabilities, dipole moments, and bond energies of the monomer within 2% of the Hartree-Fock limit values (30). Hence our basis should also be capable of yielding ionization potentials close to the Hartree-Fock limit. In Table V the resulting KT IP's of LiF and Li_2F_2 are compared with

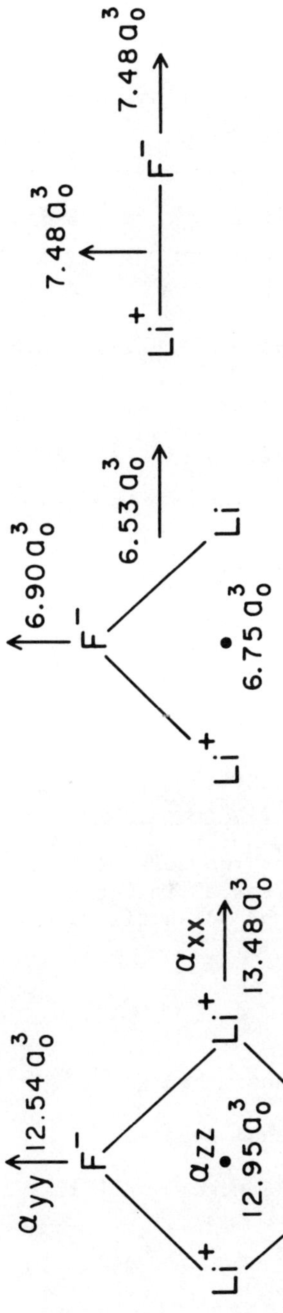

Figure 2. SCF polarizabilities of $(LiF)_2$*,* Li_2F^+ *and LiF. The geometry of* $(LiF)_2$ *and LiF are chosen to be experimental values, and that of* Li_2F^+ *is taken from* $(LiF)_2$*.*

Table V. Ionization Potentials of $(LiF)_n$ (eV)[a]

	LiF	Li_2F_2	Li_3F_3	Li_4F_4
Expt[b]	11.50(V)	12.8(V)	-	-
Expt[c]	11.8±0.5(A)	13.0±0.5(A)	12.3±0.5(A)	-
HF(1971)[d]	12.56(V)	13.18(V)	-	-
HF(1978)[e]	12.83(V)	14.09(V)	14.52(V)	14.58(V)
HF(1981)[f]	12.89(V)	14.13(V)	-	-
	12.89(V)	14.34(V)		
	13.58(V)	14.51(V)		
		14.65(V)		
		14.85(V)		
		15.38(V)		

a) The experimental and first two sets of theoretical values in this table are taken from a similar Table in Reference 13. V and A denote vertical and adiabatic values respectively.

b) From Reference 13.

c) From Reference 34.

d) From Reference 16, Koopmans' Theorem.

e) From Reference 17, Koopmans' Theorem.

f) Present study, Koopmans' Theorem. All IP's less than 30 eV are listed.

experimental and other theoretical values. The present IP's are in good agreement with those of Rupps and Ahlrichs (17), but not those of Baskin et al (16). A comparison of the basis sets used in the three studies suggests that the lack of diffuse p functions in the basis set employed on the fluorine atom is responsible for the deviation of the IP's of Baskin et al. from the other two sets of theoretical values. The close agreement between the vertical and adiabatic experimental IP's of $(LiF)_2$ is surprising since the ionic model calculations of Berkowitz et al (13) yield very different values for these two quantities. Unfortunately, there is no experimental or ab initio data available at the present time on the geometries of the ions.

The change in electronic structure of LiF upon ionization may be viewed as: $Li^+F^- \rightarrow Li^+F$. Since the SCF approximation is known to underestimate the EA of F by about 2.2 eV (12), it is at first surprising that the KT IP is too large. However it should be noted that in the KT approximation the energy difference between F^- and F is actually overestimated; at a near Hartree-Fock level SCF calculation on F^-, the binding energy of an electron in a 2p orbital is found to be 4.9 eV. Accurate (e.g., agreeing within 0.1 eV with experiment) calculations of the IP's of the alkali halides and their clusters will be difficult since large basis sets together with the inclusion of correlation are necessary to obtain accurate values of the halogen electron affinities.

The treatment of the heavier dimers and their cations at an all-electron level would be very expensive. For this reason we (36) have been exploring the utility of model or pseudopotentials (11) which account for the effect of the inner core electrons via effective potentials. In this approach one treats only the valence electrons explicitly, reducing the alkali atoms and halogen atoms to one and seven electron problems, respectively, and calculations on the heavier alkali halide dimers become comparable to that of $(LiF)_2$ in difficulty. In preliminary studies on LiCl and NaCl we have found that the SCF IP's, EA's, dissociation energies, and dipole moments resulting from the effective-potential calculations agree to within 3% with the results of all-electron calculations. Generally the errors are larger for the NaCl case, due to the neglect of the core polarization.

For LiCl, the pseudopotential calculations predict a 0.79 eV splitting between the highest lying σ and π orbitals as compared with the 0.75 eV splitting found in the orbital energies resulting from the all-electron SCF. Essentially the same splitting is found between the $^2\Sigma$ and $^2\Pi$ cation states at the ΔSCF and configuration interaction levels of treatment. While the KT, SCF, and CI treatments all give an excellent value for the splitting, it is much more difficult to obtain accurate values of the IP's. For example, for LiCl, the IP ($^1\Sigma \rightarrow {}^2\Sigma$) resulting from a singles & doubles CI using a large basis set is still 0.34 eV smaller than the experimental value (36).

Acknowledgements

We acknowledge support from the National Science Foundation and NASA Ames for this research. We also acknowledge useful discussions with Drs. Berkowitz, Gole and Hauge.

Literature Cited

1. Hauge, R.H., Kauffman, J.W., Fredin, L., and Margrave, J.L., "Matrix Isolation Infrared Studies of Metal Atom-Water Reactions", 181st National ACS Meeting, Atlanta, 1981, and references therein.
2. Hauge, R.H., Meier, P.F., and Margrave, J.L., Ber. Bunsenges. Phys. Chem., 82, 102 (1978).
3. Hauge, R.H., Kauffman, J.W., and Margrave, J.L., J. Am. Chem. Soc., 102, 6005 (1980).
4. Liu, K. and Parson, J.M., J. Chem. Phys, 68, 1794 (1974).
5. Gole, J.L. and Pace, S.A., "Boron Hydration in the Vapor Phase", J. Phys. Chem., to be published.
6. Oblath, S.B. and Gole, J.L., J. Chem. Phys., 70, 581 (1979); Comb. and Flame, 37, 293 (1980).
7. Trenary, M., Schaefer, III, H.F., and Kollman, P.A., J. Chem. Phys., 68, 4047 (1978); J. Am. Chem. Soc., 99, 3885 (1977).
8. Curtis, L.A. and Frurip, D.J., Chem. Phys. Let, to be published.
9. Kurtz, H.A. and Jordan, K.D., J. Am. Chem. Soc., 102, 1177 (1980).
10. Sakai, S., Kurtz, H.A. and Jordan, K.D., unpublished results.
11. Topiol, S., Moskowitz, J.W., Melius, C.F., Newton, M.D., and Jafri, J., "Ab Initio Effective Potentials for Atoms of the First Three Rows of the Periodic Table", Tech. Rep. Courant Inst. Math. Sci., New York University, N.Y. (1977).
12. Jordan, K. D. in "Alkali Halide Vapors: Structure, Spectra, and Reaction Dynamics", Chapter 15, Academic Press, New York, 1979, Eds. Davidovits, P., and McFadden, D.L.
13. Berkowitz, J., Batson, C.H., and Goodman, G.L., J. de Chemie Physique, 77, 631 (1980); J. Chem. Phys. 71, 2624 (1979).
14. Potts, A.W., Williams, T.A., and Price, W.C., Proc. Roy. Soc. Ser. A 341 147 (1974); Potts, A.W. and Williams, T.A., J. Chem. Soc. Far Trans II, 72, 1892 (1976).
15. Potts, A.W. and Lee, E.P.F., J. Chem. Soc. Far. Trans. II, 75, 941 (1979).
16. Baskin, C.P., Bender, C.F., and Kollman, P.A., J. Am. Chem. Soc., 95, 5868 (1973).
17. Rupp, M. and Ahlrichs, R., Theor. Chem. Acta, 46, 117 (1977).
18. Bederson, B., Kremens, R., Jaduszliwer, B., and Stockdale, J., "Electric Dipole Polarizabilities of Alkali Halide Dimers", presented at 181st National ACS Meeting, Atlanta, 1981.
19. Kahn, L.R., Hay, P.J. and Shavitt, I., J. Chem. Phys., 61, 3350 (1974).

20. Yardley, R.N. and Balint-Kurti, G.G., Mol. Phys., 31, 921 (1976).
21. Cade, P.E. and Huo, W.M., At. Data and Nucl. Data Tables, 15, 1 (1975).
22. Wahl, A.C. and Attermeyer, M., (1976), personal communication.
23. Carlsten, J.L., Peterson, J.R., and Lineberger, W.C., Chem. Phys. Lett., 37, 5 (1976).
24. Novick, S.E., Jones, P.L., Mulloney, T.J. and Lineberger, W. C., J. Chem. Phys., 70, 2210 (1979).
25. DeVreugd, C., Wijnaendt Van Resandt, R.W., Los, J., Smith, B.T. and Champion, R.L., Chem. Phys., 42, 305 (1979).
26. Jordan, K.D., Accounts of Chemical Res. 12, 36 (1979), and references therein.
27. Yoshioka, Y. and Jordan, K.D., J. Chem. Phys. 73, 5899 (1980); Yoshioka, Y. and Jordan, K.D., Chem. Phys. to be published.
28. Ebinghaus, H. Von, Naturforsch, Z., Teil. A , 19, 727 (1964).
29. Stevens, W.J., J. Chem. Phys. 72, 1536 (1980).
30. McCullough, E.A. unpublished results.
31. Hazi, A.U. "$^2\Sigma^+$ and $^2\Pi$ States of LiF^-", submitted to J. Chem. Phys.
32. Jordan, K.D., Chem. Phys. Letters, 40, 441 (1976).
33. Jordan, K.D. and Liebman, J.F., Chem. Phys. Letters, 62, 143 (1979).
34. Grimley, R.T., Forsam, J.A., and Grindstaff, Q.G., J. Phys. Chem., 82, 632 (1978).
35. Bartlett, R.J. and Purvis, G.D., Phys. Rev. A, 20, 1313 (1979).
36. Kunnathedathil, S. and Jordan, K.D., unpublished results.

RECEIVED August 26, 1981.

ALKALI METAL AND ALKALI COMPOUND VAPOR APPLICATIONS

Overview of Alkali Metal Vapor Applications

WILLIAM C. STWALLEY, MARK E. KOCH, and K. K. VERMA

University of Iowa, Iowa Laser Facility and Departments of Chemistry and Physics, Iowa City, IA 52242

The purpose of this overview is to provide a bridge between the various sorts of fundamental high temperature science, many of which are described earlier in this volume, and the areas of applications of much of this fundamental science involving specifically alkali metal vapors. Alkali metal vapors are chosen because of their unique combination of properties (Table I).

Table I. Properties of Alkali Metal Vapors

- Theoretical simplicity (M and M_2 are effectively one and two electron systems like H and H_2)
- Weakly bound valence electrons
  - low ionization potentials
  - strong visible absorption and emission
- Reversible equilibria
  - liquid $\rightleftarrows$ vapor at moderate temperatures
  - atom $\rightleftarrows$ diatomic $\rightleftarrows$ cluster
  - neutral $\rightleftarrows$ ions and electrons
- Interesting nonequilibria
  - radiatively trapped resonance states
  - ion and neutral recombination
  - near resonant energy transfer
  - ionization processes
- Nuclei (^{6}Li, ^{7}Li) for tritium breeding

The discussion here will focus on how these properties relate to the various current alkali metal vapor applications (in various stages of development) listed in Table II. It should be emphasized that this table is meant to be representative, but not comprehensive (indeed the authors would very much like to learn of further applications). A majority of these applications have been previously discussed in a more extensive review by us (1), which includes 185 references; only new references (2-32) not listed in (1) are given here. Many of the topics in Table II will be treated by the other authors in the volume, as indicated by the

0097-6156/82/0179-0397$05.00/0

Table II. Applications of Alkali Metal Vapors (with new references not already given in (1); those in this volume are indicated by an asterisk).

A.	Optical	
	1. Arc lamps (Na)	2*, 3 , 4*, 5*, 6
	2. Flash lamps (Cs, Rb, K)	-
	3. Optically pumped dimer lasers (Li_2, Na_2, K_2)	7*, 8*, 9, 10, 11
	4. Optically pumped atomic lasers (Cs, Rb, K, Na)	12
	5. Time-resolved infrared photography (K, Rb)	-
	6. Ultrahigh-Q isotropically sensitive optical filters (K, Rb, Cs)	13
	7. Optically bistable devices (Na)	-
	8. Pulse compression in optical fibers to remove distortion (Na, Cs)	14
	9. Frequency multiplication and mixing	15, 16
B.	Electrical	
	1. Thermionic convertors (Cs)	17*, 18*
	2. Thermoelectric laser energy convertors (Cs)	18*, 19, 20
	3. Alkali vapor and alkali seeded magnetohydrodynamics (K, Cs)	21*, 22*
	4. Ion propulsion (Cs)	23
	5. Vapor rectifiers (Cs)	23
	6. Sodium heat engine (Na)	24
C.	Fusion	
	1. Tritium breeding (^6Li, ^7Li)	25*
	2. First wall shield in Inertial Confinement Fusion (Li)	25*
	3. Production of negative ions for neutral beams (Na, Cs)	26, 27, 28
	4. Tritium and deuterium inventory monitoring (Li)	29, 30*
D.	Other	
	1. Coolants, e.g. using heat pipes (Li, Na, K)	-
	2. Solar tower heat transfer (Na)	31
	3. Getter (Cs)	23
	4. Alkali metal boiler topping cycle (K, Cs)	32
	5. Atomic clock (Cs)	23

asterisked references. We shall thus discuss the properties and then the applications while trying to avoid duplicating these other discussions and to provide alternate perspectives on alkali metal vapors in general.

Discussion of Properties

Theoretical Simplicity. The theoretical simplicity of alkali atoms in their ground and valence electron excited states is well known; they are well described by simple one electron "quantum defect" or "effective principal quantum number" ideas. Alkali atomic spectra traditionally follow the hydrogen atom in books on atomic spectroscopy.

The simplicity of the alkali diatomic molecules (equivalently the interactions of two alkali metal atoms) is illustrated by the work of Konowalow et al. and discussed in his contribution to this volume (33). It should be noted that because of the approximate separation of core and valence electron motion in these species, fully ab initio treatments are simpler (only valence electron excitations contribute significantly in a configuration interaction treatment) and replacement of the alkali ion cores by effective potentials becomes an accurate approximation.

Even alkali metal clusters are relatively simple species in terms of their electronic structure, as discussed by others in this volume (34). However, the nuclear motion (among many isomers for large clusters) becomes much less simple, as shown, for example, in the work of Lindsay in this volume (35).

The transition from the atom to the cluster to the bulk metal can best be understood in the alkali metals. For example, the ionization potential (IP) (and also the electron affinity (EA)) of sodium clusters Na_N must approach the metallic sodium work function in the limit $N \rightarrow \infty$. We previously displayed this (1) by showing these values from the beautiful experiments by Schumacher et al. (36, 37) (also described in this volume (38)) plotted versus N^{-1}. The electron affinity values also shown are from (39), (40) and (34) for N = 1, 2 and 3, respectively. A better plot still is versus the radius R of the N-mer, equivalent to a plot versus $N^{-1/3}$, as shown in Figure 1. The slopes of the lines labelled "metal sphere" are slightly uncertain; those shown are 4/3 times the slope of Wood (41) and assume a simple cubic lattice relation of R and N. It is clear that reasonably accurate interpolation between the bulk work function and the IP and EA values for small clusters is now possible. There are, of course, important quantum and statistical effects for small N, e.g. the trimer has an anomalously low IP and high EA, which can be readily understood in terms of molecular orbital theory (34, 42). The positive trimer ions may in fact be "ionization sinks" in alkali vapor discharges; a possible explanation for the "violet bands" seen in sodium vapor (20) is the radiative recombination of Na_3^+. Cs_3^- may be the hypothetical negative ion corresponding to EA $\simeq$ 1.2 eV which may explain subideal performance in thermionic convertors (1, 42).

Weakly Bound Valence Electrons. Figure 1 also illustrates how weakly bound the valence electrons really are in the alkali metal molecules. The corresponding Rydberg-like electronic bound states are consequently compressed in energy compared to hydrogenic species. Consequently, the strong electronic transitions begin in the visible (or even near infrared) and extend into the near ultraviolet. Hence laser technology in the visible and adjacent regions of the spectrum is well matched to the study of excitation and ionization of such species, as exemplified by a variety of contributions to this volume (7, 8, 18, 38, 43, 44, 45). Conversely, there are many important optical applications in the visible and adjacent regions, summarized in Table II and discussed below, especially the various sodium-containing lamps (2). A further corollary is that the alkali metal species are highly reactive and readily form ionic bonds, as discussed herein (45-51).

Reversible Equilibria. As noted in Table I, there are many reversible equilibria available in alkali metal vapors to accomodate external stresses and perturbations. The liquid $\rightleftarrows$ vapor equilibrium is of fundamental importance in connection with heat pipes (1) and also in many applications, e.g. lithium in fusion (25). Equilibria involving clustering (beginning with atom-atom recombination to the diatom (3, 4)) are just beginning to be understood quantitatively. Equilibria involving ionization are of major importance for the readily ionized alkali metal species (18, 19, 38). A final equilibrium which has apparently been much less studied (1) is the "Mott" metal-insulator transition which occurs for all alkali metal vapors at high density (52, 53).

Interesting Nonequilibria. Despite the above equilibria which are fairly readily established in alkali metal vapors, there are other highly nonequilibrium situations which also arise fairly readily. Foremost among these is the radiative trapping of the resonance excited states of the alkali metal atoms as in the sodium arc lamps (2). Under such circumstances, the population of the lowest np excited state may be much higher than expected from thermal equilibrium. Another mechanism for achieving a highly nonequilibrium distribution of states is ion-electron recombination to various electronically excited states; indeed population inversions and even lasing can be obtained in some cases (54). Another process of considerable interest for causing highly nonequilibrium populations is near resonant energy transfer involving neutral-neutral collisions (e.g. Na(4d) + Na(3s) $\rightarrow$ 2Na(3p) (20)) or electron-neutral collisions (e.g. e^- + Na(3p) $\rightarrow$ e^-_{SE} + Na, where e^-_{SE} is a "superelastic" electron with ~2 eV kinetic energy (18, 55)). Finally we note that ionization processes

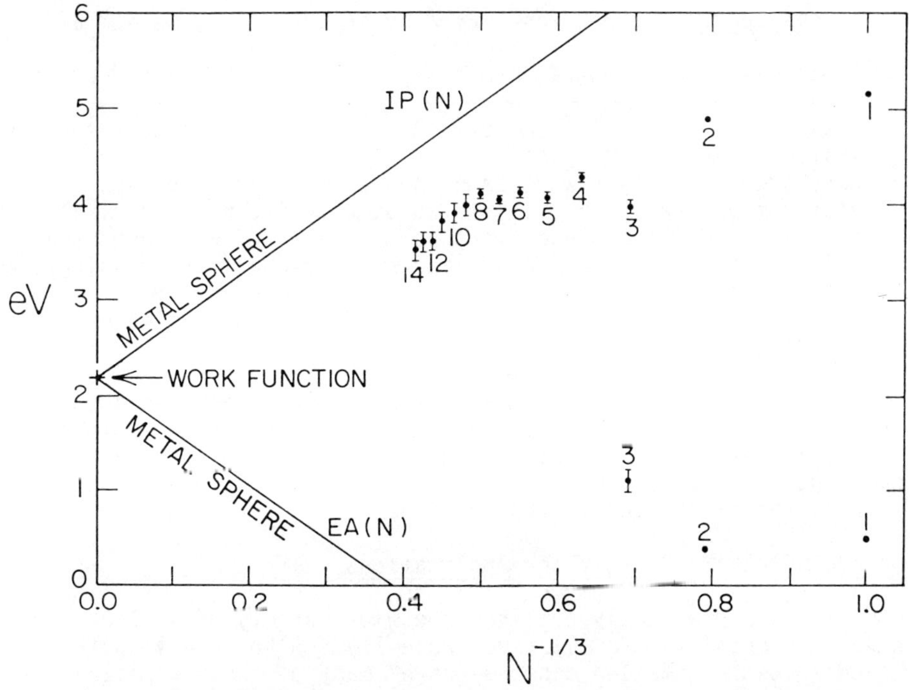

Figure 1. Ionization potential (IP(N)) and electron affinity (EA(N)), both in eV, for clusters of Na atoms Na_N plotted vs. $N^{-1/3}$ (roughly proportional to the reciprocal of the cluster radius). Also shown are bulk work function and limiting metal sphere values which the IP(N) and EA(N) curves must approach as $N \rightarrow \infty$.

can be highly nonequilibrium (18, 19), e.g. $2Na(3p) \rightarrow Na_2^+ + e^-$ with virtually no Na^+ present, despite the fact that the equilibrium ratio $[Na^+]/[Na_2^+] \simeq 10^3$ (20).

Nuclei (^6Li, ^7Li) for Tritium Breeding. For either magnetically confined or inertially confined fusion, the near term nuclear species involved will be deuterium (D) and tritium (T). While D is widely available, T must be produced and it is generally acknowledged that the nuclear reactions of neutrons with ^6Li and ^7Li should be used. For further details see (1) and (25).

Some General Properties. Finally in Table III, we summarize

Table III. Some General Properties of Alkali Metal Vapors

- Considerable energy is readily stored in the lowest np resonance excited states
- There are virtually always some ions and electrons present
- Lasers of virtually any visible wavelength incident on alkali metal vapors will produce light at a variety of wavelengths as well as some ionization (e.g. lasers at resonant and quasiresonant wavelengths)
- Low densities are easiest to understand, high densities generally of greatest interest for applications

some general properties of alkali metal vapors which are not obvious at first glance. We will not discuss these general comments in detail here, but we feel they are worth keeping in mind in considering both basic and applied studies in alkali metal vapors, such as those reported in this volume.

Update of Alkali Metal Vapor Applications

We have previously outlined the wide variety of applications of alkali metal vapors (1) (see Table II). Also we have gathered in this volume detailed discussions of many of these applications, indicated by the references with asterisks in Table II. Here we mention a few additional applications which are not covered in (1) or elsewhere in this volume.

With regard to new optical applications, Müller and Hertel (12) have developed a two-photon optically pumped Na laser, while Marling et al. (13) have developed a simple high selectivity, high sensitivity optical detector employing alkali vapors. Natatsuka and Grischkowsky (14) have demonstrated how distortion in optical fibers (pulse elongation) can be simply corrected by a compensating pulse compression in a relatively short Na vapor cell. The use of metal vapors for frequency multiplication (and also mixing) to generate coherent ultraviolet radiation is rapidly expanding field; see e.g. Vidal's review (15) and the nominally forbidden second-harmonic generation experiment of Miyazaki et al. (16).

Interesting additional electrical applications include the use of Cs for ion propulsion and to provide ionization in rectifiers (23) and the sodium heat engine concept (24).

New fusion applications include the concept of production of intense negative ion beams (27) (for neutral beam injection for heating and diagnostics in tokamaks or other magnetically confined plasmas (26, 28)) by using photodissociation to ion pairs (e.g. $NaLi + h\nu_{uv} \rightarrow Na^+ + Li^-$) in supersonic molecular beams. Another promising concept is the use of laser induced fluorescence to monitor very low tritium concentrations (as little as 10^2 T_2/cm^3) under fusion reactor conditions (29).

Other additional alkali metal vapor applications include Cs and Rb for atomic clocks (56, 57) and gettering (23), Na for solar tower heat transfer (31) and K for a boiler topping cycle (32).

It is clear that an increasing number of fascinating applications of these flexible substances will be developed and refined in the near future; we hope that this volume will make a significant contribution to such developments.

Acknowledgements

Work at Iowa on various aspects of alkali metal vapors has been supported over the years by the National Science Foundation, the National Aeronautics and Space Administration, the Department of Energy, and the Petroleum Research Fund, administered by the American Chemical Society.

Literature Cited

1. Stwalley, W. C.; Koch, M. E. Optical Eng., 1980, 19, 71.
2. Zollweg, R. J. ACS Symp. Series, 1981, XX, xxxx [this volume].
3. Woerdman, J. P. J. Mol. Structure, 1980, 61, 221.
4. Woerdman, J. P.; Eskildsen, S. S.; Rey, W. J. J. ACS Symp. Series, 1981, XX, xxxx [this volume].
5. Woerdman, J. P.; deGroot, J. J. ACS Symp. Series, 1981, XX, xxxx [this volume].
6. Gould, G. as reported in Laser Focus, March 1981, page 4.
7. Wellegehausen, B. ACS Symp. Series, 1981, XX, xxxx [this volume].
8. Man-Pichot, C. N.; Brillet, A. ACS Symp. Series, 1981, XX, xxxx [this volume].
9. Man-Pichot, C. N.; Brillet, A. IEEE J. Quantum Electronics, 1980, QE-16, 1103.
10. Verma, K. K.; Zemke, W. T.; Stwalley, W. C. J. Appl. Phys., 1981, XX, xxxx.
11. Verma, K. K.; Zemke, W. T.; Stwalley, W. C. J. Appl. Phys., 1981, XX, xxxx.
12. Müller, W.; Hertel, I. V. Appl. Phys., 1981, 24, 33.
13. Marling, J. B.; Nilson, J.; West, L. C.; Wood, L. L. J. Appl. Phys., 1979, 50, 610.

14. Nakatsuka, H.; Grischkowsky, D. Optics Letters, 1981, 6, 13.
15. Vidal, C. R. Appl. Optics, 1980, 19, 3897.
16. Miyazaki, K.; Sato, T.; Kashiwagi, H. Phys. Rev. Lett., 1979, 43, 1154
17. Britt, E. J. ACS Symp. Series, 1981, XX, xxxx [this volume].
18. Tam, A. C. ACS Symp. Series, 1981, XX, xxxx [this volume].
19. Lucatorto, T. B.; McIlrath, T. J. Appl. Optics, 1980, 19, 3948.
20. Koch, M. E.; Verma, K. K.; Stwalley, W. C. J. Opt. Soc. Am. 1980, 70, 627.
21. Hastie, J. W.; Plante, E. R.; Bonnell, D. W. ACS Symp. Series, 1981, XX, xxxx [this volume].
22. Freedman, A.; Wormhoudt, J. C.; Kolb, C. E. ACS Symp. Series, 1981, XX, xxxx [this volume].
23. Kawecki Berylco Industries, Inc. Physical Properties and Applications of Cesium, 1978.
24. Cole, T. in Report on the Symposium and Workshop on the 5 MWt Solar Thermal Test Facility, U. S. ERDA Report ALO/3701-76/1, 1976, p. 211.
25. Blink, J. A.; Krikorian, O. H.; Hoffman, N. J. ACS Symp. Series, 1981, XX, xxxx [this volume].
26. Hiskes, J. R. J. de Physique, 1979, 40, C7-179.
27. Grisham, L. R.; Stwalley, W. C., to be submitted to Rev. Sci. Instrum.
28. Grisham, L. R.; Post, D. E.; Eubanks, H. P.; Stwalley, W. C. Bull. Am. Phys. Soc., 1980, 25, 983.
29. Stwalley, W. C., "Laser Fluorescence of Li_2 and LiH: Possible Application to Fusion", unpublished.
30. Wu, C. H.; Ihle, H. R. ACS Symp. Series, 1981, XX, xxxx [this volume].
31. Report on the Symposium and Workshop on the 5 MWt Solar Thermal Test Facility, U. S. ERDA Report ALO/3701-76/1, 1976.
32. Fossil Energy Program Report, U. S. DOE Report DOE/ET-0060 (78), August 1978, p. 127.
33. Konowalow, D. D.; Rosenkrantz, M. E. ACS Symp. Series, 1981, XX, xxxx [this volume].
34. Richtsmeier, S. C.; Eades, R. A.; Dixon, D. A.; Gole, J. L. ACS Symp. Series, 1981, XX, xxxx [this volume].
35. Lindsay, D. M.; Garland, D.; Tischler, F., Thompson, G. A. ACS Symp. Series, 1981, XX, xxxx [this volume].
36. Herrmann, A.; Schumacher, E.; Wöste, L. J. Chem. Phys., 1978, 68, 2327.
37. Leutwyler, S.; Hofmann, M.; Härri, H.-P.; Schumacher, E. Chem. Phys. Lett., 1981, 77, 257.
38. Schumacher, E.; Gerber, W. H .; Härri, H.-P.; Hofmann, M.; Scholl, E. ACS Symp. Series, 1981, XX, xxxx [this volume].
39. Hotop, H.; Lineberger, C. A. J. Phys. Chem. Ref. Data, 1975, 4, 539.

40. Shepard, R.; Jordan, K. D.; Simons, J. J. Chem. Phys., 1978, 69, 1788.
41. Wood, D. M. Phys. Rev. Lett., 1981, 46, 749.
42. Laskowski, B., private communication, 1980.
43. Collins, C. B.; Lee, F. W.; Vicharelli, P. A.; Popescu, D.; Popescu, I. ACS Symp. Series, 1981, XX, xxxx [this volume].
44. Gole, J. L.; Green, G. J. ACS Symp. Series, 1981, XX, xxxx [this volume].
45. Yang, S. C.; Stwalley, W. C. ACS Symp. Series, 1981, XX, xxxx [this volume].
46. Jordan, K. D.; Kurtz, H. A. ACS Symp. Series, 1981, XX, xxxx [this volume].
47. Ault, B. S. ACS Symp. Series, 1981, XX, xxxx [this volume].
48. Hauge, R. H.; Kauffman, J.; Fredin, L.; Margrave, J. L. ACS Symp. Series, 1981, XX, xxxx [this volume].
49. Douglas, M. A.; Hauge, R. H.; Margrave, J. L. ACS Symp. Series, 1981, XX, xxxx [this volume].
50. Kauffman, J. W.; Hauge, R. H.; Margrave, J. L. ACS Symp. Series, 1981, XX, xxxx [this volume].
51. Sayer, B.; Ferray, M.; Visticot, J. P.; Lozingot, J. ACS Symp. Series, 1981, XX, xxxx [this volume].
52. Rose, J. H.; Shore, H. B.; Sander, L. M. Phys. Rev. B, 1980, 21, 3037.
53. Rose, J. H. Phys. Rev. B, 1981, 23, 552.
54. Fu-Cheng, L. Appl. Phys., 1981, 24, 299.
55. Measures, R. M.; Drewell, N.; Cardinal, P. J. J. Appl. Phys., 1979, 50, 2662.
56. Hellwig, H.; Evenson, K. M.; Wineland, D. J. Physics Today, December 1978, page 23.
57. Hellwig, H. W. Proc. IEEE, 1975, 63, 212.

RECEIVED September 3, 1981.

28

Physics and Chemistry of High Pressure Sodium Lamps

R. J. ZOLLWEG

Westinghouse Research and Development Center, Pittsburgh, PA 15235

In this review of the high pressure sodium lamp, emphasis is placed on evidence concerning the interaction of resonantly excited sodium atoms with sodium, mercury or xenon atoms, which modifies the spectral radiance of lamps to improve the color or efficacy. The influence of mercury and xenon buffer gases on the thermal and electrical conductivities and hence on lamp efficacy are also indicated.

The atomic spectra of sodium bears a unique relationship to the sensitivity of the human eye since the strong resonance lines appear in a spectral region very near the maximum eye response. This is illustrated in Figure 1. While the maximum visual response is 683 lumens per watt of energy radiated at 555 nm, sodium "D" line radiation yields about 500 lumens per watt of radiant energy. Thus sodium is currently used in three quite different types of commerical electric discharge lamps. Table I summarizes a few of the comparative parameters of these three types of lamps, the low pressure sodium discharge, the metal halide, and the high pressure sodium discharge lamps. Note the particularly wide range of sodium pressures used from millitorr to hundreds of torr. That pressure difference plus the differences in power loading and temperatures necessitates differences in discharge containing envelopes.

Low pressure sodium lamps (LPS) have the highest lamp efficacy. However, since the spectrum is essentially monochromatic, such light sources give virtually no color discrimination of colored, non-luminous objects illuminated by them. Such lamps are used mostly for such utilitarian purposes as highway, tunnel and security lighting, especially in Europe. Both metal halide (MH) and high pressure sodium (HPS) lamps do contain a variety of wavelengths in their spectra which permits color discrimination in the illuminated scene. The color rendering index (CRI) is an approximate measure of how successfully these lamps compare to a

0097-6156/82/0179-0407$05.00/0

black body source of the same chromaticity. The source then has a correlated color temperature (CCT) corresponding to the black body source.

In Table I, I have listed only two of the most widely used metal halide lamps and the "standard" high pressure sodium lamps together with typical values of their CRI and CCT. Other formulations of metal halide additives are sometimes used yielding significantly higher CRIs to 85-90 and correlated color temperatures of 5000-6000 K, matching daylight. High pressure sodium lamps with substantially higher CRI and/or higher color temperature will be discussed briefly below. At present, the usual metal halide lamp typically has better color rendering ability and a somewhat higher color temperature than the standard HPS lamp. However, the HPS lamp has an edge in lamp efficacy and in its maintenance of lamp efficacy throughout the life of the lamp. The rest of this review will be limited to the high pressure sodium lamp.

Recent reviews by deGroot et al. (1), Wharmby (2) and by McVey (3) summarize the scientific and technical aspects of the high pressure sodium lamp. Thus the present report has less need to be comprehensive and will tend instead to emphasize aspects of sodium and sodium-buffer gas interactions of interest to this symposium. We shall attempt to indicate areas where basic material data are either unknown or are inadequate.

The Spectrum

Figure 1 shows clearly how the broadened wings of the sodium "D" lines dominate the visible spectrum of the high pressure sodium lamps. Self-absorption is so strong that radiation at the undeviated line center is negligible. The broadening increases strongly and approximately symmetrically with the sodium pressure (4). Although the spectrum may appear to be accounted for by resonance broadening, most of the emitted radiation appears outside of the region of validity of the impact approximation. The latter is valid only within ~1 nm of line center for the conditions of Table I. Indeed, bumps appear on both the blue and red tails in Figure 1 which can be associated with Na-Na and Na-Hg molecules.

Mizuno et al. (5) showed that the bump on the red increases with Na and Hg partial pressures and shifts to longer wavelength if Cd is substituted for Hg. Figure 2 shows in more detail that there are 2-3 components to these red bumps arising from the Na-Hg and Na-Cd interactions. Because the ground states of the NaHg and NaCd molecules are weakly-bound van der Waals states (6), the red shift indicates that the excited molecular states formed by sodium resonance atoms and ground state Hg or Cd have higher dissociation energies than the ground states. It should be mentioned that the spectra of Figure 2 were obtained under somewhat specialized conditions to enhance the features. A sapphire arc

Table I. Types of Commercial Discharge Lamps Containing Sodium

	Low Pressure Sodium (LPS)	Metal Halide (MH)	High Pressure Sodium (HPS)
Gases	Na/Ne, 1% Ar	NaI,ScI_3/Hg/Ar (NaI,TlI,InI_X/Hg/Ar)	Na/Hg/Xe
P_{Na}(torr)	0.003	~1	50-100
Buffer	Ne/1% Ar 4-20 torr	Hg 3-5 atm.	Hg ~300 torr Xe ~140 torr
Envelope	Na-resistant glass	Quartz	Al_2O_3 {PCA Sapphire
Efficacy (lumens/watt)	≤200	≤125	≤140
CRI	-	65	20
CCT	-	3500 K	2100 K

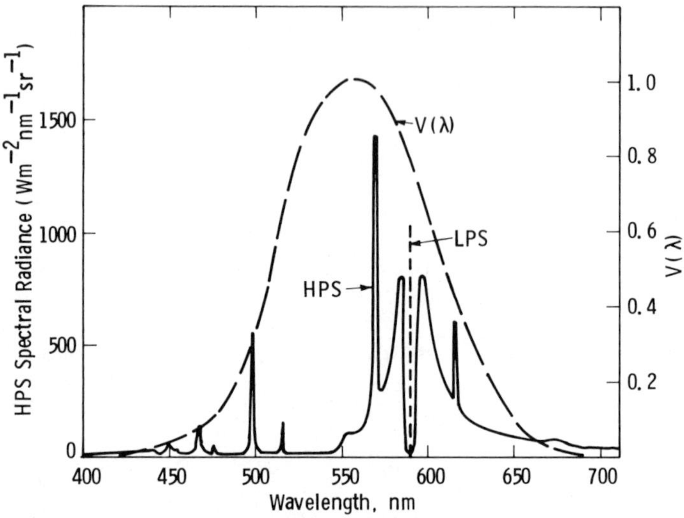

Figure 1. The measured spectral radiance of a high pressure Na (HPS) lamp (——) is compared with that of the low pressure Na (LPS) lamp (– – –) and with V(λ), the spectral sensitivity of the human eye for photopic vision.

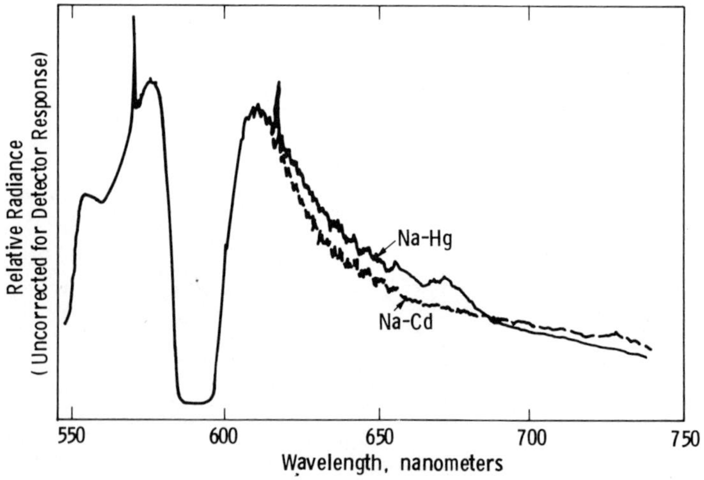

Figure 2. High resolution spectra of the Na D-line reversal from sapphire lamps operated at higher than normal Na pressures and lower than normal power loadings. Fine structure between 600 and 650 nm is the same with either Hg (——) or Cd (– – –) as the buffer gas and can be associated with vibrational transitions of the Na_2 A → X molecular system. The broader bumps at longer wavelength depend upon the buffer gas and can be related to NaHg and NaCd excimer states.

tube was used with higher Na and Hg pressures than normal. The region near the periphery of the arc core was focussed on the monochromator slit. Additional diffuse emission bands were observed in the Na-Hg arc at 465, 452.5, 446.4 and 442.2 nm. Two of these agree with absorption bands observed by Barratt (7) in Na + Hg vapor at 464.9 and 442.5 nm. The 452.5 nm band has been identified by Woerdman (8) as a collision-induced bound-free transition of Na_2 that has an intensity sensitive to the buffer gas pressure.

Also shown in Figure 2 is a considerable amount of detailed vibrational structure between about 600-650 nm that is identical in Na-Cd and Na-Hg lamps. This structure also agrees with the assignments of Singh and Rai (9) to vibrational transitions of the Na_2 $A^1\Sigma_u^+ \rightarrow X^1\Sigma_g^+$ system. This system has been studied extensively in recent years (10, 11, 12). Whittaker (13) concludes that, although both absorption and emission occur in both the red Na_2 A → X and the blue-green Na_2 B → X (476.7-492.5 nm) regions, there is net emission from high pressure sodium lamps. A paper by Woerdman and deGroot (14) at this symposium will discuss the Na_2 contribution to the prominent bump at 551.5 nm. We have also observed a broad bump at 436.3 nm at higher reservoir temperatures that, like the 452.5 nm band, has been identified by Woerdman (8) as a bound-free transition of Na_2 except that the intensity of the 436.3 nm band is insensitive to buffer gas pressure.

In his careful, high resolution studies of the HPS lamp, Whittaker (13) observed most of the expected atomic lines of Na, e.g., from the principal, sharp, diffuse and fundamental series as classically designated. Also observed were forbidden lines of the Lenard series from $^2P \rightarrow {}^2P$ and $^2F \rightarrow {}^2P$ transitions. Two forbidden lines at 552.7 and 553.2 nm are prominent on the "knee" of the blue wing reversal of the resonance lines. Many of these atomic lines are also broadened at higher lamp reservoir temperatures and buffer gas pressures. In addition atomic lines of Hg and impurity lines of Al, Ba, Ca, K, Mg and Sr were observed.

The xenon starting gas also contributes to the Na resonance line reversal and much higher than normal xenon pressures have been found to improve lamp efficacy and lumen maintenance. Figure 3 shows spectra obtained for two lamps containing xenon at different pressures with sodium but no mercury. Rather high reservoir temperatures of 812 C (1085 K) were used such that the Na partial pressures approach 400 Torr. The xenon pressures during lamp operation were approximately 140 and 2100 Torr respectively. At the higher xenon pressure it is evident that: (1) the sodium reversal is greater, (2) emission in the far red wing is much higher, (3) there is appreciably more radiation in the 552-566 nm blue wing region, (4) there is additional structure at about 557.5 and 562.5 nm, and (5) the peaks of the 568.6 and 615.9 nm self-absorbed doublets are appreciably lower. It is readily evident that the Na-Xe interaction broadens the entire sodium resonance line region at such high sodium and xenon pres-

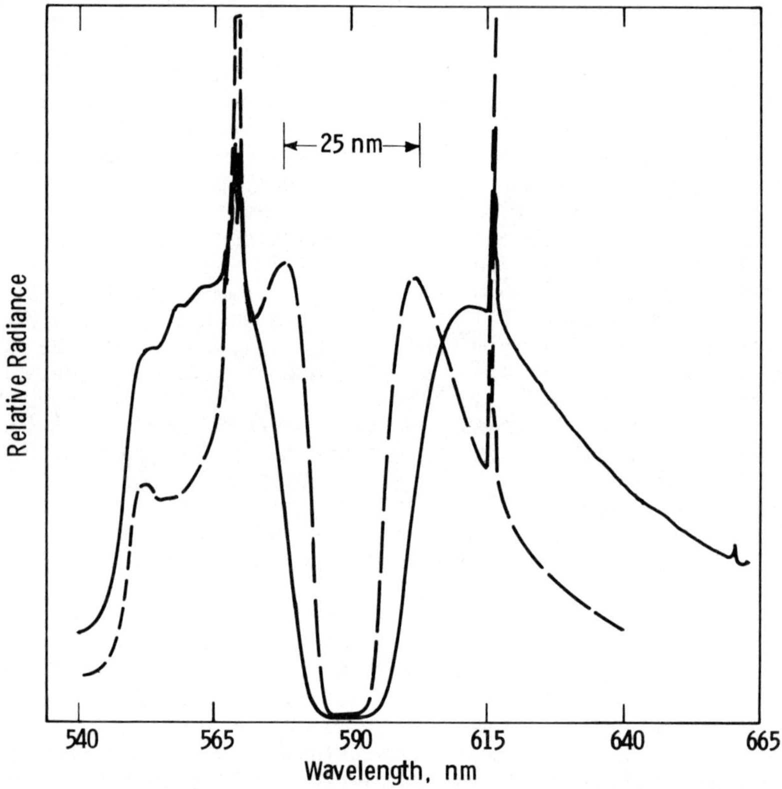

Figure 3. Dependence on Xe pressure of Na reversal is shown at higher sodium pressure. At Xe pressure of ~2100 (——) vs. ~140 (– – –) torr, separation between reversal maxima is increased, radiance in both wings increases, and additional features appear at 557.5 and 562.5 nm.

sures. The two additional bumps at 557.5 and 562.5 nm depend upon both high sodium and high xenon pressures and might arise from the self reversed satellite of NaXe* ($^2\Sigma \rightarrow {}^2\Sigma$) or from a triatomic molecule such as Na_2Xe^*. The reduced heights of the 568.6 and 615.9 nm doublets might result primarily from a lower axial arc temperature.

Although we started thinking about the sodium radiation in terms of resonance line broadening, we now see that there are components of Na_2, NaHg, NaXe and perhaps triatomic molecules as well. Much of the radiation emitted is absorbed, especially by the colder gas near the wall, so the radiation transfer problem has a considerable influence on the spectrum emanating from the lamp. The temperature profile of the arc and the arc tube wall temperature modify the layer of absorbing gas and hence the sodium reversal observed. Waymouth and Wyner (15) have emphasized that higher lamp efficacy may result from raising the arc tube wall temperature at constant diameter and power loading. The optical properties of the wall also contribute to the emitted spectra. The usual arc tube wall material, polycrystalline alumina (PCA), has rather low "in-line" transmission and scatters most of the incident radiation. That portion which is scattered back into the discharge medium and that is also in the spectral region inside or near the peaks of the self-reversed resonance lines, is likely to be absorbed. Thus we have found greater Na reversal for PCA arc tube lamps than for sapphire lamps with the same diameters, Na-Hg alloy composition and reservoir temperatures. Indeed, various authors have reported 3-8% higher efficacies for sapphire vs PCA lamps depending upon the quality of the PCA and its surface treatment (3).

With all of the complications enumerated above, how can one account for the observed HPS lamp spectra? Lamp development engineers have typically taken the simplest approximations and extended the Lorentzian, impact-approximation profile for the entire line - far beyond its region of strict validity - and combined this with radiation transfer models. Such procedures yield calculated results surprisingly close to experiment (1) provided details of the spectra are not too closely scrutinized.

A great deal of guidance on how to improve these calculations have come from the work of Gallagher and his coworkers (16) who have used laser-induced fluorescence to measure detailed line shapes for alkali metals perturbed by rare gases and determined interatomic potentials. Further aid is available from the calculated potentials of Baylis (17), Pascale (18) and Düren (19). However, the accuracy of the theoretical calculations is often insufficient for our needs. Conditions in our discharges are often at higher kinetic temperatures than those of the fluorescence measurements such that the repulsive potential curves play a more important role. We also have the complication that several broadening processes occur together, e.g., Na-Na, Na-Xe, Na-Hg, in addition to the strong self-absorption. To solve the radiation transfer problem the arc temperature profile is needed. This can be measured experimentally or calculated by solving the energy balance equation.

The Energy Balance

Calculations of the HPS lamp spectrum and of the energy balance of the arc help us to understand the importance of various controllable arc parameters and thus aid in lamp development. The assumption of local thermodynamic equilibrium (LTE) (except for radiation) has been found to be a good approximation, even for pulsed arcs, although non-LTE effects have been demonstrated in some details, e.g., in the absolute radiance at the sodium D-line reversal minima and in high frequency operation. For the ac operated arc in LTE, we solve numerically the ac energy balance equation

$$\rho\, C_p \frac{\partial T}{\partial t} = \sigma E^2 - \text{div}\,(-\kappa\ \text{grad}\ T) - u \qquad (1)$$

Ohm's Law

$$I = 2\pi E \int_0^R \sigma r dr \qquad (2)$$

and the circuit equation which connects the power source and ballast with the arc column. In equations (1) and (2) the arc current I and axial electric field E are functions of time t. The bulk material properties include the electrical conductivity σ, thermal conductivity κ, density ρ, specific heat C_p, and net radiation emission u and are all functions of temperature T. These are input parameters for the calculation as are necessary boundary conditions. Figures 4 and 5 show how the thermal and electrical conductivities of HPS lamps can be decreased with higher Hg or Xe buffer gas pressures. For temperatures below an axial arc temperature of ~4000 K, lower thermal conductivity is desirable to reduce the conduction power loss which reduces lamp efficacy. A lower electrical conductivity results in a higher impedance lamp with lower current and therefore lower electrode losses. However, both Hg and Xe buffer gases change the emitted spectrum, and too high pressures reduce lamp efficacy. The quantity $\rho\, C_p$ is also important because it has a major influence on the temperature change of the arc during the ac cycle.

The net radiation emission of a volume element of the arc is a particularly difficult parameter to determine. Because of the strong self-absorption of the arc it is not only a function of the partial pressures of the radiating species and the temperature, but also depends upon the surrounding radiating region and, in this case, the absorption will also be time-dependent. We have found that application of a modified radiation imprisonment formulation originally developed by Holstein gives us net emission coefficients in good agreement with those we derive from deGroot's arc temperature profile measurements of HPS arcs (20). deGroot *et al.* (1) show examples of the changes in axial arc

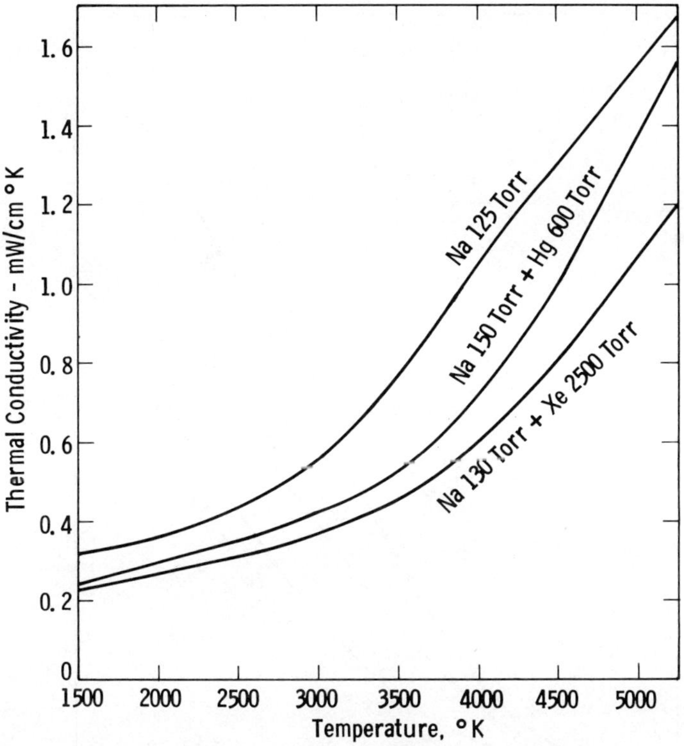

Figure 4. Calculated thermal conductivity of Na vapor and mixtures of Na vapor with Hg or Xe. Thermal conductivity is reduced significantly by the addition of Hg or Xe buffer gases.

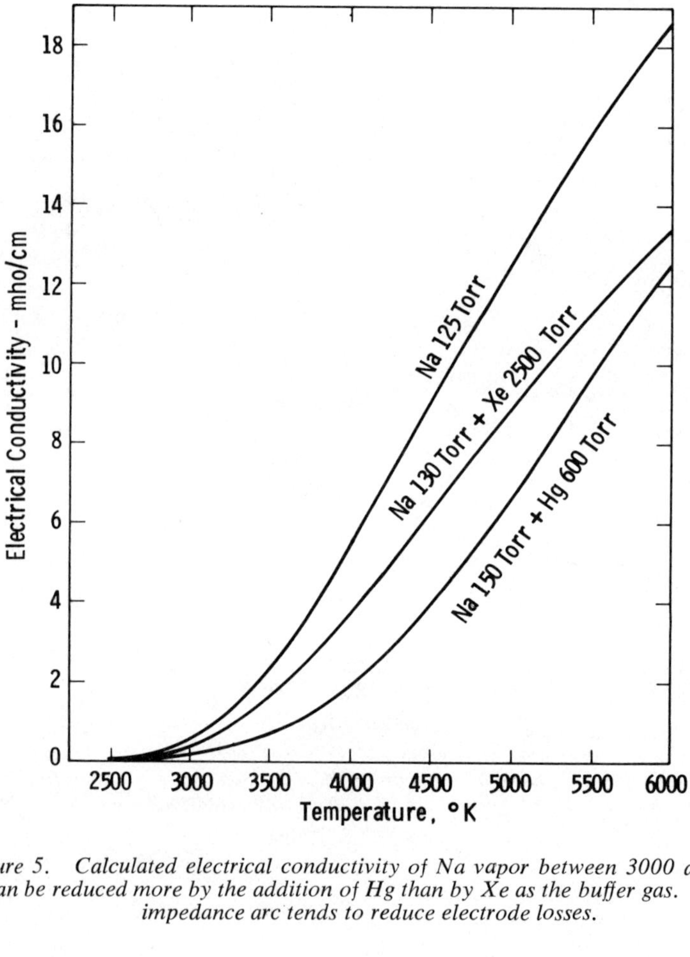

Figure 5. Calculated electrical conductivity of Na vapor between 3000 and 4500 K can be reduced more by the addition of Hg than by Xe as the buffer gas. A higher impedance arc tends to reduce electrode losses.

temperature with time during the ac cycle. They also show how the reignition voltage spikes observed for HPS lamps are modified at higher buffer gas pressures.

In order to calculate the material transport properties as given in Figures 4 and 5 we first must calculate the chemical equilibrium making use of the available thermochemical data base. Table II shows, for example, the calculated equilibrium densities of neutral and charged species near the axis (~4000 K) of a typical commercial HPS arc lamp. We then use the best available interaction data for cross sections to calculate the transport properties.

To optimize the high pressure sodium lamp is still a very formidable task (2) because of the multiplicity of parameters under the designer's control and a number of materials limitations which must be taken into account for a very long life lamp (>20,000 hours). Waymouth (21) has developed a simplified analysis that relates lamp output quantities, e.g., efficacy, radiation in various spectral regions, heat conduction losses and arc tube wall temperature, to various parameters under the designer's control. Application of the model showed that higher lamp efficacy could be attained by increasing the PCA wall temperature (now limited by Al and O vaporization) at the same arc tube diameter or by increasing the diameter but holding the wall temperature constant. Such goals could be partially achieved by substitution of an arc tube of lower total emissivity, e.g., polycrystalline yttria, for PCA. A limitation of the model is the substitution of empirical parameters for important physical characteristics of the arc and the uncertainty as to how these approximations are reflected in the numerical results. However, the development of a more rigorous model is a very formidable task which is presently limited by uncertainties of the material parameters, such as the spectral radiance in various bands including the infrared continuum, as well as the thermal and electrical conductivities.

High Pressure Sodium Lamp Problems

Because of the reactivity of sodium with quartz and other glasses an entirely new system of arc tube envelopes and seals had to be developed for HPS lamps (3). The loss of sodium during lamp life has not been completely eliminated. Therefore, an excess quantity of sodium is added to the arc tube together with mercury and the starting gas xenon. The activity of mercury in the condensed amalgam is very low ($\sim 10^{-2}$ to 10^{-3}) and, moreover, is strongly temperature dependent; hence the mercury vapor pressure is greatly reduced. Although the activities and thermodynamic properties of sodium amalgams at low temperatures have been known for over a decade, quantitative data at lamp operating temperatures and compositions have only recently been published (22). Thus the sodium and mercury partial pressures in operating

Table II. Calculated Species Densities in LTE at 4000 K for a Gas Mixture Typical of a HPS Lamp (Na 75 torr, Hg 300 torr, Xe 140 torr).

Neutral Species	Density	Charged Species	Density
Na	1.7, +17*	e^-	6.2, +15
Na_2	9.1, +12	Na^-	4.1, +12
Hg	7.2, +17	Na^+	6.2, +15
Hg_2	1.5, +14	Na_2^+	1.7, +12
HgXe	2.2, +14	Hg^+	2.2, +10
Xe	3.4, +17	Hg_2^+	6.9, +7
		Xe^+	1.9, +8

* 1.7, +17 = 1.7×10^{17} cm^{-3}.

HPS lamps depend upon the condensed phase amalgam composition and the temperature of this reservoir. These partial pressures together with the ballast (designed primarily to limit lamp current) determine the lamp voltage and power. Both the amalgam composition and the reservoir temperature may change during lamp life. The usual result is that the lamp voltage increases with lamp life until the power line-ballast system is unable to supply sufficient voltage and the lamp extinguishes. Thus HPS lamp ballasts are designed to operate lamps over a rather wide range of voltages, e.g., 80-160 V.

Changes in the amalgam reservoir temperature may result from a number of causes, such as changes in emissivity of (1) the PCA near the end of the lamp (from electrode sputtering) or (2) of the niobium end assembly or metal heat shield, or changes in the heat returned from the outer envelope when it receives a discoloring or metallic deposit. Such deposits result from (1) alumina vaporization as Al and O, (2) evaporation from support wires or welds, (3) sodium leakage from the arc tube, etc. Sodium can be lost by leaking through the seal as well as by loss into the alumina as observed by Mizuno et al. (5). Hing (23) has suggested that the loss is accelerated if calcium is present. Wyner (24) has shown that such Na loss can be accelerated by applying an electric field across the PCA surface. Other authors (25) have found most of the lost sodium in the seal region. However, Jacobs et al. (26) have related sodium loss to the electrode emission material. Thus a multiplicity of mechanisms exist for causing an increase in lamp voltage, all of which must be controlled and avoided while manufacturing lamps at rapid rates to sell at competitive prices.

Improved High Pressure Sodium Lamps

Although standard HPS lamps have high efficacy, good lumen maintenance, and the voltage rise can be controlled, they have two deficiencies in color quality. The color rendition as measured by the CRI is low and the source color as indicated by the CCT is low. McVey (2) has amply reviewed progress in this area. It has been known since 1973 (27) that higher sodium pressures could be used to increase the CRI to 85 and the CCT to 2400 K with a modest drop in the efficacy to 92 lumens per watt. Lamps approaching this goal have recently become commercially practical through improvements in ceramic-sealing techniques (28).

Higher CCT can also be achieved by lamp operation in a low duty cycle, pulsed mode at repetition rates of 500 to 2000 Hz. The arc temperature profile is appreciably modified in this mode, becoming narrower with appreciably higher peak axis temperatures. More radiation from sodium lines arising from higher excited states and from the continuum results. Disadvantages of this approach are the need for special pulsing circuitry to operate the lamps and specially constructed arc tubes to withstand fracture from the acoustic shock waves generated by the pulses.

Lamps with higher than normal xenon fill pressures have also recently been introduced (3) to improve lamp efficacy and maintenance rather than for color improvement. Reliable starting of such lamps is a problem which may require special ballasts and lamp sockets.

Conclusions

The HPS lamp is a complex device that has reached a rather advanced stage of development by largely empirical means. There remain uncertainties about the species interactions in the discharge medium which limit the accuracy of our modeling and quantitative understanding of the discharge. A number of spectral features, especially at high mercury and xenon pressures are not well understood. Better values for such material parameters would permit more rigorous lamp modeling and better assessment of the consequences of changes of parameters not accessible to convenient empirical test.

Literature Cited

1. deGroot, J. J.; van Vliet, J. A. J. M.; Waszink, J. H. Philips Tech. Rev. 1975, 35, 334.
2. Wharmby, D. O. IEE Proc. 1980, 127A, 165.
3. McVey, C. I. IEE Proc. 1980, 127A, 158.
4. Schmidt, K. Proc. 6th Int. Conf. on Ion. Phen. in Gases, 1963, Paris, p. 323.

5. Mizuno, H.; Akutsu, H.; Watarai, Y. Proc. CIE Conf. (Barcelona) 1971, Paper P-71.14.
6. Buck, U.; Pauly, H. J. Chem. Phys. 1971, 54, 1929.
7. Barratt, S. Trans. Faraday Soc. 1929, 25, 758.
8. Woerdman, J. P. Opt. Commun. 1978, 26, 216.
9. Singh, R. B.; Rai, D. K. Ind. J. Pure Appl. Phys. 1965, 3, 475.
10. Verma, K. K.; Vu, T. H.; Stwalley, W. C. J. Molec. Spectrosc. 1981, 85, 131.
11. Zemke, W. T.; Verma, K. K.; Vu, T. H.; Stwalley, W. C. J. Molec. Spectrosc. 1981, 85, 150.
12. Konowalow, D. D.; Rosenkrantz, M. E.; Olsen, M. L. J. Chem. Phys. 1980, 72, 2612.
13. Whittaker, F. L. Thorn Lighting Ltd. Tech. Rept. LRD 2005, 1972, (unpublished). See also Reference 2.
14. Woerdman, J. P.; deGroot, J. J., ACS Symposium Series 1981, XX, xxxx [this volume].
15. Waymouth, J. F.; Wyner, E. F., Paper presented at Annual Tech. Conf. of Ill. Eng. Soc., Dallas, Texas, August 1980.
16. York, G.; Scheps, R.; Gallagher, A. J. Chem. Phys. 1975, 63, 1052.
17. Baylis, W. E. J. Chem. Phys. 1969, 51, 2665.
18. Pascale, J.; Vandeplanque, J. J. Chem. Phys. 1974, 60, 2278.
19. Düren, R. J. Phys. B 1977, 10, 3467.
20. Zollweg, R. J.; Liebermann, R. W.; McLain, D. K. J. Appl. Phys., 1981, 52, 3293.
21. Waymouth, J. F. J. Ill. Eng. Soc. 1977, 6, 131.
22. Hirayama, C.; Andrew, K. F.; Kleinosky, R. L. Thermochemica Acta, 1981, 45, 23.
23. Hing, P., Paper presented at Annual Tech. Conf. of Ill. Eng. Soc., Dallas, Texas, August 1980.
24. Wyner, E. F. J. Ill. Eng. Soc. 1979, 8, 166.
25. Inouve, A.; Higashi, T.; Ishigami, T.; Nagano, S.; Shimojima, H. J. Light and Vis. Env. 1979, 3, 1.
26. Jacobs, C. A. J.; Sprengers, L.; DeVaan, R. L. C. J. Ill. Eng. Soc. 1978, 7, 125.
27. Kühl, B., paper presented at Conf. Assoc. of Public Lighting Engineers, Scarborough, 1973.
28. Bhalla, R. S.; Larson, D. A.; Unglert, M. C. J. Ill. Eng. Soc. 1979, 8, 202.

RECEIVED August 26, 1981.

Physics and Chemistry of Cesium Thermionic Converters

E. J. BRITT
Rasor Associates, Inc., Sunnyvale, CA 94086

Thermionic converters are high temperature devices which utilize electron emission and collection with two electrodes at different temperatures to convert heat into electric power directly with no moving parts. Most thermionic converters operate with a plasma of positive ions in the interelectrode space to neutralize space charge and permit electron current flow. Both the plasma characteristics and the surface properties of the electrodes are controlled by the use of cesium vapor in thermionic diodes.

Recently, surfaces with coadsorbed cesium and oxygen have been developed, which have stable work functions of about 1.0 eV. Some of these surfaces operate without loss of oxygen up to 1200 K. Use of these surfaces as advanced collectors in thermionic converters is currently being investigated. Negative ionic species of cesium and other materials can be formed by interaction with low work function surfaces present in converters. Quadrupole mass spectrographic analyses have confirmed the existence of Cs^- and other molecular negative ions in thermionic converter environments. The implications of these negative particles on converter performance is being studied.

Thermionic energy conversion is a method of converting heat directly to electricity. A metal electrode, the emitter, is heated sufficiently to emit electrons, as shown in Figure 1. The electrons cross a narrow interelectrode gap and are collected by another metal electrode, the collector. Heat is removed from the collector so that its temperature is lower than the emitter, and the electrons striking the collector cannot be returned except by

0097-6156/82/0179-0421$06.50/0

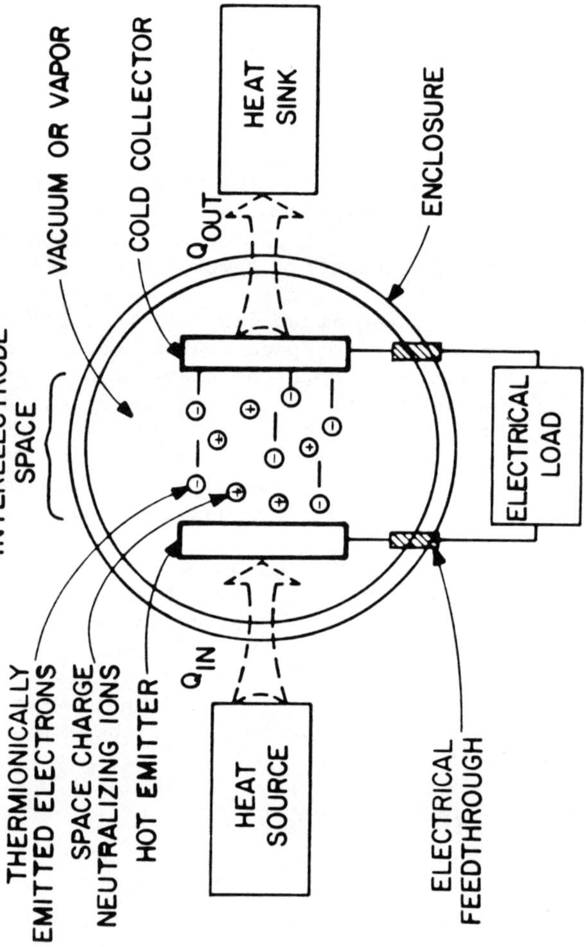

Figure 1. Scheme of thermionic diode for directly converting heat into electricity.

flowing through the external circuit. The flow of electrons constitutes an electric current which delivers power to the load.

The unique properties of cesium play a crucial role in the operation of thermionic converters. Cesium functions both as adsorbed atomic layer to produce the required work functions on the electrodes, and as a plasma medium to form Cs^+ ions which neutralize space charge in the interelectrode region. Cesium is desirable as the plasma medium because of its low ionization potential and large atomic mass. Since the surface adsorbed layers are continuously evaporating and being replenished by cesium atoms refluxing from the vapor, the surface properties are very stable. Thermionic converters have operated with no change in performance for more than 5 years.

The cycle has many similarities to a Rankine cycle which uses electrons as a working fluid. Unlike the normal Rankine cycle, however, the working fluid's "heat of evaporation", approximately the emitter work function, and its "heat of condensation", approximately the collector work function, can be varied in the thermionic converter. This feature provides the converter with great flexibility in matching the operating constraints of any particular system.

Elementary Converter Model

A simple analytical model of thermionic converter performance must be made before the impact of converter performance on system behavior can be studied. Fortunately, a very simple model of converter performance has been found to be sufficiently accurate for this purpose. The ideal thermionic diode serves as the basis for this model. Motive diagrams and converter current voltage characteristics for an ideal diode are shown in Figure 2.

In the motive diagram ϕ_E and ϕ_C, respectively are the emitter and collector work functions. They represent the potential barriers which must be overcome by electrons leaving either electrode. The output voltage of the converter is the difference between the emitter and collector Fermi levels, V.

Assuming no space charge, electron scattering between the electrodes, or back emission from the collector, the current flow from the emitter to the collector J, will be determined by the emitter temperature, T_E, the potential barrier the electrons must overcome, ϕ, and the Richardson-Dushman equation:

$$AT_E^2 \ \exp\ (^{-\phi}/kT_E) \tag{1}$$

Clearly there are two distinct regions of operation. As long as the sum of the collector work function, ϕ_C, and the output voltage, V, is less than the emitter work function, ϕ_E, the barrier to electron flow is the emitter work function, and

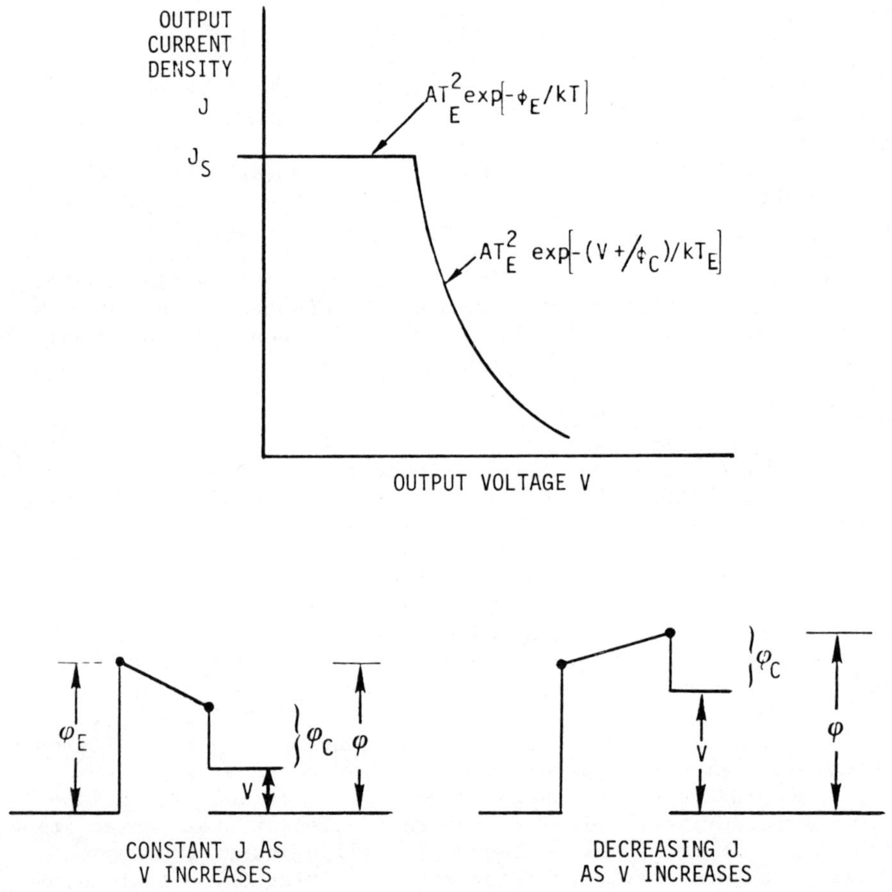

Figure 2. Current–voltage characteristics of an ideal thermionic converter. Top: J-V curve; Bottom: Electron notive diagram for two different conditions.

J is independent of output voltage. This is the "saturation region" of the JV curve and the current density is given by Eq (1) with $\phi = \phi_E$, and $J = J_S$.

As soon as the output voltage is increased to the point where $\phi_c + V = \phi_E$, the barrier becomes $\phi_c + V$ and the current density falls exponentially with further increases in output voltage.

$$J = AT_E^2 \exp\left[-(\phi_c + V)/kT_E\right] \qquad (2)$$

Two problems with this simple picture are immediately apparent. First, typical refractory metals have work functions exceeding 4 ev. Polycrystalline tungsten, for example, has a work function of 4.6 ev. To obtain a current density of 1 1 amp /cm^2 with this material would require an operating temperature of 2600 K. Second, electron space charge effects would limit current flow in a vacuum diode to less than a milliampere with a typical interelectrode gap of 254 µm (10 mils).

Both of these problems are very simply solved by adding a low pressure of cesium vapor (0.1 to 10 torr) in the interelectrode space. Such a pressure can easily be established by connecting the device to a reservoir of liquid cesium at a suitable temperature, T_R, which is typically between 500 K and 600 K. In the presence of the cesium vapor a partial monolayer of cesium will form on both the emitter and collector. The work functions of both electrodes will change in the manner shown in Figure 3 (1). By varying the cesium reservoir temperature (pressure), a range of work functions may be obtained at given emitter temperatures. The collector is operated at a low work function, in order to obtain maximum output voltage, and its work function is thus insensitive to cesium pressure changes. In this way any desired J_S, shown in Figure 2, can be obtained.

The electron space charge barrier to current flow can be neutralized by providing heavy cesium ions. Cesium is a particularly good choice because it has the lowest ionization potential of any element; and a large atomic mass 133 amu, which causes the mobility of ions leaving the plasma to be low. The ions are produced in a low voltage arc between the electrodes. In this case the power to produce the ions is reflected in a voltage drop between the emitter side of the plasma and collector side, which is called the arc drop. It typically is about one-half volt. This type of converter operation is called the ignited mode.

Since losses such as the arc drop and electron scattering are incurred when cesium is added to the converter, its performance is no longer accurately represented by the ideal diode. It has been found, however, that a converter can be well characterized simply by adding voltage loss terms in the ideal diode equation. As shown in Figure 4, the sum of these losses, includ-

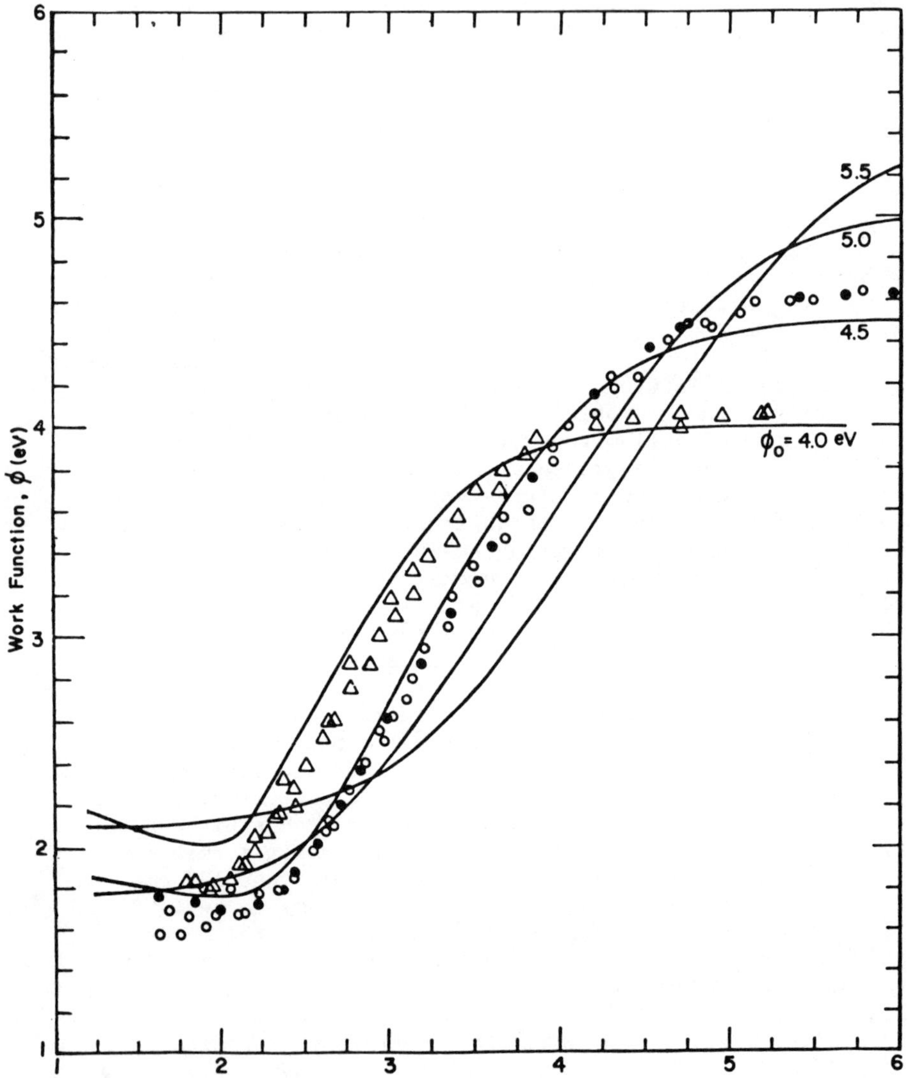

Figure 3. Calculated and experimental variation of electron work function of various refractory metals with equilibrium adsorbed Cs on surfaces. Work function is plotted vs. ratio of surface temperature to temperature of the Cs reservoir, which determines Cs coverage. Key: ——, Computed values; △, No (0 ≃ 4.1 eV); and ○, W (0 ≃ 4.6 eV).

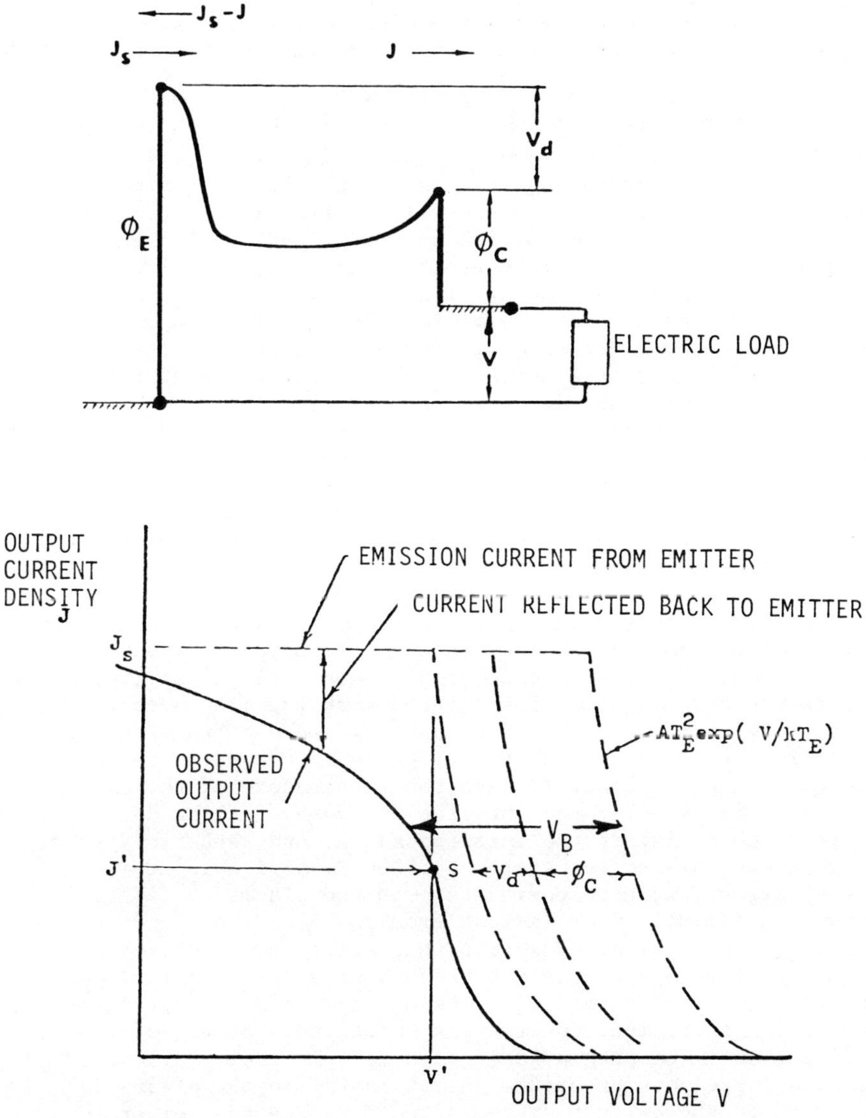

Figure 4. Characteristics of an ignited mode thermionic converter. Ignited mode motive diagram shown at top. The actual J–V curve is displaced from the ideal curve (bottom) by an amount equal to the voltage losses in the converter.

ing the collector work function, is called the barrier index, V_B. The barrier index is made up of several components as shown in Figure 4, $V_B = s + V_d + \phi$. The arc voltage drop is V_d as indicated on the electron motive diagram at the top of Figure 4. The quantity s is an equivalent voltage which accounts for the current loss due to backscattered electrons in the plasma. Actual converter performance is then represented by the equation:

$$J = AT_E^2 \exp\left[-(V + V_B)/kT_E \right] \qquad (3)$$

Reduction of any of the quantities which make up V_B are the primary approaches to improvement in thermionic performance.

The motive diagram for the potential energy of electrons in an ignited mode thermionic converter has a more complicated shape, as shown at the top of Figure 4. The presence of positive ions in the plasma creates a minimum in the electron motive inside the interelectrode gap. There are narrow collisionless sheaths (the order of a Debye length in thickness) at both the emitter and collector edges of the plasma.

Some additional understanding of the ignited mode diode may be gained by considering a fluid mechanical analog shown in Figure 5. In this analogy the electrons are represented by a two phase working fluid, and the vertical distance on the figure represents electrostatic potential, analogous to gravitational head in a fluid system. The shape of the analog is like the motive diagram for an ignited mode thermionic diode.

Electrons at ① in Figure 5 are "evaporated" from the interior of the emitter as they undergo thermionic emission. In doing this they are raised to higher potential at ② , which represents the height of the emitter work function over the emitter Fermi level. The heat input to the electrons at ① is an isothermal addition of entropy as in a Rankine cycle, so the electrons at ② still have the temperature of the emitter. They are then accelerated into the plasma by falling down ~0.8 volts of potential in the emitter sheath, and rethermalyzed at ③ to a much higher temperature, ~3000 K. Following this, the electrons are transported through the plasma to ④ by a density gradient. Heat is lost from the electron fluid in the plasma in the form of radiation, hot electrons, and excited atoms, that escape to the electrode boundaries. Collisional impacts of hot electrons with cesium atoms produce multistep ionization sufficient to create a spontaneous plasma of $\sim 10^{13}$ ions per cm^3. The plasma generation is relatively efficient since a large portion of the energy lost from the plasma is recycled to the emitter. These effects reduce the temperature of the electrons at ④ to ~2200 K, and thus they can only surmount a much smaller sheath at the collector edge of the plasma.

The electrons must be cooled from a temperature of ~2200 K at ④ to the temperature of the collector (~900 K) before they

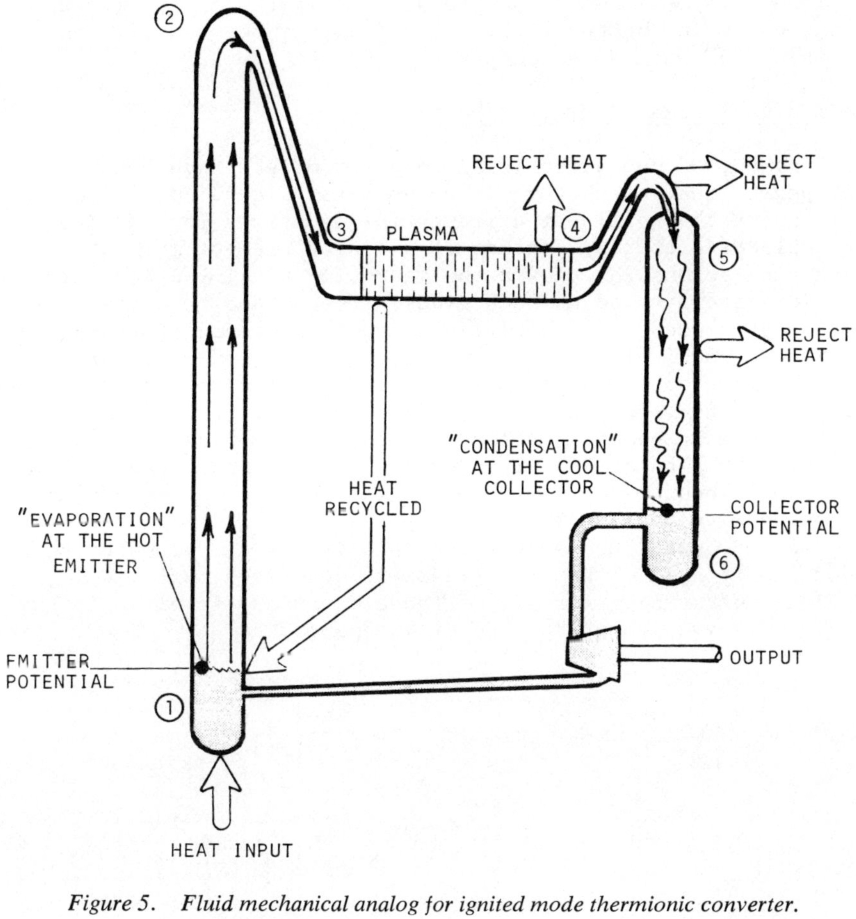

Figure 5. Fluid mechanical analog for ignited mode thermionic converter.

are "condensed" into the collector. Waste heat is given off irreversibly as the electron temperature is reduced. This leads to some of the converter inefficiencies. The potential of the collected electrons is lowered from ⑤ to ⑥ by an amount representing the collector work. This step also produces reject heat, but the energy loss is necessary to prevent the electrons from being back emitted. Finally, the electrons are returned to the emitter by an external circuit delivering the output power to a load. This is shown analogously in Figure 5 as a liquid turbine operating on the potential difference between ⑥ and ①, which represents the output voltage.

Detailed Analytical Converter Model

A number of one dimensional computer models have been developed to analyze thermionic converters. These numerical models solve the nonlinear differential equations for the thermionic plasma either by setting up a finite element mesh or by propagating across the plasma and iterating until the boundary conditions are matched on both sides. The second of these approaches is used in an analytical model developed at Rasor Associates. A highly refined "shooting technique" computer program, known as IMD-4 is used to calculate converter characteristics with the model (2).

Five simultaneous equations are solved numerically in order to carry out calculations with IMD-4. Two transport equations, Eqs 4 and 5 shown below describe the currents for electrons and ions respectively. The total energy transport Q_e is described by Eq. 6. Production of electron-ion pairs is described by Eq. 7. Production of heat within the plasma is described by Eq. 8. Definition of symbols for all of these equations is given in the nomenclature section at the end of the paper.

$$J_e = -\mu_e \left[kT_e \frac{dn}{dx} + enE + (1 + k_e^T) nk \frac{dT_e}{dx} \right] \tag{4}$$

$$J_i = -\mu_i \left[kT_i \frac{dn}{dx} - enE + (1 + k_i^T) nk \frac{dT_i}{dx} - eR_{ie} \right] \tag{5}$$

$$Q_e = J_e \left[(5/2 + k_e^T) kT_e + V \right] - K_e \frac{dT_e}{dx} \tag{6}$$

$$\frac{dJ_e}{dx} = \frac{dJ_i}{dx} = e(Sn_a n - \alpha n^3) \tag{7}$$

$$\frac{dQ_e}{dx} = (V - V_i)\frac{dJ_e}{dx} \tag{8}$$

The ion drag term R_{ie} is

$$R_{ie} = -g_1(\frac{kT_e}{e}\nabla n + nE) + (g_2 - g_1)\frac{nk}{e}\nabla T_e \tag{9}$$

Here g_1 and g_2 are functions of the electron collision frequency as evaluated by Wilkins and Gyftopolous (3).

All terms of the transport equations are retained and included in the solution. This is significant because both thermal diffusion effects and the ion drag affect the calculated performance. Boundary conditions for these equations have electron retaining sheaths at the edges of the plasma. Electrode area ratios and electron reflectivities are included in the boundary conditions also. Electron back emission from the collector is in the collector side boundary conditions, but ion emission from the emitter has been neglected.

The computational model operates by specifying a value for J, the output current density, which specifies a position on the volt-ampere characteristic. The output voltage, V_o, is the value to be calculated.

Values are also specified for the emitter temperature T_E, the collector temperature T_C, the reservoir temperature T_R, the cesiated emitter work function ϕ_E, the collector work function ϕ_C, and the interelectrode spacing d.

The saturation and back-emission current densities are calculated by the Richardson-Dushman equation. The shooting method involves guessing those values which are unknown in the boundary conditions at one side, integrating over to the other side using a standard stepping technique, and iterating on the unknown values until the boundary conditions at the far side are satisfied within a specified degree of accuracy. This method is useful when the differential equations to be integrated are strongly nonlinear. In this model, integration is begun at the collector edge and proceeds toward the emitter.

A flow chart which shows the calculational logic for solution of the IMD-4 computer program is given in Figure 6. The calculation contains three nested interation loops. The inner most loop adjusts the value of the electron temperature at the collector edge of the plasma T_{eC}, by checking the continuity of heat flux. The second loop, which adjusts the value of the plasma density at the collector edge n_C, iterates until the electron current at the emitter side has the correct value. During each

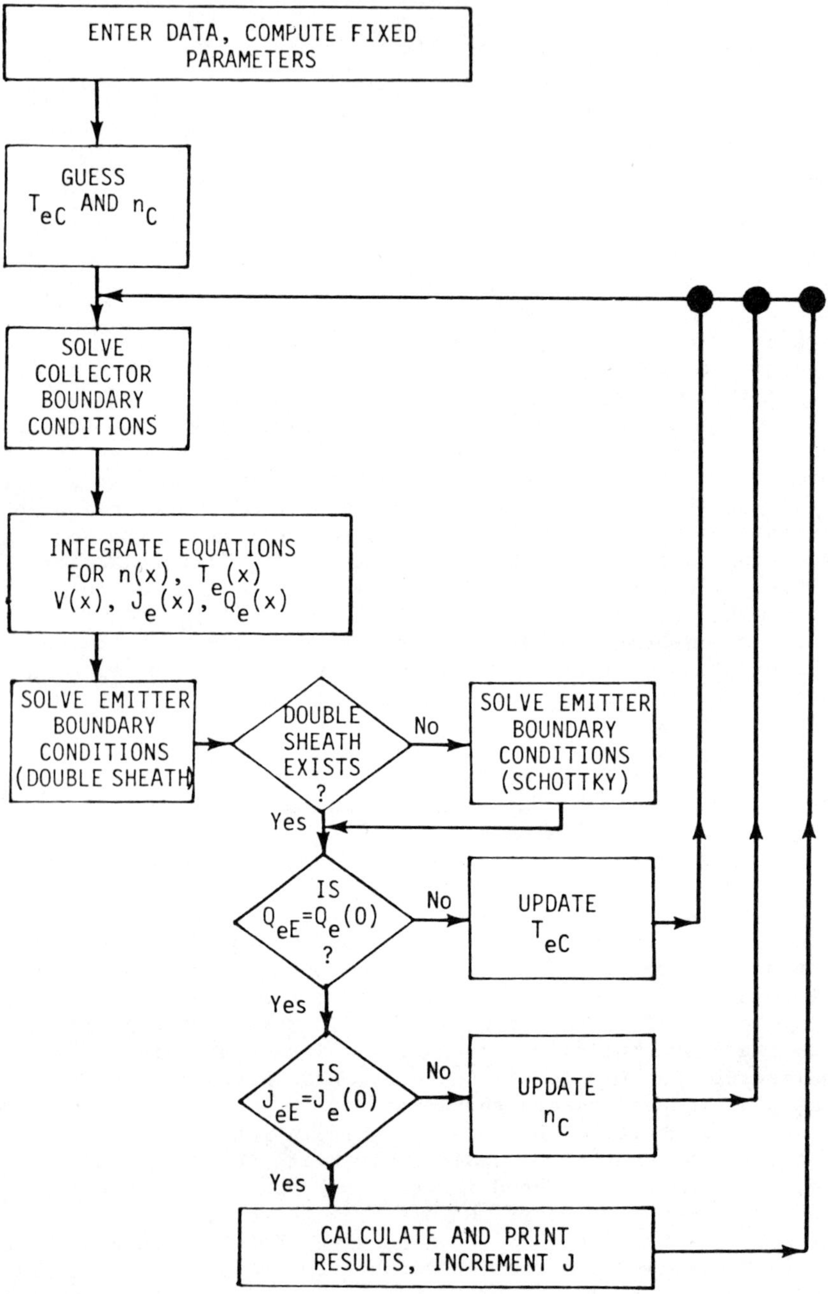

Figure 6. Flow chart for the computer program logic in IMD-4 analytical converter model.

iteration of the plasma density n_C , the inner loop on T_{eC} must be reconverged. There is also a small loop on the emitter side boundary condition which is executed during every shooting attempt. This loop tests for the existence of a double sheath and then matches the boundary conditions at the emitter edge. Finally both values of n_C and T_{eC} achieve sufficiently small errors; and the results are calculated, plotted and printed for a given J-V point. The next point on the J-V curve is then calculated by incrementing the current density.

An example of the results obtained with the IMD-4 model is given in Figure 7 which shows a computed I-V curve (solid line) compared with an experimental curve (dashed line). The calculated motive diagram at various points along the curves is also shown in the figure as indicated. For points (1) to (4) in the obstructed region of the J-V curve, there is a double sheath barrier at the emitter edge of the plasma. The emitter sheath in this region of the characteristic is approximately constant in size, equal to about 0.81 to 0.83 volts. The collector sheath does not vary much either in this region. It remains at about 0.26 to 0.3 volts. Points (5) to (7) are in the quasi-saturation region of the J-V curve, and monotonic sheaths are present on both sides of the plasma. Schottky effects on the emitter are included in the calculation in this region which, along with the ion currents, give the slope to the characteristic in the quasi-saturation both the emitter sheath and the collector sheath increase in size as the current is raised and the plasma becomes more dense.

Converter Performance

Calculated Power and Efficiency. The simplified analytical models of thermionic characteristics have been used to project the converter efficiency and power density with the barrier index as a parameter. These projections are shown in Figures 8 and 9 as functions of the emitter temperature. The dashed lines in these two figures are for a constant current density of 10 A/cm^2. If the current density is adjusted to maximize the efficiency at each temperature, the calculated performance is represented by the solid lines. Typical present generation thermionic converters operate with V_B near 2.0. Ignited mode converters in laboratory experiments have demonstrated practical operation with $1.85 \leq V_B \leq 1.90$. Other laboratory devices with auxiliary sources of ions and/or special electrode surfaces have achieved $V_B \leq 1.5$, but usually not under practical operating conditions.

Hardware Experience. The efficiency of cylindrical thermionic converters can be determined accurately because the thermal input to the emitter is easy to measure. The efficiency of actual converters is shown in Figure 10. The lines for

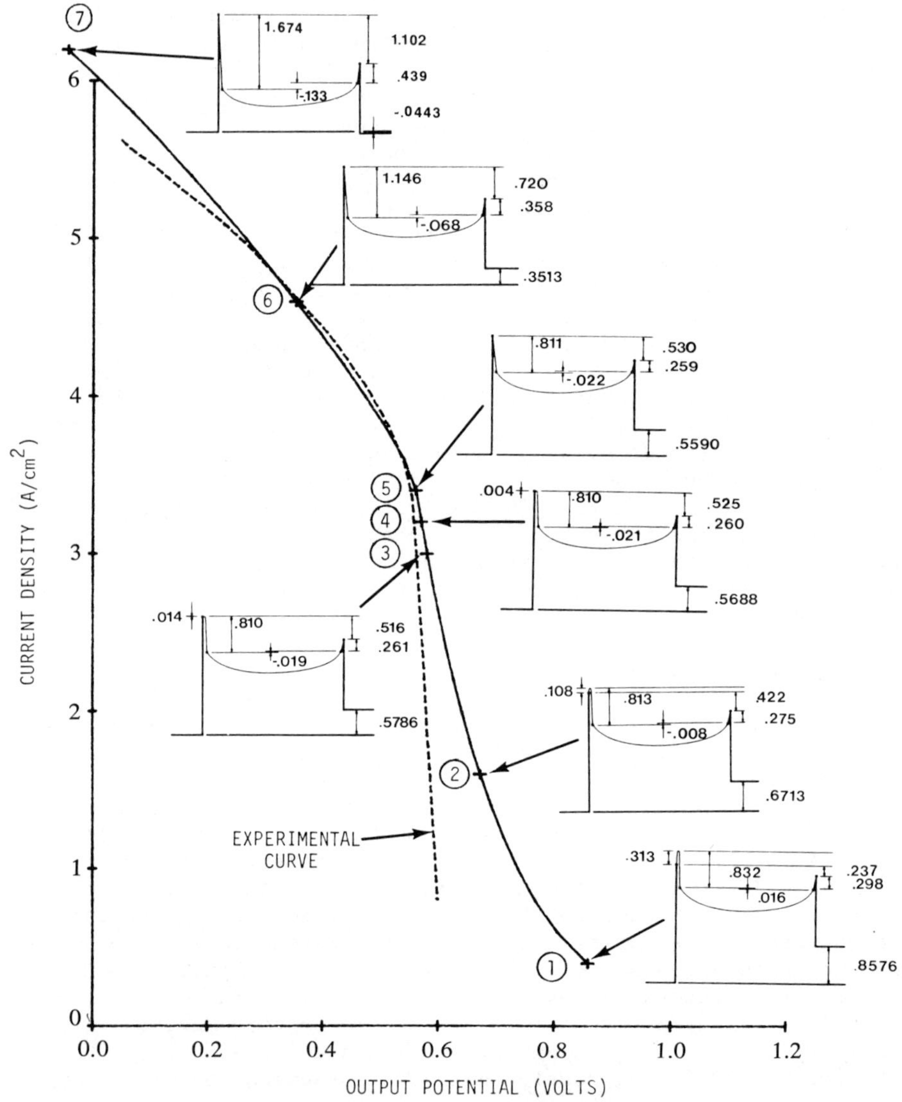

Figure 7. Comparison between calculated results from IMD-4 and experimental data from a Planar converter. T_E, 1700 K; T_C, 773 K; T_R, 567 K; ϕ_E, 2.654 eV; ϕ_C, 1.560 eV; and d, 10 ml; Key: – – –, experimental curve.

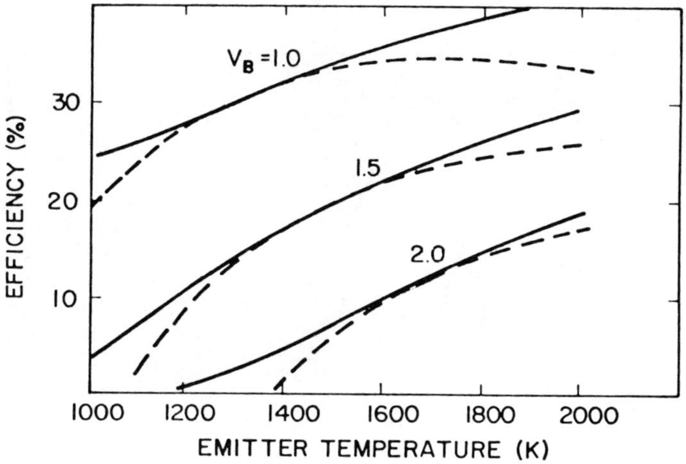

Figure 8. Projected converter efficiency. Key: ——, maximum efficiency; and – – –, 10 amp/cm².

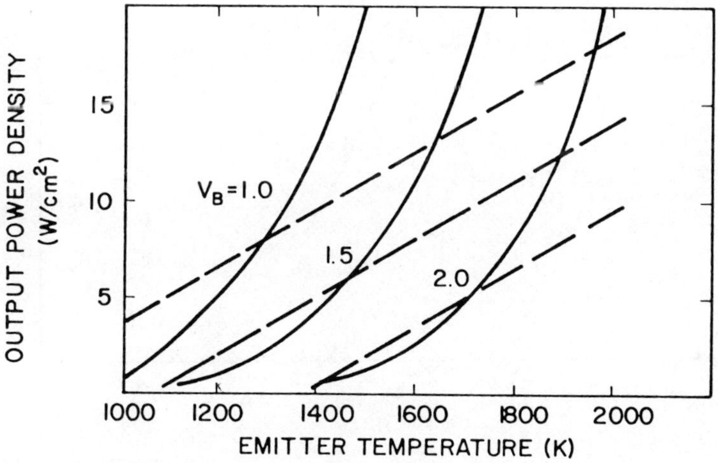

Figure 9. Projected converter output power density.

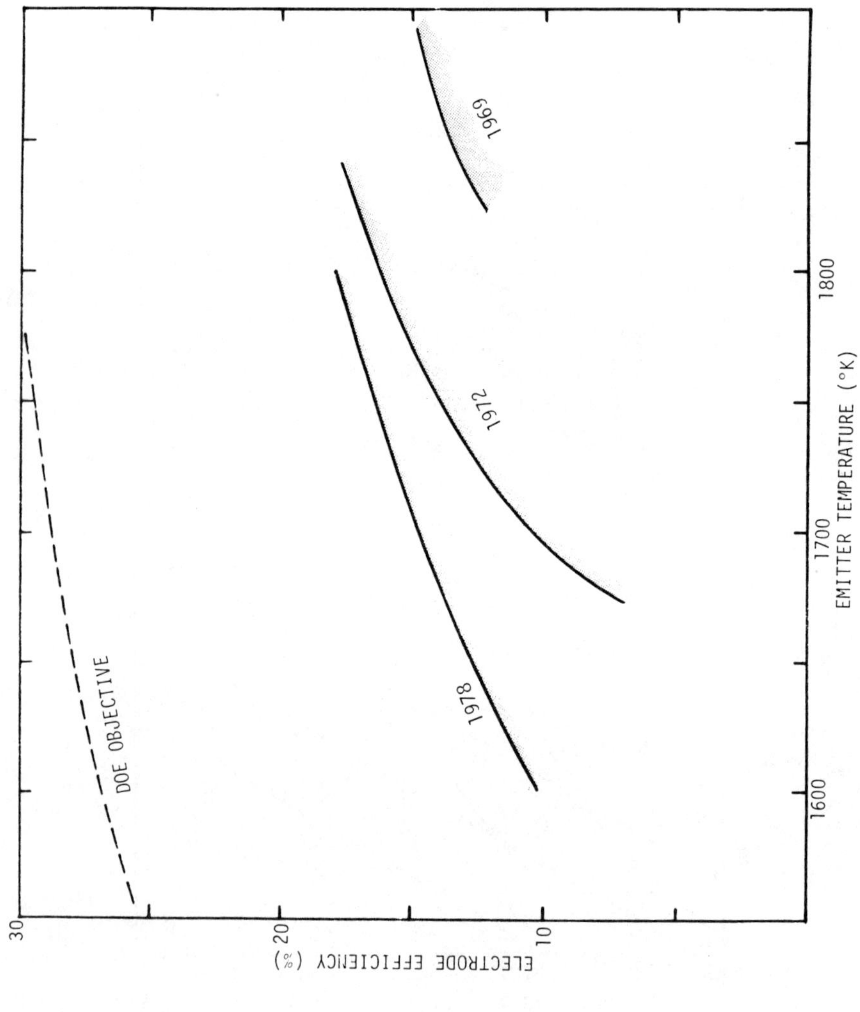

Figure 10. Progress in increasing efficiency of thermionic converters at lower emitter temperatures.

1969 and 1972 are converters built by General Atomic Company for the nuclear space power effort. A group of converters with specially structured (roughened) electrode surfaces, was built by Rasor Associates and produced the results labeled 1978 in Figure 10. The performance gains have been made by reducing the emitter temperature while maintaining the same efficiency.

A complementary plot showing output power density is given in Figure 11. On this figure, the earlier converters are all grouped in the shaded region labeled 1965-1978. The 1978 group are the ones built by Rasor Associates, which were mentioned in the previous paragraph. An unusual converter known as the "cold seal converter", (4) with gas buffered heat pipe containment of the cesium vapor, produced the 1979 point on Figure 11.

There are no inherent degradation mechanisms to limit the life of thermionic converters. The lifetime of test devices has usually been related to damage from the environment of the heat source. Converters with nuclear fuel are affected by fisson product swelling, which distorts the emitter. Flame heated converter lifetimes are controlled by the durability of the hot shell, which protects the emitter from the combustion atmosphere.

The record holder for converter life is LC-9, a converter built for NASA by General Atomic as part of the in-core nuclear space reactor program. LC-9 operated with perfectly stable performance for over five years at an emitter temperature of 1970 K. As shown in Figure 12 (5), LC-9 had an electrode efficiency of 17%, and generated 8 W/cm^2 of output power (80 KW/m^2). The converter was still performing stably when tests were terminated for programmatic reasons. This test illustrates well the long life capability of the thermionic converter process.

Recent converter hardware development has concentrated on flame heated devices for terrestrial topping cycle applications. Significant progress has been made with chemical vapor deposited silicon carbide as the oxidation protection hot shell. Flame heated converters have now been operated by Thermo Electron Corporation with emitter temperatures over 1700 K for 7000+ hours (6).

Prospects for Performance Improvement. Performance improvement in thermionic converters is being pursued by suppressing arc voltage drop in the plasma and by reducing the size of the collector work function. The voltage loss in a spontaneous ignited mode plasma is about 0.5 volt. Typical cesium converters behave as though the collector work function is about 1.5 ev, yet laboratory surfaces can be produced with $\phi_C \simeq 1.0$ ev. Reduction of either V_d or ϕ_C in practical converters would lead to a substantial improvement in converter performance, since the output voltage of a typical ignited cesium diode is only about 0.5 volt.

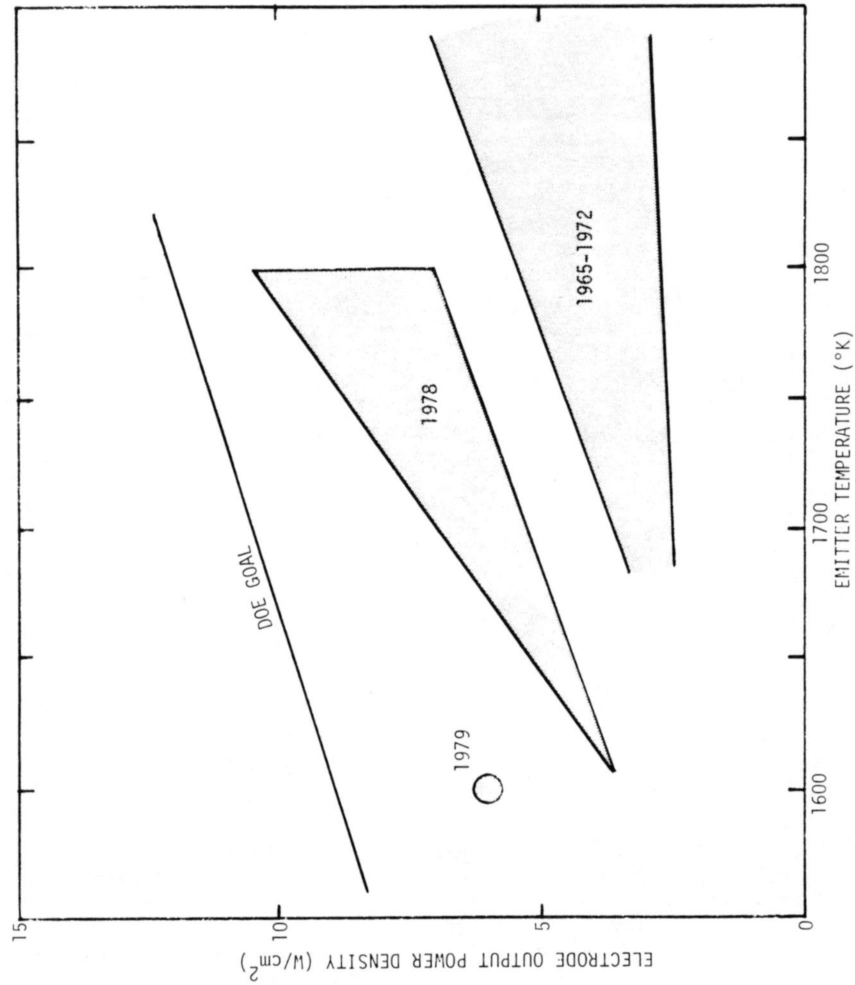

Figure 11. Progress in increasing output power density of thermionic converters at lower emitter temperatures.

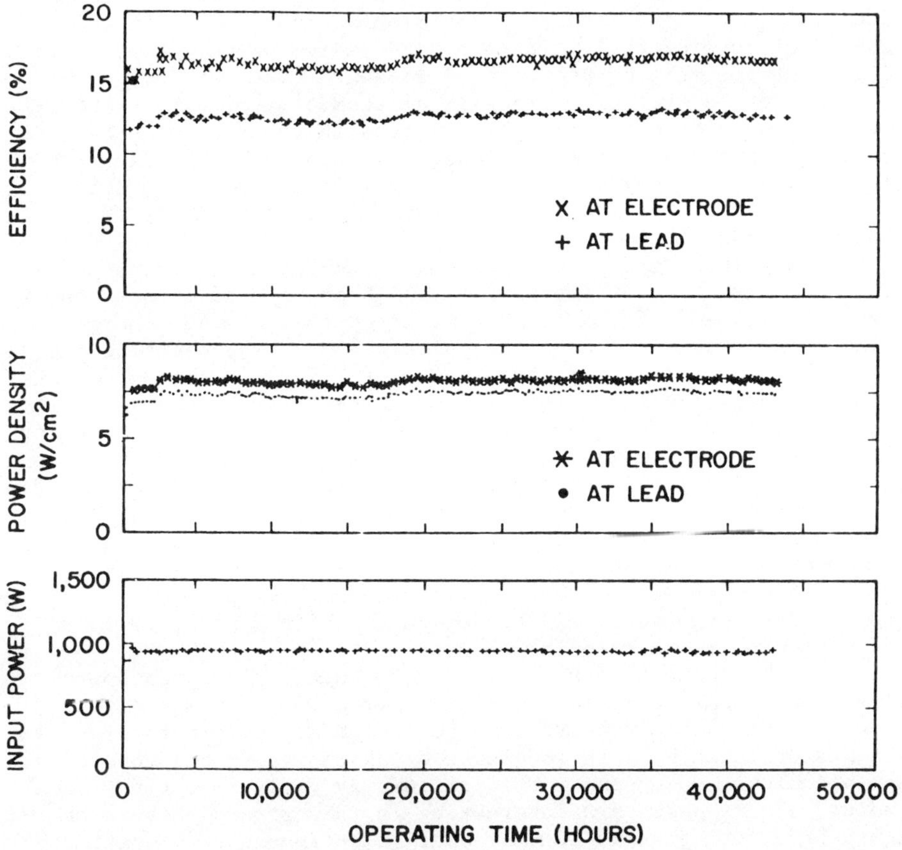

Figure 12. Operating history of converter test LC-9 at General Atomic Company. Stable performance was monitored for > 5 years with no failure.

Recently, two types of surfaces with coadsorbed cesium and oxygen have shown promise for low work function operation in thermionic converters. The first type of surface is made by co-deposition of a "thick layer" (~ 30 Å) of cesium. As shown in Figure 13 (7), if the proportions of cesium and oxygen are properly controlled, a work function as low as 1.0 ev can be obtained. The substrate material does not affect the low ϕ_C obtained with a thick C_s - O layer. This type of surface could potentially be maintained in a thermionic converter by an equilibrium mixture of cesium, oxygen, and cesium oxide. Experiments to demonstrate this in operating diodes are underway (8).

The second type of surface is an atomic layer of oxygen strongly bound to a tungsten 110 surface in the form of a 2-D oxide. This surface will not lose oxygen up to a temperature of 1200 K; and when cesium is adsorbed on it with proper coverage, a work function near 1.0 ev can be obtained. An example (9) is shown in Figure 14.

In previous attempts to use low ϕ_C surfaces in cesium thermionic diodes, the surface performs with an effective work function of ≥ 1.5 ev. It has been postulated that a space charge layer of negative ions near the collector may be preventing the low work function surfaces from being utilized. Negative Cs ions and other negative ions with smaller masses have been detected by a quadrupole mass spectrometer attached to a thermionic converter (10). Atomic cesium and cesium dimers have too small a value of electron affinity (11) to produce a serious space charge layer. But near the surface there may be a large concentration of trimers and other atomic clusters. It is very possible that these larger particles could become negative ions in sufficient quantities to affect converter performance. Further research and understanding is needed to determine whether negative ions are a problem in thermionic converters.

Another route to performance improvement is through operation with plasma conditions different from the cesium ignited mode. Considerable research has been done on converters with plasmas produced by a third electrode discharge or pulsed diodes. Operation with gases other than cesium may yield a better plasma medium. Inert gases are interesting in this regard because the Ramsauer effect results in very small electron-neutral scattering at thermionic energies. Converters with other types of alkali metal vapors are also of interest. The associative ionization of excited sodium atoms to produce dimer ions, (12) is particularly promising for a thermionic converter plasma.

Thermionic conversion is a technology that needs, and can immediately use, research on high temperature properties of alkali metals. Electron transport properties of alkali vapors and characteristics of atomic clusters are particularly important. Improved understanding in these areas could lead to performance improvements that would more than double the output power density and efficiency of cesium ignited mode thermionic converters.

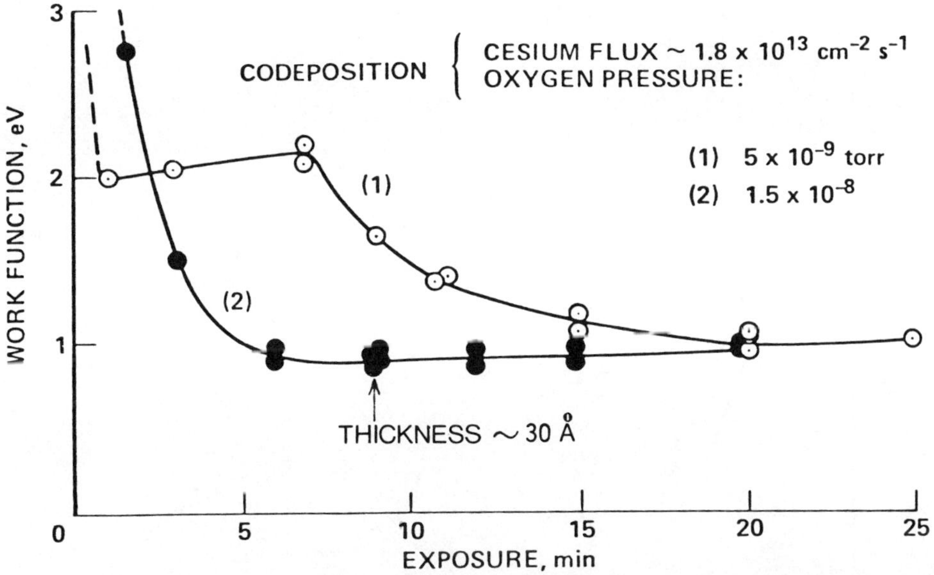

Figure 13. Low work function of a thick layer of CsO obtained by codeposition of Cs and O.

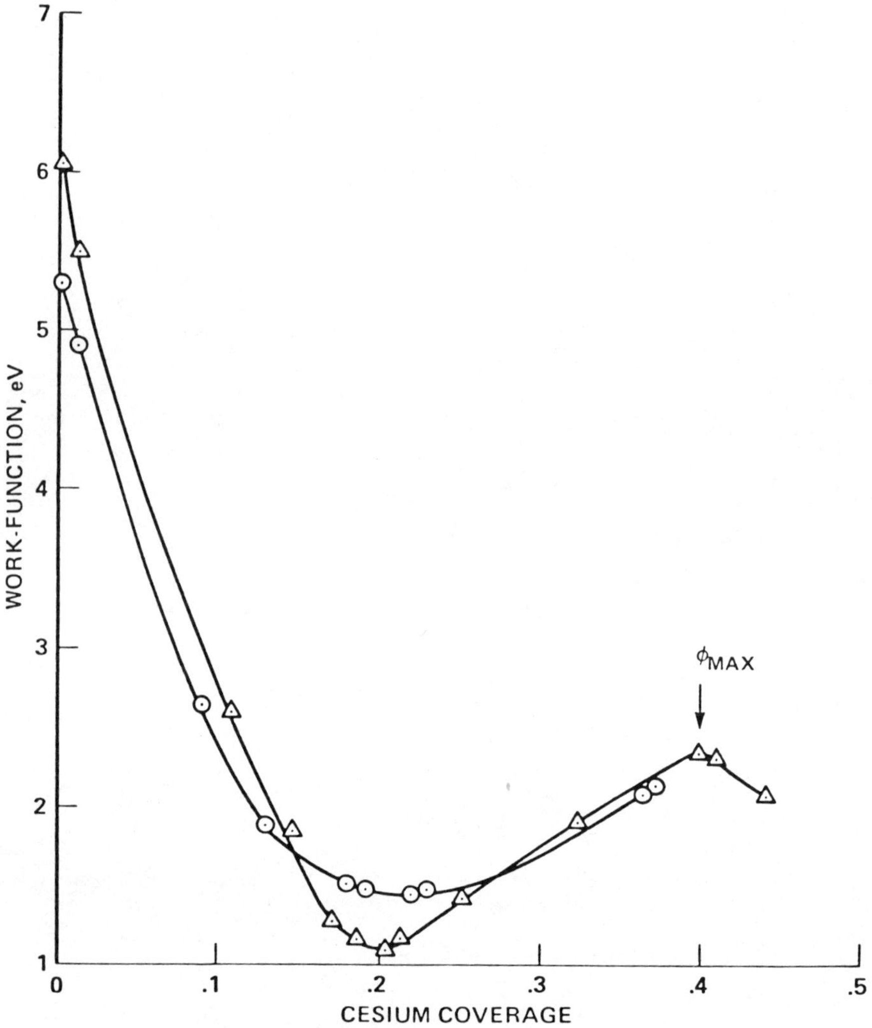

Figure 14. Low work function obtained by Cs adsorbed on an atomic layer of annealed 2-D oxide on a W(110) surface (–△–) at 295 K. Work function of Cs on a clean W(110) (–○–) surface is shown for comparison.

NOMENCLATURE

Symbol		Units
d	= interelectrode gap width	cm
e	= electron charge	coulombs
E	= electric field	volts/cm
g_1	= function used in evaluating the ion drag term	-dimensionless-
g_2	= function used in evaluating the drag term	-dimensionless-
J_e	= net forward electron current density	A/cm^2
J_{eE}	= net forward electron current density at the emitter sheath-plasma interface (computed from boundary conditions)	A/cm^2
J_i	= net forward ion current density	A/cm^2
J	= net current density through the diode	A/cm^2
k	= Boltzmann's constant	erg/°K
k_e^T	= electron thermal diffusivity	-dimensionless-
k_i^T	= ion thermal diffusivity	-dimensionless-
K_e	- electron thermal conductivity	watt/cm K
n_a	= atom density	cm^{-3}
n_C	= charged particle density at the collector plasma-sheath interface	cm^{-3}
n	= charged particle density in the plasma	cm^{-3}
Q_e	= total electron energy flux	$watts/cm^2$
Q_{eE}	= electron energy flux at the emitter sheath-plasma interface (calculated from the boundary conditions)	$watts/cm^2$
R_{ie}	= ion drag term	$volt/cm^4$

S	= ion source coefficent	cm^3/sec
T_e	= electron temperature	°K
T_i	= ion temperature	°K
T_R	= cesium reservoir temperature	°K
T_{eC}	= electron temperature at collector sheath-plasma interface	°K
T_E	= emitter temperature	°K
T_C	= collector temperature	°K
V	= electron potential	volts
V_d	= arc drop	volts
V_B	= barrier index	volts
V_o	= output voltage	volts
V_i	= cesium ionization potential	volts
α	= recombination coefficient	cm^6/sec
μ_e	= electron mobility	cm^2/volt-sec
μ_i	= ion mobility	cm^2/volt-sec
ϕ_E	= field-free emitter work function	volts
ϕ_C	= collector work function	volts

Literature Cited

1. Rasor, N. S. and Warner, C., JAP, 1964, 35, 2589-2600.
2. Britt, E. J. and McVey, J., "Advanced Thermionic Energy Conversion, Joint Highlights and Status Report"; Rasor Associates., COO-2263-16, NSR 2-16, 1979.
3. Wilkins, D. R. and Gyftopoulos, E. P., J. Appl. Phys., 1966, 37, 3533-3540.
4. Smith, M. D., Manda, M. L., and Britt, E. J., "Utilization of Low Temperature Insulators and Seals in Thermionic Con verters", 15th Intersociety Energy Conversion Engineering Conference, Washington, 1980.
5. "Thermionic Converter and Fuel Element Testing Summaries at Gulf General Atomic Company, " GULF-GA-C12345, California, 1972.
6. Goodale, D. B., Reagan, P. Miskolczy, G., Lieb, D. and Huffman, F. N.; "Characteristics of CVD Silicon Carbide Thermionic Converters", 16th Intersociety Energy Conversion Engineering Conference, Atlanta, GA, 1981.
7. Papageorgopoulos, C. A. and Desplat, J.-L.; Surface Science, 1980, 92, 119.
8. Hansen, L. K. and Woo, H.; "Thermionic Converters with a Thick Cesium Oxide Collector", IEEE International Conference on Plasma Science, Madison, Wisconsin, 1980, 75.
9. Desplat, J.-L. and Papageorgopoulos, C. A. Surface Science, 1980, 92, 97.
10. Laskowski, B. "Lowest Autodetaching States of Cs^-", IEEE International Conference on Plasma Science, Madison, Wisconsin, 1980, 44.
11. Hansen, L. K. and Woo, H. "Positive and Negative Ions from a Plasma Diode", IEEE International Conference on Plasma Science, Madison, Wisconsin, 1980, 45.
12. Koch, M. E., Verma, K. K. and Stwalley, W. C., J. Opt. Soc. Am., 1980, 90, 627.

RECEIVED August 26, 1981.

30

Plasma Formation in Alkali Metal Vapors by Quasi-Resonant Laser Excitation

ANDREW C. TAM

IBM Research Laboratory, San Jose, CA 95193

Quasi-resonant laser light, which couples the resonant excited state with a higher excited state in an alkali atom, is useful for efficient breakdown and plasma formation in the alkali metal vapor. For example, we have used a cw quasi-resonant laser beam of 100 mW power to produce a Cs plasma with an electron density of 3×10^{14} cm^{-3}, and Stwalley and co-workers have succeeded in producing a Na plasma by using an even weaker cw beam, of power 1 mW. Sharp thresholds exist for the atomic density and the laser power to form such plasmas, and collisions involving excited states and involving electrons play major roles in producing and maintaining such plasmas. The electron temperature in such a plasma can be measured by observing radiative recombination continua, or by measuring atomic line emissions. Electron density is obtained by measuring plasma line broadening, and can also be estimated by using current probes. We have also used pulsed dye lasers to effectively cause Quasi-Resonant Laser-Produced Plasma (QRLPP), thus providing a new method to generate and study shock and sound waves in vapors, and to investigate molecular destruction, diffusion, and repopulation dynamics. QRLPP should have important applications because it is a highly efficient mechanism for producing a localized plasma; it should also be a fairly general technique applicable to many gas systems.

The interaction of light with alkali metal vapors has been extensively investigated, because of the strong coupling with light, and the simplicity of alkali atoms and molecules. This paper is concerned with a particular optical interaction: efficient plasma generation by light of suitable wavelengths. Light can produce ionization either by direct photo-ionization, or by collisional processes. An example of the latter is associative ionization or Hornbeck-Molnar process (1,2), for example

$$Cs(8D) + Cs(6S) \rightarrow Cs_2^+ + e \quad (1)$$

Equation (1) is possible because of the binding energy of the molecular ion, Cs_2^+, which is 0.64 eV (3). Similar associative ionization processes exist for other alkali metal vapors. While Eq. (1) is energetically possible only for a sufficiently high

0097-6156/82/0179-0447$05.00/0

excited state, ionization may also be possible when only the lowest (resonant) excited state is produced, in high enough density such that their mutual collisions become probable. Thus, Yamada and Okuda (4) discovered that a Cs resonance lamp produces detectable amount of ionization in an irradiated Cs vapor.

Lasers, of course, are much better light sources to study optical ionization. The first study of multiphoton breakdown in an alkali vapor was performed by Okuda et al. (5,6), who focused a pulsed Ruby laser beam into a Cs vapor. With the advent of tunable dye lasers, the possibility of efficient plasma-production by a laser beam tuned to the resonance line of an alkali atom was recognized (Fig. 1a), as first proposed by Measures (7), and experimentally demonstrated by Lucatorto and McIlrath (8,9). This highly efficient method of "resonant laser-produced plasma" (RLPP) has been exploited to perform spectroscopy of alkali ions (8,9). Theoretical modelling has been advanced (10,11) to explain the RLPP mechanism. However, a resonant beam is not the only light that strongly couples with an alkali vapor. If excited states are already present in the vapor (as in a discharge), then light that couples two excited states may also interact strongly with the vapor, and efficient energy deposition in the vapor (leading, for example, to a doubling of the initial electron density) may be possible, as suggested by Oettinger and Dewey (12). On the other hand, alkali molecules (dimers) have very broad absorption bands, and excitation of the dimer may result in the production of the lowest excited atomic P state through dissociation (13) or excitation transfer (14). An example of dissociation is:

$$Cs_2(X^1\Sigma^+) + \text{photon (e.g.,6010Å)} \rightarrow Cs_2^* \rightarrow Cs(6P) + Cs(6S) + \text{energy} \quad (2)$$

Afterwards, the 6010Å light strongly excites the Cs(6P) state to a higher excited state Cs(8D),

$$Cs(6P_{1/2}) + \text{photon (6010Å)} \rightarrow Cs(8D_{3/2}) \quad (3)$$

Equations (2) and (3) suggest that a 6010Å laser beam, for example, may effectively couple energy into a neutral vapor. Collins et al. (15,16) first observed that multi-photon ionization of Cs through such types of hybrid excitation (molecular excitation followed by atomic excitation) is possible in a comparatively dilute Cs vapor (atomic density of 3×10^{15} cm^{-3}). Tam and Happer (17) extended Collins et al.'s work into much higher atomic densities (10^{17} cm^{-3} and higher), and observed that very efficient plasma generation in Cs or Rb vapors is possible with a suitable hybrid excitation (for example, reactions (2) and (3)), using a focused cw laser beam of intensity ~0.1W. For the Cs case, we have observed plasma production for the following dye laser wavelengths: 5839, 5845, 6010, 6035, and 6213Å (using R6G dye), and 8761Å (using HITC dye). For the Rb case, we have also observed plasma production at laser wavelengths of 5724 and 6071Å. These laser wavelengths all couple the resonant excitated state A* of the alkali atom with some higher excited state A** (Fig. 1b) through strong dipole-allowed transitions. We shall call such an excitation a "quasi-resonant" laser excitation, and the resultant plasma (18) a "quasi-resonant laser-produced plasma" (QRLPP), because the laser is, in a sense, resonant with the vapor, since A* is readily produced by dimer excitations, and long-lived due to radiation trapping. QRLPP has also been recently observed in other alkali vapors such as Na (by Koch, Verma, and Stwalley in 19), and Li (by Lucatorto and McIlrath in 11), and should also be observable in other vapors (18) such as the alkaline earth metal vapors, where RLPP has been recently observed (20,21).

Threshold Conditions

As reported earlier (18), the QRLPP effect is a threshold process, requiring sufficiently high atomic density N and laser intensity I. The threshold values depend on various parameters like laser detuning and bandwidth, focusing optics, and the particular excited state involved; also, the threshold N and threshold I values are correlated, with the increase of one allowing a decrease of the other. Typical threshold values in Cs are $N=10^{17}\ cm^{-3}$ and $I=10^4 W\ cm^{-2}$ when a 6010Å laser is used to excite the $Cs(8D_{3/2})$ state. Threshold values when other states are excited or when other alkali vapors are used can be quite different; for example, Stwalley and co-workers (19) have reported that the threshold values in a sodium vapor are $N \sim 10^{16}\ cm^{-3}$ and $I \sim 10^2 W\ cm^{-2}$ when a focused cw dye laser at 5688 or 5683Å is used to excite the Na(4D) state.

The existence of thresholds for the QRLPP mechanism may be understood by considering Fig. 2. Denote the ground, resonant excited, and higher excited atom by A, A*, and A**, with the corresponding densities by N, N_1, and N_2, respectively. At low N, the laser excited A** atom decays mainly radiatively to form various lower states A' (Fig. 2a). At high enough N (but low I), collisional ionization begins to be significant (Fig. 2b). The collisional ionization may be associative ionization

$$A^{**} + A \rightarrow A_2^+ + e \tag{4}$$

or collisional ion pair formation,

$$A^{**} + A \rightarrow A^+ + A^- \tag{5}$$

The first condition of plasma formation due to quasi-resonant laser excitation is that collisional ionization (4) or (5) be sufficiently fast (compared to radiative decay rate of A**),

$$N\,\alpha_i\,\tau_2^{rad} \gtrsim 1 \tag{6}$$

where α_i is the ionization rate coefficient when A** collides with A, and $\tau_2{}^{rad}$ is the radiative lifetime. For the Cs case cited above, α_i is estimated (2) to be $10^{-10}\ cm^3 sec^{-1}$, and $\tau_2{}^{rad}$ is 10^{-7} sec; hence, the threshold density condition is $N \gtrsim 10^{17}\ cm^{-3}$, by Eq. (6).

The threshold laser intensity condition may be estimated as follows. At low I, but high enough N, ionization occurs, but the electrons produced (for example, by Eq. (4)) generally have energy less than 1 eV, and are ineffective to produce A* from A. However, at high enough I, with a corresponding large density N_2, "electron heating" by superelastic collisions with A** becomes possible (Fig. 2c)

$$A^{**} + e \rightarrow A^* + e' \tag{7}$$

Here, e and e' represent slow and fast electrons, respectively, and the superelastic collision may preferentially produce the A* resonant excited state because A**—A* is a strong dipole transition. (Note that inelastic cross sections for electron-atom collisions are generally larger for dipole-allowed transitions compared to dipole-forbidden transitions.) Since the ground state density N is high, the electron excitation of A following Eq. (7) is fast:

$$A + e' \rightarrow A^* + e \tag{8}$$

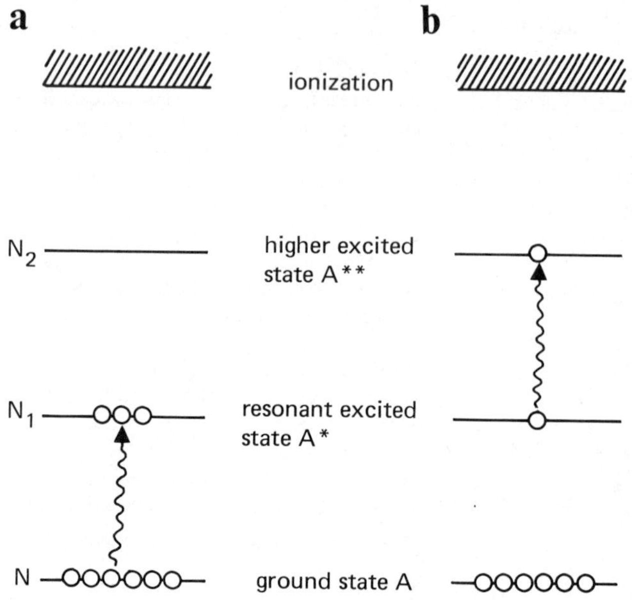

Figure 1. Scheme of efficient laser-production of a plasma by a, resonant excitation (RLPP); and b, quasi-resonant excitation (QRLPP) (18).

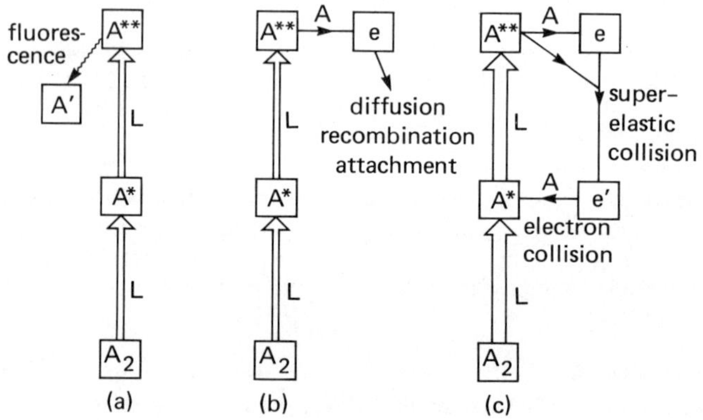

*Figure 2. Diagram of the existence of thresholds for N and I in QRLPP. Key: A, ground state atom; A_2, ground state molecule; and L, laser excitation. a: At low N, radiative decay of A** dominates. b: At high N but low I, A** collisionally produces an electron which causes no further excitation. c: At high N and high I, the electron produced can undergo superelastic and inelastic collisions, effectively completing a self-propagating loop.*

Adding Eqs. (7) and (8), we have the following net reaction:

$$A^{**} + A \rightarrow 2\,A^{*} \qquad (9)$$

which means that for each A* atom removed by laser excitation, two A* atoms will be returned by the "electron-catalysed" reaction (9). Reaction (9) may also occur directly when there is "near resonance", i.e., energy of A** is close to twice the energy of A*, as is the case for A** being $Cs(6D_{3/2})$ where the cross-section for the direct reaction (9) is measured (22,23) to be very large,being 1.5×10^{-14} cm^2; a similar situation exists for A** being Na(4D) state (24,25). Indeed, for the case of Na(4D) being excited, Stwalley and co-workers (19) have reported extremely high efficiency of plasma production in a Na vapor, because merely 1 mW of CW light coupling the Na(3P) and Na(4D) states is needed for the plasma production, indicating that if reaction (9) is directly possible, plasma-production may be greatly faciliated. For the general case of an indirect, electron-assisted reaction (9), near-resonance is not needed, but the energy of A** plus thermal energy must exceed twice the energy of A*. Reaction (9) represents one way to trigger a run-away situation, because it allows the laser to regenerate its own lower state faster than the rate it consumes (true when each absorption causes, on the average, the regeneration of more than 1 A* atom); thus a large N_1 can be built up by the laser starting from a very small N_1, as noted earlier (22). Thus for the electron-catalysed reaction (9) to be effective, the electron-heating collision (7) must be sufficiently fast, i.e.,

$$N_2\, v_e\, \sigma_{e2}\, \tau_e \gtrsim 1 \qquad (10)$$

where v_e is the average electron velocity, σ_{e2} is the average superelastic collision cross section, and τ_e is the electron lifetime determined by recombination, diffusion, or attachment collisions. For the Cs case cited above, Eq. (10) leads (18) to an estimated threshold laser intensity of the order of 10^4W cm^{-2}, which agrees in order of magnitude with our experiment, and is some ten orders of magnitude lower than the conventional laser breakdown threshold due to multiphoton ionization and inverse bremsstrahlung. By monitoring the fluorescence, we observed that a large increase in N_1 occurs at the breakdown threshold for QRLPP, in agreement with the above discussion, as shown in Fig. 3.

Diagnostic Methods

Optical diagnostic techniques are convenient and popular for alkali metal plasmas, because the strong and well-understood emission lines and continua can be easily observed in the visible and near-visible region.

The emission spectrum observed from a Cs QRLPP is shown in Fig. 4. The electron temperature can be derived from such a spectrum by two traditional methods used for Cs discharge plasmas: (a) Radiative continuum due to recombination of free electrons with atomic ions can be observed, as seen (Fig. 4) in the 4000-5200Å region when the recombination product is a Cs(6P) excited state, and in the 5500-6200Å region when the recombination product is a Cs(5D) excited state. The shape of the continuum gives the electron temperature T_e (26), because it is determined by the Boltzmann factor $\exp(-h\nu/kT_e)$ where $h\nu$ is the photon energy and k is the Boltzmann constant. (b) Atomic line emission intensities (see Fig. 4) also provide an estimate of T_e if equilibrium excited state densities are assumed, i.e., $N_j \sim g_j \exp(-E_j/kT_e)$, where N_j is the density of the jth excited atomic state of statistical

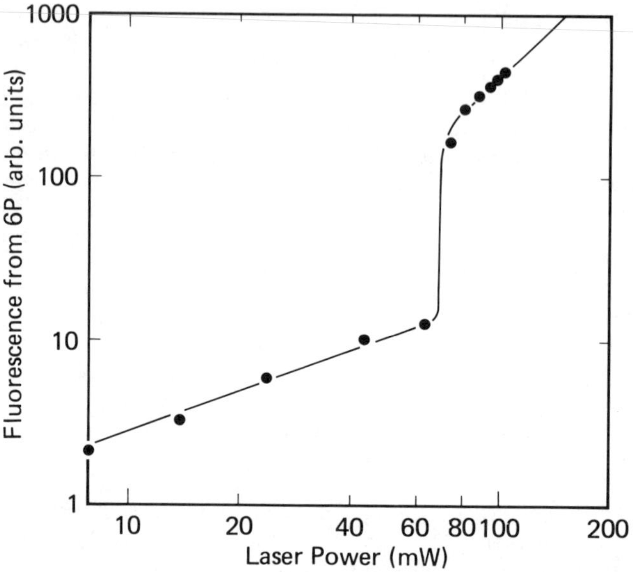

Figure 3. Observed fluorescence (6P → 6S) from a Cs vapor at 420°C as a function of the incident laser power (cw laser at 6010.5 Å). The strong increase of 6P population at the QRLPP threshold (about 70 mW power) can be clearly seen.

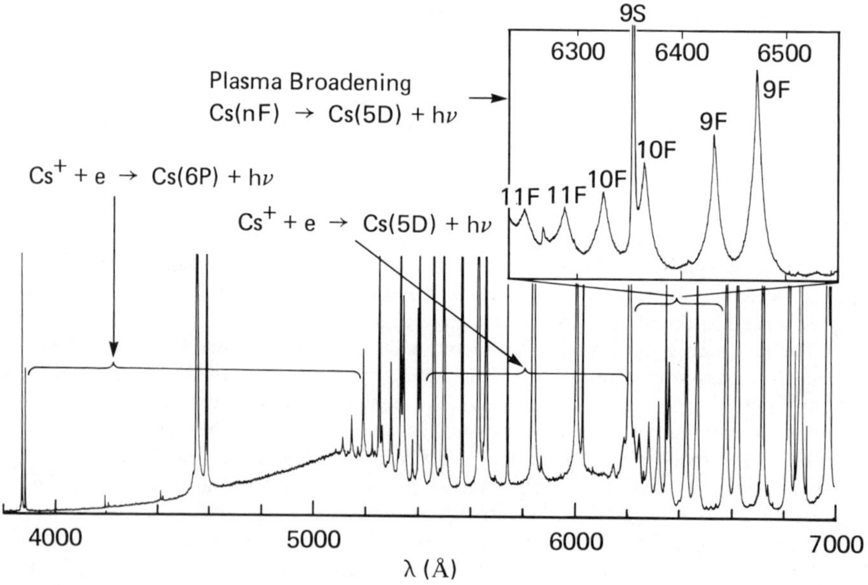

Figure 4. Visible emission spectrum of a Cs QRLPP, excited by a focused cw laser of 170 mW power tuned to 6010.5 Å. The Cs atomic density is 3×10^{17} cm^{-3}, (the saturated density at 410°C).

weight g_j and energy E_j. However, to use atomic emission intensities to derive N_j, the integrated emission intensity of each line must be measured, and the radiative decay rate must be known. Furthermore, local thermodynamic equilibrium (26,27) must be completely satisfied for method (b) to be applicable, but need only be partially satisfied for method (a) to be applicable. Thus, method (b) tends to be more difficult or less reliable than method (a), which is to be preferred if atomic radiative recombination continuum is observed. By analyzing the 4000-5200A recombination continuum (which is more conspicuous than the 5500-6200A continuum in Fig. 4), we obtain an electron temperature of 0.2 ± 0.02 eV. Method (b) has also been applied to the line emission intensities indicated in Fig. 4, but the resultant T_e is less accurate. We obtain $T_e \approx 0.16$ eV from an Arrhenius plot of the 9S through 12S emission lines, and $T_e \approx 0.1$ eV from the intensities of the 9D through 13D emission lines. These discrepancies are partially due to the uncertainties of the linewidths of the emission lines from high lying excited states, and this is more serious for a D state than for an S state at approximately the same energy from the ground state.

The emission spectrum of Fig. 4 also provides the electron density through plasma line broadening (27) measurements. This is particularly convenient for the emission lines originating from the Cs(nF) excited states, because nF is nearly degenerate with nG excited state, so that large Stark broadening occurs due to the micro-electric-fields in the plasma, and emission linewidths vary linearly with the electron density N_e. For the line broadening shown in Fig. 4, N_e is found to be 3×10^{14} cm^{-3}.

The spatially averaged electron density $\overline{N}_e$ can be estimated by using current probes positioned close to the QRLPP but without intercepting any laser light. The current collected (Q electrons/sec) is given by

$$Q = \overline{N}_e \, u_e \, A \tag{11}$$

where u_e is the electron drift velocity depending on the ratio of the electric field E to the atomic density N (28), A is the area of charge collection, and current due to the positive ions is neglected. In our previous experiment (17), we have $Q=1\times10^{14}$ electrons/sec for E=2V/cm and $N=2\times10^{17}$ cm^{-3} corresponding to $u_e=1\times10^3$ cm/sec (28), and A is taken as the projected area of the plasma, i.e., 2×10^{-2} cm^2. Using Eq. (11), we obtain $\overline{N}_e = 5\times10^{12}$ cm^{-3} which is smaller than N_e as determined by line broadening because $\overline{N}_e$ is averaged over the space between the electrodes, while N_e is obtained at the center of the plasma spark.

Naively, we may expect T_e and N_e in the QRLPP to increase as the incident cw laser power increases. Experimentally, we have found that T_e and N_e increase very slowly as I increases; instead, the volume of the QRLPP increases much more visibly as I increases, i.e., the QRLPP lengthens from the point where the laser is focused towards the incident laser beam. As an example, the length of the QRLPP is observed to increase from about 3 mm to about 6 mm when the laser power is increased from 100 mW to 200 mW, with the T_e increase being less than 10%. These observations indicate that QRLPP tends to expand as more energy is deposited into it. This is the basis of shock wave generation by transient QRLPP generated by pulsed lasers, described in the next section.

We have previously noted (17) that the values of T_e=0.2 eV and $N_e=3\times10^{14}$ cm^{-3} satisfy the Saha equilibrium condition:

$$N_e^2/N = 3.0\times10^{21}\, T_e^{1.5} \exp(-3.894/T_e) \tag{12}$$

This indicates that the QRLPP in our Cs case is in local thermodynamic equilibrium with T_e=0.2 eV. At such an electron temperature, the positive ions should mainly be Cs^+ (Cs_2^+ density is smaller by about 2 orders of magnitude), as shown by Agnew and Reichelt (26). Indeed, in our case, the readily observed e-Cs^+ radiative recombination continua is indicative of the high Cs^+ concentration. However, in the Na QRLPP observed by Stwalley and co-workers (19), no e-Na^+ radiative continuum was observed, and they suggested that the ionic species are mainly Na_2^+ in their case. The matter needs further clarification.

Transient QRLPP by Pulsed Lasers

We have shown that pulsed dye lasers are also quite capable of generating transient QRLPP, similar to the cases of RLPP reported (8,9,20,21). We have used a flashlamp pumped dye laser (Candela LFDL-1), with output pulses of 1Å bandwidth, 1 μsec duration, and energy up to 20 mJ at 10 Hz repetition rate. The pulsed dye laser is typically set at 6010Å to excite QRLPP in a Cs vapor. The time development of the absorption of the incident laser pulse can be monitored by measuring the transmitted pulse with a fast photodiode. We have observed that the earlier part of the incident laser pulse is not much absorbed; as the incident laser intensity increases, the QRLPP threshold is reached and breakdown occurs, causing a sudden drop in the transmitted light.

Apart from the fact that the breakdown is sudden, not much is known about the time-development of the transient QRLPP. Detailed time-resolved absorption studies, similar to the investigation of Lucatorto and McIlrath (11), would provide valuable information on the transient dynamics, but this has not been done for the QRLPP case. Instead, we have examined the transient fluorescence from several Cs excited states, as shown in Fig. 5. Each fluorescence line observed from the transient QRLPP exhibits a fast decay component of about 1.5 μsec time constant, and a slow decay component of about 10 μsec time constant (depending on the particular line observed). The fast decay time constant of 1.5 μsec agrees very well with the previously observed "electron response time" of 1.8 μsec in a cw QRLPP (29). The two-component decay of the excited states in the transient QRLPP is not presently clearly understood, but it is quite similar to the results reported by Lam and Schearer (30) for a laser-excited afterglow in a K/Kr vapor mixture.

The fast deexcitation after the laser pulse is focused into the vapor may cause the generation of an acoustic blast wave or shock wave, whose magnitude can be quite large. In a typical experiment, 1 mJ of pulsed laser energy is absorbed in a vapor volume of 10^{-3} cc containing 10^{14} atoms. Thus, the initial total thermal energy in the volume is 10^{13} eV, while the laser irradiation causes the thermal energy to increase by about 1 mJ or 6×10^{15} eV in about 1 μsec. This fast heating behaves like an exploding wire or a fast meteor, creating a cylindrical blast wave. We found that (31) the shock wave propagation can be probed by a weak cw probe laser beam (from a HeNe or Kr laser) that is parallel to but displaced from the pulsed laser. The arrival of the acoustic pulse causes a well-defined transient deflection and absorption signal in the probe beam, with a well-defined delayed time t from the laser pulse. By moving the probe beam using a micrometer translation stage, a graph of the separation R versus t can be plotted, as given in Fig. 6. The data of Fig. 6, never previously observed for any metal vapor, are characterized by two regions. In the near region (close to the line blast source), the blast wave propagates supersonically, slowing down as it propagates, because the wave is weakening. After propagating

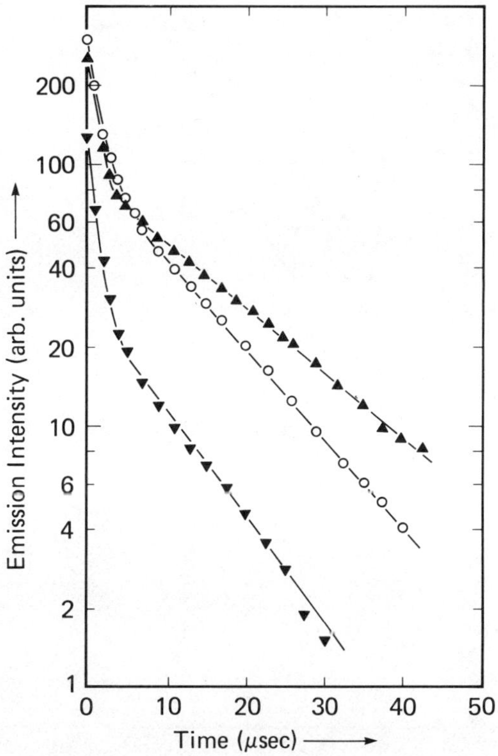

Figure 5. Observed time-dependence of emission lines from a transient Cs QRLPP excited by a pulsed laser beam at 6010.5 Å, of energy about 5 mJ and duration 1 μsec. The Cs density is 1.5×10^{17} cm^{-3}. Key: ▲, 7D → 6P; ○, 9D → 6P; and ▼, 7P → 6S.

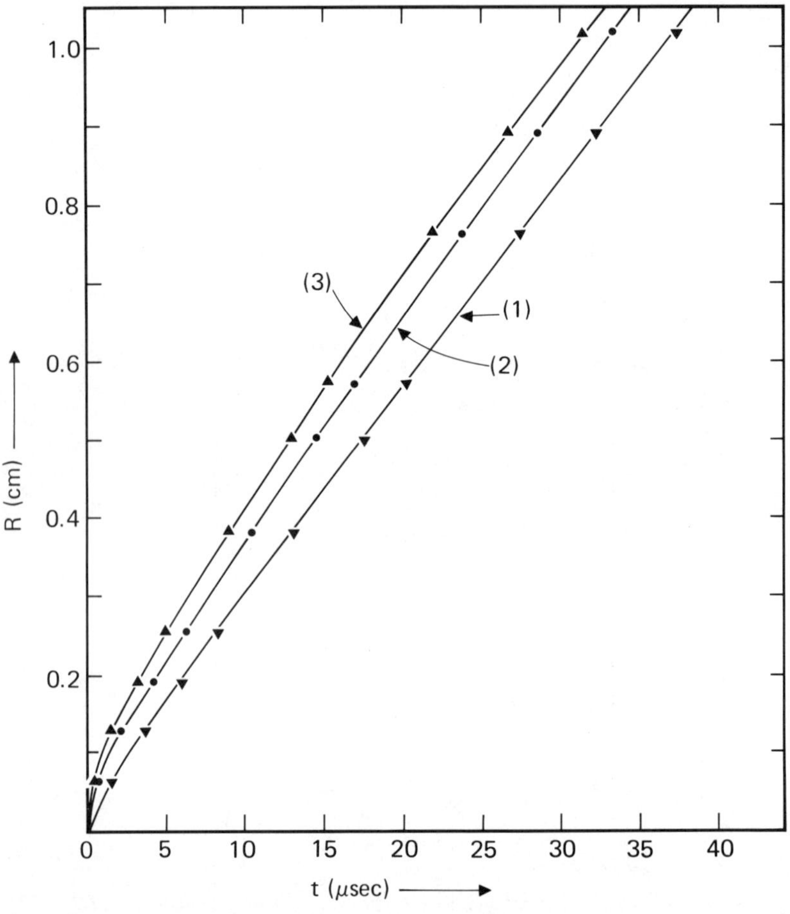

Figure 6. Observed relation between propagation distance (R) and arrival time (t) for a Cs vapor at 402°C. Key: R, displacement between the pulsed laser (producing a transient QRLPP) and the probe cw laser beam. The laser pulse energy is 106, 876, and 2420 μJ for curves (1), (2), and (3), respectively. The corresponding energies absorbed by the Cs vapor are 21, 464, and 2001 μJ, respectively. For details of comparison of the present data with blast wave theories (Eq. 13), see Ref. (31).

through a distance of about 1 cm (depending on the pulsed laser energy and on the atomic density), the blast wave decays to a sound wave and propagates at a constant sound velocity afterwards. The data of Fig. 6 can be fitted quite well by a general formula for shock propagation proposed by Vlases and Jones (32):

$$R^2 = c_o^2 t^2 + R_o c_o t \quad (13)$$

where c_o is the (infinitesimal-amplitude) sound velocity, and $R_o = [E_o/1.75\rho_o c_o^2]^{1/2}$ is the "range" of the blast wave (i.e., for $c_o t << R_o$, we have $R^2 \approx R_o ct$, which is the strong shock limit; while for $c_o >> R_o$, we have $R \approx c_o t$, which is the weak sound wave limit). Here E_o is the energy deposited in the blast source per unit length and ρ_o is the particle density. For more details, see Ref. 31. The good fit, shown in Fig. 6, is to our knowledge the first experimental verification of Eq. (13) in the extreme weak shock limit, when transition to sound occurs. This transition corresponds to a change from a nonlinear R-t dependence to a linear R-t dependence. In fact, the linear portion of Fig. 6 gives the sound velocity in a metal vapor, which had also never previously been measured because of experimental difficulties associated with the generation and detection of acoustic waves in such a hot corrosive metal vapor. We have used the linear portion of the R versus t data to obtain sound velocities in saturated Cs vapors at different temperatures. Our velocity measurements represent a new and useful method to probe thermodynamic properties (such as dimer/monomer ratio) of metal vapors (see Ref. 31).

Since the QRLPP has a fairly high electron temperature, the normal equilibrium between the Cs_2 molecules and the Cs atoms is affected in the QRLPP. This means that the bound Cs_2 (singlet) density is decreased while the unbound Cs_2 (triplets) density is increased due to the QRLPP. We have indeed observed that the transient destruction of molecular state by the pulsed QRLPP can be measured by positioning the cw probe beam close to or in coincidence with the pulsed laser beam. The transient absorption signals observed in the probe beam, with long transient time constants (for example, on the order of 1 msec) provide a new method to study diffusion dynamics in a vapor system (33).

Discussions and Future Applications

The QRLPP method represents a highly efficient method to produce plasmas in local thermodynamic equilibrium in metal vapors of densities exceeding 10^{16} cm^{-3}. The threshold intensity for QRLPP is many orders of magnitude lower than that required for multiphoton and inverse-bremsstrahlung plasma formation. Although efficiency of QRLPP may not be as high as RLPP, QRLPP has the advantage that highly-localized plasma spark formation is possible (17), and generally, more laser wavelengths are available for QRLPP than for RLPP for each vapor system.

QRLPP is a general plasma-formation mechanism, already demonstrated for Cs, Rb, Na, and Li vapors. It should also be useful for other vapors or vapor mixtures, e.g., alkaline earth metal vapors.

Possible applications of QRLPP include the following: (1) The QRLPP can be quite localized if a focused laser beam is used. Such a small, isolated "point-source" of plasma may be useful for providing ion beams for ion-beam microscopy, ion-beam lithography, doping, bombardment, or surface etching. It may also be useful (as suggested by Falk, (34)) to study and optimize a Cs thermionic energy converter by causing local increase in electron density by QRLPP at various locations from the electrodes, and observing the corresponding change in the performance of the converter. (2) The QRLPP possesses the unusual behavior of inducible optical absorption under appropriate conditions (29). This property of stronger abosorption for stronger incident light is useful for automatic noise-reduction of a laser beam (29) or for automatic attenuation of intense laser radiation (35). (3) Transient QRLPP provides an ideal source for blast wave and acoustic wave generation in a gas system. Besides being useful for new acoustic studies in corrosive metal vapors as indicated above, the QRLPP acoustic source may provide a new acoustic diagnostic tool for plasma and laser fusion studies. (4) Other applications of QRLPP include (a) basic spectroscopy, e.g., spectroscopy of ions, as done by Lucatorto and McIlrath (8,9); (b) quantum electronics, e.g., identifying lasing transitions in a recombining plasma, as shown by Silfvast, Szeto and Wood (36); (c) energy conversion, e.g., the thermoelectric laser energy conversion done by Thompson et al. (37); (d) afterglow kinetics studies, e.g., afterglow investigation in a K and Kr vapor mixture excited by a violet pulsed laser done by Lam and Schearer (30); and (e) plasma wave studies, e.g., the investigation of waves and instabilities in plasmas studied by Ikezi et al. (38).

Acknowledgments

The earlier part of this work was performed in Columbia University, in collaboration with W. Happer, and was continued at Bell Laboratories. The recent experiments on transient QRLPP have been performed in IBM San Jose Research Laboratory, in collaboration with W. Zapka, W. Imaino, and K. Chiang. I am grateful to the above mentioned colleagues for their important contributions. I also thank W. C. Stwalley of University of Iowa, T. Falk of Calspan Corp., and B. H. Schechtman of IBM for their valuable encouragement and suggestions. The useful comments on this manuscript by R. Balanson, K. Chiang, and W. Imaino of IBM, and T. B. Lucatorto of NBS are greatly appreciated.

Literature Cited

1. Manus, C. Physica 1976, 82C, 165.
2. Dobrolezh, B.U.; Klyucharev, A. N.; Sepman V. Yu. Opt. Spectrosc. 1975, 38, 630.
3. Marr, G. V.; Wherrett, S. R. J. Phys. B: Atom. Molec. Phys. 1972, 5, 1735.
4. Yamada, J.; Okuda, T., J. Phys. Soc. Jpn 1973, 35, 881.
5. Okuda, T.; Kishi, K.; Swada, K. Appl. Phys. Lett. 1969, 15, 181.
6. Yamada, J.; Okuda, T. Jpn J. Appl. Phys. 1975, 14, 903.
7. Measures, R. M. J. Quant. Spectrosc. Radiat. Transf. 1970, 10, 107.
8. Lucatorto, T. B.; McIlrath, T. M. Phys. Rev. Lett. 1976, 37, 428.
9. McIlrath, T. J.; Lucatorto, T. B. Phys. Rev. Lett. 1977, 38, 1390.
10. Measures, R. M.; Drewell, N.; Cardinal, P. J. Appl. Phys. 1979, 50, 2662.
11. Lucatorto, T. B.; McIlrath, T. J. Appl. Optics 1980, 19, 3948.
12. Oettinger, P. E.; Dewey, Jr., C. F. AIAA Journal 1970, 8, 880.
13. Brom, J. M.; Broida, H. P. J. Chem. Phys. 1974, 61, 982.
14. Lam, L. K.; Fujimoto, T.; Gallagher, A. C.; Hessel, M. M. J. Chem. Phys. 1978, 68, 3553.
15. Collins, C. B.; Johnson, B. W.; Mirza, M. Y.; Popescu, D.; Popescu, I. Phys. Rev. A 1974, 10, 813.

16. Collins, C. B.; Johnson, B. W.; Popescu, D.; Musa, G.; Pascu, M. L.; Popescu, I. Phys. Rev. A. 1973, 8, 2197.
17. Tam, A. C.; Happer, W. Optics Commun. 1977, 21, 403.
18. Tam, A. C. J. Appl. Phys. 1980, 51, 4682.
19. Koch, M. E.; Verma, K. K.; Stwalley, W. C. J. Opt. Soc. Am. 1980, 70, 627.
20. Skinner, C. H. J. Phys. B: Atom. Molec. Phys. 1980, 13, 55.
21. Jahreiss, L.; Sandeman, R. J.; Huber, M. C. E.; Burgess, D. D.; Wheaton, J. E. G., EGAS Conference, Pisa, September 1980 (unpublished).
22. Yabuzaki, T.; Tam, A. C.; Hou, M.; Happer, W.; Curry, S. M. Optics. Commun. 1978, 24, 305.
23. Tam, A. C.; Yabuzaki, T.; Curry, S. M.; Hou, M.; Happer, W., Phys. Rev. A. 1978, 17, 1862.
24. Allegrini, M.; Alzetta, G.; Kopystynska, A.; Moi, L.; Orriols, G. Optics Commun. 1976, 19, 96; and 1977, 22, 329.
25. Stwalley, W. C. private communication.
26. Agnew, L.; Reichelt, W. H. J. Appl. Phys. 1968, 39, 3149.
27. Griem, H. R. "Plasma Spectroscopy"; McGraw-Hill, N.Y., 1964.
28. Saelee, H. T.; Lucas, J. J. Phys. D: Appl. Phys. 1979, 12, 1275.
29. Tam, A. C. Appl. Phys. Lett. 1979, 35, 683.
30. Lam, L. K.; Schearer, L. D. Chem. Phys. Lett. 1978, 60, 130.
31. Tam, A. C.; Zapka, W.; Chiang, K.; Imaino, W., submitted to Applied Optics.
32. Vlases, G. C.; Jones, D. L. Phys. Fluids, 1966, 9, 478.
33. Zapka, W.; Tam, A. C.; to be published.
34. Falk, T., private communication.
35. Weiss, P. Appl. Phys. Lett. 1977, 30, 261.
36. Silfvast, W. T.; Szeto, L. H.; Wood III, O. R. Appl. Phys. Lett. 1977, 31, 334.
37. Thompson, R. W.; Manista, E. J.; Alger, D. L. Appl. Phys. Lett. 1978, 32, 610.
38. Ikezi, H.; Mima, K.; Nishikawa, K.; Inutake, M. Phys. Rev. Lett. 1976, 36, 794.

RECEIVED August 26, 1981.

31

Optically Pumped Alkali Molecule Lasers

B. WELLEGEHAUSEN

University of Hannover, Institute of Applied Physics, 3000 Hannover, West Germany

By optical excitation with argon and krypton laser lines, continuous laser oscillation on A → X and B → X transitions of Li_2, Na_2 and K_2 molecules can be achieved. These dimer lasers show such interesting features as multiline emission, extremely low threshold pump intensities and forward-backward amplification asymmetry. Basic principles, operating conditions and applications of these lasers will be discussed. The dimer lasers operate between bound electronic states, resulting in the emission of discrete lines. To achieve tunable laser oscillation, continuous emission bands from bound-free transitions have to be considered. Some possibilities for alkali dimers are outlined and recent spectroscopic investigations on UV excited diffuse bands are reported.

An active field of spectroscopic investigation presently is molecular metal vapors and vapors of metal-compounds, such as the alkalis (1-6), the alkaline earth (7,8) and the metal halides (9,10). Furthermore, these materials are also of increased interest for the development of new laser systems. A nice example of how new spectroscopic results have stimulated new laser developments and how these new lasers have lead to special spectroscopic techniques and application is given by the optically pumped homonuclear diatomic lasers, which have been developed since 1976 (11 -19). It is the aim of this article to review features, operating conditions and possible applications of these dimer lasers, with special regard to alkali dimer systems and to outline how interesting laser systems with alkali molecules might be developed.

0097-6156/82/0179-0461$06.50/0

The saturated vapors of many elements contain moderately high concentrations of dimer molecules; some vapors or gases even consist only of dimer molecules. The longest known and widely used lasers involving diatomic molecules are obviously the hydrogen and nitrogen lasers (20,21), which can be excited by short pulsed discharges. The laser transitions in dimer molecules are transitions between rotational-vibrational levels of different electronic states. At low pressure, individual rotational-vibrational levels have to be excited, which requires a selective excitation mechanism. For the heavier dimer molecules with dense rotational-vibrational states, this selectivity is difficult to achieve with discharge excitation. Therefore the light dimer systems nitrogen and hydrogen are the only presently important discharge pumped systems. The requirements of selective excitation, however, can be ideally fulfilled with optical exitation, especially if laser radiation is used. Consequently, with the development of strong pulsed and cw primary laser systems, laser pumping techniques have become more and more attractive for new secondary laser systems. In 1976 the first realization of a pulsed optically pumped Na_2 dimer laser system was reported (11) and in 1977 this system could also be operated cw by excitation with an argon ion laser (13). Figure 1 surveys the present cw dimer systems, their spectral emission ranges and dominant pump laser lines. Figure 2 shows a typical laser cycle within the energy level diagram of a dimer molecule. With a suitable pump frequency, ω_p, a rotational-vibrational level in an excited electronic state is populated. The fluorescence or laser emission starts again from this level. Consequently dimer lasers are three level laser systems. For laser purposes, those transitions are important which have high Franck-Condon factors and which end in rotational-vibrational levels high enough above the pump level. Assuming a negligible population density in the lower laser level 2 ($n_2 \approx o$), the maximum small signal cw gain for a laser transition is approximately given by $g = \sigma_d I_p \alpha_p L/A_3 \hbar \omega_p$, where σ_d is the cross section for stimulated emission, α_p the absorption coefficient for the pump radiation (Intensity I_p), L the amplification length and A_3 the total loss rate of the upper laser level, including spontaneous emission and possible collisional deactivation. For low pressures, A_3 is mainly determined by the natural lifetime of typically

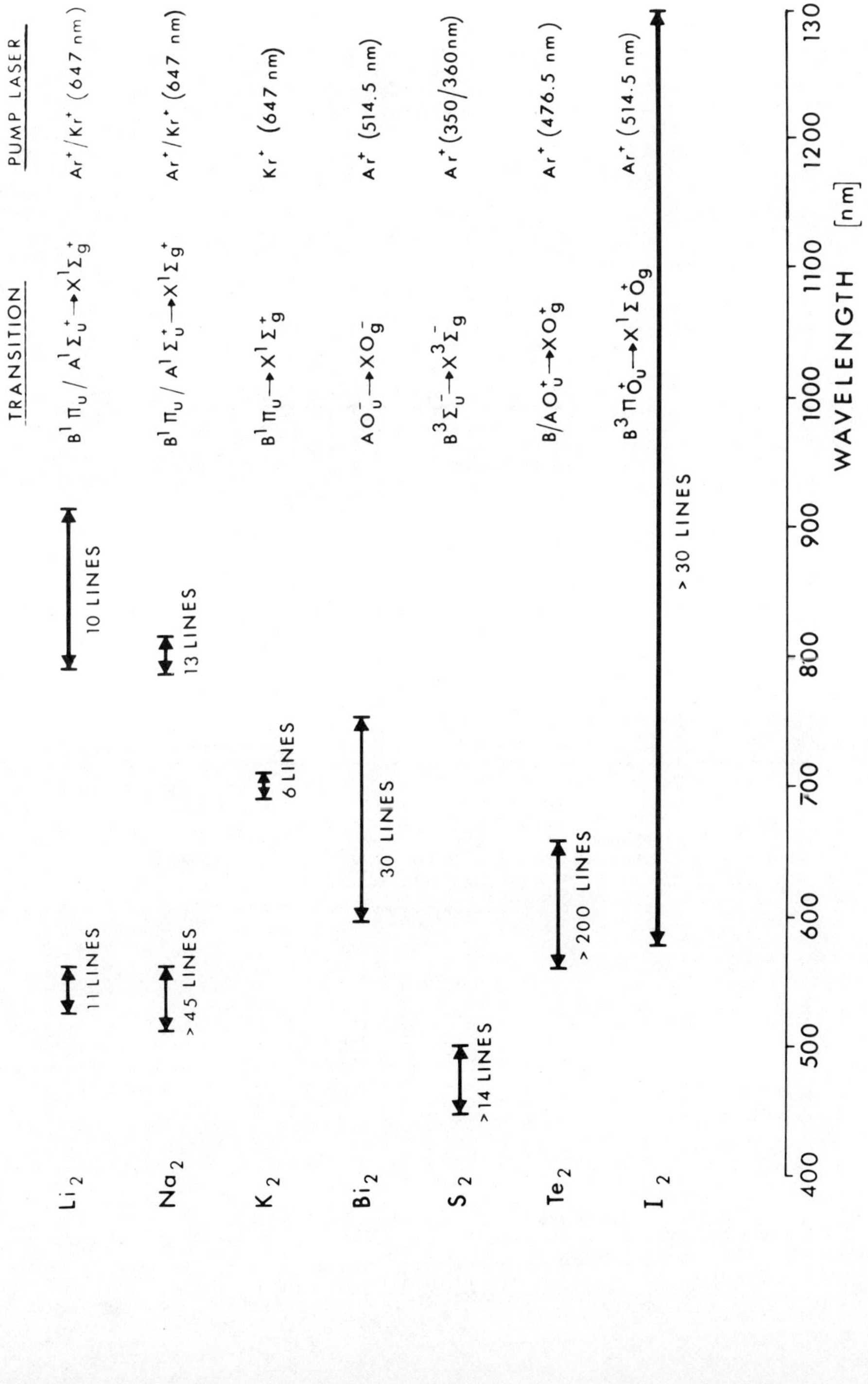

Figure 1. Spectral ranges of cw dimer laser lines.

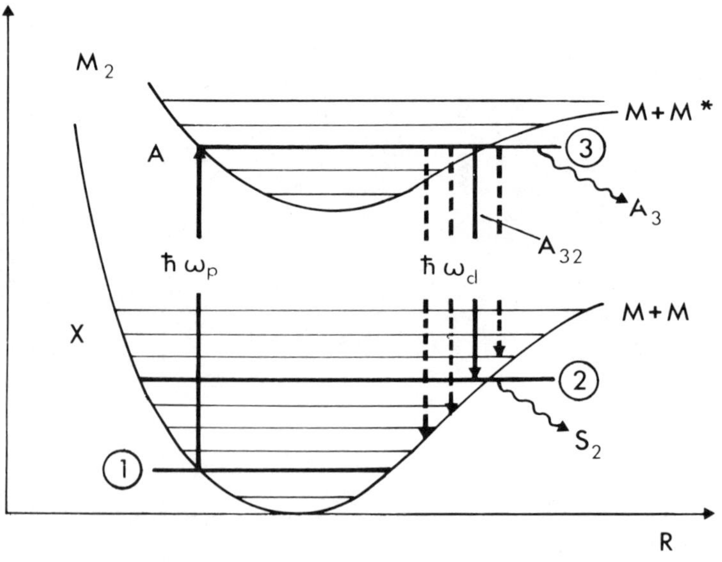

Figure 2. Energy level diagram of a dimer molecule M_2 including pump and laser transitions. A_{32}, spontaneous transition rate; A_3 and S_2, total loss rates of upper and lower laser level.

10^{-8}s. For $\lambda_p = 500$ nm, L = 10 cm, $\alpha_p = 10^{-1}cm^{-1}$ and $\sigma_d = 3 \cdot 10^{-12}cm^2$ for Na_2 (16), a gain of 10% sufficient for laser operation, is obtained with only 1 Wcm^{-2} pump intensity. Such low pump intensities, which can easily be achieved by focussing a laser with only 1mW output power, have indeed been observed for the Na_2 system. In order to maintain the cw-population inversion density, however, the lower laser level must be rapidly depopulated. As this level is metastable, it represents a bottleneck in the laser cycle, which may prevent cw laser oscillation. Cw oscillation therefore requires a relaxation rate S_2 of the lower laser level at least larger than the spontaneous transition rate A_{32} which populates this level at threshold. The rate S_2 is $\sigma_2 p\sqrt{8/\pi\mu KT}$, where p is the pressure of the collision partner, σ_2 the corresponding cross section, T the absolute temperature and μ the reduced mass for the collision process. For the Na_2-Na system σ_2 values of about $10^{-14}cm^2$ have been measured (16). With $A_{32} \approx 10^7 s^{-1}$ for strong transitions, the condition $S_2 > A_{32}$ then requires a minimum Na pressure of about 10^{-2} mbar, which is much less than the Na vapor pressure of typically 9 mbar (corresponding to T $\approx$ 800K) for laser operation. The condition $S_2 > A_{32}$ can in principle always be fulfilled by a sufficient buffer gas pressure. However, with increasing buffer gas pressure also the loss rate of the upper laser level and with it the threshold pump intensity will increase as well. Therefore, under certain circumstances cw oscillation will not be possible even for very high pump intensities. Laser oscillation can also be prevented by internal losses due to self absorption. The probability for self absorption increases for the heavier alkali dimer molecules, especially if higher lying electronic states are excited, since the absorption spectra of these dimers extend throughout the visible and near infrared spectral range. This is probably why up to now no laser oscillation could be obtained for Rb_2 and Cs_2 dimers, and why the laser oscillation in K_2 is much weaker as for Na_2 and Li_2 systems. A typical set-up for a linear dimer laser system is shown in Figure 3. The metal dimer molecules are generated in a heatpipe with Brewster angle windows. The pump laser radiation is focussed collinearly into the heatpipe and the resonator consists of two focussing elements, a plane output coupling mirror and a suitable tuning element, which allows selection of an individual line. Instead of a middle lens a concave mirror together with a folded cavity can be used. Two types of excitation of the dimer molecules, with

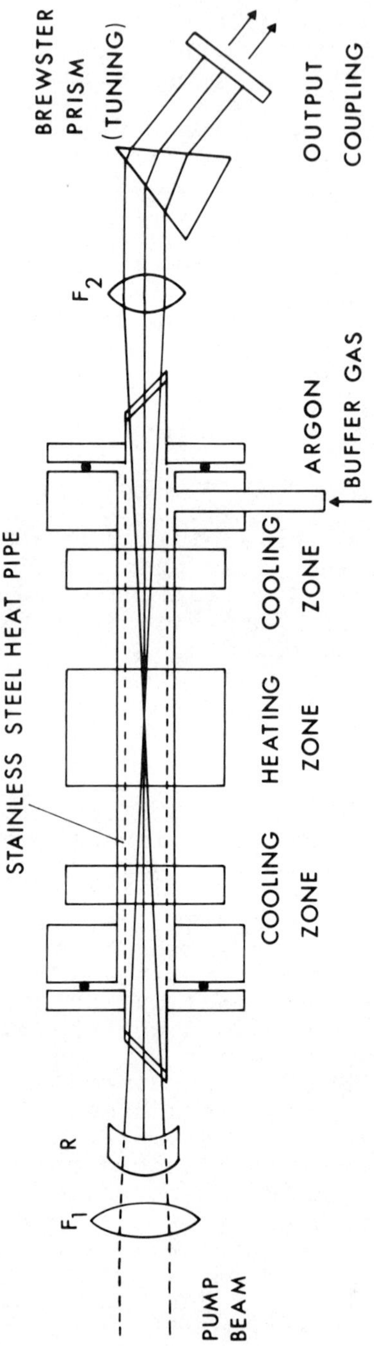

Figure 3. Dimer laser setup with linear resonator.

a multimode or a single mode laser, have to be distinguished. For multimode excitation, the total spectral width of the pump laser (5-7 GHz for an argon ion laser) is 3 or 4 times broader than the Doppler width of the molecular absorption lines. Consequently, only part of the pump power can be used for a single transition; in addition normally several transitions are excited simultaneously. For a laser system without a tuning element, this multimode excitation therefore gives multiline emission, with lines belonging to different fluorescence series. (See Figure 4). For multimode pumping the total output power can be high, with more than 10% conversion efficiency. On the other hand, the single line output power from a resonator with an internal tuning element is low, due to the inefficient pumping. Single frequency excitation of an individual pump transition is much better suited for single line operation. Figure 5 shows a complete fluorescence progression from single frequency excitation of the $X^1\Sigma_g^+$ (3,43) → $B^1\Pi_u$ (6,43) transition with an argon ion laser at 488 nm. The progression extends from v'' = o up to v'' = 17. Laser oscillation was observed for the strongest transitions with v''≫ 9. Without a tuning element, all laser lines share the same upper laser level population, and consequently strong competition effects can be observed, and normally only the strongest lines (v'' = 13, 14) will oscillate simultaneously. With a tuning element, a single line can use the whole population, resulting in high output power values and good conversion efficiencies. In suitably designed low loss resonators, it should also be possible to achieve laser oscillation for weak transitions and for transitions which are still closer to the pump transition.

Up to now, the dimer laser system has been described alone in terms of population inversion between suitable energy levels, and for this description the condition $S_2 > A_{32}$ is indeed the only necessary condition for cw laser oscillation, as long as the thermal population density in the lower laser level remains negligibly low. However, as this optically pumped laser system is a coherently excited three level system, the coherent emission can also be described as stimulated Raman scattering, which is resonantly enhanced by the common level 3 of the pump and laser transitions. This coupled two photon or Raman process does not require a population inversion between levels 3 and 2 and introduces qualitatively new aspects which appreciably influence and change the normal laser behaviour. For a detailed and deeper description of the coherently excited three level dimer

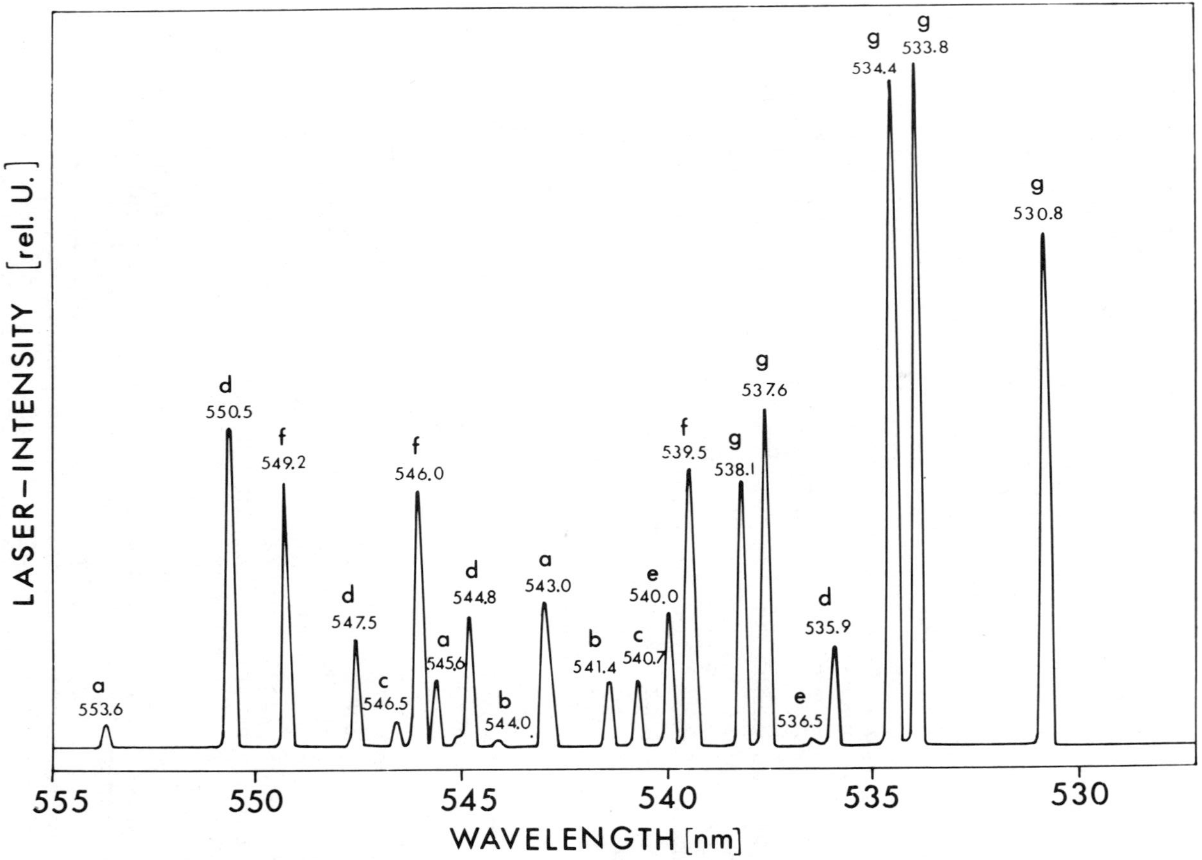

Figure 4. Multiline laser spectrum of Na dimer laser on $B^1\Pi_u \rightarrow X^1\Sigma_g^+$ transitions, excited by 472.7-nm Ar laser line (multimode). Pump power 1 W, all lines output power 70 mW. Pump transitions $(v'',J'') \rightarrow (v',J')$: a, (12,21) → (27,22); b, (12,48) → (29,49); c, (8,65) → (22,64); d, (6,83) → (20,82); e, (4,66) → (15,65); f, (3,80) → (14,81); and g, (1,37) → (9,38).

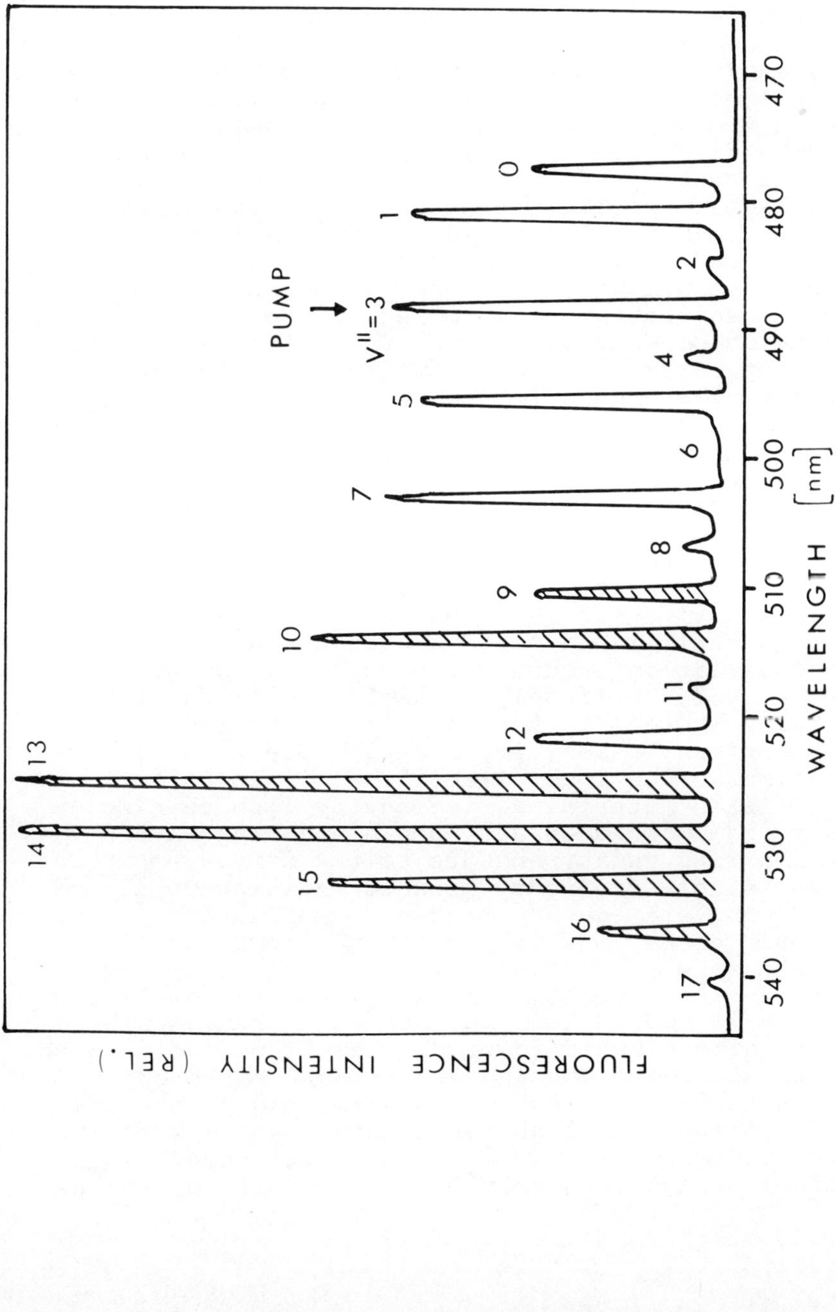

Figure 5. Na_2 $B^1\Pi_u$ *(6,43)* $\rightarrow$ $X^1\Sigma_g^+$ *(v″, 43) fluorescence progression. Pump transition,* $X^1\Sigma_g^+$ *(3,43)* $\rightarrow$ $B^1\Pi_u$ *(6,43) (488 nm). Key:* ▨ *, laser lines.*

laser system see References 16, 22, 23 and further References therein. We will restrict our discussion here to a simple phenomenological description and discuss consequences for operation and applications of these laser systems. For multimode excitation, the whole Doppler profile is excited, and the laser emission again comes from a Doppler broadened ensemble and consists of many modes. For narrowband single frequency excitation, however, only a sub Doppler molecular ensemble is excited. This is schematically shown in Figure 6. For a given direction of the pump beam with frequency ω_p, only molecules around a velocity component u_p with respect to the pump beam direction will be excited. Therefore a hole, whose width is given by the natural linewidth, is burned into the Doppler broadened population density distribution of the lower pump level. The upper laser level population density then consists of a molecular ensemble with velocity components around u_p. The emission from such a sub Doppler ensemble alone will change the laser behaviour. For a pump frequency $\omega_p = \omega_{13}+k_p u_p$, the center frequencies of the emissions in the forward and backward direction are given by $\omega_d^{\pm} = \omega_{32} \pm k_d u_p$ with ω_d^+ being the center forward frequency and k_p and k_d the wave vectors of the pump and the emission frequencies. For off resonant pumping $\omega_p \neq \omega_{13}$, the material therefore has an anisotropic emission frequency characteristic which, for example, can be used to decouple optical systems, as has been pointed out by Skribanowitz (24). In case of population inversion, laser oscillation around ω_d^+ or ω_d^- will be possible. For $u_p \neq o$, ω_d^+ will only be amplified in the forward direction, whereas in the backward direction ω_d will see absorption losses, for example, due a residual population in the lower laser level 2. It is obvious that for this material a ring cavity with running waves is now more adequate than a standing wave cavity. In such a ring cavity and due to the narrow emission profile, single frequency forward oscillation around ω_d^+ and single frequency backward oscillation around ω_d^- should occur. However, both emissions have to share the same population inversion, so that they can not exist simultaneously. As long as pure population inversion is considered, neither the forward nor the backward direction is favored. This situation changes, if we consider the coupled absorption of photons ω_p and emission of photons $\omega_d^{\pm}$ in a two photon or Raman type process. The two photon transition 1 → 2 has the highest probability for $\Delta\omega_{12} = \omega_p - \omega_d^{\pm}$ being close to the resonance frequency $\omega_{12} = \omega_{13} - \omega_{32}$ of the two

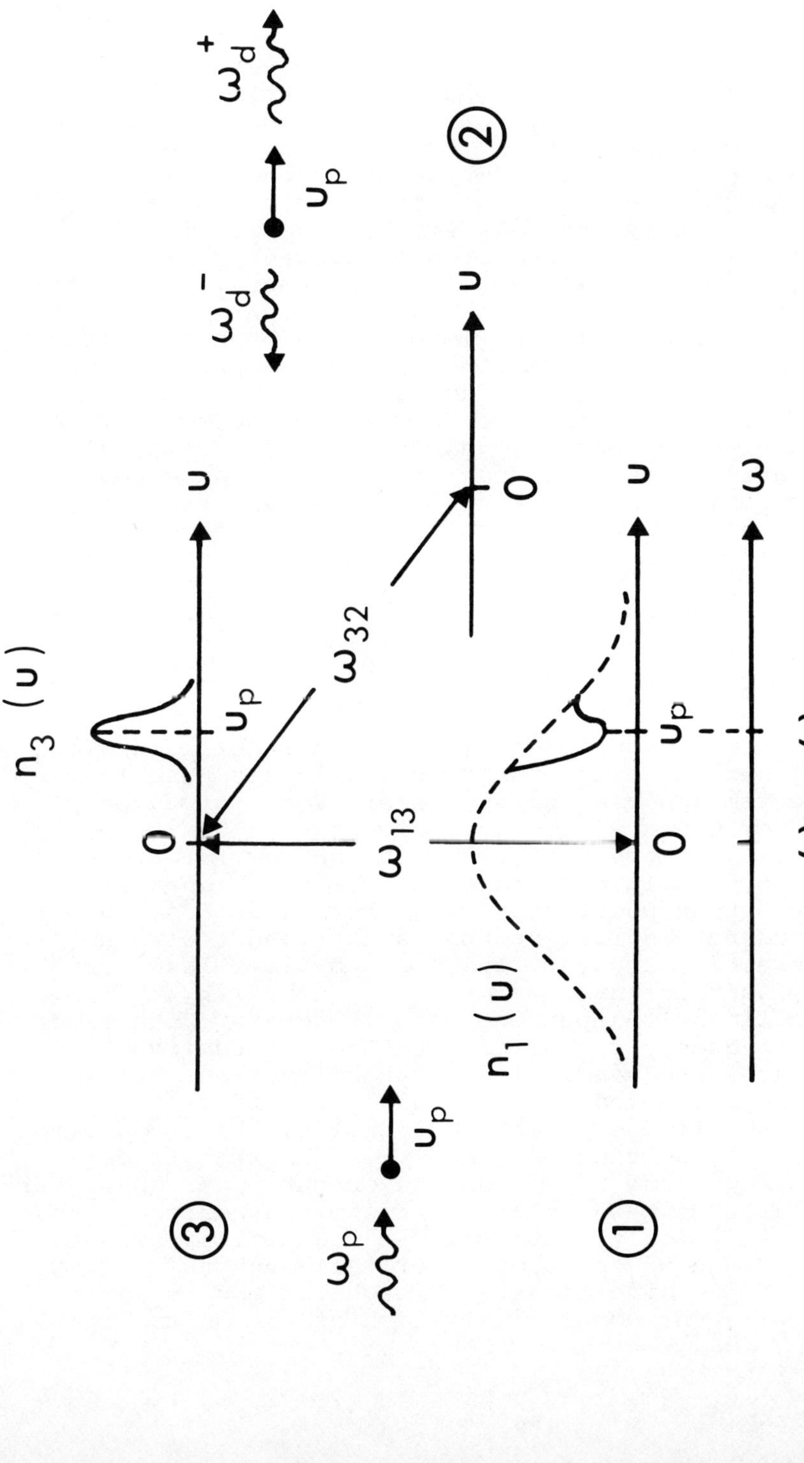

Figure 6. Sub-Doppler excitation–emission scheme.

photon transition. Inserting the expressions for ω_p and $\omega_d^{\pm}$ we obtain $\Delta\omega_{12} = \omega_{12} - k_p u_p \cdot (1 \mp k_d/k_p)$. As k_d/k_p is close to 1, it can immediately be seen, that for a variation of u_p, equivalent to considering all velocity components that may contribute to the process, the forward resonance frequencies ω_d^+ practically do not change ($1 - k_d/k_p \approx 0$; compensation of the Doppler effect), whereas the backward resonance frequencies ω_d^- are greatly shifted due to the Doppler contribution ($\approx 2k_p u_p$). The total amplification profile, which is given by averaging over the contributions of all velocity components, therefore is narrow with a high peak in the forward direction and broad with a flat maximum in the backward direction. As a consequence, the two photon process introduces a forward-backward amplification asymmetry which strongly favors the forward frequency ω_d^+. In a ring cavity, unidirectional and single frequency oscillation around ω_d^+ can therefore be expected, without using unidirectional devices and frequency selective elements to restrict the mode structure. This is fully confirmed by experimental observations. Backward oscillation so far could only be observed for multimode pumping, where the forward-backward amplification asymmetry is small, and under suitable conditions for high intensity off resonant single mode pumping (25).

Figure 7 shows a schematic of a stabilized dimer ring laser which is pumped by a slightly tunable single frequency argon ion laser. For a given pump frequency ω_p, a mode of the ring cavity has to be stabilized at the maximum of the narrow forward amplification profile, in order to obtain a high and stable output power and avoid thermal drifts. This is achieved by controlling the cavity length with a piezoelectric crystal. The optimum dimer laser frequency is then given by $\omega_d^+ = (\omega_p - \omega_{13})\, k_d/k_p + \omega_{32}$ and is linearly dependent on ω_p. Therefore the dimer laser frequency can simply be tuned by tuning the pump laser frequency, if the stabilization circuit is adequately coupled to the tuning circuit of the pump laser. By tilting an internal etalon, the argon pump laser could be tuned by about $\pm$ 2 GHz. The linear change $\Delta\omega_d^+ = \Delta\omega_p k_d/k_p$ and the output power characteristic are shown in Figure 8. In this experiment, the tuning range of the dimer laser was limited by the tuning range of the pump laser. Although the tuning range of the dimer laser of roughly 4 GHz is more than twice the Doppler halfwidth of the molecular

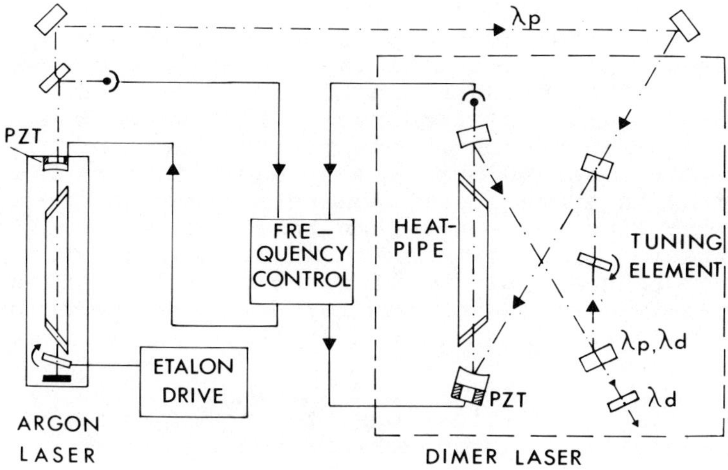

Figure 7. Tunable dimer ring laser scheme.

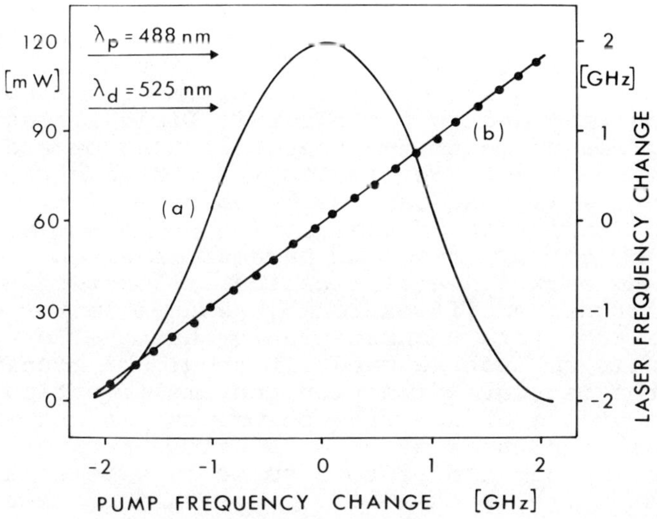

Figure 8. a: Na_2 laser tuning curve (——). Pump power, 1.5 W; length of vapor zone, 10 cm; and 780 K. b: Laser frequency change versus pump frequency change (●).

transition, population density effects still play an important role as can be seen from absorption experiments, and it would therefore be interesting to tune the dimer laser further away from resonance into the pure Raman region. Such experiments, where the dimer laser is pumped by a tunable dye laser, are presently underway. The broad tuning capabilities and the forward-backward amplification asymmetry are the most important Raman aspects. Recently, Brillet and Man-Pichot (26) observed pure Raman laser lines in Li_2 excited by an argon laser, however, attempts to tune these lines have not been performed so far. The demonstration of pure Raman tunability over extended frequency ranges would be of great practical and also principal importance for the further development and understanding of these molecular three level laser systems.

Ring laser investigations, tuning experiments and measurements of forward-backward amplification profiles have been performed for sodium dimer B → X transitions. Detailed results and a comparison with theory are given in Reference 16. Small signal forward profiles for different pump intensities and zero detuning of the pump laser are shown in Figure 9. For low pump intensities a narrow profile is obtained, which splits at higher pump intensities due to the dynamic Stark effect (27,28). The splitting is proportional to the dipole transition moment for the pump transition and to the electric field strength. Splitting measurements can therefore also be used to determine transition moments. In the backward direction a splitting can not be observed (Figure 10) due to the influence of the uncompensated Doppler effect. For the experimental parameters given here, the maximum backward amplification is lower than the maximum forward amplification by about a factor of 20. It is therefore understandable that the forward direction is so much favored. It should be mentioned that in the backward direction the amplification contribution due to the two photon process is negligibly small (16). The observation of a strong backward amplification can therefore also be considered as a proof that a real population inversion is achieved and that the condition $S_2 > A_{32}$ is fulfilled. The splitting of the forward profile would lead to undesired frequency jumps and deviations from the exact resonance frequency and therefore limit the use of

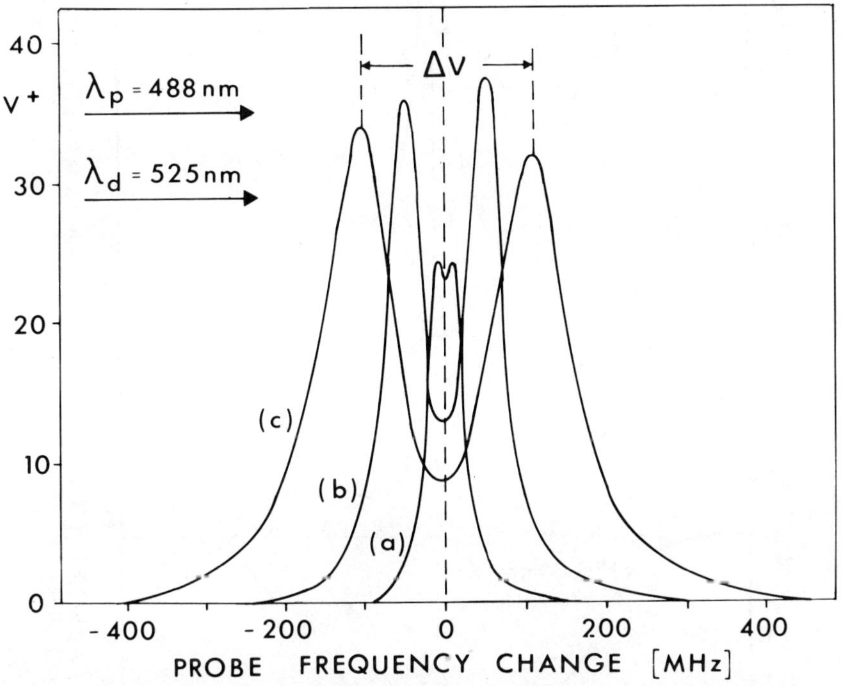

Figure 9. Forward amplification profiles for different pump intensities at vapor zone of 10 cm and 780 K. $\lambda_p = 488$ nm; $\lambda_d = 525$ nm. Curve a, 85 W cm^{-2}; curve b, 430 W cm^{-2}; curve c, 2150 W cm^{-2}.

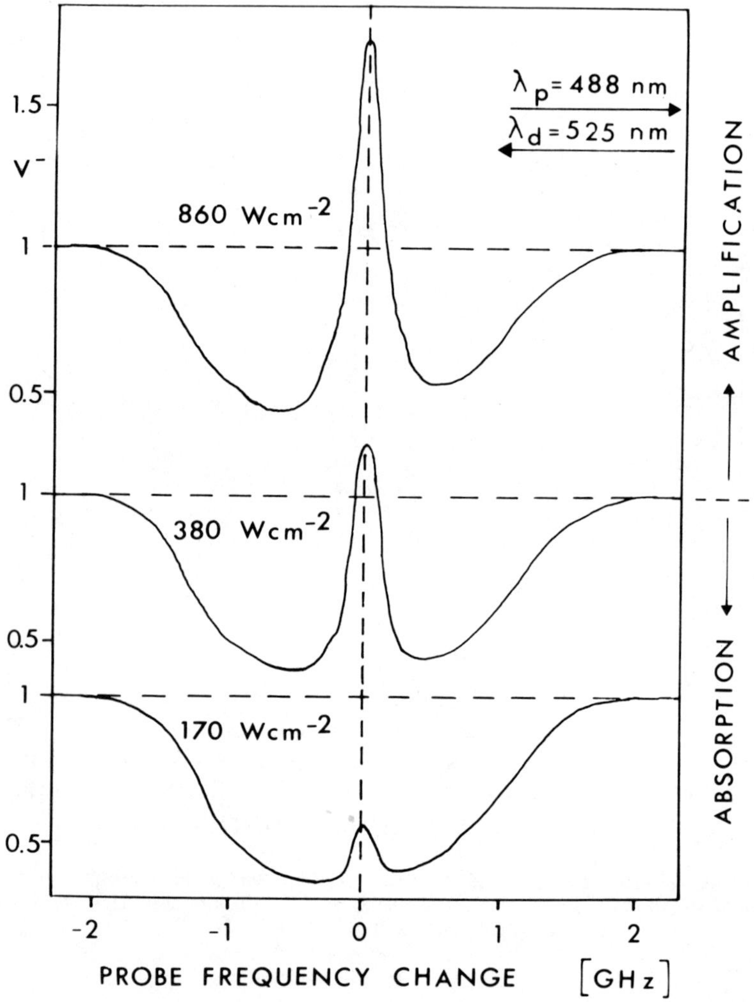

Figure 10. Backward amplification profiles for different pump intensities at vapor zone of 10 cm and $\lambda_p = 488$ nm; and $\lambda_d = 525$ nm; 780 K.

the dimer ring laser as a slightly tunable single frequency system. Fortunately, the splitting is dependent on the intensity of the dimer laser field and vanishes at relatively low dimer laser intensities (Figure 11). In a normal dimer laser system, operating well above threshold, the splitting will therefore not be present and by tuning the pump laser a smooth tuning curve without frequency jumps as shown in Figure 8 will be obtained. The disappearence of the splitting is a further manifestation of the influence of the two photon processes. For multimode excitation the situation is quite different (Figure 12). Each mode of the pump laser gives forward-backward amplification profiles which are frequency shifted according to the mode frequencies. The superposition of all these profiles yields a broad profile with some residual structure in the forward direction (Figure 12a), corresponding to the pump laser modes. However, the envelope and the maxima of the forward-backward profiles are almost identical. Due to the broad profiles the dimer laser will also oscillate in several modes. Since the first succesful operation of a cw dimer laser a lot of data concerning operation conditions, gain, output power and laser lines has been evaluated. Some data for alkali dimer lasers are summarized in Table I, but as this field is still developing, improvements and new results are expected in the future, especially for the less intensively investigated Li_2 and K_2 lasers.

Considering briefly applications, these dimer lasers are obviously ideal suited for special spectroscopic and kinetic investigations on the dimer molecule itself, such as measurements of transition moments, relaxation rates and precise determination of molecular constants. A more detailed discussion of these special applications is given in References 29,30 and in contributions by J.L. Kinsey and R.W. Field presented at this conference. Other possible applications are closely related to the specific features of these dimer lasers, such as low threshold pump intensities, high gain, good conversion efficiency, simple single mode operation and multiline emission with lines throughout the visible and near infrared spectral range. Furthermore these lasers are easy to operate and the laser material is stable and can be handled in long lived closed cells or heatpipes. These systems might therefore be of interest in all cases where fixed laser lines are needed, such as for frequency standards, as a reference system for technical measurements or just as a simple and inexpensive conversion system. For all applications, where only fixed laser

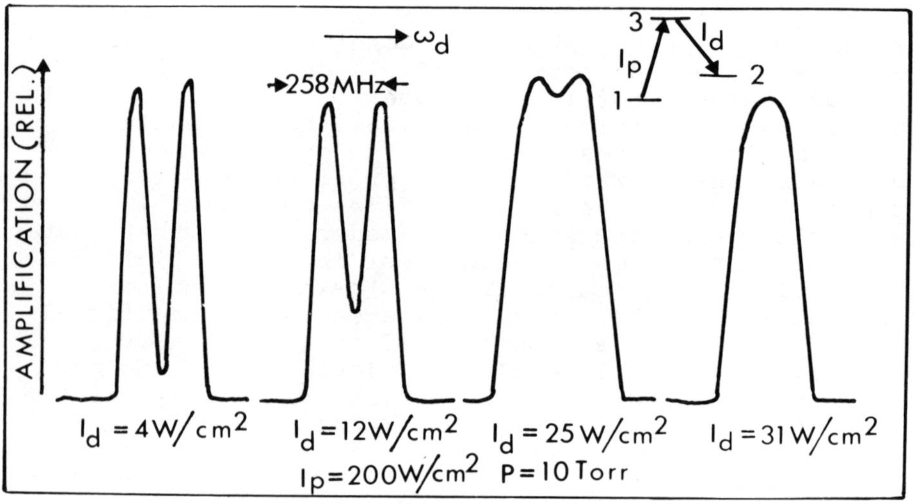

Figure 11. Relative forward amplification profiles for different dimer laser intensities I_d and fixed pump intensity I_p.

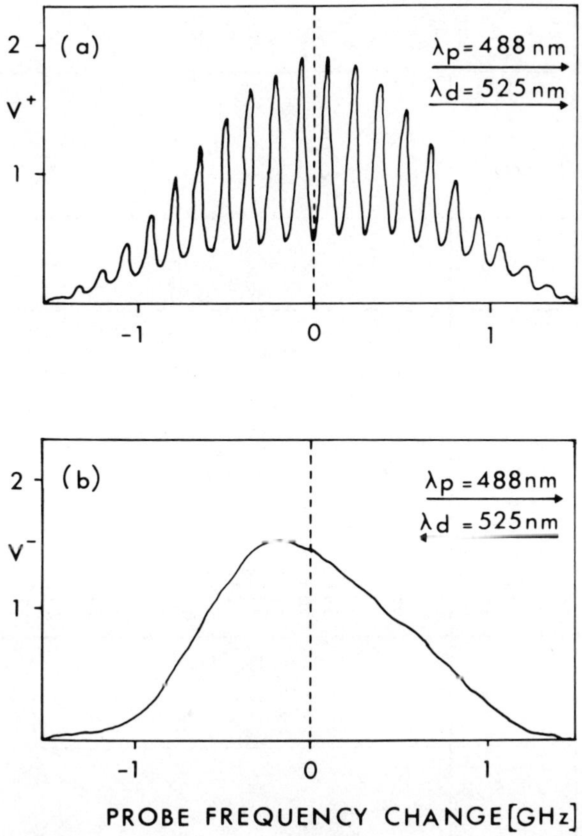

Figure 12. Amplification profiles for multimode pumping at vapor zone of 10 cm and 780 K. Key: a, forward profile; b, backward profile; $\lambda_p = 488$ *nm; and* $\lambda_d = 525$ *nm.*

Table I. Data of CW Alkali Dimer Lasers

	PUMP			LASER		GAIN b)	THRESHOLD c)	OPERATION
	TRANSITION	LINE [nm]	POWER [W]	LINES [nm] a)	POWER [mW]	[cm^{-1}]	[mW]	COND. c)/ REMARK:
Na$_2$	$X^1\Sigma_g^+ \rightarrow A^1\Sigma_u^+$ multimode Kr^+	647.1	1	793.0 799.3 800.4 801.8 802.7 803.0 806.5 806.9 807.6 808.2 810.1 813.6 814.8	20		< 200	temperature T ∼ 700 - 800 K
								length of vapor zone L ∼ 10 - 20 cm
	$X^1\Sigma_g^+ \rightarrow B^1\Pi_u$ multimode Ar^+	457.9	1	556.4 558.7 559.7 561.9	3		< 15	
		465.8	1	545.4 545.9 548.6 549.2 560.2 562.3 564.3	8		< 30	linear resonator
		472.7	1	530.8 532.9 533.8 534.4 535.9 536.5 537.6 538.1 539.5 540.0 540.7 541.4 543.0 544.0 544.8 545.6 546.0 546.5 547.5 549.2 550.5 553.6	70	∼0.05	< 20	most multimode power data not optimized; buffer gas pressures up to 200 mbar possible
		476.5	2.5	531.7 533.9 537.6 542.1	25		< 70	
		488.0	2.5	525.1 528.9 532.6 534.0 534.8 535.7 536.3 537.5 541.1	20		< 100	

Table I. Data of CW Alkali Dimer Lasers

Continued

Na_2	$X^1\Sigma_g^+ \rightarrow B^1\Pi_u$ singlemode Ar^+	488.0 (3.43) → (6.43)	2.0	510.0 /513.7 525.1 /528.9 532.6 /536.3	30 /30 200 /150 190 /130	~0.2	< 20	ring resonator
		476.5	1.2	533.9 /537.6	30 /20			lowest threshold < 1 mW
		472.7	1.0	530.8 /533.8 534.4 /538.1	20 /50 40 /30			optimum output coupling ~ 30%
Li_2	$X^1\Sigma_g^+ \rightarrow A^1\Sigma_u^+$ multimode Kr^+	647.1	1	791.3 826.4 845.4 884.6 886.5 896.3 903.7 904.7 906.8 912.2	30	-	< 200	T ~ 1300 K L ~ 5 - 10 cm system not optimized
		457.9	1	550.2 558.4	4		< 50	
	$X^1\Sigma_g^+ \rightarrow B^1\Pi_u$ multimode Ar^+	472.9	1	550.0 552.2 558.2 560.4	3		< 100	
		476.5	2	533.6 542.3 551.0	15	~0.05	< 80	
		488.0	2	531.9 559.8	1		< 200	
K_2	$X^1\Sigma_g^+ \rightarrow B^1\Pi_u$ multimode Kr^+	647.1	1	696.1 697.5 698.5 699.5 703.7 708.9	3	< 10^{-3}	< 200	T ~ 600 K L ~ 20 cm system not optimized

a) Wavelength accuracy ± 0.2 nm; b) determined from maximum output coupling; c) typical values lines are needed, these dimer lasers are without doubt as well suited as cw dye lasers, which operate in almost the same spectral region, which use the same pump lasers but which have to be stabilized at the desired frequency and are more complicated to handle. It can therefore be expected that in the future these dimer lasers will be more intensively used.

Regarding future laser developments which make use of alkali metal vapors, several interesting possibilities exist, which all allow for tunable lasers. The Raman aspects of the dimer systems have already been pointed out, where in this case cw operation is of special interest. In the atomic vapors, broadly tunable pulsed Raman conversion has recently been demonstrated (31,32). Another interesting possibility makes use of the atomic alkali vapors to realize a tunable two photon laser (33,34). More conventional methods to achieve new tunable laser systems include bound-free molecular transitions with more or less broad continuous emission bands. Such bands are obtained from excimer molecules such as NaXe, which have no stable ground state, or by considering bound-free transitions in the triplet manifold of stable homonuclear or heteronuclear alkali dimers (35-37). Alkali noble gas excimer systems have long been regarded as potential powerful laser systems (38), but up to now no laser oscillation could be achieved. Problems that exist for the direct production of excited alkali noble gas molecules may however be circumvented by recently developed techniques which start from alkali-halide noble gas mixtures and use photodissociation of the alkali-halides to produce excited alkali atoms and alkali noble gas molecules. For a further discussion see References 9, 39. The use of excimer-like transitions in stable molecules is an extremely interesting alternative as it would allow, in principle, operation at low pressures and thus avoid problems associated with high pressure and high temperature in alkali noble gas systems. To achieve the necessary population inversion between bound-free transitions in the triplet system of the alkali dimers, direct optical excitation from the singlet ground state is not possible. Therefore discharge excitation or optical pumping connected with collisional energy transfer have to be considered, and indeed many diffuse bands have first been discovered in discharges (40-41). In the sodium dimer system, for example, a violet diffuse band that originates from certain not yet identified C-levels and

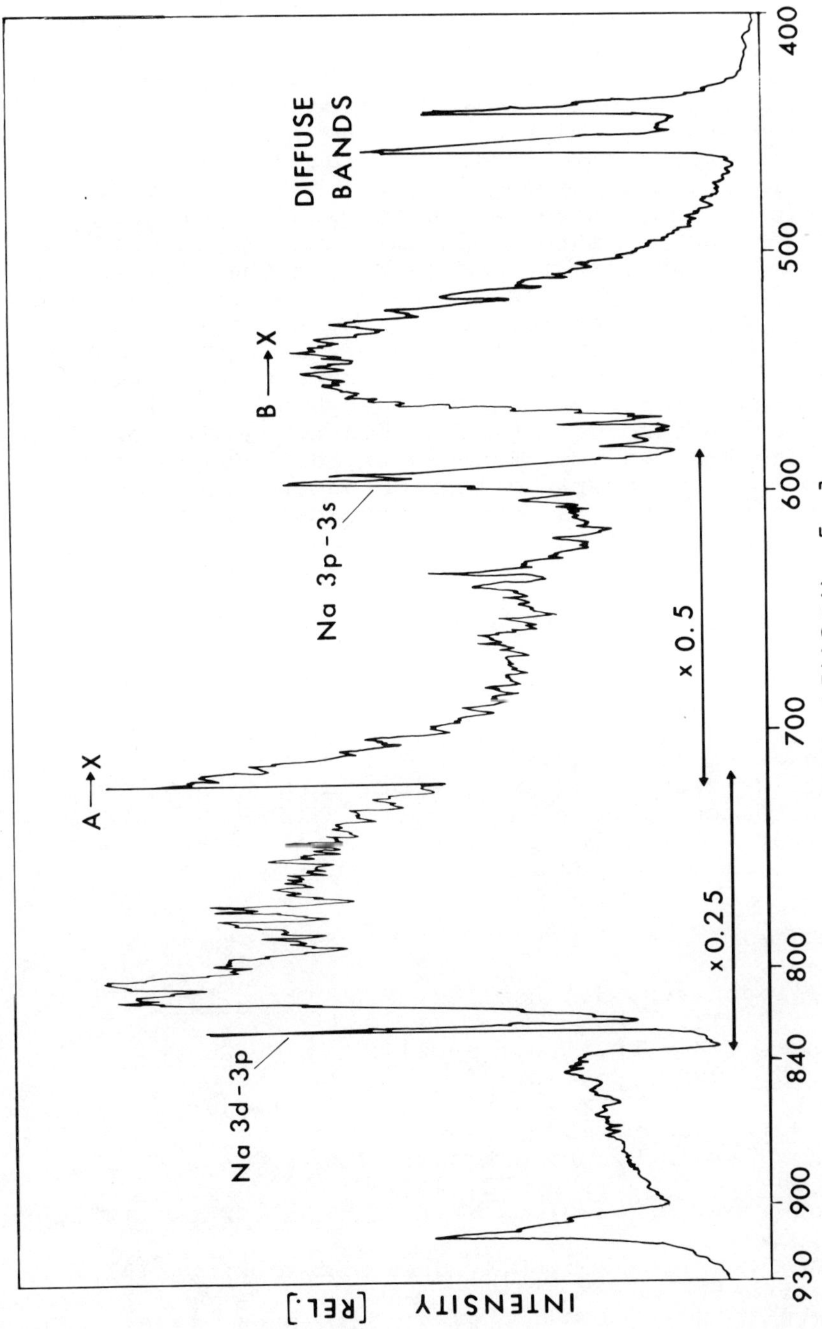

Figure 13. Na–Na_2 fluorescence spectrum excited by broadband excimer laser radiation at 351 nm, 780 K, and Ar buffer gas pressure of 12 mbar.

ends in the lowest repulsive $^3\Sigma_u^+$ state (42) can be obtained upon excitation of singlet C -levels either with UV radiation or in a two step process with visible radiation (43-44). A typical UV excited spectrum (351 nm excimer laser) from Na-Na_2 vapor is shown in Figure 13. Besides the violet diffuse bands around 438 nm and 453 nm, emission from the molecular A and B band and from atomic lines is obtained. In addition we see some unidentified structure in the range of 840 nm to 930 nm, which increases with temperature and buffer gas pressure. This emission may result from bound-bound transitions between higher lying electronic states or even correspond to the bound-free $^3\Sigma_g^+ \rightarrow {}^3\Sigma_u^+$ transitions recently discussed for laser purposes by Konowalow (45). However, at present, these are speculations and more details have to be worked out. Similar violet or blue diffuse bands and infrared emissions can also be obtained upon UV excitation of Li-Li_2 and K-K_2 vapors. Besides the discussed optically pumped cw laser oscillation on the A → X and B → X bands of Li_2, Na_2 and K_2, optically pumped pulsed laser oscillation in the infrared between higher lying bound-bound electronic states in Na_2 has recently been observed (46). However, up to now laser oscillation on a bound-free transition of a stable molecule has never been observed. Going back to laser history, one of the first proposed tunable laser systems was the molecular hydrogen system on the bound-free $^3\Sigma_g^+ \rightarrow {}^3\Sigma_u^+$ transition (47). The alkali dimers can be considered as visible molecular hydrogen and the study of corresponding transitions is also of principal importance. Thus alkali vapors, although being investigated for quite some time, still offer challenging problems for spectroscopists and laser physicists.

Literature Cited

1. Tellinghuisen, J.; Hoeller, M.B. Chem.Phys. 1980, 50, 301-11.
2. Verma, K.K.; Vu, T.H.; Stwalley, W.C. J. Mol. Spectr. 1981, 85, 131-49.
3. Zemke, W.T.; Verma, K.K.; Vu, T.; Stwalley, W.C. J. Mol. Spectr. 1981, 85, 150-76.
4. Konowalow, D.D.; Rosenkrantz, M.E.; Olson, M.L. J. Chem. Phys. 1980, 72, 2612-15.
5. Konowalow, D.D.; Olson, M.L. J. Chem. Phys. 1977, 67, 590 - 2.
6. Wine, P.H.; Melton, L.A. Chem. Phys. Lett. 1977, 45, 509-12
7. Sakurai, K.; Broida, H.P. J. Chem Phys. 1976, 65, 1138-45.

8. Vidal, C.R. J. Chem. Phys. 1980, 72, 1864-74.
9. Ehrlich, D.J.; Osgood, Jr. R.M. IEEE J. QE 1980, 16, 257-67.
10. Maya, J. IEEE J. QE 1979, 15, 579-94.
11. Henesian, M.A.; Herbst, R.L.; Byer, R.L. J. Appl. Phys. 1976, 47, 1515-18.
12. Leone, S.R.; Kosnik, K.G. Appl.Phys.Lett. 1977, 30, 346-8.
13. Wellegehausen, B.; Shahdin, S.; Friede, D.; Welling, H. Appl. Phys. 1977, 13, 97-9.
14. Welling, H.; Wellegehausen, B. "Laser Spectroscopy III", Springer Series in Optical Sciences 7, New York, 1977, p. 365 - 9.
15. Koffend, J.B.; Field, R.W. J.Appl. Phys. 1977, 48, 4468 - 72.
16. Wellegehausen, B. IEEE J.QE 1979, 15, 1108-30.
17. Wellegehausen, B.; Stephan, K.H.; Friede, D.; Welling, H. Opt. Commun. 1977, 23, 157-61.
18. Wellegehausen, B.; Friede, D.; Steger, G. Opt. Commun. 1978, 26, 391-5.
19. West, W.P.; Broida, H.P. Chem. Phys. Lett. 1978, 56, 283-5.
20. Hodgson, R.T. Phys. Rev. Lett. 1970, 25, 494-7
21. Godard, B. IEEE J.QE 1974, 10, 147-53.
22. Hänsch, T.; Toschek, P. Z.Physik, 1970, 236,213-44.
23. Beterov, I.M.; Chebotayev, V.P. "Progress in Quantum Electronics", Pergamon Press, New York, 1974, p. 1 - 106.
24. Skribanowitz, N.; Feld, M.S.; Francke, R.E.; Kelly, M.J.; Javan, A. Appl. Phys. Lett. 1971, 19, 161-4.
25. Luhs, W.; Wellegehausen, B. to be published.
26. Man-Pichot, C.N.; Brillet, A. IEEE J.QE 1980, 16, 1103 - 8.
27. Delsart, C.; Keller, J.C. J. de Phys. 1978, 39, 350-60.
28. Feneuille, S.; Schweighofer, M.G. J. de Phys. 1975, 36, 781 - 86.
29. Koffend, J.B.; Field, R.W.; Guyer, D.R.; Leone,S. R. "Laser Spectroscopy III" Springer Series in Optical Sciences 7, New York 1977, p. 382 - 93.
30. Koffend, J.B.; Goldstein, S.; Bacis, R.; Field,R. W.; Ezekiel, S. Phys. Rev. Lett. 1978, 41, 1040-44
31. Cotter, D.; Hanna, D.C.; Tuttlebee, W.H.W.; Yaratich, M.A. Opt. Commun. 1977, 22, 190-4
32. Wyatt, R.; Cotter, D. Appl. Phys. 1980, 21, 199-204.
33. Nikolaus, B.; Daozhong, Z.; Toschek, P.E. "Proc. German Phys. Soc." Münster 1981, 420 - 1.
34. Frölich, D.; Krökel, D. to be published.

35. Breford, E.J.; Engelke, F. Chem. Phys. Lett. 1978, 53, 282 - 7.
36. Vasilakis, A.; Bhaskar, N.D.; Happer, W. J. Chem. Phys. 1980, 73, 1490 - 3.
37. Kato, H.; Noda, C. J. Chem. Phys. 1980, 73, 4940-5
38. Gallagher, A. "Excimer Laser", Topics in Applied Physics 30, Springer Verlag, New York 1979, p. 135 - 174.
39. Shahdin, S.; Ludewigt, K.; Wellegehausen, B. IEEE J. QE 1981, 17, 1276 - 80.
40. Rebbeck, M.M.; Vaughan, J.M. J. Phys. B: Atom Mol. Phys. 1971, 4, 258 - 60.
41. Woerdman, J.P.; de Groot, J.J. to be published Chem Phys. Lett.
42. Woerdman, J.P. Opt. Commun. 1978, 26, 216 - 8.
43. Kopystynska, A.; Kowalczyk, P. Opt. Commun. 1979, 28, 78 - 80.
44. Allegrini, M.; Moi, L. Opt. Commun. 1980, 32, 91 - 5.
45. Konowalow, D.D.; Julienne, P.S. J. Chem. Phys. 1980, 72, 5815 - 18.
46. Müller, W.; Hertel, I.V. Appl. Phys. 1981, 24, 33 - 8.
47. Houtermans, F.G. Compte rendu Society Suisse de Phys. 1960, 33, 597 - 8.

RECEIVED August 26, 1981.

Physical Studies of Optically Pumped Dimer Lasers

C. N. MAN–PICHOT and A. BRILLET

Laboratoire de l'Horloge Atomique, Equipe de Recherche du CNRS, Associée à l'Université Paris-Sud, Bât. 221 - Université Paris-Sud, 91405-Orsay, France

We report on saturated absorption experiments in Na_2, realized with a tunable and stabilized argon laser. These experiments provide both spectroscopic and physical results, which help in understanding the behavior of optically pumped alkali dimer lasers. We briefly describe a new double resonance experiment which enables us to study the gain lineshapes of the dimer laser and to demonstrate the backward-forward gain competition.

Since 1977, optical pumping of dimer molecules such as I_2, Bi_2 and Te_2 and especially alkali metal dimers Li_2, Na_2 and K_2, has provided us with a new family of visible and near infrared CW lasers (1). The usual pumping scheme is the following: dimer molecules are pumped by an argon or krypton laser from a low lying rovibrational level of the X ground electronic state (0) to one of the A or B state levels (1), which may result in a population inversion between this A or B level and high lying X state levels (2).

These systems are interesting from different viewpoints:

- for the spectroscopist, because they allow a very precise determination of the transition wavelengths, resulting in improved molecular constants, and also because it becomes possible to study higher electronic levels, connected with level 1 or 2 (2).
- for the laser scientist, dimer lasers are 3 level systems interacting with two nearly resonant laser fields. They provide an excellent opportunity for the experimental study of a number of related effects, such as the dynamic Stark effect, the Raman effect, multiphoton effects resulting in gain without population inversion, gain anisotropy, etc. (1).
- these lasers are interesting in themselves because of their large number of lines, high efficiency and ease of single frequency operation. In particular,

0097-6156/82/0179-0487$05.00/0

they seem to be well suited for applications such as optical frequency (or wavelength) standards since the active medium is a quiet vapor, which should result in a very narrow emission linewidth (3).

In this paper, we report on two experiments which enabled us to improve our understanding of Na_2 and $^7Li_2$ lasers. In particular, we obtained assignment of the levels involved in the laser cycles, and evaluation of their lifetimes, relaxation rates and mechanisms, and transition dipole moments. We will also give preliminary results on the experimental study of gain lineshapes, which will ultimately be compared with a theoretical model.

Assignment of Na_2 and Li_2 Laser and Pump Lines

In a first step to understand the physics of these dimer lasers, we measured precisely (10^{-2} Å or better) and simultaneously the dimer laser and the argon pump laser wavelengths, using our lambdameter (4). This led to the unambiguous assignment of most laser lines, many of them being observed for the first time (5). In particular, when pumping Li_2 with the 4966 Å argon laser line, we could observe a peculiar laser oscillation at 5049 and 5133 Å, which can be understood only as pure Raman lines, for the following reasons: i) the energy difference (324 cm^{-1}) between the pump line and the 5049 line is too small to allow a population inversion between levels (1) and (2); ii) there was no measurable absorption of the pump laser; iii) the most likely assignment, taking into account the Franck-Condon factors, is Q(10)3-3 for the pump line and Q(10)3-4 and Q(10)3-5 for the emission lines. This corresponds to a detuning of 1 cm^{-1} (or 6 Doppler widths) for both the pump and laser lines.

This Raman gain is interesting, since it extends the tuning range of optically pumped lasers, provided the pump laser itself is tunable and powerful enough.

The identification work has been completed recently by saturated absorption experiments. All the coincidences between the argon laser lines and Na_2 and Li_2 absorption lines have been measured with an accuracy of 0.002 cm^{-1} and identified in terms of the $B^1\Pi_u$-$X^1\Sigma_g^+$ systems using the set of molecular constants given by Kusch and Hessel (6) and Hessel and Vidal (7), for Na_2 and $^7Li_2$, respectively. The results are shown in Table I. We have included some transitions which were not measured in saturated absorption because their line center is out of the argon laser tuning range, but which we could nevertheless localize and identify by observing their (broader) linear absorption and because their pumping results in dimer laser emission.

Saturated Absorption Experiments

The saturated absorption experiments were used not only for the identification work described above, but also for the

Table I. Coincidences between transitions in the $B^1\Pi_u$-$X^1\Sigma_g^+$ Systems of Na_2 and $^7Li_2$ and argon laser lines. Dimer laser emission has been observed from all the underlined transitions. The accuracy is 5 x 10^{-4} Å for all resolved lines. Δ is the difference in Å between measured vacuum wavelength and the calculated wavelength from the constants of References 6 and 7.

Na_2			$^7Li_2$		
Measurement	Transition	Δ[a]	Measurement	Transition	Δ[a]
4580.6468	Q(43)28-7	-.0008	4580.6456	R(8)9-2	-.0042
			4580.6600	Q(5)9-2	-.0015
4728.1682	R(37)9-1	-.0012	~4728	R(31)8-3[c]	
4728.1820[b]	R(9)13-4	.0043	4728.2258	P(62)8-1	.0192
	Q(63)18-6	-.0090			
4728.1985[b]	Q(36)12-3	.0025			
	Q(45)17-6	.0017			
4766.1824	Q(98)9-0	.0046	4766.1753	R(44)8-3	.0040
4766.1871	Q(90)25-9	.0077	4766.1757	R(60)8-2	-.0040
4766.1920	P(48)17-7	.0070	4766.1957	Q(24)4-1	-.0048
4766.1965	P(13)10-3	.0095	4766.2115	P(63)14-3	-.0138
4766.2142	P(28)6-0	.0071	4766.2258	P(62)9-2	-.0022
4881.2054	R(41)10-6	.0018	4881.2052	Q(55)14-6	.0021
4881.2136	R(128)23-8	-.0026	4881.2082	Q(36)12-7	-.0176
4881.2173	Q(74)16-9	-.0010	4881.2452	R(30)2-1	-.0036
4881.2225	P(99)15-7	-.0004			
4881.2351	R(8)25-16	.0085			
4881.2433	Q(43)6-3	.0083			
4966.4517	P(29)8-7	.0088	4966.4576	Q(47)0-0	-.0210
4966.4557	P(44)7-6	.0095	4966.4775	R(78)12-3	-.0044
4966.4709	R(13)19-15	.0070	4966.4979	P(70)9-4	-.0023
4966.4857	R(84)10-7	.0070			
5146.7217	R(151)0-2	.0008			
5146.7293	P(65)9-12	.0075			
5146.7353	Q(49)11-4	.0070			
5146.7441	R(73)13-15	-.0057			
5146.7583	R(115)7-9	-.0027			
5146.7656	P(76)3-7	-.0028			

[a] Difference between the measured wavelength and the wavelength computed from (6).
[b] Unresolved transitions.
[c] Line center falls out of the argon tuning range.

evaluation of the linewidth, the relaxation rates and other processes involving the levels (1), and of the transition dipole moments of the (0)-(1) transitions. It is important to know all these parameters in order to construct a detailed theoretical model for the dimer laser.

Experimental Techniques. A block diagram of the experimental set-up used for saturated absorption experiments is shown in Figure 1. The argon laser is a commercial 4W tube in a home made cavity. This cavity is made of three Invar rods, decoupled from the tube in order to avoid vibrations. Line selection is made with a prism, and single frequency operation is obtained with a Michelson interferometer. The laser can be frequency locked to a stable Fabry-Perot resonator with a double servo-loop acting on a fast PZT for line narrowing and on a galvo-plate for wide tunability. This results in a linewidth of less than 10 KHz and a continuous tunability of 6 GHz.

The argon laser beam is split into a strong saturating beam and two weaker probe beams, counterpropogating in a ring shaped saturated absorption interferometer containing the Na or Li heat pipe. The saturating beam is amplitude-modulated at 10 KHz with an acousto-optic modulator, and the difference between the 2 probe beams is synchronously detected. Beam geometries are carefully matched and controlled, since the knowledge of these geometries is necessary in order to understand quantitatively the power broadening curves. Wavelength calibration is provided by the lambdameter.

Results. By varying the temperature and the pressure inside the heat-pipe, we could measure the broadening rates of the transitions due to M_2-M or M_2-A collisions (M = Na, Li), since we used argon as the buffer gas in both heat pipes (Figure 2). In the working range of 1-10 Torr, the effect of velocity changing collisions was predominant as we could verify both from the nonlinear pressure broadening rate above 10 Torr and from the shape of the collision broadened signals (Lorentzian curve above a Gaussian background).

The power broadening curves allow one to determine the saturation intensity:

$$I_S = \frac{\hbar^2 \gamma^2}{2c \, \varepsilon_0 \, \mu^2} \qquad (1)$$

(where γ is the natural linewidth) and then to evaluate the dipole moment μ of the transition. The determination of I_S is not exactly straightforward, since the laser beams are Gaussian beams and not plane waves. There is no simple analytical formula giving the saturated absorption lineshape at high intensity in a Gaussian beam, but one can obtain numerical results. Figure 3 shows a comparison of experimental results with the theoretical model of Salomon (8).

By extrapolating the broadening curves to zero power and

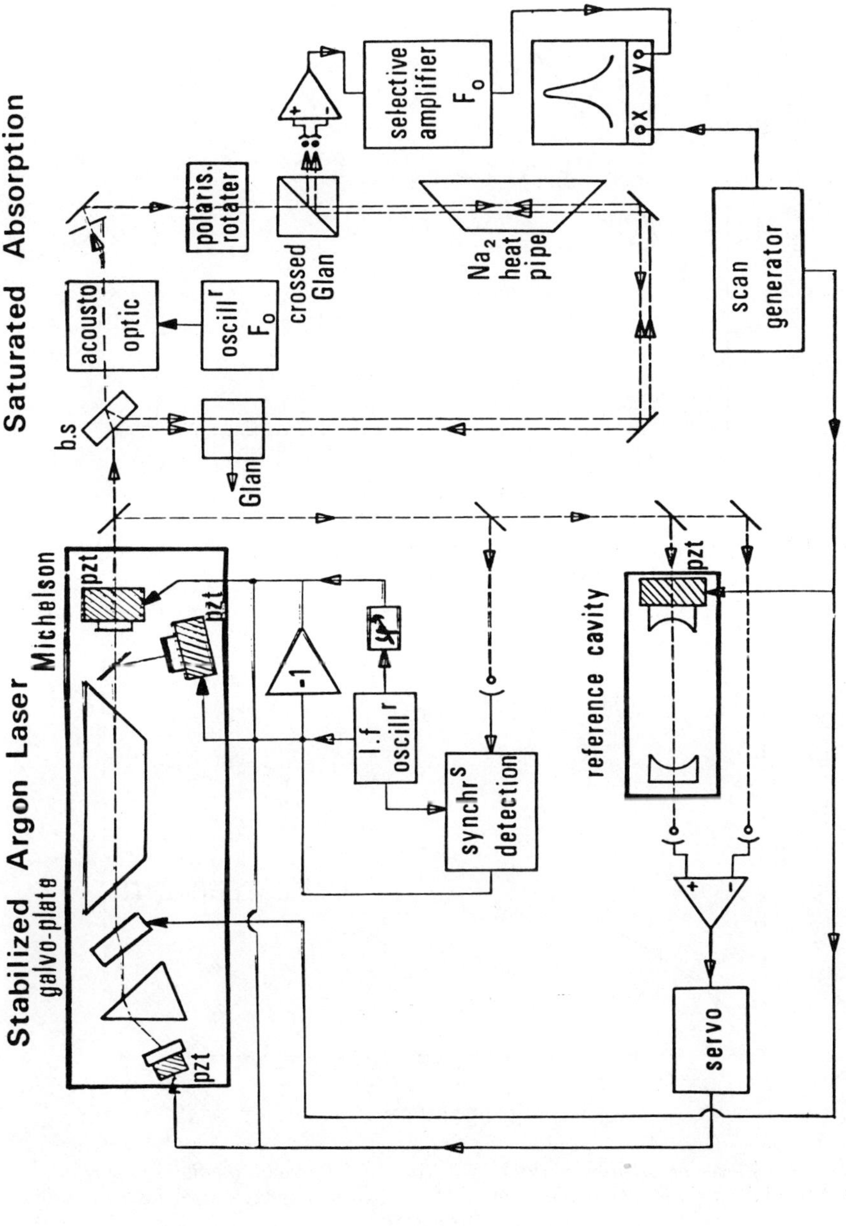

Figure 1. Block diagram of saturated absorption experiment.

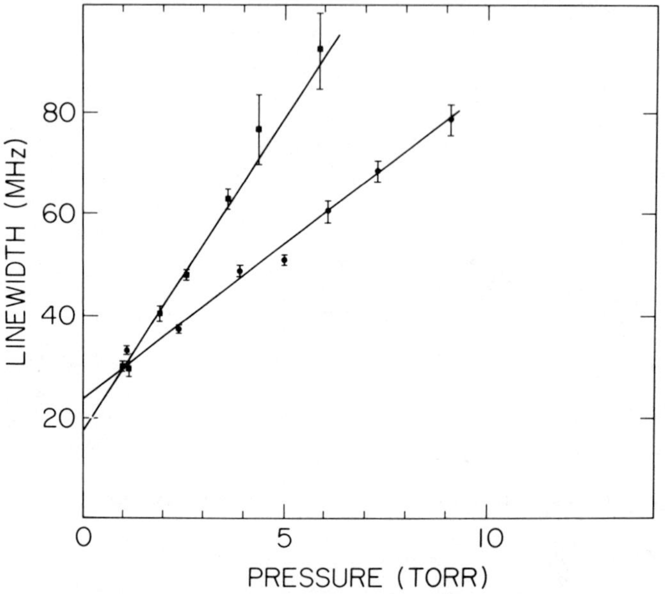

Figure 2. Pressure broadening rates of the P(99)15-7 Na_2 line. Key: (■) Na_2-Na collision; and (●) Na_2–Ar collision.

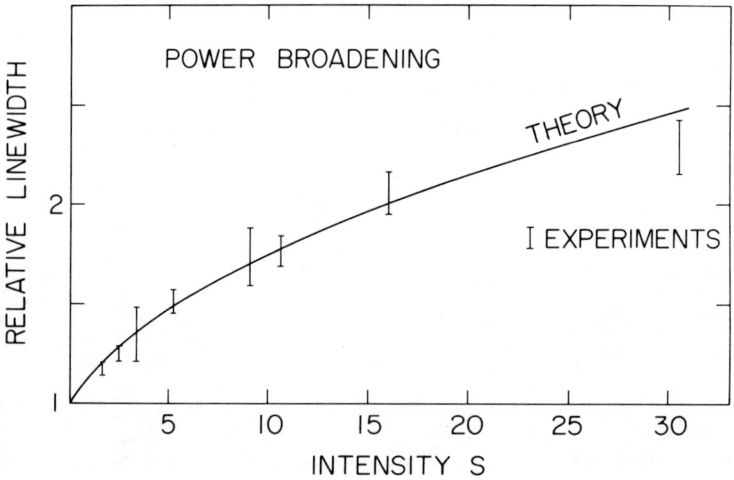

Figure 3. Power broadening of P(99)15-7 Na_2 line, fitted with the theory (8) (rate equation approximation, Gaussian beams). $S = 1$ corresponds to an Ar laser intensity of 54 mW/mm².

zero pressure, one can get a value of the natural linewidth of the transition which in this case should be the natural linewidth of level (1).

As an example, Table II shows a summary of the results obtained for the P(99)15-7 transition in Na_2 and the Q(47)0-0 transition in Li_2.

Table II. Line broadening parameters for the following $B^1\Pi_u$-$X^1\Sigma_g^+$ transitions: P(99) of the (15-7) band of Na_2; and Q(47) of the (0-0) band of $^7Li_2$.

Transition	M_2 =	Na_2 P(99)15-7	Li_2 Q(47)0-0
Collisional Braodening Rate (FWHM) (MHz/Torr)	M_2-M	25.0	18.6
	M_2-A	19.4	30
Natural Line-width, γ (MHz)		18	28
Transition Dipole Moment, μ (amp-second-meter)		$(9.1 \pm 4) \times 10^{-30}$	$(2.3 \pm 1) \times 10^{-29}$

The collisional broadening rates are in good agreement with those observed in Na_2 by Tsai (9). The natural linewidths are larger than what one would expect from the lifetime measurements of Demtröder et al. (10). The additional broadening (7 MHz for Na_2 and 20 MHz for Li_2) could result from an unresolved hyperfine structure or from the Zeeman effect produced by the oven heating wire.

Gain Lineshape Measurements

We are presently performing gain measurements in Na_2. The experimental set-up is close to the saturated absorption set-up, except that the probe beams are replaced by two beams from a tunable dye laser parallel to the argon beam, one of them propagating in the same direction (forward) and the other one in the opposite direction (backward). With this set-up, we can measure the gain lineshapes in both directions, when the dye laser is tuned across the (1)-(2) transitions. Figure 4 shows an interesting example of these lineshapes, in a case where the argon laser is close to resonance on the (0)-(1) transition. As expected, the gain lineshape is broader and weaker in the backward direction. Furthermore, we observe an additional feature: as a result of the competition between forward and backward waves, the backward gain profile shows a hole in the presence of forward

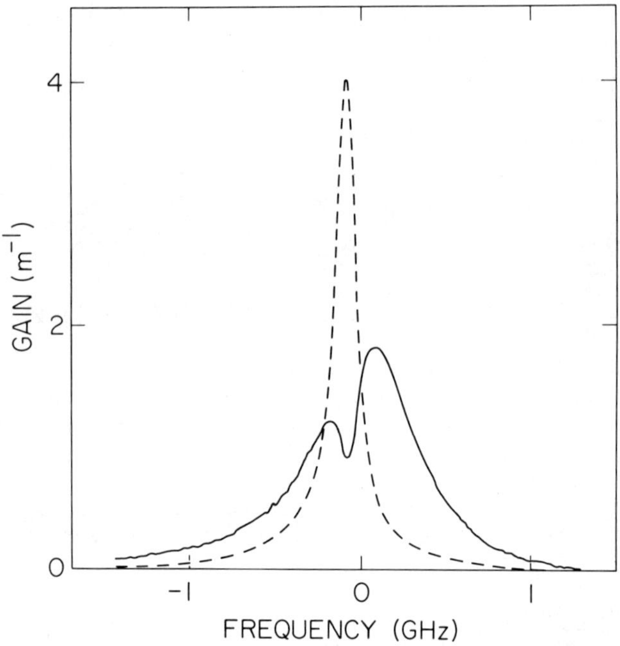

Figure 4. Gain lineshapes on P(43)10-20 in Na_2 probed by dye laser at 5389 Å showing the hole in backward gain resulting from forward–backward competition. Key: ——, backward; and – – –, forward.

amplification. This effect explains why one never could observe backward laser oscillation, even though the backward gain may be larger than the cavity losses.

Further experiments are underway with this set-up, in order to study the Raman gain lineshapes and their saturation behavior, and to compare these results with a theoretical model.

Literature Cited

1. Wellegehausen, B. IEEE J. of Quant. Elec. 1979, QE-15, 1108.
2. Carlson, N. W.; Kowalski, F. V.; Teets, R. E.; Schawlow, A. L. Opt. Commun. 1979, 29, 302.
3. Drullinger, R.; Brillet, A.; Dallarosa, J.; Hall, J. L. J. Opt. Soc. Am. 1978, 68, 635.
4. Cachenaut, J.; Man, C.; Cerez, P.; Brillet, A.; Stoeckel, F.; Jourdan, A.; Hartmann, F. Revue de Physique Appliquée, 1979, 14, 685.
5. Man-Pichot, C.; Brillet, A. IEEE J. of Quant. Elec. 1980, QE-16, 1103.
6. Kusch, P.; Hessel M. M. J. Chem. Phys. 1978, 68, 2591.
7. Hessel, M. M.; Vidal, C. R. J. Chem. Phys. 1979, 70, 4439.
8. Salomon, C. Thèse de 3e Cycle, 1978, Paris XIII.
9. Tsai, J. Thesis, 1980, Stanford.
10. Demtröder, W.; Stetzenbach, W.; Stock, M.; Witt, J. J. Mol. Spectrosc. 1976, 61, 382.

RECEIVED August 26, 1981.

33

Use of Lithium in Fusion Reactors

J. A. BLINK and O. H. KRIKORIAN
Lawrence Livermore National Laboratory, Livermore, CA 94550

N. J. HOFFMAN
Energy Technology Engineering Center, Canoga Park, CA

The use of lithium as a solid compound, a pure melt, or a molten alloy is required for tritium breeding in at least the first generation of fusion reactors. Three fusion reactor concepts are discussed with emphasis on material selection and material compatibility with lithium. Engineering details designed to safely handle molten lithium are described for one of the example concepts. Tritium recovery from the various breeding materials is reviewed. Finally, two aspects of the use of molten Li-Pb alloys are discussed: the solubility of hydrogen isotopes, and the influence of the alloy vapor on heavy ion beam propagation.

Fusion research offers the potential to use the nuclear binding energy of light elements such as hydrogen. Release of fusion energy requires heating the nuclei to about 10 keV (10^8 K) and confining the plasma for sufficient time for fusion to occur in a reasonable fraction of the nuclei. The quality of confinement (density - confinement time product) and the plasma temperature are the figures of merit for the fusion process. There are two approaches to achieve an acceptable quality of confinement. In magnetic confinement fusion (MCF), a plasma with density of about 10^{14} nuclei per cm^3 (10^{-5} times the density of air) must be confined for periods of many seconds. In inertial confinement fusion (ICF), the plasma confinement time is determined by the inertia of the heated nuclei ($\sim 3 \times 10^{-11}$ seconds); the corresponding density is about one thousand times the normal liquid density of the fuel or 12 times the density of lead.

* Work partially performed under the auspices of the U. S. Department of Energy by Lawrence Livermore National Laboratory under contract number W-7405-ENG-48.

0097-6156/82/0179-0497$11.50/0

The fusion reaction least difficult to initiate is the deuterium-tritium (DT) reaction which releases a 14.1 MeV neutron and a 3.5 MeV alpha particle. However, because neutrons activate the reactor structure, other fusion reactions have been considered. These reactions are either neutron free, or they produce fewer and less energetic neutrons. The required quality of confinement for these more desirable fusion reactions is much higher than for DT, and it is not yet clear if it will be achieved. Hence, fusion reactor designers have concentrated on the DT reaction for at least the first generation of fusion plants.

Deuterium is present in ordinary sea water at a concentration of one atom per 3300 water molecules, and it can be inexpensively extracted. Tritium is radioactive with a 12.3 y half life, and there is no abundant natural source. However, tritium can be produced in a fusion reactor by absorbing the neutrons in lithium (Figure 1). Natural Li is 92.5% ^{7}Li and 7.5% ^{6}Li. The ^{7}Li nucleus can absorb a fast (above 3 MeV) neutron to produce a tritium nucleus, an alpha particle, and a slower neutron. A moderated neutron can be absorbed by a ^{6}Li nucleus to produce a tritium and an alpha. Neutronic calculations indicate that a thick sphere of natural lithium could breed about 1.8 tritium atoms for each tritium atom burned in a fusion reaction (1). Structure and portions of the volume left open for fueling or driver beams reduce the 1.8 tritium breeding ratio. If the ratio falls below 1.0, it may be increased by addition of a neutron multiplier such as Be or Pb, and by isotopically enriching the Li in ^{6}Li.

Lithium and its compounds may be used in fusion reactors in either liquid or solid form. Liquid Li is an excellent coolant with low density and viscosity, and with high heat capacity and thermal conductivity (Table 1). Consequently, it is used in many designs as a combined breeding material and coolant. However, hot molten Li can react violently with water or air under certain conditions. Hence, either strict engineering design must preclude large scale Li - air or water reactions, or another form of Li must be used. Both approaches have been studied.

Molten Li-Pb solutions of several compositions have been suggested for fusion reactors. These include $Li_{99}Pb_1$ which is pure Li contaminated with 1 a/o Pb from ICF target debris, and $Pb_{83}Li_{17}$ which is a eutectic composition with chemical properties similar to Pb. A phase diagram of the Li-Pb system is shown in Figure 2.

Solid phase lithium compounds such as Li_2O, $LiAlO_2$, Li_2ZrO_3, or Li_4SiO_4 have been proposed for fusion reactors. An example is the STARFIRE Tokamak blanket conceptual design in which a packed bed of highly porous alpha-$LiAlO_2$ is used to absorb neutron energy and breed tritium (4) (Figure 3).

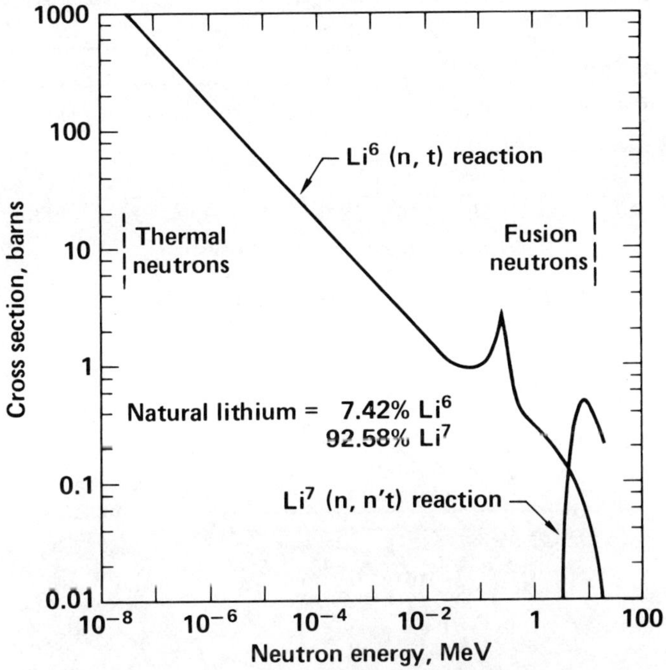

Figure 1. Tritium producing cross sections of Li.

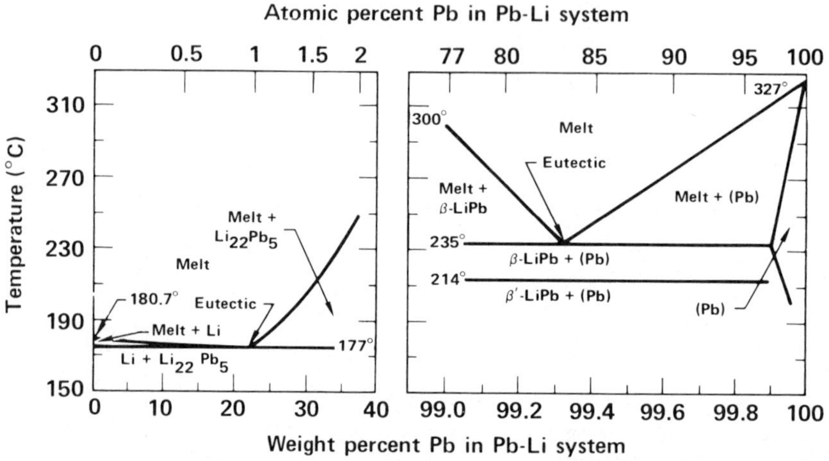

Figure 2. Phase diagram of the LiPb system.

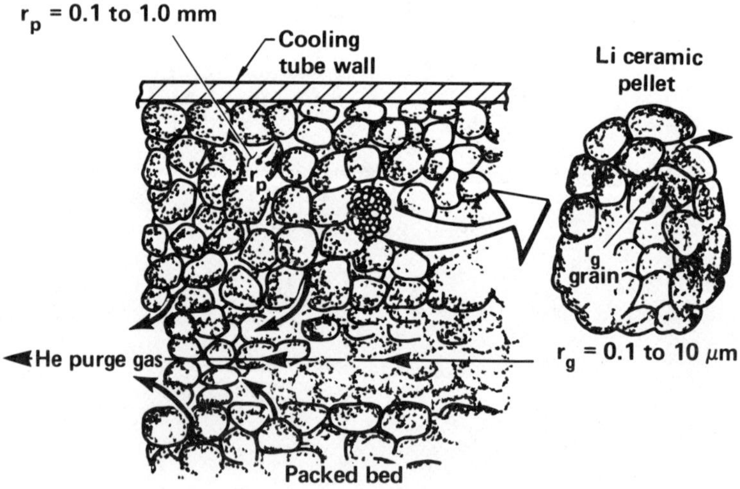

Figure 3. The STARFIRE tritium breeding blanket (4).

Table I. Comparison of Coolant Properties

	Li	Na	H_2O
Density, kg/m^3	480	830	840
Viscosity, MPa-s	0.36	0.24	0.12
Melting point, ^oC	181	98	0
Boiling point, ^oC	1347	883	100
Heat capacity, J/(kg K)	4170	1260	4570
Thermal conductivity, W/(m•K)	50	67	0.65

• Temperature dependent Li and Na properties evaluated at 500^oC (Ref.2).

• Temperature dependent water properties evaluated at 13.7 MPa (2000 psi) and 230^oC (440^oF) (from Ref. 3).

The tritium diffuses to the grain boundaries, migrates through the grain boundaries to the particle surface, and then percolates through the packed bed to low pressure helium purge channels. Heat is removed by pressurized water flowing in a stainless steel tube network in the packed bed.

The next sections describe three reactor studies with emphasis on the lithium-structure compatibility. HYLIFE is a liquid metal wall (LMW) ICF reactor considered here for electricity production. It has also been adapted to fissile fuel production (5). The Tandem Mirror Reactor (TMR) Cauldron Blanket Module is an MCF concept designed to produce hydrogen. The TMR Heat Pipe Blanket Module is designed to produce either hydrogen or electricity. All three studies emphasize materials compatibility with lithium. Tritium recovery techniques and two aspects of lead-lithium liquids are also discussed.

The HYLIFE Design (6,7)

Design Concept. The High Yield Lithium Injection Fusion Energy chamber design (HYLIFE) uses 175 jets of 0.2 m diameter to absorb the energy from 1800 MJ - 1.5 Hz ICF pulses

(Figure 4). The free falling jets are grouped into an annular 50% packing fraction array with a 0.5-m inner radius (Figure 5). The array has an effective thickness of 0.74 m, and it shields the 5-m-radius first structural wall from x-rays, debris and 95% of the neutron energy. Due to this shielding, the structural wall is predicted to survive for 30 years without replacement. The jets are arranged to allow two-sided illumination of the target by two-36 m^2 arrays of mirrors set back 60 m from the chamber center (Figure 6). The jet array is completely disassembled and reestablished between fusion pulses (Figure 7). Only ∿ 4% of the 1800-MJ fusion yield is converted to kinetic energy of liquid or gas; hence, the wall can be designed to survive the fatigue loading of the 10^9 pulses that occur in a 30-year lifetime.

Several other ICF reactor concepts use liquid metal walls (LMW). These include the Los Alamos National Laboratory (LANL) wetted wall concept, the Bechtel concept called EAGLE (which uses a lithium spray in the chamber), the Lawrence Livermore National Laboratory (LLNL) concept called JADE (which uses a fiber-metal structure to control liquid metal flow), and the German/University of Wisconsin concept called HIBALL (which uses carbide "socks" to control liquid metal flow).

HYLIFE Lithium Flow. The lithium jets are injected into the 8-m-high HYLIFE chamber with 9.5 m/s initial velocity (72.2 m^3/s flow rate). Additional coolant in the neutron reflector increases the flow to 86.2 m^3/s. The mixed-mean temperature rise in the lithium is 18 K, and the peak lithium temperature is 500°C. Eleven recirculation pumps, each with 7.8 m^3/sec (124,000 gpm) capacity, return the lithium to the top of the vessel. About 9.8 m^3/s is diverted from the flow loop to four Li-Na intermediate heat exchangers which in turn drive twelve steam generators. The plant gross electric power is 1236 MW_e; 135 MW is used to drive the 4.5 MJ - 5 % efficient laser, 95 MW is used elsewhere in the plant, and 1006 MW is the net power.

Lithium-Structure Compatibility. One of the critical chemistry problems of HYLIFE is the compatibility of structural alloys with the molten liquid of the jet array. Two candidate liquid metals are lithium and $Pb_{83}Li_{17}$. High-Z metal (such as lead from target debris) will enter the liquid metal and may affect the compatibility. The structural alloy selected in the HYLIFE study is 2¼ Cr-1 Mo, a ferritic steel. The carbides usually present in this steel are M_3C (cementite) and M_2C, where M is primarily Fe. Both of these carbides are unstable in lithium. M_3C is usually present as platelets within pearlite, the eutectoid structure in pearlitic steel. The most common microstructure for the 2¼ Cr-1 Mo steel is large grains of ferrite with small islands of pearlite. M_2C is present as a fine spray of precipitate within large ferrite grains. Lithium

Figure 4. The HYLIFE ICF reactor concept.

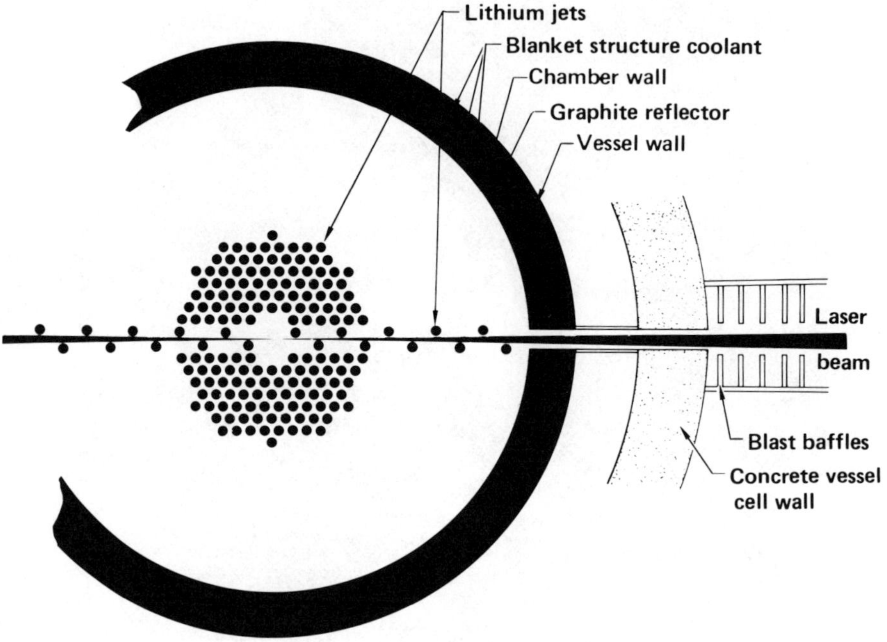

Figure 5. HYLIFE midplane cross section.

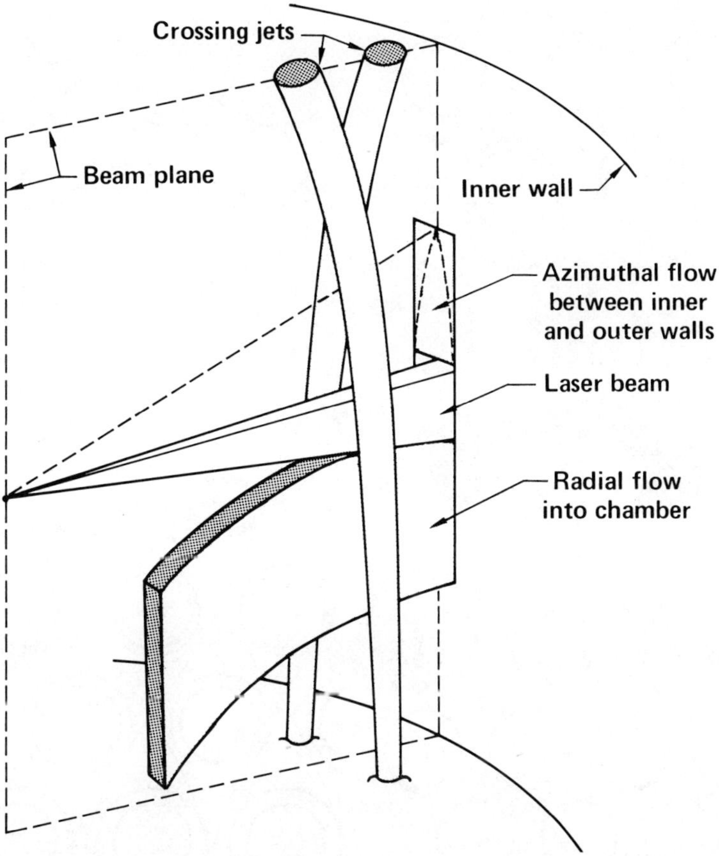

Figure 6. HYLIFE driver-beam port protection.

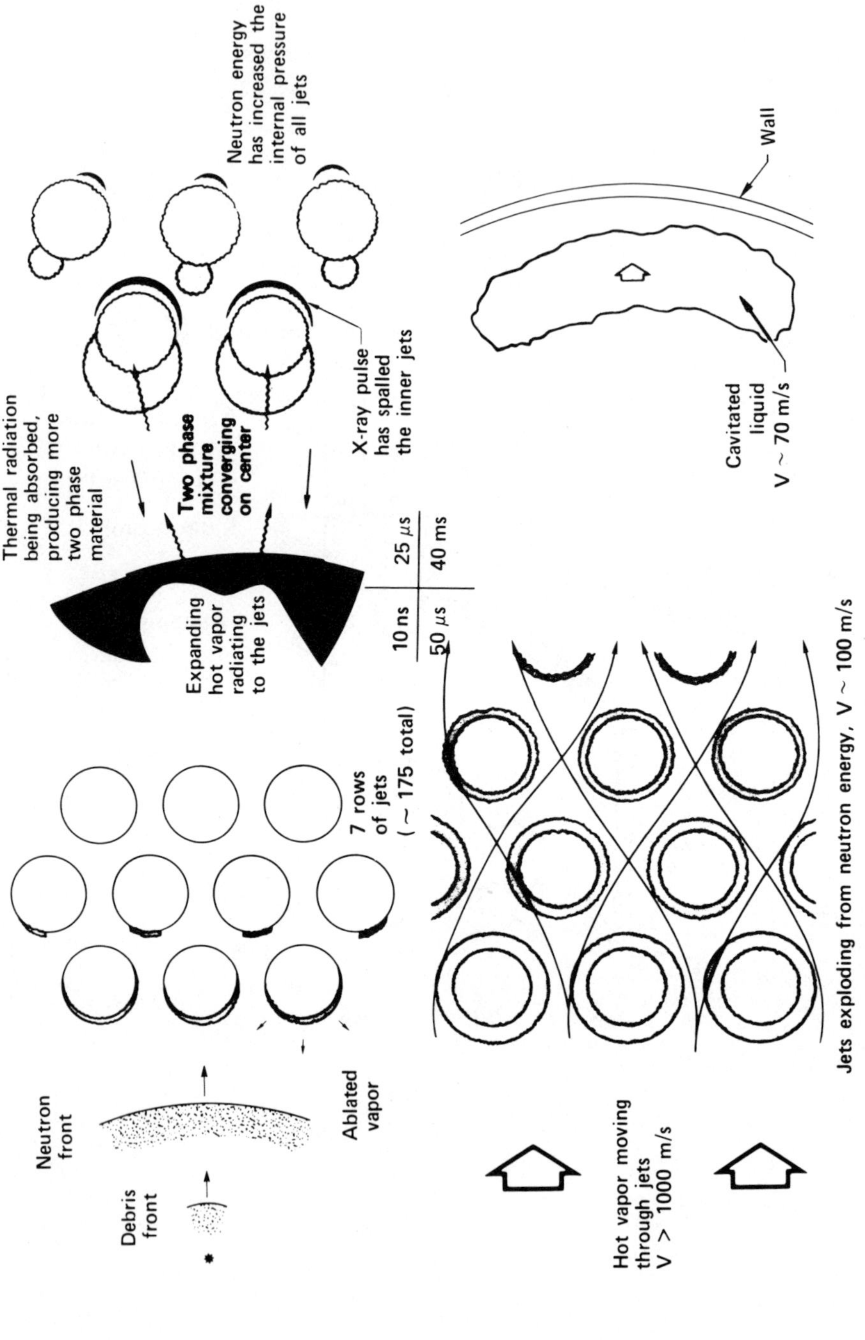

Figure 7. HYLIFE jet array response to a fusion pulse.

will attack the fine spray of exposed M_2C and the platelets of M_3C which intersect the surface. Heat treatments have been developed which convert the unstable M_3C and M_2C to M_7C_3, $M_{23}C_6$, and M_6C which are Cr-rich carbides that are stable in lithium (Figure 8) (8).

If 2¼ Cr-1 Mo steel is welded, a region of steel near the molten weld is heated high enough to convert the pearlite plus ferrite to austenite. A continuous film of M_3C can form around the austenite grains (Figure 9). As this hot heat-affected-zone (HAZ) cools, the austenite decomposes to ferrite and pearlite, but a ghost of the prior austenite grain boundary remains. This ghost boundary is made up of the film of M_3C and possibly other impurities deposited in the austenite boundary. Since such a film may be continuous in a weld HAZ, lithium reaction along the film can be catastrophic; therefore, post weld heat treatment could be required.

The M_3C and M_2C carbides are stable in $Pb_{83}Li_{17}$ melt; however, this high lead melt may create liquid metal embrittlement problems. Indeed, crack propagation of 2¼ Cr-1 Mo steel could possibly be accelerated by either liquid. Since fatigue damage is a major concern to the HYLIFE concept, studies have begun relating the number of cycles required to propagate a crack a certain distance to both temperature and strain rate. The general trends of such a relationship are shown in Figure 10.

The stress level in an alloy may also affect the corrosion rate in lithium. Failures of bellows and sharp cornered channels have often been noted in lithium systems. This is contrary to considerable experience in sodium and to recent experiments at the Colorado School of Mines (Figure 8), and the subject is still controversial. High nitrogen content (> 100 ppm) in lithium certainly increases the corrosion rate and might be a prerequisite for stress-related increases in the corrosion rate. In addition, the role of hydrogen in lithium is believed to be critical in all these compatibility studies, but no research has yet begun on this topic.

Liquid metal walls in ICF reactors will become contaminated with target debris. In addition to posing compatibility problems, these contaminants affect pumping power, pump life, and radioactive inventory. Thus, methods of removing contaminants, particularly high-Z elements, from liquid lithium are required. For example, no economical method to remove Pb to below 1 a/o (23 w/o) in Li has been devised.

HYLIFE Safety Engineering. In order for an accident in a liquid lithium cooled fusion reactor to endanger the general public, the accident must involve large scale reaction of hot molten lithium with air, water, or concrete. Further, the reaction heat must be coupled to radioactive components either directly in the flame, or indirectly via heated gas. Therefore, fusion reactor designers try to preclude even a remote possibility that such reactions and heat transfer can take place.

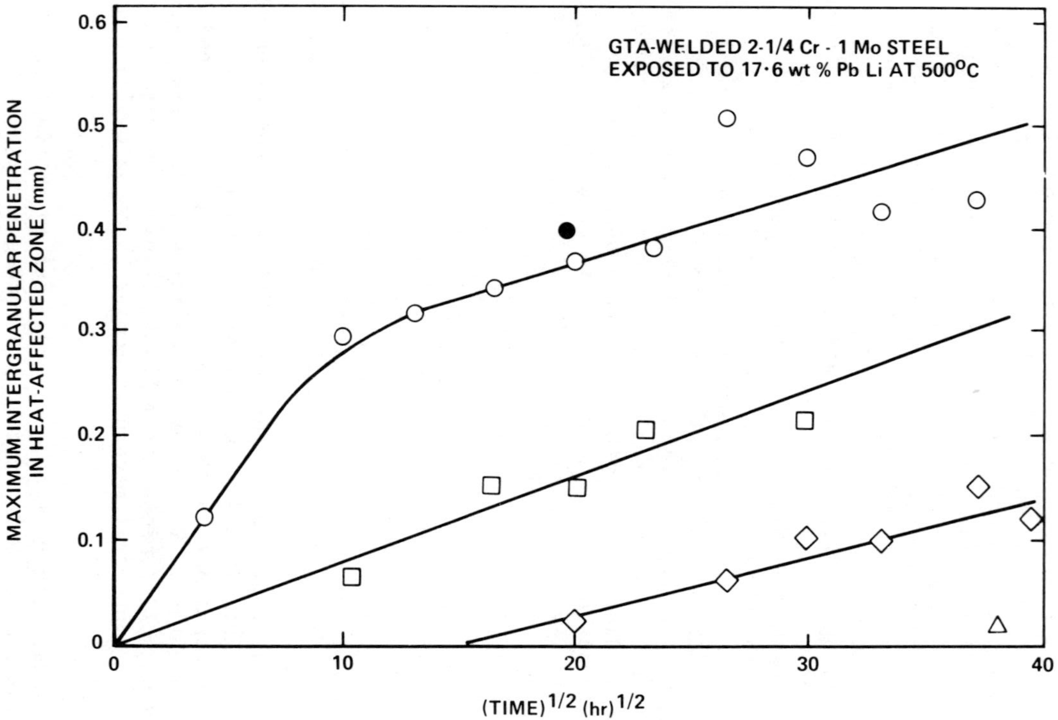

Figure 8. Corrosion rate is a function of the 2¼ Cr-1 Mo heat treatment. Key for no post weld H.T.: ○, unstressed coupon; ●, 18 ksi applied stress; for post weld H.T.: ◇, 10 hr, 760°; □, 10 hr, 710°C; Nb-stabilized parent metal △, no post weld H.T.

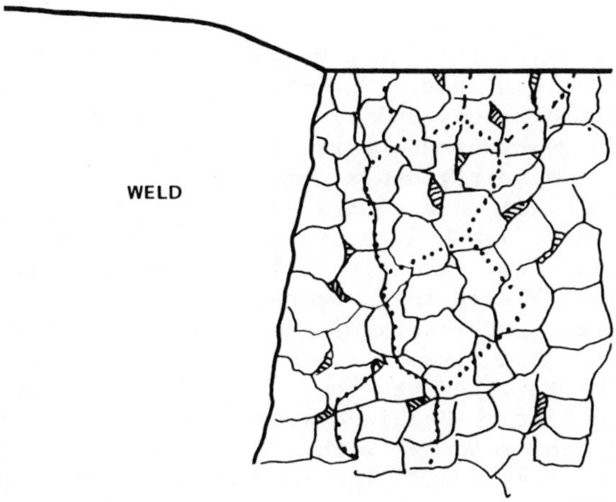

Figure 9. Microstructure of 2¼ Cr-1 Mo steel. Key: [symbol] *, ferrite grain peppered with M_2C;* [symbol] *, pearlite (platelets of alternating M_3C and ferrite) and* [symbol] *, ghost boundary of prior austenite grain boundary.*

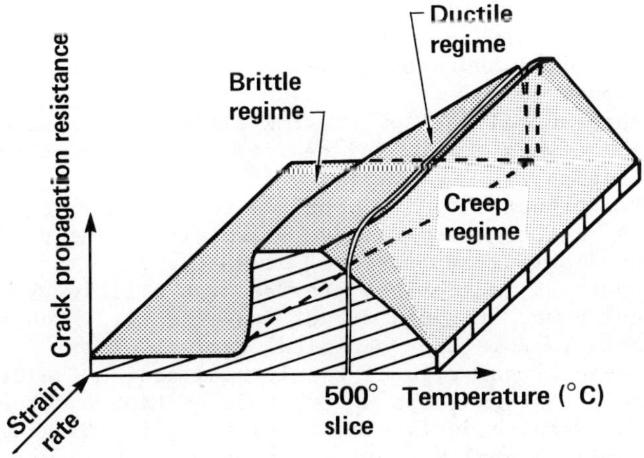

Figure 10. Relationship of crack propagation resistance to strain rate and temperature. Thermally stabilized 2¼ Cr-1 Mo in high-N Li. Stress intensity factor, $K = 20.0$; and fatigue load ratio, $R = 0.0^+$.

As an example of intrinsic safety features, consider the HYLIFE design. The HYLIFE reactor room has no external walls or roof; it is surrounded by other rooms and covered with a crane loft. The reactor room contains an inert gas (no air). The entire Li inventory can be drained in minutes in case of an air leak into the room; such a leak would require hours before lithium combustion is possible. There are no water or steam components in the reactor room, and all concrete is steel lined. The lithium loop is everywhere sub-atmospheric pressure (1 Pa to 80 kPa or 10^{-2} to 620 Torr); hence, small leaks will be inward. Large leaks will fall on a sloped floor that drains to tall narrow tanks containing hollow graphite spheres that would float above spilled lithium. Additional inert gas injection capability will be available in these areas.

These design features eliminate almost all pathways to a serious accident, but elimination of the lithium chemical energy would be even better. Hence, the ideal DT fusion reactor would utilize lithium that is combined with other elements to produce a fluid that is not reactive with air or water, but still retains the low density and high heat capacity of pure lithium. Various lead-lithium solutions have been considered, but none fully satisfy these criteria.

The TMR Cauldron Blanket Module

Design Concept. Thermochemical cycles for hydrogen production lend themselves well to use of a Tandem Mirror Reactor (TMR) heat source with a liquid Li blanket design. This linkage of a thermochemical hydrogen cycle to a TMR is favorable for several reasons: (9)

- the central cell of the TMR allows the design of relatively simple and readily accessible blanket modules,
- the process heat and maximum temperature requirements for a thermochemical cycle are within the performance available from liquid Li blankets, and
- advanced TMR designs provide high efficiency electrical power (through direct converters), which is desirable in some thermochemical cycles.

A cross-sectional view of the lithium-sodium Cauldron Blanket Module that has been under study at LLNL and the University of Washington is shown in Figure 11. Although the module resembles a pool boiler, it is substantially more complicated due to the non-symmetric geometry, the exponential energy generation in the fluid contained within the module, and also due to MHD effects on the convective mixing of the liquid pool. The Li-Na liquid metal in the Cauldron Blanket Module acts as a neutron moderator, heat transfer fluid, and tritium producer. Heat is removed by vaporizing the sodium. The void fraction

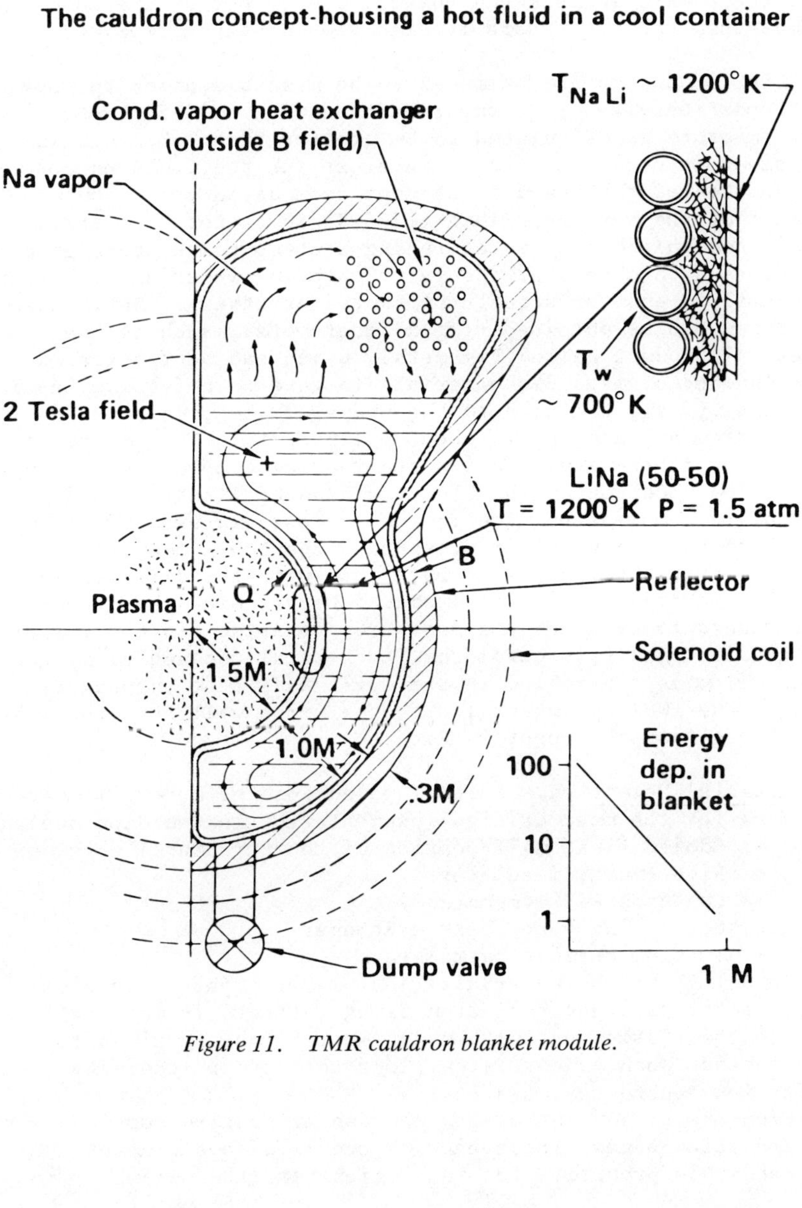

Figure 11. TMR cauldron blanket module.

created in the liquid pool by the vaporizing sodium is an inherent limitation of the Cauldron concept.

The Na vapor, traveling at vapor velocities of roughly 8-10 m/s at 1200 K, condenses on the heat exchanger tubes in the dome (the condensing vapor heat exchanger, CVHX) and returns as liquid droplets to the pool, thus completing the cycle. In the dome, the CVHX transfers the thermal energy out of the module to various chemical processors located some distance from the reactor.

Lithium and Na are believed to be miscible under the operating conditions, and tritium production with a 50-50 Li-Na atomic mixture is calculated to be greater than 1.0. Lithium-potassium mixtures are also of interest for the Cauldron concept.

Success of the Cauldron concept depends in part upon containment of the very hot fluid (1200 K) in a cool container. This is accomplished by establishing a steep temperature gradient across a thickness of insulator material intervening between the hot fluid and the structural container itself. Satisfactory insulation can be obtained from a fiber-metal, such as the commercial product called Feltmetal, which can be fabricated from a number of metallic materials to give an effective thermal conductivity, k_e, tailored to the needs of the problem. In this particular case, k_e is about 0.175 W/m•K. Two important features of fiber-metal are its substantial compressive strength and its low shear strength. The Cauldron application loads the material in compression, due to the hydrostatic head of fluid and due to the sodium vapor pressure. The 500 K temperature gradient across the insulator material, with the differential expansion it creates, is accommodated by allowing the fiber-metal fibers to slide on one another. To establish the temperature gradient and to control the fiber-metal conductivity, the coolant flow is controlled in the first wall heat exchanger (FWHX). The FWHX is integral with the first wall structure, as may be seen from the inset in Figure 11.

Material Selection. Initial selections of materials have been made for the most critical elements in the Cauldron design:

- Cooled first wall portion of the structural envelope
- Fiber-metal insulator
- Cauldron wall (membrane)
- Condensing vapor heat exchanger
- Tritium permeation membranes

In selecting the materials (10), major emphasis is given to neutron activation and radiation damage effects in materials close to the plasma and further considerations are given to other factors such as corrosion, long-term creep strength, fabrication technology, and cost.

From the standpoint of low neutron activation combined with good radiation damage resistance, V and Ti alloys present the best materials prospects for the region near the plasma. The current choices for the first wall are V-10% Ti and V-20% Ti, since both of these alloys have been shown to suppress helium gas bubble formation. Further, triton impact should not cause problems of localized tritium buildup and blistering, since tritium is known to permeate very rapidly through V and its alloys. The technologies of producing V alloy sheet and tubing and of fabrication by electron beam welding are well established and present no problems as long as interstitial O, N, and H are kept at low levels. The current cost of V is high (∿ $440/kg), but V resources are extensive and cost reductions can be expected for quantity use. Even at $440/kg, the cost is not excessive for the TMR first wall (∿ $10M). Corrosion should not be a problem if liquid Na or helium is used as a coolant. If organic liquids (polyphenyls) are used, the coolant would need to be continuously purified to avoid carbon deposition in the tube passages due to radiation damage.

Beta-Ti alloys are under consideration as a backup material for the first wall. Although radiation damage information is sparse on these alloys, it is encouraging that hydrogen diffusivity has been shown to be substantially higher than in the alpha-Ti alloys. For the cooled structural wall outside of the first wall region, a broad range of conventional low-cost alloys (e.g., Fe-Ni based alloys) are available, but have not yet been specified in the initial scoping study.

For the fiber-metal insulator, the beta-Ti alloys, Ti-13V-11Cr-3Al, Ti-2.5Al-16V, and Ti-7Al-4Mo are the current selections, based on existing commercial alloys. Improved Ti alloys or use of a Nb alloy layer next to the Cauldron wall may be necessary to avoid creep at the highest temperatures. A helium sweep gas would be used to remove tritium from the porous fiber-metal channel.

For the material of the Cauldron wall which is near the plasma, and in contact with the liquid Li-Na pool and the resultant vapors, neutron activation remains of major importance, and radiation damage effects are somewhat reduced (no triton impact). Corrosion resistance is a critical concern throughout the Cauldron wall, and long-term creep (in spite of the low loading stresses) needs to be considered. The same commercial beta-Ti alloys were selected for the Cauldron wall as for the fiber-metal insulator. The corrosion resistance of pure Ti and of all the alloying constituents except Al are expected to be adequate in high temperature liquid Li, and the corrosion of Al is believed to be negligible because of its low diffusion rate through Ti. Also, liquid Li is usually more corrosive than Na, so that the beta-Ti alloys are expected to have good corrosion resistance in the Li-Na mixture. A definitive statement cannot be given on the long-term creep strengths of beta-Ti alloys, due to lack of any direct data under the low-stress, high-temperature conditions encountered here. The LLNL workers are optimistic, however, in view of the fact that short-term tensile yield strengths (0.2% deformation) are quite high (225 MPa or 33 ksi) at 1200 K for the MST-881 Ti alloy, Ti-8Al-8Zr-1(Nb+Ta). Fabrication and welding of both alpha and beta-Ti alloys into large complex structures using electron beam welding techniques is a well-established technology in the aerospace industry. Tritium retention in the titanium alloys is not expected to be a problem because of the high temperatures involved.

The condensing vapor heat exchanger is sufficiently removed from the plasma region that neutron activation and radiation damage are no longer significant. The important materials considerations here are long-term creep resistance, high thermal conductivity, corrosion resistance to the Cauldron vapors and circulating heat exchange fluids, and ease of fabrication. The molybdenum alloy, TZM, has been selected because of its outstanding behavior for the first three criteria above. Small-scale (up to ∿ 3 cm dia.) welding of TZM for heat pipes and laboratory test capsules is well established and should be directly translatable to heat exchanger fabrication, although a larger scale technology has not yet been established. Certain developmental grades of Nb and Ta alloys would also be suitable for the heat exchanger. In-800 and In-800H also become acceptable if temperatures are held below 1100 K.

Tritium permeation membranes can be applied in several places depending upon the design specifics of the Cauldron and the heat transport loop, i.e., tritium may be removed directly from the Cauldron dome, from a processing slip-stream off the Cauldron dome, from the fiber-metal channel, and from a liquid Na, liquid K, or gaseous helium heat transport loop. The V, Nb, and beta-Ti alloys have been selected as the preferred membrane materials because of a combination of high hydrogen permeability and good corrosion resistance to a Li-Na environment. Nb alloys are preferred at the highest temperatures where high strengths and low creep rates are required.

A summary of the material selection considerations is presented in Table 2.

The TMR Heat Pipe Blanket Module

A conceptual design study is currently underway (11) at LLNL and the University of Washington on a Heat Pipe Blanket Module (Figure 12) to provide process heat at temperatures of about 800-1000 K. Since work on this concept has only recently begun, details cannot yet be provided on operating parameters. Process heat provided by this blanket design should have utility for either hydrogen production by thermochemical cycles or electrical power production. Both liquid Li-Pb and pure liquid Li are under consideration for the breeder-moderator region, which is about 1 m thick. Gravity-assisted heat pipes ∿ 5-cm diameter and ∿ 3-m long remove heat and extract tritium from the liquid metal. The heat pipe enclosure will probably use two

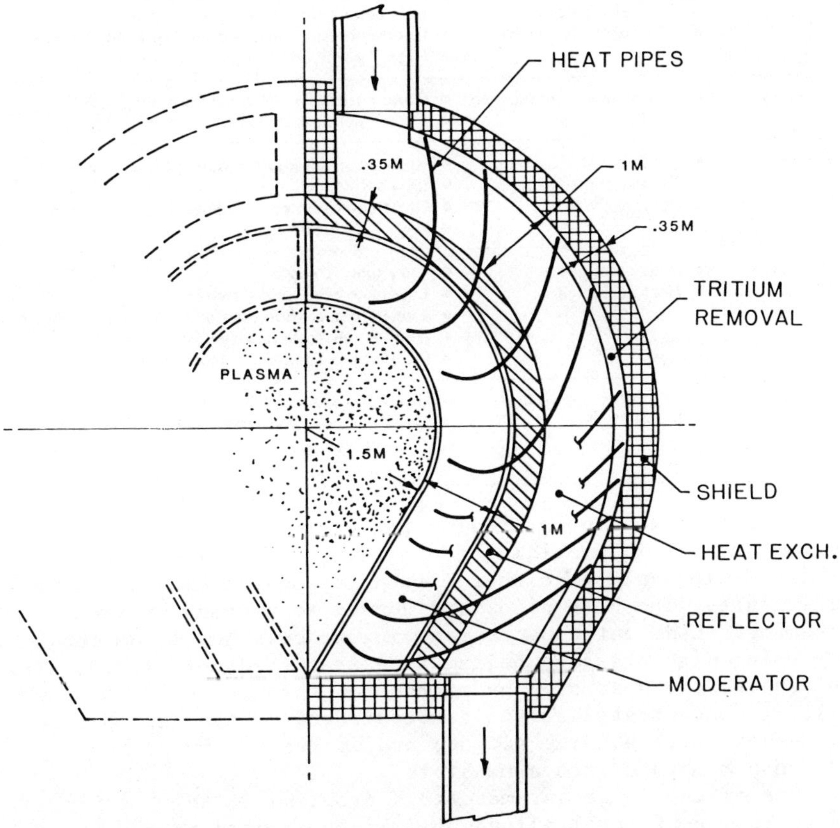

Figure 12. TMR heat pipe blanket module.

Table II. Initial Materials Selections for the Cauldron Blanket Module

Material	Application	Pros and Cons
V alloys	• First wall • Feltmetal insulator • T permeation membrane	• Good resistance to radiation damage • Low activation • Good corrosion resistance to liquid Li, Na, K • High T permeation rates
Nb alloys	• T permeation membrane	• Good corrosion resistance to liquid Li, Na, K • High T permeation rates
β-Ti alloys and Ti-V alloys	• First wall • Feltmetal insulator • Cauldron wall	• Resistance to radiation damage is probably good • Low activation • Good corrosion resistance to liquid Li, Na, K • High T diffusion rates
Mo alloys (TZM)	• Heat pipes • Heat exchangers	• Very high strength • Excellent heat conductivity • Excellent corrosion resistance to liquid Li, Na, K • Good small-scale fabricability • Problems in large-scale fabrications
In-800	• Transport piping for liquid Na, K	• Good creep strength to 1100 K • Good corrosion resistance to liquid Na, K

metals, with TZM alloy providing for the main structure and with a metal such as Nb providing for tritium permeation. The working fluid would be either K or Na depending upon operating temperature. The heat pipes are curved to prevent neutron streaming. LLNL anticipates removing process heat from the heat pipe using high pressure helium as a heat transport fluid. Tritium enters the heat pipe by permeation through the heat pipe wall, is concentrated at the condenser end of the heat pipe due to the heat pipe pumping action, and is removed by permeation through a membrane into a manifold.

One of the important materials problems in this design is the corrosion of metal alloys when Li-Pb is used as the breeder-moderator. The dual wall structure used in the Cauldron Blanket Module should apply equally well here, so that V can be used for the plasma first wall to provide structural strength with minimum activation. The choice of a material for the wall membrane in contact with liquid Li-Pb becomes difficult. Ti alloys cannot be used because of reaction with Pb. Pure iron and iron-rich alloys appear marginal because of a solubility of Fe in Pb of $\sim 10^{-5}$ atom fraction at 900 K (12). Use of Ni or Cr as alloying constituents in iron do not improve the situation, since Ni and Cr have solubilities in Pb of 0.03 and $\sim 10^{-4}$ atom fraction at 900 K (13). The refractory metals Nb, Ta, Mo, and W should resist corrosion in Li-Pb, and although Nb and Mo are good candidates for both the wall membrane and the heat pipe, they present activation problems if used in the vicinity of the first wall. Further work is needed to make a proper selection for the wall membrane in contact with Li-Pb.

If pure Li is used as the breeder-moderator, Ti and a number of other materials can be used, as discussed for the Cauldron Blanket Module. However, the tritium activity in Li drops markedly in the absence of Pb, and the blanket temperature will need to be raised to alleviate the tritium removal problem.

Tritium Extraction

One of the key chemical problems associated with lithium in fusion reactors is extraction of the tritium that has been either generated in or trapped by lithium. A figure of merit for tritium extraction is the blanket tritium inventory. Very large inventories require excessive start-up inventory and are a potentially large radioactive effluent in the event of a catastrophic accident.

For solid blankets such as the STARFIRE concept discussed in the introduction, diffusion of tritium through the solid is the limiting factor. A blanket tritium inventory of 0.7 to 8.2 kg in a new blanket has been estimated (4). However, radiation could cause stoichiometry changes, grain growth, tritium-trapping, and sintering; these effects could increase the blanket tritium inventory to hundreds of kilograms (4).

If the tritium breeding medium is not in direct contact with the plasma exhaust (i.e., in all but some LMW ICF reactor concepts), tritium must be recovered from the plasma exhaust, a gas. In addition, solid breeders, and some liquid breeders, use a gas to remove tritium from the breeder, and then process the gas. The practical prospects of using a vapor phase slip stream to extract tritium from a liquid Li breeder were significantly enhanced due to the discovery by Ihle and Wu (14) in mass spectrometer studies that high concentrations of tritium-bearing species are expected in the vapor phase. Ihle and Wu's work was actually on deuterium, but some 30% higher gas-phase concentrations are expected with tritium. When Ihle and Wu's results are extrapolated to a somewhat higher temperature of 850°C and assumed to be applicable to tritium at an atom fraction of $x_T = 10^{-5}$, the LiT and Li_2T pressures are of comparable magnitude at 2.6 mPa (2 x 10^{-5} Torr), and exceed the T_2 partial pressure by a factor of 40. Thus, the vapor phase tritium concentration is substantially increased over T_2 alone, and extraction of the tritium compounds through permeation membranes may be practical.

Ihle and Wu, in extrapolating their results, also found that the concentration of tritium in the gas phase should exceed that in liquid Li at temperatures above ∿ 1200 K, reaching a value of x_T(gas phase)/x_T(liquid phase) ≈ 6 at 1600 K. This led them to suggest distillation as a means of tritium recovery.

Several methods have been proposed to extract tritium from molten Li or molten Li alloys. These include:

- Fractional distillation
- Cold trapping
- Gettering
- Diffusion through a large permeable window
- Diffusion into heat pipes
- Molten salt extraction

In all of these methods, concurrent removal of corrosion products could increase the corrosiveness of the liquid metal downstream from the extraction equipment. This factor has not been evaluated for most schemes, but it has caused problems in some lithium loops that use getters.

Distillation. An azeotrope has been reported in the lithium-tritium phase diagram at ∿ 1000°C (Figure 13). Reference 15 estimates the azeotrope concentration at 2,000 wppm tritium while Reference 14 estimates 3 wppm of deuterium (corresponding to ∿ 4 wppm tritium). If the high value is correct, a dilute tritium solution will have a vapor that is less rich in tritium than is the melt. As the vapor is removed, both the vapor and the melt would increase in tritium concentration until a constant boiling point mixture at the azeotrope concentration is achieved. If the lower azeotrope concentration is correct, fractional distillation could be effective. However, these arguments are complicated by the presence of other tritium vapor phases such as LiT.

Cold Trapping. There is considerable uncertainty in the concentration to which tritium can be cold trapped from Li (Figure 14). The work of Natesan and Smith (Figures 8 and 9 of Reference 16) can be used to estimate concentrations of 1800 wppm protium and 2700 wppm tritium at a cold trap temperature of 195°C. Katsuta (17) estimates a protium concentration of 200 wppm at a cold trap temperature of 203°C. Veleckis (18) estimates protium and deuterium concentrations of 75 and 135 wppm at a cold trap temperature of 195°C. The corresponding tritium concentration would be ∿ 200 wppm. Thus, Veleckis (200 wppm tritium) and Natesan (2700 wppm) bracket the available estimates. In a 1630 m^3 Li blanket (such as HYLIFE), 200 wppm and 2700 wppm tritium concentrations result in 160 and 2200 kg blanket tritium inventories, respectively. (The actual tritium inventory may be reduced by about 25% due to the presence of deuterium, but this effect has been ignored here.)

Gettering. The gettering approach to tritium extraction applies to processing of either liquid metals or gas streams

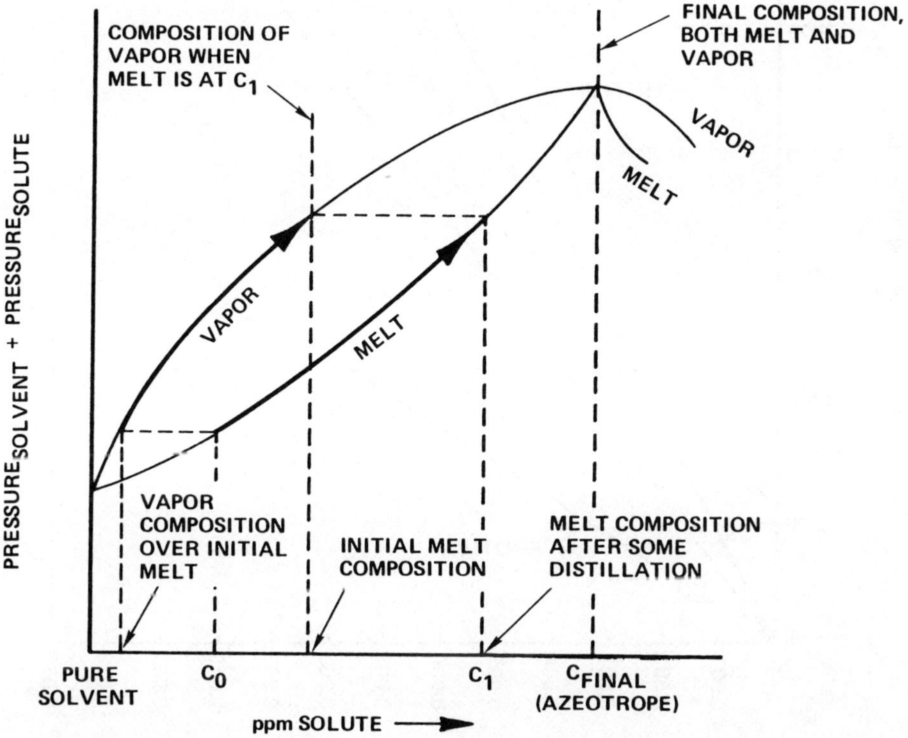

Figure 13. Phase diagram for a dilute solution with an azeotrope. Diagram can be used qualitatively to predict the distillation behavior of T (solute) in Li (solvent).

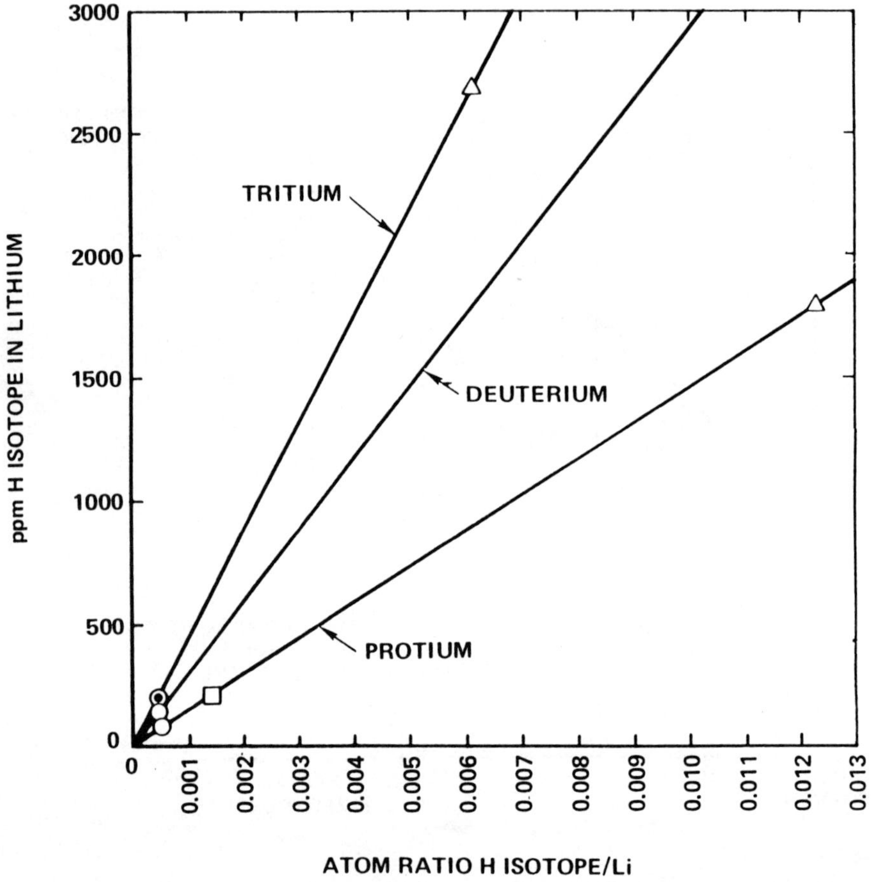

Figure 14. Concentrations to which H isotopes can be cold trapped from Li. Key: △, cold trapped at 195°C (16); □, cold trapped at 203°C (17); ○, cold trapped at 195° (18); and ⊙, estimated value if cold trapped at 195°C (18).

containing tritium. The stream to be processed is passed through a bed of porous metal that serves as the getter. After sufficient exposure, the getter bed is regenerated by heating under vacuum or in a carrier flow stream to recover the tritium. There is very little information on the use of getters for tritium extraction. Yttrium (19) and cerium (20) have shown a limited degree of success as getters, and then only at relatively low temperatures (i.e., below 350°C).

Diffusion Through the Primary Heat Transfer Loop. An equation for the permeation of a gas through a metal is

$$F = C\ A\ \Delta\sqrt{P}/t \quad (1)$$

where F is the gas flux, C is the permeability constant, and $\Delta\sqrt{P}$ is the difference of the square roots of the tritium pressures on the high and low pressure sides of the diffusing medium which has area, A, and thickness, t.

For lithium temperatures of ∿500°C, the liquid metal heat exchanger requires ∿ 3.3 to 8.0 m^2 of area per MW_t (10,600 to 25,600 m^2 for a 1-GW_e plant). The wall thickness will be about 0.65 mm to 1.3 mm for a 2¼ Cr-1 Mo heat exchanger. For a 30% tritium burn fraction and a 1.75 tritium breeding ratio, 19.5 mg/sec of tritium are added to a 1-GW_e HYLIFE plant's liquid metal wall, and the same amount must diffuse out. The permeability constant for tritium diffusing through 2¼ Cr-1 Mo steel is (21)

$$C = 206 \times 10^{-4}\ \frac{\text{mg}}{\text{s}\cdot\text{m}\cdot\sqrt{\text{Pa}}}\ \exp[-4706/T(K)]$$

$$= 8.81\ \frac{\text{cm}^3(\text{STP})\cdot\text{mm}}{\text{h}\cdot\text{cm}^2\cdot\sqrt{\text{atm}}}\ \exp[-4706/T(K)]$$

At T = 773 K, C = 4.68 x 10^{-7} mg/(s·m·$\sqrt{\text{Pa}}$) = (0.20 cm^3·mm/h·$cm^2\sqrt{\text{atm}}$). For the range of heat exchanger thicknesses and areas,

$$1.1 < \Delta\sqrt{P} < 5.1\sqrt{\text{Pa}}\ (0.092 - 0.44\sqrt{\text{Torr}}) \quad (2)$$

If the low pressure side is considered to have essentially a zero tritium pressure,

$$1.1 < P < 27\ \text{Pa}\ (0.0084 - 0.20\ \text{Torr}) \quad (3)$$

These pressures correspond to tritium concentrations of 2900 and 14,000 wppm (Figure 15) or to a tritium inventory in a 1630 m^3 liquid lithium blanket of 2300 to 11,000 kg. (The pressure-concentration relationship shown in Figure 15 is that of Ref. 16.) If the low pressure side is liquid sodium cold trapped to 4 mPa (3 x 10^{-5} Torr), the high pressure side ranges from 1.3 to 27 Pa (0.0095 to 0.20 Torr), i.e., 3100 to 14,000 wppm, or 2400 to 11,000 kg in a 1630 m^3 liquid lithium blanket. Therefore, using a secondary fluid that can be cold trapped to a lower pressure than Na (such as NaK which can be cold trapped to 30 nPa or 2 x 10^{-10} Torr) does not significantly improve the situation.

If a much larger area heat exchanger is used, considerable improvement is possible. For example, if the 25,600 m^2 heat exchanger has its tubes reduced in radius and coiled, perhaps 100 times more area could result. Using the 0.65 mm thick 2¼ Cr-1 Mo, and cold trapped NaK on the secondary side results in a primary tritium pressure of 0.12 mPa or 8.7 x 10^{-7} Torr (29 wppm or 23 kg in a 1630 m^3 liquid lithium blanket).

An alternate method of reducing the blanket tritium inventory is to use a more permeable heat exchanger material such as Nb which has a permeability constant (22) of 2.6 x 10^{-4} $mg/(s{\cdot}m{\cdot}\sqrt{Pa})$ = 110 $cm^3(STP){\cdot}mm/(h{\cdot}cm^2{\cdot}\sqrt{atm})$ at 500°C (Figure 16). For the original heat exchanger areas and thicknesses, and for a cold trapped NaK secondary, tritium pressure ranges from 4.4 to 89 μPa (3.3 x 10^{-8} to 6.7 x 10^{-7} Torr), i.e., 6 to 26 wppm, or 5 to 20 kg inventory for a 1630 m^3 liquid lithium blanket).

These calculations are uncertain, however. The permeation equation may pass through a regime proportional to pressure rather than root pressure. Additionally, the rate controlling factor may not be permeation, but rather the rate at which tritium atoms skating about on the low pressure surface find other atoms and leave the surface as gaseous molecules.

Heat Pipe Scheme. The principle involved in tritium removal with heat pipes has been described by Lee and Werner (23,24). The method is applicable mainly to liquid metal breeder-moderators. Tritium at the ppm level in the liquid metal breeder enters the heat pipe through a permeation membrane (i.e., a metal such as Nb or V that permits rapid transport of tritium through the lattice). The tritium is then picked up by the rapid sweeping action of the heat pipe vapor (e.g., Na or K), and is carried to the heat rejection end of the heat pipe where the tritium is entrapped and compressed by the continuous impingement of the heat pipe vapor. With a tritium concentration of a few ppm in the liquid metal breeder (corresponding to ∿ 1 μPa or 10^{-8} Torr in 500°C Li), tritium concentrations up to ∿ 130 Pa (1 Torr) are expected in the heat rejection end of the heat pipe, where the tritium can then be extracted through a metal permeation membrane. Because of the selective sweeping and concentrating action of heat pipes for the non-condensible hydrogen isotope containing gases (when metal membranes are used), heat pipes present a very significant potential as a tritium recovery method. Development of further designs and materials for this application would be very useful.

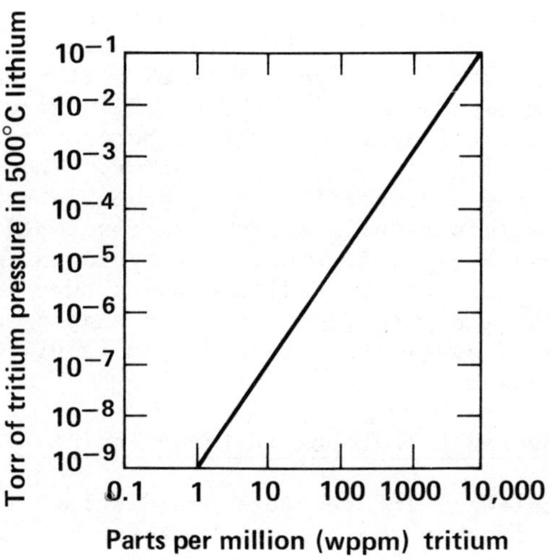

Figure 15. Relationship of T pressure to T concentration in 500°C Li.

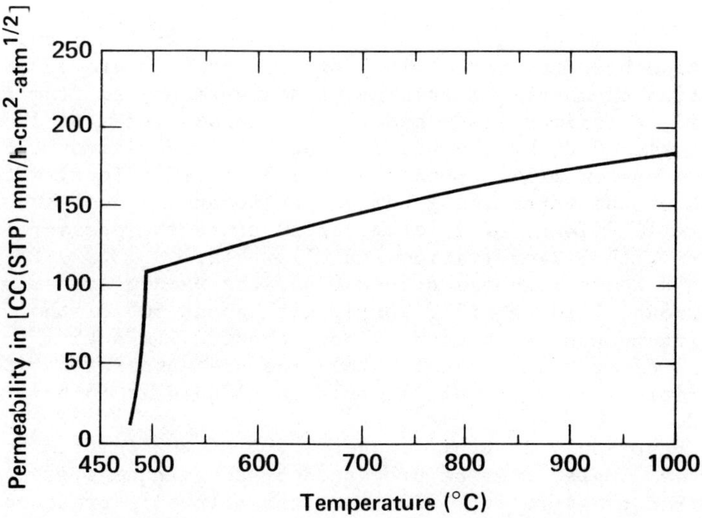

Figure 16. Permeability of Nb to T.

Molten Salt Extraction. Molten lithium or lithium alloy is placed in contact with a molten solution of LiF-LiCl-LiBr. This lithium halide solution is 22 mol % LiF, 31 mol % LiCl and 47 mol % LiBr, melting at 445°C. Hydrogen isotopes are preferentially extracted from the metallic solution into the lithium halide melt. The immiscible solutions of molten halide and molten metal are separated, with the metallic liquid, carrying less than 1 wppm total hydrogen isotope burden, returning to the liquid metal circuits of the fusion reactor. The halide melt is processed to remove hydrogen isotopes, probably by electrochemical evolution. This technique has been developed (25) by Calaway and co-workers at Argonne. The hydrogen isotopes released at the anode of the cell are swept away in molecular form by argon gas emerging from a bubbler-shaped anode. This technique could remove tritium to 0.5 wppm (a 0.4 kg inventory in a 1630 m^3 blanket).

Solubility of Hydrogen Isotopes in Lead-Lithium Liquids

Liquid Metal Wall ICF reactors are distinct from other fusion reactors in that the lithium breeding blanket is in direct contact with the plasma exhaust. Thus, the solubility of hydrogen isotopes in the liquid metal will determine whether tritium must be recovered from either the liquid loop or the vacuum system, rather than both systems.

Hydrogen Pressure Ranges for Liquid-Only or Vapor-Only Tritium Recovery for LMW ICF Reactors. For a 2700-MW_f ICF reactor with a 600-m^3 chamber, a 30% tritium burn fraction and a 1.75 tritium breeding ratio, 19.5 mg/s of tritium are added to the reaction chamber. If tritium is to be recovered from only the liquid, a maximum loss through the vacuum system of 1% (0.2 mg/s) is assumed to be acceptable. With a laser driver (26), the vacuum system might operate at $\sim$ 1.3 Pa (10^{-2} Torr) with about 10% of the vapor being pumped per second. Then, about 3.5 mPa (2.6×10^{-5} Torr) of tritium, or 10 times the pressure associated with 0.2 mg tritium, is allowable.

With a heavy-ion-beam driver (26), the vacuum system might operate around 0.13 Pa (10^{-3} Torr), with about 50% of the vapor being pumped per second. Then, about 0.7 mPa (5×10^{-6} Torr) of tritium, or twice the pressure associated with 0.2 mg tritium, is allowable if only the liquid is to be processed.

For vapor-only tritium recovery, 99% of the tritium production (or 19.3 mg/s) must be processed by the vacuum system, and the required pressure is $\sim$ 100 times the allowable pressure for liquid-only recovery. Therefore, with a laser driver, tritium vapor pressure between 3.5 and 350 mPa (2.6×10^{-5} and 2.6×10^{-3} Torr) requires tritium recovery from both the liquid-circulation and the vacuum systems. For the heavy-ion-beam driver, the corresponding pressure range is 0.7 to 70 mPa (5×10^{-6} to 5×10^{-4} Torr). The actual tritium vapor pressure will depend on both the vacuum system design and the solubility of tritium in the LMW.

Startup Inventory in a LMW ICF Reactor. Efficient tritium recovery cannot begin until the tritium concentration builds up in the liquid. For a 30% burn fraction, a start-up tritium inventory of about 1.4 kg is required for each day of start up. For a 1630-m^3 system of $Li_{99}Pb_1$ about 14 hours is required to reach 1 wppm (1 kg) tritium. For a 900-m^3 system of $Pb_{83}Li_{17}$, about 5 days are required to reach 1 wppm (7 kg tritium).

Hydrogen Solubility. As of December 1980, no data have been generated that directly give the hydrogen isotope pressure above $Pb_{83}Li_{17}$ liquid in the 400 to 500°C temperature range. However, data at other temperatures and concentrations can be extrapolated to the needed temperatures and compositions. At a given composition, Sievert's constant (K_s) is related to temperature by

$$\log K_s = m/T + b \quad (4)$$

where m and b are composition-dependent constants. If K_s vs composition is known at two temperatures, the constants m and b can be calculated at each composition and K_s can be estimated for all temperatures and compositions.

The data points of Ihle, Neubert, and Wu (27), for the solubility of deuterium in several Pb-Li compositions are shown in Figure 17. The solid lines at temperatures of 677 and 767°C are interpolations between data points. The gap in the 677°C curve is due to the presence of a solid phase. The data have been extended to the deuterium-in-lithium data from Argonne (28) and the hydrogen-in-lead data from Opie and Grant (29). The hydrogen-in-lead data were converted to deuterium-in-lead data by assuming that the ratio of Sievert's constants for hydrogen and deuterium is constant (independent of the Pb-Li composition) and by using the Argonne data for hydrogen and deuterium in pure lithium. The adjustment was relatively small, with a 9% increase at 767°C and a 20% increase at 477°C.

Using the curves at 677 and 767°C and Eq. (4), curves were constructed at 477 and 577°C. However, at Pb concentrations above 50 a/o, the 677°C curve is quite uncertain. Figure 18 presents a second plausible 677°C curve and the resulting lower-temperature curves.

For a "pure lithium" LMW, Pb target debris can build up to ∿ 1 a/o before cold trapping of binary Pb-Li compounds is effective. If deuterium is removed from this liquid to a concentration of 1 wppm (4.5×10^{-6} atom fraction of deuterium in

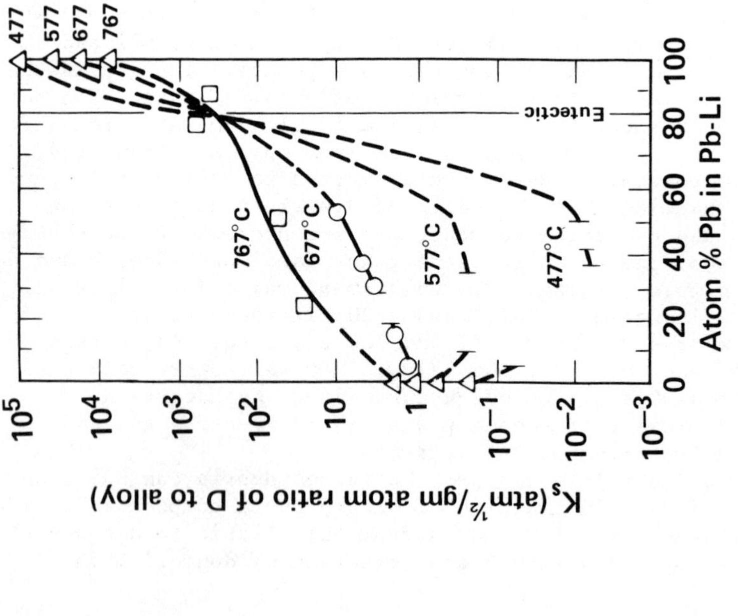

Figure 17. Sievert's constant for solubility of D in Pb–Li.

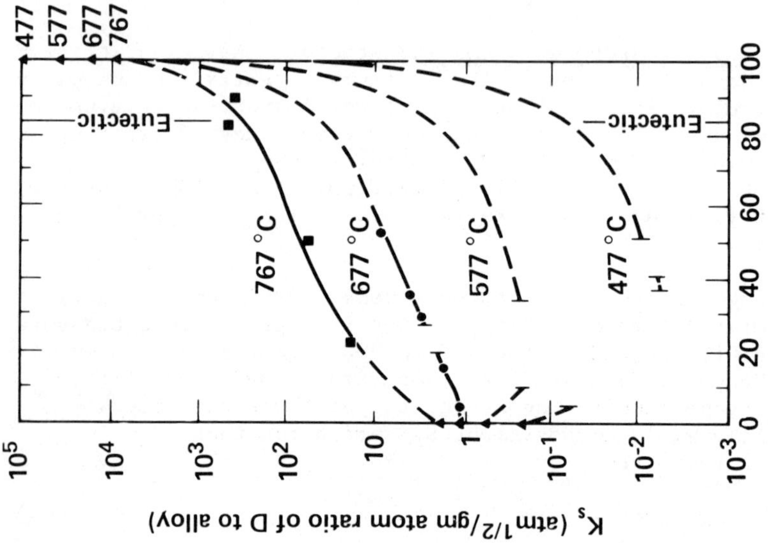

Figure 18. A second possible extrapolation of the solubility data of D in Pb–Li.

alloy), the diatomic deuterium pressure at 477°C is ∿ 70 nPa (5×10^{-10} Torr), and hydrogen isotopes need not be recovered in the vacuum system for either laser or heavy-ion-beam drivers.

For the $Pb_{83}Li_{17}$ LMW, no conclusion can be reached from the data. The extrapolation of Figure 17, and liquid recovery of hydrogen isotopes down to 1 wppm each (8.7×10^{-5} atom fraction of deuterium in alloy) produce a diatomic deuterium pressure at 477°C of ∿ 4 μPa (3×10^{-8} Torr), and no vacuum-system hydrogen-isotope recovery is required. However, if the Figure 18 extrapolation is used, and if deuterium is extracted from the vapor to a pressure of 350 mPa (2.6×10^{-3} Torr) for a laser driver or 70 mPa (5×10^{-4} Torr) for a heavy-ion-beam driver, the corresponding deuterium concentrations in the liquid are 0.034 and 0.015 wppm, respectively, and no liquid recovery of hydrogen isotopes is required. To resolve this issue, researchers at Argonne National Laboratory (ANL) and elsewhere are conducting appropriate experiments. Initial ANL data indicate that tritium recovery will only be required from the vapor phase over $Pb_{83}Li_{17}$ (30).

Heavy Ion Beam Propagation in Pb-Li LMW ICF Reactors

For a heavy-ion-beam driven, liquid-metal-wall ICF reactor, the liquid metal wall (LMW) composition and temperature must not preclude transport and focus of the heavy ion beam on the target. For a near-vacuum system, (∿ 10^{14} cm^{-3}) both the vapor pressure and the stripping effectiveness of the chamber gas must be known before it can be determined if ballistic propagation of the beam is possible. These conditions have been considered for $Li_{99}Pb_1$ and for $Pb_{83}Li_{17}$ liquid metal walls (31).

Li and Pb Partial Pressures. The vapor pressure above a Pb-Li solution is

$$P = \gamma_{Li} N_{Li} P_{Li} + \gamma_{Pb} N_{Pb} P_{Pb} \qquad (5)$$

where N_{Li} and N_{Pb} are the atom fractions of lithium and lead in the liquid, and P_{Li} and P_{Pb} are the vapor pressures of the pure elements at the given temperature. The temperature-and-composition dependent functions, γ_{Li} and γ_{Pb}, are activity coefficients; they describe the ideality of the solution. If the activity coefficients are 1.0, the solution is "ideal" (Raoult's Law). If they are greater than 1.0, the solution has "positive deviation from ideality" and a higher than expected vapor pressure. If they are less than 1.0, there is a "negative deviation" and lowered vapor pressure. In the case of the Pb-Li system, experimental data (32,33,34) indicate a profound negative deviation from ideality, probably due to conversion from metallic to ionic bonding. The nonideality of the Pb-Li system is dramatically shown in a plot of electrical resistivity (35) as a function of composition (Figure 19).

The vapor pressures of pure Pb and Li are shown as a function of temperature in Figure 20 and Tables 3 and 4. Vapor pressure of the pure lead is several hundred times less than that of pure lithium in the 400-500°C temperature range. (Higher temperatures are discouraged by a decrease in mechanical properties of ferritic steels and by Li corrosion; lower temperatures are discouraged by brittleness of ferritic steels and by thermal efficiency considerations.)

For $Li_{99}Pb_1$, the Li activity coefficient, γ_{Li}, is about unity. However, data (32,40) at 527 and 659°C indicate that the lead activity coefficient is less than 2×10^{-6}, so that the Pb-component partial pressure is ten orders of magnitude less than that of the Li component for $Li_{99}Pb_1$.

For high-concentration Pb solutions, such as $Pb_{83}Li_{17}$, the activity coefficients must be carefully evaluated. Demidov's γ_{Pb} data (40) at 527°C were used to determine activity coefficient values which ranged from 0.932 at 77 a/o Pb to 1.0 above 88.3 a/o Pb. If γ_{Li} is again of order 1, the Li-vapor component would dominate the Pb vapor by a factor of 50. However, as is shown below, γ_{Li} is less than 10^{-3} at temperatures below 500°C, and the Pb component dominates the vapor.

The Li-component activity coefficient, γ_{Li}, can be determined from electromotive force (emf) data taken by Saboungi (32) at temperatures of 497, 539, 596, and 659°C in Pb-Li mixtures with Pb atom fractions above 45%. In Figure 21, γ_{Li} is shown as a function of (1/T) at five liquid compositions. The lines in Figure 21 are least-square fits to allow extrapolation to the temperature region of interest to ICF reactor designers, 400 to 500°C.

Since the emf data did not involve direct pressure measurements, experimental verification involving a direct or indirect pressure reading of Li and Pb above a binary solution was sought. Knudsen cell work by Ihle, Neubert, and Wu (33) provide data that are also shown in Figure 21. Additional Knudsen cell work by Neubert, Ihle, and Gingerich (34) is shown in Figure 22 which plots the Pb and Li partial pressures above $Pb_{50}Li_{50}$ liquid as a function of (1/T). Figure 22 also shows the earlier Knudsen cell work for Li, the emf data, and least-square fits to the emf data. The emf data for Pb uses, at all temperatures, the γ_{Pb} value of 0.4 shown in Fig. 3 of Reference 32 for a temperature of 659°C. If a value of 1.0 is used for γ_{Pb} at all temperatures, the emf extrapolation falls in the middle of the higher temperature Knudsen cell Pb data.

The total Pb and Li vapor pressure over 497°C Pb-Li liquid is shown in Figure 23 as a function of Pb concentration. The pressure above $Pb_{83}Li_{17}$ (1.6 mPa or 1.2×10^{-5} Torr) is a factor of ~ 50 lower than ideal and a factor of ~ 270 lower than the pressure of pure Li at the same temperature (0.43 Pa or

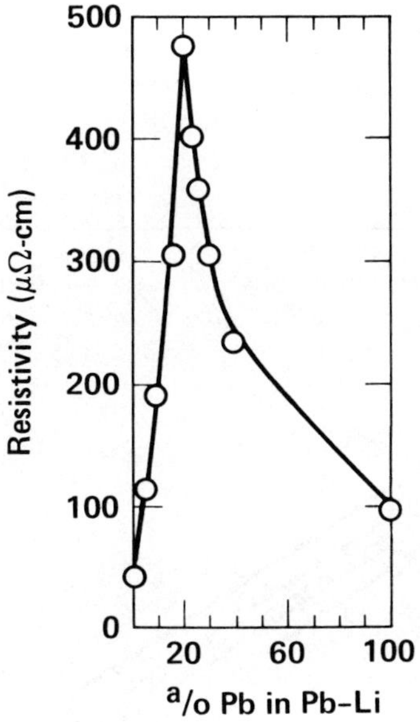

Figure 19. Electrical resistivity of Pb–Li alloys at 800°C.

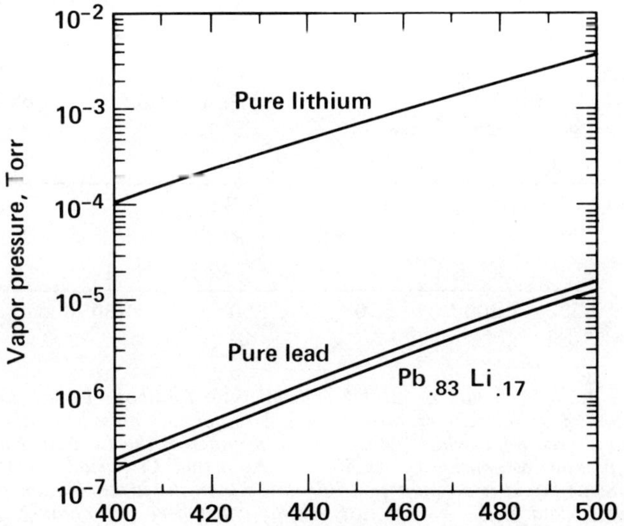

Figure 20. Vapor pressure of pure Pb and pure Li as a function of temperature.

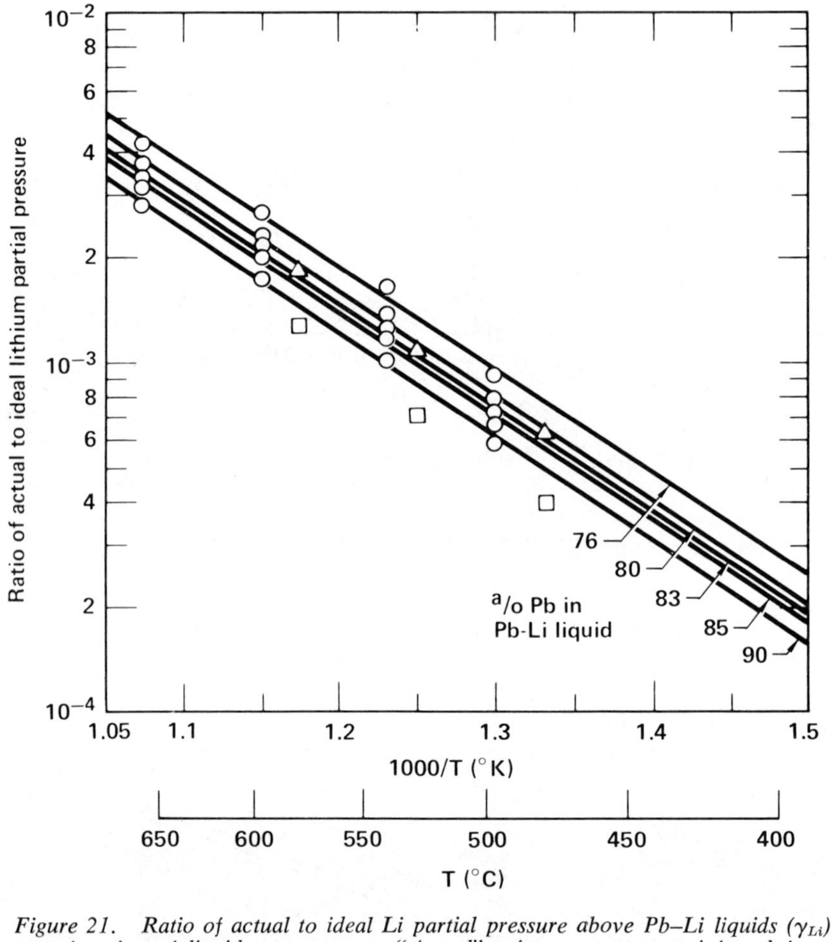

Figure 21. Ratio of actual to ideal Li partial pressure above Pb–Li liquids (γ_{Li}) as a function of liquid temperature. "Actual" refers to pressures inferred from electromotive force data, while "ideal" is the product of the Li atom fraction and the vapor pressure of pure Li. At 500°C, the actual Li partial pressure above high-Pb eutectic liquid is a factor of 1400 lower than the ideal solution prediction. Key: ○, EMF data; and ——, fit to EMF data; Knudsen cell data: △, 80% Pb; and □, 90% Pb.

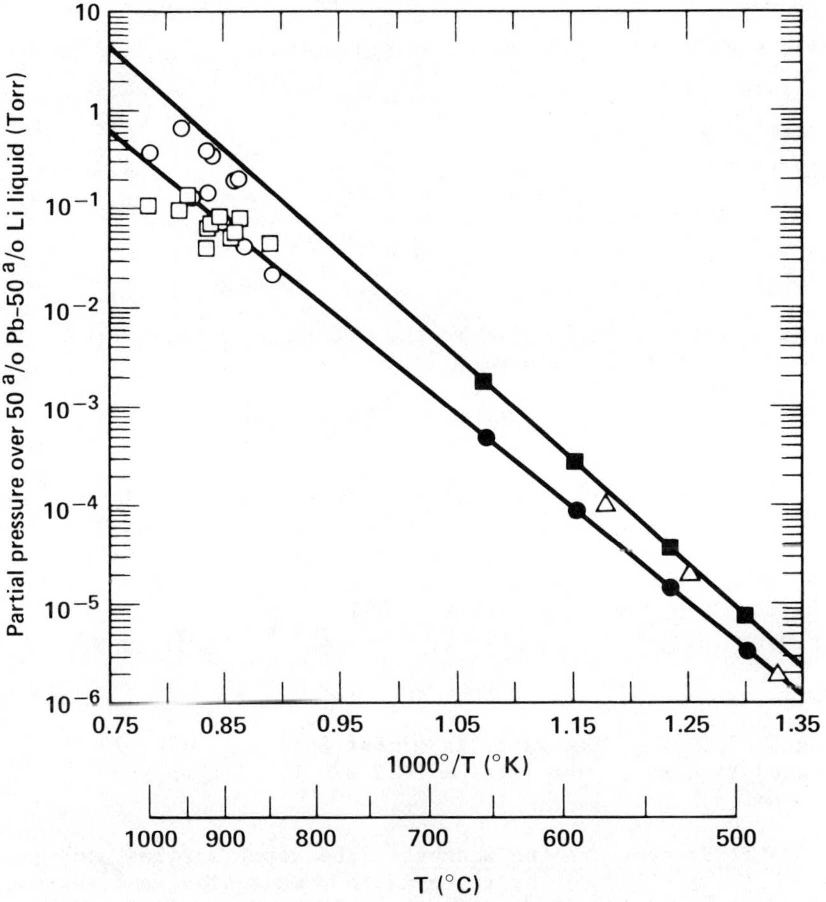

Figure 22. Partial pressures of Li and Pb above $Pb_{50}Li_{50}$ liquid from EMF and Knudsen cell data. Key for EMF data (2): ■, Li; ●, Pb; and ——, fit to EMF data; and for Knudsen cell data: □, Li (4); ○, Pb (4); and △, Li (3).

Table III. Range of Vapor Pressures in the Literature for Pure Lithium

Source	Temperature (°C) 400	500
Cowles & Pasternak (36)	13.6 mPa	473. mPa
Nesmeyanov (37)	12.1	433.
Hultgren (38)	11.4	423.
JANAF (39)	11.2	392.
Fit to average of Refs. 37,38,39	11.4 mPa (8.57×10^{-5} Torr)	401. mPa (3.02×10^{-3} Torr)

Notes: 1) An equation fit to the "Douglas" curve (Fig. 6 of Ref. 36) was used in this article.

$$\ln P_{Li}\ (Pa) = 23.111 - \frac{18{,}444}{T(K)}$$

2) The data from Hultgren, Nesmeyanov, and JANAF were averaged at several temperatures between 327 and 627°C. An equation fit to the averages is

$$\ln P_{Li}\ (Pa) = 23.054 - \frac{18{,}526}{T(K)}$$

3.2×10^{-3} Torr). The vapor is almost pure Pb ($\sim$96 a/o Pb) compared to almost pure lithium ($\sim$ 2 a/o Pb) for an ideal solution.

Total Pressure in the Chamber. The vapor species are Li, Li_2, Pb, Pb_2, LiPb, He, hydrogen isotope molecules, and gaseous lithium hydride (including deuteride and tritide). The significant species are Li, Pb, He, and hydrogen isotope molecules.

For the $Li_{99}Pb_1$ LMW, the total chamber pressure ranges between 0.03 and 0.9 Pa (2×10^{-4} Torr and 7×10^{-3} Torr) as liquid temperature increases from 400 to 500°C (Figure 24). Monatomic Li is the primary condensible vapor. Diatomic Li partial pressure is small compared to monatomic Li pressure. The Pb partial pressure is at least a factor of 600,000 lower than

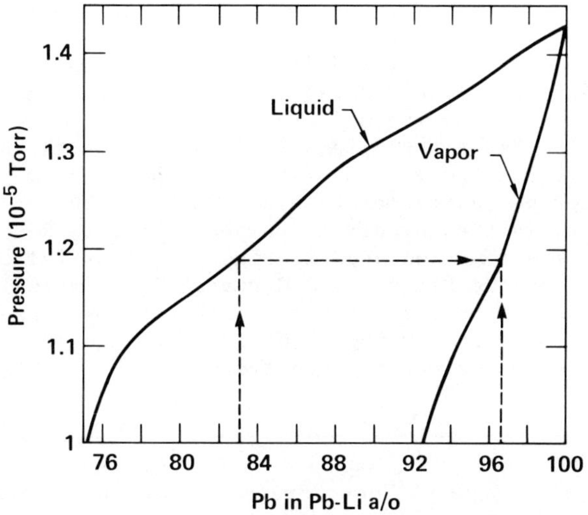

Figure 23. Vapor pressure and composition as a function of liquid composition at 497°C.

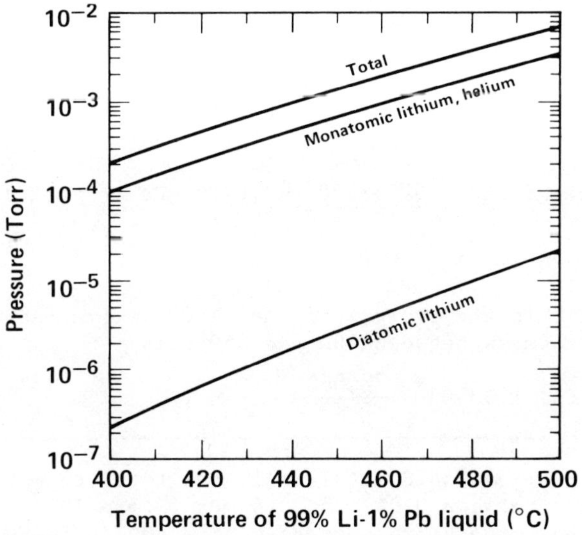

Figure 24. Partial pressures of vapor constituents above $Li_{99}Pb_1$ liquid. He pressure is assumed to equal the sum of the Li, Pb, and H pressures because detailed vacuum system design has not been addressed. Monatomic Li and He are the primary constituents. ^{P}Hydrogen isotopes $<10^{-9}$ Torr at 1 wppm in liquid $P_{Pb} < 10^{-12}$ torr.

Raoult's law predicts due to the nonideality of Li-Pb solutions; hence it is negligible. If the hydrogen isotope concentration in the liquid metal is maintained below 1 wppm by chemical processes outside the chamber, and if the time for gaseous hydrogen to dissolve into the liquid metal is much less than the interpulse time, then the hydrogen isotope partial pressure is also negligible. The He partial pressure depends on both the He generation rate and the vacuum system design. Helium partial pressure cannot be easily maintained much below the Li vapor pressure since, as the chamber is pumped to remove He, more Li vaporizes. The assumption used in figure 24 (partial pressure equals the sum of the Li, Pb, and H_2 partial pressures) should be tested by vacuum pump designers. Although lower pressures are unlikely, considerably higher He pressures could well be required by vacuum engineering constraints.

Table IV. Range of Vapor Pressures in the Literature for Pure Lead

Source	Temperature (°C) 400	500
Nesmeyanov (37)	0.0256 mPa	2.00 mPa
JANAF (39)	0.0265	2.09
Hultgren (38)	0.0316	2.33
Fit to average of Refs. 37,38,39	0.0278 mPa	2.13 mPa
	(2.09×10^{-7} Torr)	(1.60×10^{-5} Torr)

An equation fit to the average of the three references at several temperatures between 400 and 550°C is

$$\ln P_{Pb}\ (\mathrm{Pa}) = 23.03 - \frac{22,560}{T(K)}$$

For a $Pb_{83}Li_{17}$ liquid metal wall, the total chamber pressure ranges between 0.047 and 3.6 mPa (3.5×10^{-7} and 2.7×10^{-5} Torr) as temperature increases from 400 to 500°C (Figure 25). The Li partial pressure is negligible. The H_2 partial pressure depends on the solubility as discussed above; two possible solubility cases are plotted. The lines labeled "Case 1" correspond to high hydrogen solubility. In this case, H_2 pressure is negligible as it is with the other liquid metal. Helium pressure is then also treated as before (set equal to the total of the Pb, Li, and H_2 pressures). The lines labeled "Case 2" correspond to low hydrogen solubility, the more probable case according to initial ANL experiments (30). In this case, H_2 is removed from the chamber through the vacuum system, and the sum of the H_2 and He pressures are set equal to the sum of the Pb and Li pressures. Again, vacuum system design is required. In either case, hydrogen solubility has almost no effect on the total chamber pressure.

Equivalent Lithium Pressure in the Chamber. In order to compare the effects of the two liquid metals on heavy ion beam transport, a method of converting the pressures of each element to effective pressures of a single element is needed. According to Gillespie (41), Pb vapor pressure (and pressures of other elements) can be converted to equivalent Li vapor pressure if it is assumed that stripping of +1 ions in a heavy ion beam to a higher ionization state is the main beam degradation effect. When Gillespie's multipliers of 1/3, 4/9, 1/2, and 113 are applied to the partial pressures of H_2, He, Li_2, and Pb, respectively, each pressure is converted to equivalent (monatomic) Li pressure. Figure 26 shows the results for both fluids. For $Li_{99}Pb_1$, the equivalent pressure for stripping is dominated by Li and He; these components are shown with dashed lines. Two cases are shown for $Pb_{83}Li_{17}$; the lower line uses the minimum He pressure set by the Pb-Li vapor pressure, and the upper line uses the higher He pressure calculated for the other fluid. The higher He pressure seems more likely based on vacuum pumping design considerations.

Figure 26 can be used to compare three cases: $Li_{99}Pb_1$, $Pb_{83}Li_{17}$ with high He pressure, and $Pb_{83}Li_{17}$ with low He pressure. To make the comparison, draw a horizontal line at the maximum allowable lithium pressure. (This value will eventually be provided by the beam propagation community; current estimates range from 13 to 130 mPa or 10^{-4} to 10^{-3} Torr.) Then, compare the liquid metal temperatures for the three cases. For example, at 66 mPa (5 x 10^{-4} Torr) allowable, the maximum liquid metal temperatures are 431, 450, and 472°C for the three cases.

Recently, it has been suggested (42) that beam stripping might be significantly decreased if the beam ion and background gas species were identical. If so, Pb would make an excellent candidate for a heavy ion beam. As shown earlier, the vapor above $Pb_{83}Li_{17}$ is ∿96 a/o Pb (considering only the Pb and Li). If the stripping effectiveness of lead is much less than 100, He will be the dominant component of the background gas with respect to beam propagation. In addition, a ^{208}Pb heavy ion beam has high mass, Pb may also be a target material, and Pb has a very low neutron activation probability. However, isotope separation will be required to separate ^{208}Pb from natural lead.

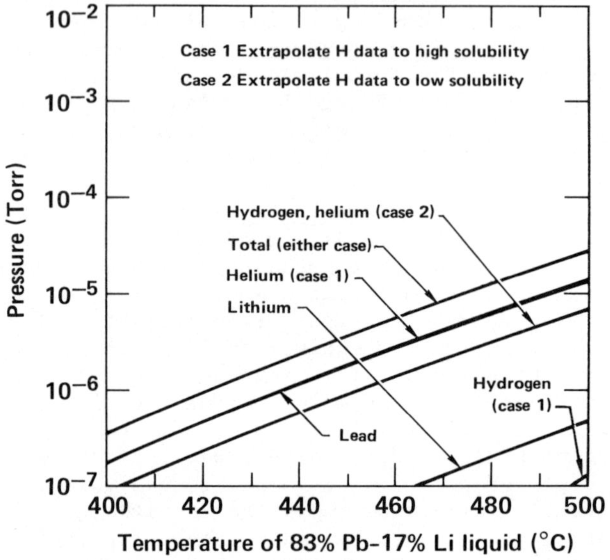

Figure 25. Partial pressures of vapor constituents above $Pb_{83}Li_{17}$ liquid. H solubility is still undetermined, both extremes are plotted. Pb and He are primary constituents.

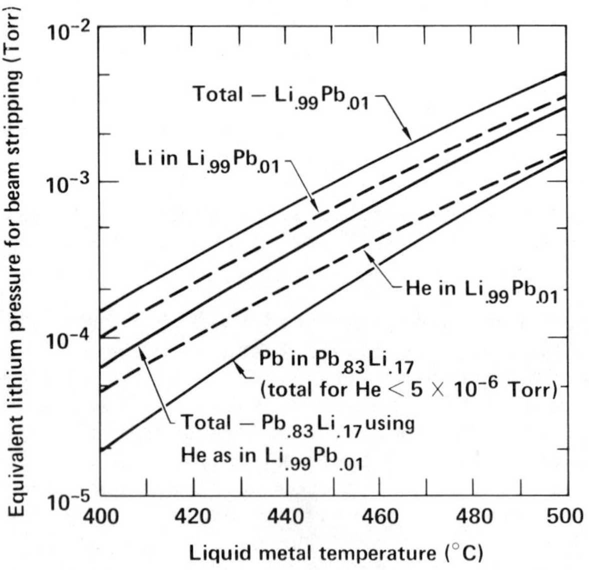

Figure 26. Equivalent Li pressures for heavy ion beam stripping above $Li_{99}Pb_1$ and $Pb_{83}Li_{17}$ liquids.

The lead stripping multiplier of 113 that was used in Figure 26 is the maximum predicted theoretically, and a value of one-half that is not unlikely, according to the theoreticians. However, German experimental work (43) using U^{+8} ions interacting with nitrogen background gas indicates that the theoretical maximum cross section predictions may be too low. If these surprising experimental results are confirmed, the Pb-stripping multiplier for U^{+1} could be as high as 400. The data of Figure 26 are cross plotted in Figure 27 as a function of the Pb-stripping multiplier. Clearly, the allowable equivalent liquid metal temperature is most sensitive to the allowable lithium vapor pressure, even though the pressure range plotted is less than one-half the range of estimates made by the beam propagation community. The lead-stripping multiplier is also very uncertain, but it is of secondary importance in determining the maximum allowable temperatures for the two liquid metals.

Comparison of $Pb_{83}Li_{17}$ and $Li_{99}Pb_1$ as Liquid Metal Walls. A comparison of $Pb_{83}Li_{17}$ with $Li_{99}Pb_1$ can be made by using the difference in allowable temperature based on heavy ion beam propagation. If the $Pb_{83}Li_{17}$ has a 30°C higher liquid temperature, the net plant efficiency is increased by about 0.5% and about 20 MW_e additional power is produced (44). In addition, the higher neutron energy multiplication associated with lead-lithium could produce about 150 MW_e additional power in a 1000 MW_e-net power plant (45). However, the higher density and lower heat capacity of $Pb_{83}Li_{17}$ require about 130 MW_e additional pumping power (46) for the LMW and about 100 MW_e for the primary loop (47). Hence, from a rough power flow viewpoint, using $Pb_{83}Li_{17}$ results in a loss of 60 MW_e net power. A full evaluation of $Pb_{83}Li_{17}$ must also consider, on the negative side, increased flow loop structural requirements, increased fluid radioactivity, plating of activated lead on plant surfaces after a spill, and larger primary heat exchangers; and, on the positive side, additional recovered pump heat and elimination of liquid metal fire hazards. At this point, both fluids are of interest to LMW fusion reactor designers.

Summary

The use of lithium in fusion reactor designs has been described with emphasis on materials compatibility, safety, and tritium extraction. Several chemical issues remain unresolved, including:

- Compatibility of structural alloys with Li-Pb liquids
- Extraction of high-Z target debris from Li
- Chemical physics of heavy ion beams and background gas
- Tritium gettering from the gas phase

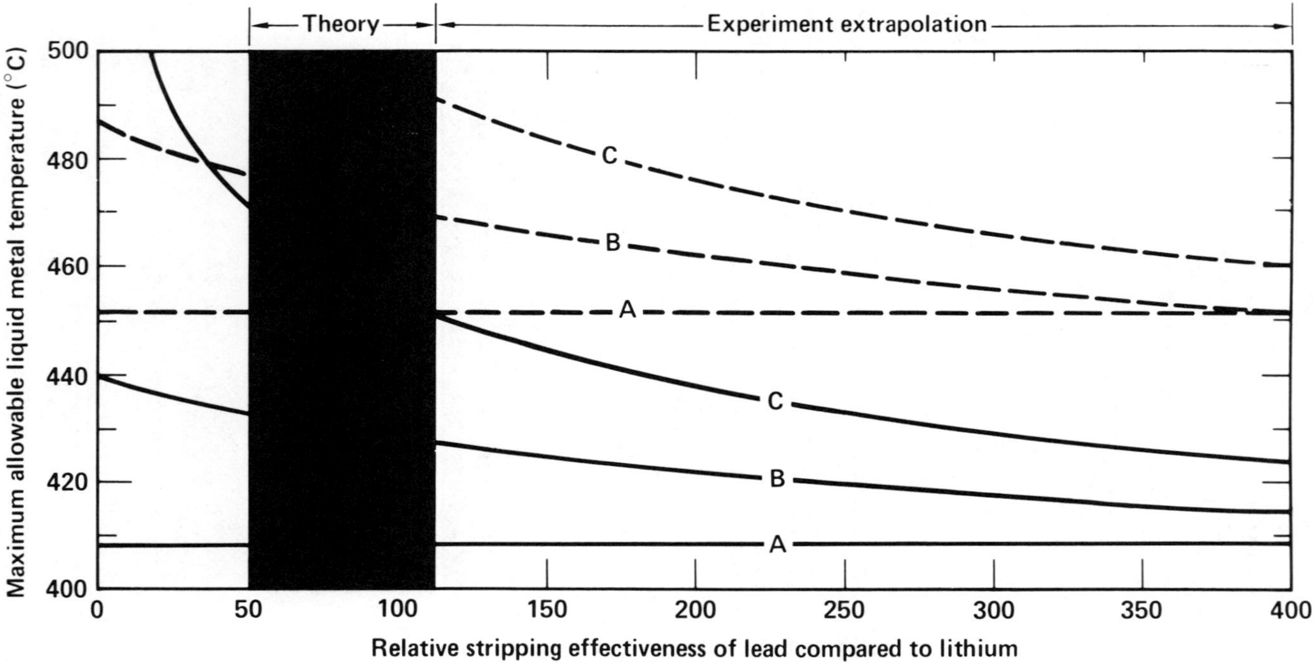

Figure 27. Maximum allowable LMW temperature is more sensitive to allowable equivalent Li pressure than to the stripping effectiveness of Pb. The shaded band is the range predicted by theory. Key: – – –, 10^{-3} Torr Li allowable; ——, 2×10^{-4} Torr Li allowable; A, $Li_{0.99}$ $Pb_{0.01}$; B, $Pb_{0.83}$ $Li_{0.17}$, He ~ $10^{-4} - 10^{-5}$ torr; and C, $Pb_{0.83}$ $Li_{0.17}$, He ~ 10^{-6} torr.

- Effects of radiation on tritium inventory in solid breeders
- Accurate hydrogen concentration measurement in Li-Pb liquids
- Sieverts constant as a function of temperature and Pb concentration in Li-Pb liquids
- Effects of intense magnetic fields on boiling and heat pipes.

Literature Cited

1. Meier, W. R. Nucl. Tech. 1981, 52, 170.
2. Meier, W. R.; Maniscalco, J. A. "Liquid Metal Requirements for Inertial Confinement Fusion"; Lawrence Livermore National Laboratory: Livermore, CA, UCRL-80424, 1977.
3. El-Wakil, M. M. "Nucl. Power Eng."; McGraw-Hill: New York, NY, 1962; p 534.
4. Smith, D. L.; Clemmer, R. G.; Jankus, V. Z.; Rest, J. "Analysis of In Situ Tritium Recovery From Solid Fusion-Reactor Blankets"; Proc. 4th ANS Top. Mtg. Tech. Controlled Nucl. Fusion: King of Prussia, PA, 1980.
5. Maniscalco, J. A.; Berwald, D. H.; Engdahl, J. C.; McKone, T. E.; Whitley, R. H.; Allen, W. O.; Massey, J. V.; McGrath, R. T. "Laser Fusion Driven Breeder Design Study - Final Report"; U.S. DOE Contract DE-AC08-79DP40-11; TRW, Inc.: Redondo Beach, CA, 1980.
6. "Laser Program Annual Report--1978"; Lawrence Livermore National Laboratory: Livermore, CA; UCRL-50021-78, 1979; pp 9-1 to 9-75.
7. "Laser Program Annual Report--1979"; Lawrence Livermore National Laboratory: Livermore, CA; UCRL-50021-79, 1980; pp 8-1 to 8-118.
8. Anderson, T. "An Evaluation of the Corrosion Resistance of 2¼ Cr-1 Mo Steel in a Lithium-Lead Liquid"; M.S. Thesis, Dept. of Metallurgical Engineering, Colorado School of Mines: Golden, CO, 1980.
9. Werner, R. W "The Fusion-Synfuel Tie Producing Hydrogen With the Tandem Mirror Reactor"; presented at 15th Intersociety Energy Conversion Engineering Conference: Seattle, WA, 1980.
10. Krikorian, O. H. "Materials Considerations for the Coupling of Thermochemical Hydrogen Cycles to Tandem Mirror Reactors"; Proc. 4th ANS Top. Mtg. Tech. Controlled Nucl. Fusion: King of Prussia, PA, 1980.
11. Werner, R. W.; private communication: Lawrence Livermore National Laboratory; Jan. 1981.
12. Shunk, F. A. "Constitution of Binary Alloys, Second Supplement"; McGraw-Hill: New York, NY, 1969.
13. Elliott, R. P. "Constitution of Binary Alloys, First Supplement"; McGraw-Hill: New York, NY, 1965.

14. Ihle, H. R.; Wu, C. H. "Experimental Determination of the Partial Pressures of D_2, LiD, and Li_2D in Equilibrium With Dilute Solutions of Deuterium in Liquid Lithium"; Proc. 8th Symp. Fusion Tech.; Leeuwenhorst Congress Centre: Noordwijkerhout (The Netherlands), 1974: published by Commission of European Communities Directorate General; Scientific and Tech. Information and Information Management; Luxembourg, 1974.
15. Watson, J. S. "Evaluation of Methods for Recovering Tritium From the Blankets or Coolant Systems of Fusion Reactors"; Oak Ridge National Laboratory: Oak Ridge, TN, ORNL-TM-3794, 1972.
16. Natesan, K.; Smith, D. "Effectiveness of Tritium Removal From a Lithium Blanket by Cold Trapping Secondary Liquids, Na, K, and NaK"; Nucl. Tech. 1974, 22.
17. Katsuta, H.; Ishigai, T.; Furukuwa, K. "Equilibrium Pressure and Solubility of Hydrogen in Liquid Lithium"; Nucl. Tech. 1977, 32.
18. Veleckis, E.; Yonco, R. M.; Maroni, V. A. "Solubility of Lithium Deuteride in Liquid Lithium"; J. Less-Common Metals, 1977, 55.
19. Buxbaum, R. E.; Johnson, E. F. "The Use of Yttrium for the Recovery of Tritium From Lithium at Low Concentrations"; Princeton Plasma Physics Laboratory, Princeton University: Princeton, NJ, PPPL-1548, 1979.
20. Singleton, M. F.; Folkers, C. I.; Griffith, C. M. "Assessment of Uranium and Cerium as Hydriding Materials for Hydrogen Isotopes in Flowing Argon"; ANS Annual Meeting: New York, NY, 1977.
21. Selle, J. E.; Personal Communication, Oak Ridge National Laboratory: Oak Ridge, TN, 1977.
22. Webb, R. W. "Permeation of Hydrogen Through Metals"; Atomics International: Canoga Park, CA, NAA-SR-10462, 1965.
23. Lee, J. D.; Werner, R. W. "Concept for a Gas-Buffered Annular Heat Pipe Fuel Irradiation Capsule"; Lawrence Livermore National Laboratory: Livermore, CA, UCRL-50510, 1968.
24. Werner, R. W. "The Generation and Recovery of Tritium in Thermonuclear Reactor Blankets;" Lawrence Livermore National Laboratory: Livermore, CA, UCID-15390, 1968.
25. Calaway, W. F. "Electrochemical Extraction of Hydrogen From Molten LiF-LiCl-LiBr and its Application to Liquid-Lithium Fusion Reactor Blanket Processing"; Nucl. Tech., 1978, 39, 63.
26. "Laser Program Annual Report--1979"; Lawrence Livermore National Laboratory: Livermore, CA, UCRL-50021-79, 1980, p 8-68.
27. Ihle, H. R.; Neubert, A.; Wu, C. H. "The Activity of Lithium, and the Solubility of Deuterium in Lithium-Lead Alloys"; Proc. 10th Symp. Fusion Tech.: Padova, Italy, 1978, p 639.

28. Veleckis, E.; Yonco, R. M.; Maroni, V. A. "The Current Status of Fusion Reactor Blanket Thermodynamics;" Argonne National Laboratory: Argonne, IL, ANL-78-109, 1979.
29. Opie, W. R.; Grant, N. J. "Solubility of Hydrogen in Molten Lead"; J. Metals 1951, p 244.
30. Cafasso, F.; Veleckis, E.; Private Communications, Argonne National Laboratory, Argonne, IL, 1980 and 1981.
31. Hoffman, N. J.; Blink, J. A.; Darnell, A. "Properties of Lead-Lithium Solutions"; Proc. 4th ANS Top. Mtg. Tech. of Controlled Nucl. Fusion: King of Prussia, PA, 1980.
32. Saboungi, Marie Louise; Marr, Jane; Blander, Milton. "Thermodynamic Properties of a Quasi-ionic Alloy From Electromotive Force Measurements: The Li-Pb System," J. of Chem. Phys. 1978.
33. Ihle, H. R.; Neubert, A.; Wu, C. H. "The Activity of Lithium, and the Solubility of Deuterium in Lithium-Lead Alloys"; Proc. 10th Symp. Fusion Tech.: Padova, Italy, 1978, p 639.
34. Neubert, A.; Ihle, H. R.; Gingerich, K. A. "Thermodynamic Study of the Molecules BiLi and PbLi by Knudsen Effusion Mass Spectroscopy," J. Chem. Phys. 1978.
35. Nguyen, V. T.; Enderby, J. E. "The Electronic Structure of Lithium-Based Liquid Semiconducting Alloys"; Philisophical Mag. 1977, 35, 4, p 1013-1019.
36. Cowles, J. O.; Pasternak, A. D. "Lithium Properties Related to Use as a Nuclear Reactor Coolant"; Lawrence Livermore National Laboratory: Livermore, CA, UCRL-50647, 1969.
37. Nesmeyanov, A. N. "Vapor Pressure of the Elements"; Academic Press: New York, NY, 1963.
38. Hultgren, R.: et al. "Selected Values of Thermodynamic Properties of Metals and Alloys"; John Wiley and Sons Inc.: New York, NY, 1963.
39. Stull, D. R.; Prophet, H. "JANAF Thermodynamic Tables, Second Edition"; National Bureau of Standards: Washington, D.C., 1971.
40. Demidov, A. I.; Morachevskii, A. G.; Gerasimenko, L. N. "The Thermodynamic Properties of Liquid Li-Pb Alloys;" Translated from Elektrokhimiya, 9, #6, 848-851, (translation, June 1973; original October 1972).
41. Gillespie, G.; private communication: Physical Dynamics, 1980. For convenience, uranium was used as the heavy ion beam in these calculations although a non-fissionable ion will almost certainly be the eventual choice for a HIB fusion reactor.
42. Kim, Yon-Ki; private communication: Argonne National Laboratory; Jan. 1981. Kim said that for identical beam and background species, the beam stripping cross section will decrease to an extent difficult to theoretically predict.

43. Mueller, R.; private communication; Gesellschaft fur Schwerionenforschung, Darmstadt, West Germany; Nov. 7, 1980.
44. "Laser Program Annual Report-1979;" Lawrence Livermore National Laboratory: Livermore, CA, UCRL-50021-79, 1980, 8-78 to 8-83. Steam cycle III in Fig. 8-79 was used.
45. Meier, W.; private communication; Lawrence Livermore National Laboratory, 1980. The ratio of thermal to fusion power for an 80 v/o Pb_4Li - 20 v/o steel liquid metal wall is 1.29 (1-D calculation) as compared to 1.16 for the original HYLIFE (2-D calculation).
46. Monsler, M. J.; Meier, W. R. "A Conceptual Design Strategy for Liquid-Metal-Wall Inertial Fusion Reactors; Lawrence Livermore National Laboratory: Livermore, CA, UCRL-84881, 1980, 28.
47. Ref. 44; Figs. 8-82 and 8-84.

RECEIVED August 26, 1981.

34

Alkali Vapor Transport in Coal Conversion and Combustion Systems

J. W. HASTIE, E. R. PLANTE, and D. W. BONNELL

Center for Materials Science, National Bureau of Standards, Washington, DC 20234

Alkali metal-containing vapor species are ubiquitous in coal conversion and combustion systems. These species originate from coal mineral and atmospheric impurities (organic and inorganic) and from ceramic construction materials. Alternatively, they are present as additives, such as with potassium seeding for MHD or with bulk glass as a particle absorbing medium, or with dolomite in fluidized bed systems. Alkali vapor transport over representative slag, glass, and simple halide, hydroxide, and sulfate systems is discussed in relation to materials and process limitations in coal-supported energy systems. Problems associated with molecular-level vapor transport measurements are also considered.

Vapors containing alkali metal species have diverse implications to high temperature processes (1). Potential new applications of alkalies in combustion systems include--their vapor phase catalytic action in smoke reduction (2, 3), their liquid phase catalysis of coal gasification (4), and their role as electron sources for magnetohydrodynamic (MHD) combustion systems (3). In most combustion systems, however, their presence is undesirable. This is particularly true in fossil energy systems.

More efficient coal utilization can be realized with combined power plant cycles. For instance, the post combustion gases of a conventional combustor or an advanced MHD system can be further utilized to drive a gas or steam turbine. However, the sustained durability of downstream turbine or heat exchanger components requires minimal transport of corrosive fuel impurities. Control of mineral-derived impurities is also required for environmental protection. For the special case of open cycle-coal fired MHD systems, the thermodynamic activity of potassium is much higher in the seeded combustion gas (plasma) than in common coal minerals and slags. This results in the loss of plasma seed by slag absorption and is of critical concern to the economic feasibility of MHD.

0097-6156/82/0179-0543$14.50/0

Empirical experience with conventional coal-fired power plants has indicated minerals containing alkali metal (Na, K), sulfur- and chlorine-bearing species to be the most aggressive fuel components leading to fire-side or hot corrosion (5). Species containing these elements appear to act synergistically in degrading alloy or ceramic materials. The mechanisms by which such species are released from their mineral source, transported, and deposited are not known, though the literature contains numerous speculative schemes (see Ref. 3, p. 216). Rational development of new control strategies, such as gas clean-up or the use of fuel additives, requires a clear understanding of the role played by the active fuel impurities. For instance, new control systems based on scavenging (e.g., absorption of alkali by glass or other oxide media) or chemical modification of the active inorganic impurities will need as design criteria information, such as species identity, concentration profiles, dew points, thermodynamic reactivity, nucleation and absorption rates and diffusivities. Such data will also be pertinent to minimization of seed-slag interaction in MHD systems.

Previous attempts to define the mode of release and transport of fuel impurities have largely been unsuccessful, owing mainly to a lack of knowledge concerning species identities. This has resulted from the inability of molecular specific measurement techniques to function under the combined aggressive conditions of high temperature, high pressure, and high chemical reactivity. We have developed several new measurement techniques ideally suited to such conditions. The first of these techniques is a High Pressure Sampling Mass Spectrometric method for the spatial and temporal analysis of flames containing inorganic additives (6, 7). The second method, known as Transpiration Mass Spectrometry (TMS) (8), allows for the analysis of bulk heterogeneous systems over a wide range of temperature, pressure and controlled gas composition. In addition, the now classical technique of Knudsen Effusion Mass Spectrometry (KMS) has been modified to allow external control of ambient gases in the reaction cell (9). Supplementary to these methods are the application, in our laboratory, of classical and novel optical spectroscopic methods for *in situ* measurement of temperature, flow and certain simple species concentration profiles (7). In combination, these measurement tools allow for a detailed fundamental examination of the vaporization and transport mechanisms of coal mineral components in a coal conversion or combustion environment.

As a long-term objective, we aim to define the mechanisms by which inorganic fuel impurities (particularly K, Na, Cl, S, and heavy metals) and additives (e.g., K in MHD) are released to or removed from the environment, transported in a gas stream and deposited in cooler or chemically less reactive regions. To meet this objective, we are addressing the following basic tasks:

(a) Measurement of species vaporization rates and related thermodynamic functions for well-characterized salt, oxide, mineral, slag, glass, and ash samples under controlled gas conditions and as a function of temperature, time, gas composition and total pressure.

(b) Development, from the basic data derived from task (a), together with auxiliary literature thermochemical data, of computer-based models for prediction of release or retention of alkali and other inorganic components under actual coal combustion, gasification, or MHD conditions.

(c) Validation of models developed from task (b) through comparisons with large-scale test data.

This paper summarizes the status of this activity. Emphasis is given to systems showing unusual behavior, or where additional lines of research are revealed. A more comprehensive presentation of data for many of the systems considered here may be found in the cited references.

Combustion Systems of Interest

Alkali, in the form of Na and K-containing species, can lead to a dramatic reduction in the durability of metal (alloy) and ceramic reactor components through a complex process known as hot corrosion (5). Examples of energy systems where this process occurs include, coal-fired utility boilers, turbines, gasifiers, MHD generators, and pressurized fluidized bed combustors (PFBC). In such cases, the alkali enters the vapor phase by vaporization from coal minerals, dolomite (10) or limestone, as in sulfur-scrubbing processes, or from air-ingested salt particles (e.g., in marine environments). Hardesty and Pohl (11) have recently reviewed the major problem areas and data limitations relating to the properties of coal mineral matter and ash.

Even a minor amount of alkali vapor transport can be significant, as revealed by the turbine tolerance level of 0.02 ppm alkali needed for corrosion control in pressurized fluidized bed combustors (12). If we consider only the alkali halide content of the dolomite component, this tolerance level would require an alkali-scrubbing efficiency of better than 99.9999 percent for PFBC. Even if corrosion (alkali) resistant materials were available, uncontrolled alkali vapor transport would still lead to unmanageable deposits on cool downstream components. For instance, under typical coal gasifier conditions, a species partial pressure as low as 10^{-6} atm would lead to vapor transport and deposition in metric ton quantities on an annual basis.

The need for a basic understanding of alkali vapor transport in fossil energy systems can be appreciated when we consider the diversity of conditions such as temperature, pressure, chemical composition, and time scale, present in existing and developing fossil fuel technologies. Table I summarizes some typical process conditions.

Table I

Typical Coal-Conversion and Combustion Systems

Process	Temperature K	Pressure atm[e]	Comments
	Conventional Steam Plants		
	1000-1500	1	
	Pressurized Fluidized Bed Combustion		
	1200	1-10	2% excess O_2, 300 ppm SO_2
	Coal Gasification[a,b,c]		
Cogas	570-2250	3-6	entrained/slag
Koppers-Totzek[d]	1750-2100	0	entrained/slag, medium btu
Bi-gas	1200-1920	66-100	entrained/slag
Atgas	480-1750	0-0.3	molten iron
Kellogg	1100-1480	27-80	molten salt
CO_2 Acceptor	1100-1370	10-20	fluidized (dolomite)
Hygas	920-1300	66-100	fluidized, high btu, high S
Synthane	370-1260	33-66	entrained/fluidized, high btu
Lurgi	870-1260	20-33	fixed bed, low btu
	Magnetohydrodynamics		
Open cycle--hot walls	1500-3000	1-10	1% K seed, fuel rich or stoichiometric

[a]Comparable conditions exist in ammonia plant secondary reformers, e.g., 1060K, 13 to 20 atm, 28 percent H_2, 50 percent H_2O, 6.4 percent CO, and 4 percent CO_2.
[b]After Crowley (14).
[c]For technological status, see Vorres (15) and Lenzer and Laurendeau (16).
[d]In commercial operation.
[e]1 atm = 1.01325×10^2 kpascal.

Coal Mineral Characteristics

Typically, coal contains about 10 wt.% mineral matter. There is a recognized need for improved understanding of coal mineral transformations and slag-forming processes (13). The chemical form of alkali and halogen in coal is of considerable importance to the mode of alkali vapor transport. Analyses of Gluskoter and Ruch (17) suggest that halogen is present in two forms, NaCl and organic. Also, most of the potassium is present in a halogen-free highly bound form, such as for the mineral illite, or other potassium alumino-silicates. A number of coal combustion systems utilize limestone and dolomite additions for sulfur removal. These materials provide an additional source of alkali. Alkali (Na + K) contents of 0.05 to 1 wt.% are usual and the predominant mineral form is the chloride (18).

Laboratory simulation of coal combustion indicates several modes of mineral decomposition (19). Submicron-size particles tend to be derived from vaporization with subsequent homogeneous and heterogeneous condensation. These particles are rich in silica (SiO vapor transport) but with large enhancements of trace metals, including alkalies, Cd, As, and other heavy metals. Various mechanisms have been suggested concerning the combustion history of the alkali components (20). The principal alternatives are, that NaCl is vaporized during combustion and is not incorporated into silicate minerals--at least in the initial combustion phase--or, that NaCl reacts with the ash thereby lowering the alkali activity and hence the extent of vapor phase alkali transport. This latter statement remains qualitative pending the determination of alkali activities for coal minerals and slags in combustion atmospheres.

Corrosion by Alkalies

Alkali vapor transport and deposition is a well-known, though poorly understood, factor in the corrosion or fouling of alloys and ceramics, both in established and developing technologies. Problem areas include oil-fired glass melting operations (21), blast furnaces, boilers, turbines, coal gasification (22), MHD (23, 24, 25) and coal-fired pressurized fluidized beds (26).

In general, the corrosive effects of alkali deposits result from the high solubility of ceramic and oxide coatings (e.g., for alloys) in molten alkali sulfate, carbonate, chloride, or vanadate deposits. This solubility results from the high stability of Na (or K)-Al-silicates, or similar oxide phases. Formation of these silicates, for instance, leads to a volume increase and loss of structural integrity in ceramic materials (27, 28). In alloy systems fluxing can also occur and this results in greatly increased oxidation rates (5). Even when hot corrosion is not a problem, alkali deposits can lead to fouling and thermal barrier effects. For instance, in secondary naphtha reformers, oxide deposits containing Na_2O (21 wt.%), K_2O (3 wt.%), plus SiO_2, Al_2O_3 and CaO, lead to fouling of waste heat boiler tubes (29).

Alkali vapor transport and deposition places severe limitations on ceramic materials for MHD generator walls and electrodes. Here, the corrosive action of K_2SO_4 and K_2CO_3-containing liquids appears to be the major problem. In the combined presence of potassium seed and coal slag, the rate of electrochemical corrosion of MHD ceramic electrodes increases by two to three orders of magnitude. Alloy corrosion at intermediate temperatures (900 to 1200 K) also can be related to formation of a liquid Na_2SO_4 phase, which prevents formation of a protective Cr_2O_3 scale and greatly enhances corrosion of Co- and Ni-Cr alloys (30).

The thermodynamics of corrosive alkali salt-oxide interaction is not well established. In an assessment of research needs for materials in coal conversion, the need for carbonate-silicate melt studies, including activity and phase equilibrium measurements, was stressed (31). The lack of thermodynamic data for fused salts, and their reactions with oxides and alloys leading to models of hot corrosion, was also indicated.

Hot corrosion of Ni-base turbine alloys by Na_2SO_4 and K_2SO_4 is strongly dependent on the Na_2O activity in the salt (32). Further activity measurements of this type are needed over a wider range of conditions. Raymon and Sadler (22) have reviewed evidence for reactions involving alkali vapors and refractory lining materials for coal gasifiers. They indicate a pressing need for studies of alkali attack in reducing atmospheres at pressures up to 30 atm and temperatures to 1500 K.

Alkali Vapor Transport Modeling

Despite the incomplete state of a thermodynamic data base and limited mechanistic insight, several attempts to model alkali vapor transport in reactive atmospheres have been made. The increased sophistication of modeling efforts in recent years is demonstrated by the following examples:

(a) Coal gasification (33).

(b) Glass furnace corrosion (34).

(c) Pressurized fluidized bed combustion (35).

In example (a), the gas composition was modeled assuming ideal solution phases and neglecting known complex vapor species, such as K_2SO_4, K_2CO_3 and alkali chlorides. These serious limitations resulted from the non-availability of oxide solution-activity data, accurate vapor species thermodynamic functions, and the inability of existing computer codes to handle non-ideal solution multiphase, multicomponent equilibrium computations.

The more recent work of example (b) modeled Na vapor transport by including NaOH, NaCl and Na_2SO_4 as vapor species, with the major uncertainties arising from neglect of solution non-ideality and inaccurate thermodynamic functions for Na_2SO_4 (8).

In example (c), many of the limitations represented by (a) and (b) were resolved; the greatest uncertainty resulted from the highly approximate nature of the alkali-silicate activity data. For this system, the data base requirements are more critical since, in pressurized fluidized bed combustion, the alkali tolerance levels for downstream turbine operation are of the order of 0.02 ppm. Spacil and Luthra (35) compared their thermochemical predictions with observed combustion gas stream alkali concentrations. Here, Na_2SO_4, NaCl, NaOH, and the K-analogues were included as significant molecular species in the thermochemical model data base. Albite and sanidine were assumed to represent the coal ash alkali-getter substrates. Fair, but encouraging, agreement between calculated and observed gas phase alkali concentration was obtained.

Experimental Methods

The primary experimental methods used in this study are the Knudsen Effusion Mass Spectrometric (KMS) and Transpiration Mass Spectrometric (TMS) methods, as described elsewhere [KMS, (9); TMS, (8)]. Both are modulated molecular beam methods with phase sensitive detection, and they allow for accurate measurement of gaseous and condensible species concentrations. The basic differences between the KMS and TMS methods are, the upper pressure limits of 10^{-4} and one atm, respectively, and the upper limit gas residence times of about 0.04 and 20 sec, respectively. Thus, the TMS method will more closely approach the equilibrium condition for systems exhibiting non-equilibrium behavior. As the TMS method is still relatively novel, a brief description is given here.

The Transpiration Reactor. The TMS facility consists of a transpiration reactor mounted in one of two available multi-chambered (two or four), differentially pumped, vacuum systems with a quadrupole mass filter (cross-beam) located in the high vacuum stage. Essential features of the reactor include: a sample container or boat, a boat carrier, a thermocouple for temperature measurement, a carrier gas inlet system, and a gas extraction system or probe, as shown in Figure 1. The boat carrier allows for boat removal from the reactor without need for a complete disassembling of the transpiration system. Molecular beam sonic probes are, typically, conical nozzles with design details determined by reasonably well established gas dynamic criteria. However, for highly reactive systems, we also found it desirable to develop a more robust capillary probe, at the possible expense of sampling fidelity. For the present study, all of these components were fabricated from platinum metal. The reactor is usually operated at total pressures of 0.2 to 1.0 atm, using N_2 or Ar as a carrier gas and temperatures up to about 1700 K.

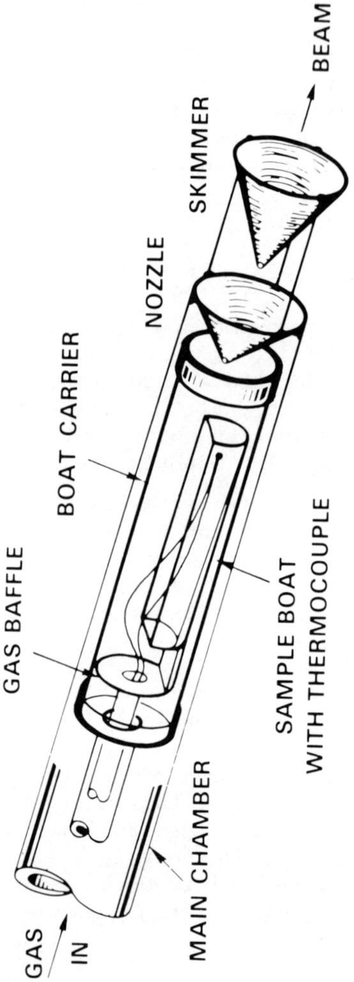

Figure 1. Transpiration reactor showing internal details.

Partial Pressure Determination. For the KMS method, conversion of mass spectral ion intensities to species partial pressures is made through the basic relationships,

$$P_i = k_i I_i^+ T \quad (1)$$

and

$$k_i = \frac{GA_i (2\pi R/M_i)^{\frac{1}{2}} m_i}{ac \sum I_i^+ \, t(T)^{\frac{1}{2}}} \quad (2)$$

where, for species i; P_i is partial pressure, I_i^+ the corresponding ion intensity, T the temperature, k_i an instrument and system sensitivity constant, G a gravimetric factor, A_i an isotope abundance factor, R the gas constant, m_i the weight loss, t the time, c the Clausing factor, a the orifice area, and m_i the molecular weight (or an average value for Knudsen effusion), as described elsewhere (9). In some cases, k_i for different species, but arising from the same experimental configuration, can be inter-related through known ionization cross sections, σ_i, and transmission/detector efficiencies, S_i, i.e.,

$$k_i \propto (\sigma_i \cdot S_i \cdot A_i)^{-1} \quad (3)$$

as discussed elsewhere (8). The degree of agreement for the k_i's obtained from weight loss [expression (2)] or cross section data [expression (3)] provides a useful check on the internal consistency of the partial pressure data.

Relation (1) can also be applied to TMS data. The sensitivity factor k_i is obtained by several independent methods, thereby providing a good test of internal consistency in the data. The basic relationships are,

$$k_i = A_i \, R \, \frac{n}{V} \, \Delta t / \sum I_i^+ t \quad (4)$$

and

$$k_i = k_j \left(\frac{\sigma_j}{\sigma_i}\right)\left(\frac{S_j}{S_i}\right)\left(\bar{S}\right)^{-1} \quad (5)$$

where, for species i and transport gas j (N_2 or Ar); n is the number of moles of substrate transported, V the volume of gas transported, Δt the transport time interval, and $\bar{S}$ an instrument gas scattering correction [$\bar{S}$ = 0.6 to 1.0; (8)]. The reference or transport gas sensitivity factor k_j is readily obtained from equation (1), since the pressure of transport gas is known from an external manometric determination.

Single Component Systems

In complex combustion systems, alkali vapor transport can occur as the metal or as molecular species, such as NaCl, $(NaCl)_2$, NaOH, $(NaOH)_2$, Na_2SO_4, $NaSO_x$ (x = 2,3), $NaPO_x$ (x = 2,3) and, possibly, other yet-to-be established postulated species, such as Na_2CO_3 and $NaCl{\cdot}2H_2O$. Potassium-containing systems show an analogous behavior. Species identities, and their basic thermodynamic functions, are usually best established by vaporization studies over the single component systems. Considerable uncertainty and literature disagreement has existed for most alkali containing systems. We have obtained new thermodynamic vaporization data for the most important systems in this category.

$NaCl(\ell)$ and $Na_2SO_4(\ell)$ Vaporization. These systems are the best established and they serve, primarily, as test cases for the TMS method. The principal TMS results for liquid NaCl may be summarized as follows. Additional detail may be found elsewhere (8). For the reaction,

$$NaCl(\ell) = NaCl(g)$$

$$\log P_{NaCl}\ (atm) = 4.85\ (\pm\ 0.3) - 8820\ (\pm\ 200)/T$$

$$\Delta H_v\ (1300) = 40.4 \pm 0.9\ (42.7 \pm 3.5)\ \text{kcal/mol, and}$$

$$\Delta S_v\ (1300) = 22.2 \pm 1.4\ (24.2 \pm 0.5)\ \text{cal/deg mol}$$

where T is in Kelvin. These second-law data compare favorably with the JANAF (36) evaluation of previous literature data, indicated in parentheses. Our second-law data, obtained by both TMS and KMS, for the dimerization reaction,

$$2NaCl(g) = (NaCl)_2(g)$$

give

$$-\ \Delta H_d\ (1300) = 46.4 \pm 0.7\ (47.1 \pm 3.5)\ \text{kcal/mol, and}$$

$$-\ \Delta S_d\ (1300) = 30.2 \pm 0.6\ (29.7 \pm 1.0)\ \text{cal/deg mol}$$

which compare favorably with the JANAF (36) values, indicated in parentheses. Thus, the vaporization data for this system are well established.

The $Na_2SO_4(\ell)$ system has historically been difficult to characterize, due largely to containment problems. However, from our TMS and KMS second-law data (8), and other recent literature results, a quantitative thermodynamic description of this system is now possible. For the major vaporization process,

$$Na_2SO_4(\ell) = 2Na + SO_2 + O_2$$

$$\Delta H_v\ (1550) = 69.8 \pm 3\ (71.7 \pm 0.6)\ \text{kcal/mol, and}$$

$$\Delta S_v\ (1550) = 27.8 \pm 3\ (29.7)\ \text{cal/deg mol}$$

with the JANAF (36) data given in parentheses. For the secondary reaction,

$$Na_2SO_4(\ell) = Na_2SO_4(g)$$

our data are consistent with those of Kohl et al. (37), who give,

$$\Delta H_v\ (1267) = 65.9 \pm 3\ \text{kcal/mol, and}$$

$$\Delta S_v\ (1267) = 22.3 \pm 2.8\ \text{cal/deg mol}$$

K_2SO_4 Vaporization. Potassium sulfate is a potentially important constituent in coal burning systems, being formed by reaction of ash/slag potassium with combustion SO_2. The solubility of K_2SO_4 (and Na_2SO_4) in coal ash is small, but reaction can occur to form K_2O (and Na_2O) dissolved in the ash with release of SO_2 and O_2. However, the deposition of K_2SO_4 (and Na_2SO_4) in hot corrosion processes indicates that K_2O (and Na_2O) is only partially removed by interaction with coal ash. This could result from slow kinetic processes, competitive chemical reactions, transport of the sulfate as gaseous molecules, or by a series of chemical reactions involving alkali-containing gaseous molecules which lead to a net transport of K_2SO_4 (or Na_2SO_4) across a temperature/concentration gradient.

The thermodynamics of K_2SO_4 and Na_2SO_4 condensed phases are generally considered to be well known and are tabulated by JANAF (36). Thermodynamic data for the gaseous molecules are much less certain, however.

Using separate gravimetric Knudsen effusion and KMS techniques, we have obtained data for the heat of vaporization to form $K_2SO_4(g)$, and detailed results will appear elsewhere (38). From thermal data, the dissociation pressure, as represented by the reaction,

$$K_2SO_4(s,\ell) = 2K(g) + SO_2(g) + O_2(g) \qquad [1]$$

can be readily calculated. The only other reaction of comparable importance is that of the sublimation or vaporization process;

$$K_2SO_4(s,\ell) = K_2SO_4(g) \qquad [2]$$

A number of alternative vaporization processes have previously been suggested for K_2SO_4. However, one or more of the postulated reaction products [$K_2O(\ell,g)$, $SO_3(g)$] have a sufficiently positive free energy of formation so that the proposed alternate reaction paths lead to an insignificant partial pressure of evaporation products compared to reactions [1] and [2]. Most of the previous K_2SO_4 vaporization measurements are thought to be in error because of container material reactions or, possibly, creep of liquid K_2SO_4 from the metal container (usually a platinum metal).

In the present study, weight loss measurements of $K_2SO_4(\ell)$ were made using a thermobalance equipped with a Pt Knudsen cell. Combining these data with the known dissociation pressures for reaction [1] leads to partial pressure data for reaction [2], as shown in Figure 2. Complementary KMS data were also obtained. For comparison we have also indicated, in Figure 2, smoothed data extrapolated from the torsion effusion measurements of Lau, et al. (39), and data from two mass effusion measurements made by Efimova and Gorokhov (40). The absolute pressure data from these three independent recent investigations is in unusually good agreement, but some significant differences still exist (38). These pressure data are about 400 percent greater than the corresponding values of Ficalora et al. (41).

KOH(ℓ) and KCl(ℓ) Vaporization. The thermodynamic stability of KOH in the vapor phase can be obtained from the Second and Third Law analyses of KOH(ℓ) vaporization as the thermodynamic functions for the liquid phase are reasonably well established. JANAF (36) has evaluated the various disparate sets of KOH vaporization data but with considerable uncertainty. Much of the difficulty associated with obtaining reliable thermodynamic data for this system arises from its reactivity with container materials, the presence of carbonate impurity, and the coexistence of dimers and monomers. Previous studies have also been hampered by decomposition to K and H_2O. In the present work, using the TMS technique, we have suppressed this decomposition by addition of H_2O to the carrier gas.

We have obtained extensive data for the KOH(ℓ) and KCl(ℓ) systems, which will be presented in a formal publication elsewhere (42). Representative data for KOH(ℓ) are presented here in comparison with other recent results not considered by JANAF (36). Species partial pressure data are summarized in Figure 3. Note that the KOH species data are in good agreement with JANAF (36), as might be expected. However, there is no agreement between workers regarding the $(KOH)_2$ species, except that the relative amounts of dimer to monomer found in the present study agree quite well with the KMS results of Gusarov and Gorokhov (43). When the monomer and dimer partial pressures are summed, the total pressures are about a factor of two greater than the JANAF (36) data.

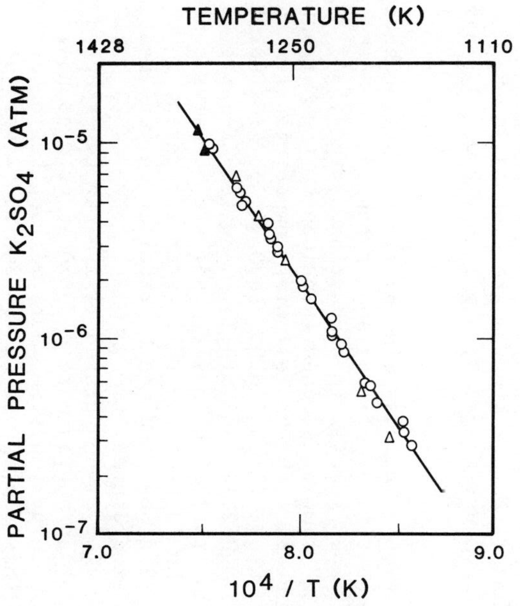

Figure 2. Partial pressure mass effusion data for molecular K_2SO_4 over K_2SO_4(l) (38). Key: ○, present study; △, Ref. 39; and ▲, Ref. 40.

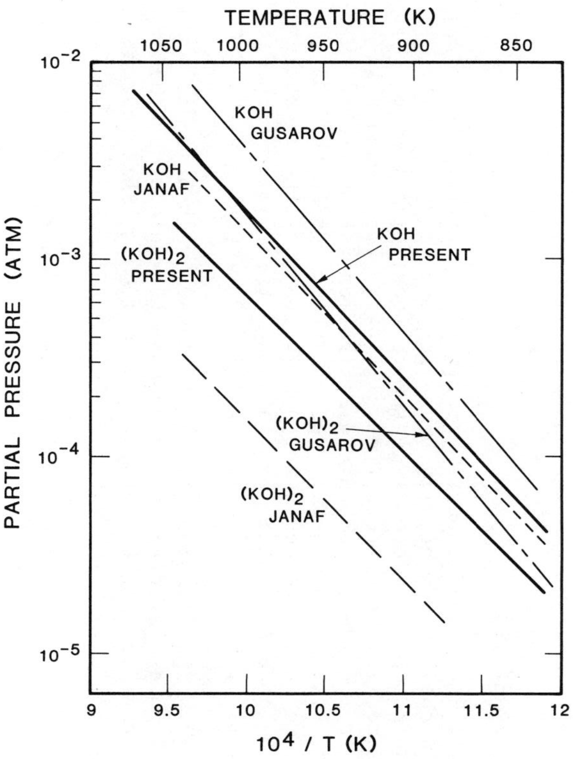

Figure 3. Partial pressure data for KOH and $(KOH)_2$ over liquid KOH, obtained by TMS. Key: ——, present study; — — —, Ref. 43; and – – –, Ref. 36.

With regard to the KCl(ℓ) system, the KCl species data were found to be in good agreement with JANAF (36). However, the dimer species $(KCl)_2$ was found to have an appreciably greater enthalpy of formation than the JANAF (36) value.

Coal Mineral Systems

Alkali Benzoate/Carbonate Vaporization. Part of the alkali content in coal is organically bound and the benzoate salts, NaCOOØ and KCOOØ, have been selected to model alkali release from such a state e.g., see Stewart et al. (44). For Na and K in coal, typical organic/inorganic distribution ratios lie in the range 2 to 9, and 0.02 to 0.1, respectively. Using the TMS approach, we have obtained comprehensive data on the mechanism of Na and K release from their benzoates and the related carbonate-char systems. The detailed results will be reported elsewhere (45).

The principal findings are as follows. On heating to ∿ 800 K, sodium and potassium benzoates decompose to yield a carbonate plus char residue. At temperatures of ∿ 1000 K, this residue reacts to yield Na in the vapor phase according to the reaction,

$$Na_2CO_3(c) + 2C(c) = 2Na + 3CO$$

and similarly for potassium. Comparison of species partial pressures with those of multicomponent equilibrium calculations indicate that this reaction is at, or near, equilibrium, a surprising result for what is formally considered a solid-solid interaction at these relatively low temperatures. Thus, the organic alkali components of coal can be a significant source of alkali in the gas phase. Future studies are planned in the presence of reactive combustion gases, e.g., SO_2, H_2O, CO_2, H_2, and HCl.

Illite Vaporization. The clay mineral illite is frequently found as a mineral constituent of coal and is considered a major source of potassium vapor species in combustion systems. A "Beaver's Bend illite" sample was used (provided by the Morgantown Energy Technology Center) with the following composition for the major components, in wt.%: Al_2O_3(26.0), Fe_2O_3(4.4), SiO_2(60.2), K_2O(7.4), Na_2O(0.2), MgO(2.1), and S(0.1). A water analysis, carried out by heating to 1300 K in air, indicated a water content of 7.3 percent. It should be noted that water is present in the illite structure as OH groups and can be expelled only by heating to relatively high temperatures.

The melting behavior of dehydrated illite can be expected to be similar to that for an approximately 30-60-10 wt.% composition in the Al_2O_3-SiO_2-K_2O ternary system where melting begins at about 1300 K. From this analogy we can assume that over the temperature range of the present experiments (∿ 1540 to 1950 K), illite (dehydrated) will be present as a reasonably homogeneous liquid.

During the initial heating period, significant release of Na and K to the vapor phase was noted over the temperature interval 1300 K to 1500 K. The predominant vapor species were SO_2 (∿ 10^{-3} atm), Na (∿ 10^{-6} atm), K (∿ 10^{-6} atm), and possibly NaOH (∿ 10^{-6} atm). The mass spectral signals for the latter species were less certain, being detected as signal/noise ratios of about two. These initial data also tended to show time dependence in the form of pressure bursts and isothermal signal decay indicative of sample inhomogeneity. Also during this initial period, the O_2 pressures were in excess of the reaction [3] level owing to reduction of Fe_2O_3 to Fe_3O_4.

Following this initial vaporization phase the principal reactions were,

$$K_2O\ (soln.) = 2K(g) + 1/2\ O_2(g) \qquad [3]$$

and

$$SiO_2\ (soln.) = SiO(g) + 1/2\ O_2(g) \qquad [4]$$

At the higher temperatures, Fe_3O_4 contributed to a small partial pressure of Fe(g), of a magnitude similar to the SiO(g) pressure. Representative vaporization data, obtained using the KMS and TMS methods, are given in Figure 4 (see also Table II). A more detailed account of this study is given elsewhere (46). Not all the O_2 pressure data points appear in the figure for reasons of clarity. The O_2/K pressure ratios varied from about 2 to 1/2 during the experiment. This behavior is not consistent with the vaporization stoichiometry indicated by reactions [3] and [4]. Since the K pressure is about 25 times greater than for SiO, reaction [4] is a negligible source of O_2, and most of the O_2 should result from K_2O decomposition according to reaction [3]. For this to be true, the O_2 pressure should be only about 1/4 of the K pressure and have the same temperature dependence as K. The variation of O_2/K pressure with time may be due to excess O_2 dissolved in the illite or, more likely, to iron oxide decomposition. This time-dependent oxygen activity could also account for the K pressure differences between the KMS and TMS methods, as shown in Figure 4.

The Soda-Lime-Silica System

Background. A commercially common soda-lime-silica glass has been considered as an absorbing medium for removing fly ash particulates in combustion gas streams (47). However, a possible limitation with this application is the release of alkali from the glass into the gas stream. Glass also has some common features

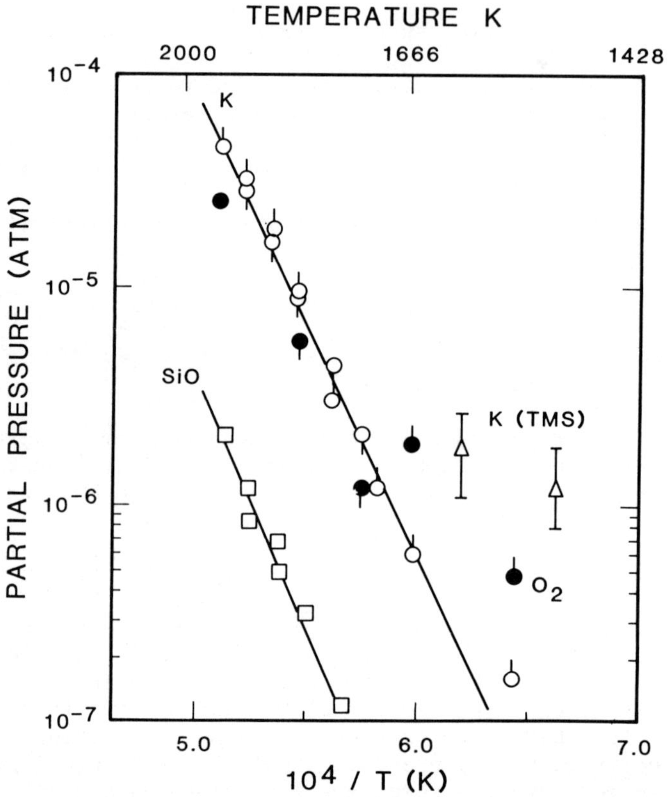

Figure 4. Partial pressure data for K, O_2, and SiO over illite. Key: △, TMS method with N_2 atmosphere; □, ○, KMS under vacuum vaporization conditions; and ●, O_2 with ticks indicating increasing (up) or decreasing (down) temperature chronology.

Table II

Summary of Alkali Vaporization Data for K-Containing Complex Oxide and Coal Slag Systems

System[a]	Phases[a,b,e]	Temperature[c] K	K-Pressure[d,e] atm A	K-Pressure[d,e] atm B	Comments[e]
K_2O-SiO_2	Solution 44 to ~ 5[K_2O]	1300-1800	4.721	15624	Pressures higher than earlier data (56) because of orifice effect.
			C = 9.32		
K_2O-Al_2O_3	$KAlO_2 \rightarrow K_2O \cdot 5Al_2O_3$[f]	1300-1800	5.489	15036	P_K independent of bulk composition (56).
	$K_2O \cdot 9Al_2O_3 \rightarrow Al_2O_3$[f]	1500-1800	7.487	21453	
K_2O-ZrO_2	$K_4Zr_5O_{12}$-ZrO_2	1100-1450	4.524	12873	(57).[g]
K_2O-Fe_2O_3	$KFeO_2$-$K_2O \cdot 6Fe_2O_3$[f]	1150-1450	4.524	12873	Phase boundary composition slightly dependent on T.
	$K_2O \cdot 6Fe_2O_3$-Fe_3O_4	1200-1600	7.028	16750	Data adjusted for constant K/O_2 (58).
K_2O-Al_2O_3-SiO_2 ($KAlSiO_4$)[h]	$KAlSiO_4$, $K_2O \cdot 9Al_2O_3$, $KAlSi_2O_6$	1400-1650	8.722	20923	Probably not equilibrated because of slow condensed phase kinetics (59, 60).
K_2O-Al_2O_3-SiO_2[h]	Possibly $KAlSi_2O_6$, Al_2O_3, $K_2O \cdot 9Al_2O_3$ ~ 22[K_2O]	1600-1900	7.068	20763	Evidence of melt formation.
	Phases uncertain ~ 17[K_2O]	1600-2000	6.350	20693	
	Phases uncertain ~ 11.5[K_2O]	1750-2150	4.667	19800	
	Liquid + $3Al_2O_3 \cdot 2SiO_2$, 10.8[K_2O]	1800-2150	4.464	20424	Decreasing T-chronology used to minimize composition change.

Table II. Continued

System[a]	Phases[a,b,e]	Temperature[c] K	K-Pressure[d,e] atm A	B	Comments[e]
Simplified Western Slags K_2O-CaO-Al_2O_3-SiO_2	Solution + $KAlSiO_4$ at higher K_2O content	1400-1900	5.228 C = 7.887 exp-2, D = 1.741 exp-3, E = 1.825 exp-2	16794	Composition range, [K_2O] 3-14, [CaO] 17-36, [Al_2O_3] 14-36, [SiO_2]-balance. Standard dev. ~ 30 percent, maximum dev. ~ 100 percent. Additional data for 20[K_2O] compositions have been obtained (60-64).
Real MHD Channel Slag (K_1): [K_2O] 19.5, [Al_2O_3] 12.1, [Fe_2O_3] 14.3, [CaO] 3.8, [MgO] 1.0, [SiO_2] 46.8, [Na_2O] 0.5.	$KAlSiO_4$ major + minor unidentified crystalline phase + slag	1600-1800	5.564	16650	Illinois no. 6 coal, UTSI test sample. Initially high pressure of K due to presence of K_2SO_4 and K_2CO_3 (0.2 percent S). Equation corresponds to [K_2O] ~ 14 (65).
Synthetic Western Channel slag: [K_2O] 22.7, [Al_2O_3] 24.6, [Fe_2O_3] 5.3, [CaO] 9.3, [MgO] 3.3, [SiO_2] 34.8.	$KAlSiO_4$ + glassy phase	1500-1700	6.818	18472	[K_2O] 18.9-17.6 (66).
Synthetic Eastern Channel slag: [K_2O] 23.6, [Al_2O_3] 25.5, [Fe_2O_3] 12.5, [CaO] 1.8, [MgO] 0.6, [SiO_2] 36.0.	$KAlSiO_4$ + glassy phase	1600-1750	6.831	19056	23.3-22.1 [K_2O]. Linear portion of AB, BC curves in Figure 9 (66).
Illite: [K_2O] 7.4, [Al_2O_3] 26.0, [Fe_2O_3] 4.4, [MgO] 2.1, [SiO_2] 60.2.	Liquid	1550-2085	6.286	20642	(46).

Table II. Continued

System[a]	Phases[a,b,e]	Temperature[c] K	K-Pressure[d,e] atm A	K-Pressure[d,e] atm B	Comments[e]
Low Melting Synthetic Slag (K_2): $[K_2O]$ 8.7, $[Al_2O_3]$ 11.1, $[Fe_2O_3]$ 12.0, [CaO] 13.9, [MgO] 7.9, $[SiO_2]$ 47.3.	Liquid	1470-1820	6.231	17863	$6[K_2O]$. Initial $P_{O_2} \geq P_K$, but later fell to the level of K_2O (see text) (65).

[a]Initial compositions, given in weight percent (wt.%). Note, actual analyses not normalized to 100 percent. Fe_2O_3-containing systems usually converted to Fe_3O_4 with loss of O_2 under vacuum or neutral atmosphere vaporization conditions.

[b]Phases indicated are solid.

[c]Temperature range over which experimental data were taken.

[d]Results obtained by KMS. See comments for applicable compositions. Potassium partial pressure data can be represented, to a good approximation, by the empirical form,

$$\log P\ (\text{atm}) = A - B/T + C + D - E \quad ,$$

where A is an entropy term and B is enthalpy; C, D, and E are the factors by which the concentrations (normally in wt.%) of K_2O, CaO, and Al_2O_3, respectively, are to be multiplied. The factor C in the K_2O-SiO_2 system is to be multiplied by the mole fraction squared. P refers to pressure (atm) of elemental K, as the only K-species present. T is temperature in Kelvin. Other data for individual compositions may be found in the reports cited (see comments). Note, similar data-fits are obtained for compositions expressed in mole %.

[e]Square brackets indicate weight percent composition. Exp-2 denotes, for instance, 10^{-2}.

[f]Exact crystalline phase compositions uncertain.

[g]Pressure could be low by a factor of two because of vapor unsaturation (large orifice). Also, K_2CO_3 contamination present.

[h]All compositions constrained to mole ratio $SiO_2/Al_2O_3 = 2$.

with coal slag, and the basic thermodynamic data derived from glass-combustion gas studies will benefit our basic understanding of slag-gas interactions.

Description of alkali release, or retention, by glass requires accurate Na_2O activity data at various temperatures and glass compositions. However, thermodynamic activity data for glass systems are surprisingly sparse. No critical analysis, for instance of the type represented by JANAF Thermochemical Tables (36), has been made on the available data. Glass activity data for a common system can vary by several orders of magnitude, or more, depending on the measurement method used.

In the present study, we have utilized the KMS and TMS methods to obtain vaporization and activity data for a glass of initial composition (wt.%): $Na_2O(17)$, $CaO(12)$, and $SiO_2(71)$. The sample, obtained from the Ceramic, Glass and Solid State Science Division (NBS), was prepared by melting together the appropriate mixture of Na_2CO_3, $CaCO_3$, and SiO_2 in a platinum container in air to about 1800 K until all visible CO_2 was expelled, except for a residual Na_2CO_3 impurity level as discussed below. This glass is similar to those suggested as being useful for removing fly ash particulate matter from fluidized bed coal combustion systems. In addition, the binary silicate system (no CaO present) is one that has been extensively studied by a variety of methods and reliable Na_2O activity data are available. There is also limited information on the effect of CaO additions on the Na_2O activity (49). Thus, indirect comparisons can be made with activity data for the ternary system.

KMS Measurements. Partial pressures for Na and O_2 were obtained under vacuum vaporization conditions using the KMS technique. Following an initial heating period, these pressures followed the stoichiometry of the reaction,

$$Na_2O \text{ (soln.)} = 2Na(g) + 1/2\ O_2(g) \qquad [5]$$

During the initial heating period, an excessive amount of Na, as well as CO_2, was observed. Initially the CO_2 pressures approximated those for O_2 but decreased to a negligible level ($CO_2/O_2 < 0.1$) in subsequent experimental runs. Two processes appeared to be controlling the release of CO_2,

$$Na_2CO_3 \text{ (soln.)} = 2Na + 1/2\ O_2 + CO_2 \quad \text{and} \qquad [6]$$

$$Na_2CO_3 \text{ (soln.)} = Na_2O \text{ (soln.)} + CO_2 \qquad [7]$$

with the former predominating in the early and lower temperature phase of the experimental runs, and the latter at higher temperatures and later observation times.

Figure 5 summarizes Na partial pressure curves for two experimental runs. Experimental details have been presented

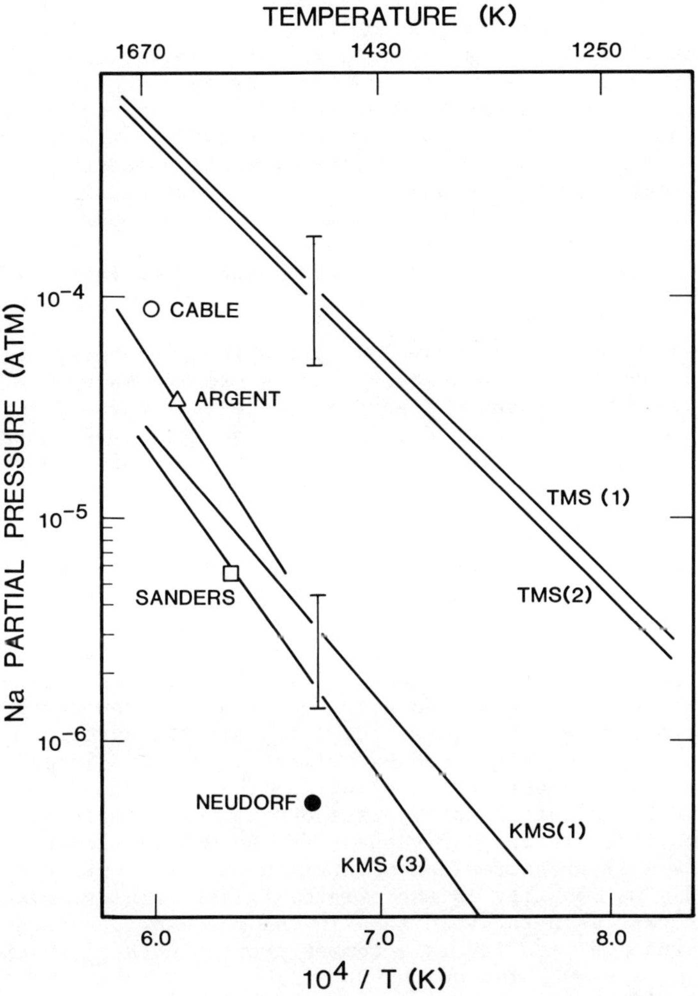

Figure 5. Comparison of glass melt Na partial pressure data obtained by various workers for compositions (wt %) similar to Na_2O(17), CaO(12), and SiO_2(71). KMS(1), 17.0–16.7 and KMS(3), 15.6–13.3 wt %. TMS(1), 17.0 and TMS(2), 16.9 wt %. Key: ●, Ref. 49; □, *Ref.* 51; △, *Ref.* 52; *and* ○, *Ref.* 50.

elsewhere (48) and the data points closely followed the curves given in Figure 5. The vertical bar represents the maximum conceivable error that could have arisen with the KMS (run 1) data. Several data sets from the literature are indicated for comparison. Neudorf and Elliott (49) measured Na_2O activities in the binary silicate solution, as well as the effects of CaO on the Na_2O activity, using an emf method. We have extrapolated their data to our experimental conditions based on the effects of Na_2O and CaO content on the Na_2O activity. The data point of Cable and Chaudhry (50) was obtained by a classical transpiration method under conditions where surface segregation effects were negligible. Similarly, the data point of Sanders et al. (51) represents a stirred-melt transpiration experiment where surface depletion is also unlikely. The curve of Argent et al. (52) represents Knudsen effusion mass spectrometric data (without beam modulation).

TMS Measurements. Glass vaporization in a N_2 atmosphere was monitored using the TMS technique. Representative Na-partial pressure curves are given in Figure 5. Note that these partial pressures are more than an order of magnitude greater than those obtained by the KMS technique. Ideally, both sets of data should coincide. We believe that the explanation for this apparent discrepancy is as follows.

Under the conditions of the KMS experiments, the rate of alkali removal was about an order of magnitude greater than for TMS. This may be seen by the calculated composition change in the respective glass samples, i.e., KMS(3) final wt.% Na_2O = 13.3 and TMS(2) = 16.9 as compared with the initial composition of 17.0 wt.%. During the initial phase of each type of experiment, excessive amounts of CO_2 and Na were released. Only when about one percent of the glass Na_2O was depleted did the excess CO_2 and Na become negligible in the KMS experiments. As this level of alkali depletion was never reached during the TMS experiments, we believe that these latter data correspond to the anomalously high Na pressures found in the early phase of the KMS experiments. These high alkali pressures can be attributed to the presence of unreacted Na_2CO_3 impurity in the original glass samples, even though care was taken to avoid this in the glass preparation. Residual carbonate impurity is a common problem with glass experimentation e.g., see Cable and Chaudhry (50).

We calculate from the time-integrated CO_2 and excess Na signals that the initial concentration of impurity Na_2CO_3 was 0.45 wt.%. From the relative amounts of Na, O_2, and CO_2 released prior to vaporization from the silicate itself, two types of impurity-related vaporization processes appear to be present, as represented by reactions [6] and [7]. Comparison of the TMS and KMS data indicates that the activity of Na_2O (solution), produced by reaction [7], is substantially greater than that for the silicate-bound Na_2O characteristic of the pristine glass. Apparently, the Na_2O produced in situ by carbonate decomposition is not readily incorporated into the silicate matrix, at least on the time scale of the vaporization measurements. Formally, we can consider Na_2CO_3 as a solute, in a metastable glass solution, and with an activity defined by reaction [6], as discussed in the following section. Alternatively, one could argue that under the higher vaporization rate conditions typical of KMS, surface depletion of alkali led to the relatively low Na-pressures observed, e.g., see Cable and Chaudhry (50). However, no significant isothermal time dependent vaporization was noted on the several minute time scale of individual KMS measurements. Also, the KMS(1) pressures are greater than the stirred-melt data of Sanders et al., (51) which would not be the case in a surface-alkali depleted system.

Na_2CO_3 in Glass. We have interpreted the anomalously high alkali vapor pressures of the TMS experiments described above in terms of impurity Na_2CO_3 decomposition in a glass solution. By monitoring the release of CO_2, and integrating over time, we have determined the mole fractions of Na_2CO_3 present at the various measurement temperatures and times. Hence, from the observed partial pressures for reaction [6], and the corresponding reference state values (36), we can calculate Na_2CO_3 activity coefficient data, as shown in Figure 6. These data appear to be thermodynamically reasonable and tend to support the alkali carbonate impurity interpretation of alkali vaporization differences between the KMS and TMS experiments (see above). We can likewise argue that the data of Cable and Chaudhry (50), shown in Figure 5, also appear to suffer from this impurity problem, even though they also took precautions to eliminate residual carbonate during the glass synthesis process.

Future studies should be pursued under controlled doping conditions and in atmospheres containing CO_2 and O_2. The known synergistic effect of CO_2 on O_2-solubility in silicate melts at very high gas pressures has, in fact, been interpreted in terms of Na_2CO_3 formation in solution. Effects of this type could significantly enhance alkali vapor transport in practical combustion systems.

Complex Oxide Systems and Slags

Background. As part of a program of systematic studies on potassium-containing slag systems, vaporization data have been obtained for a series of synthetic-binary, ternary, quarternary and sexternary oxide mixtures, as well as an actual MHD coal-slag sample. Virtually no experimental thermodynamic activity data exist for these systems. Even for the most studied relatively simple K_2O-SiO_2 system, existing data is extremely crude and incomplete. Comparison of our activity data for this system with those reported by Charles (53), as based on the moist-atmosphere

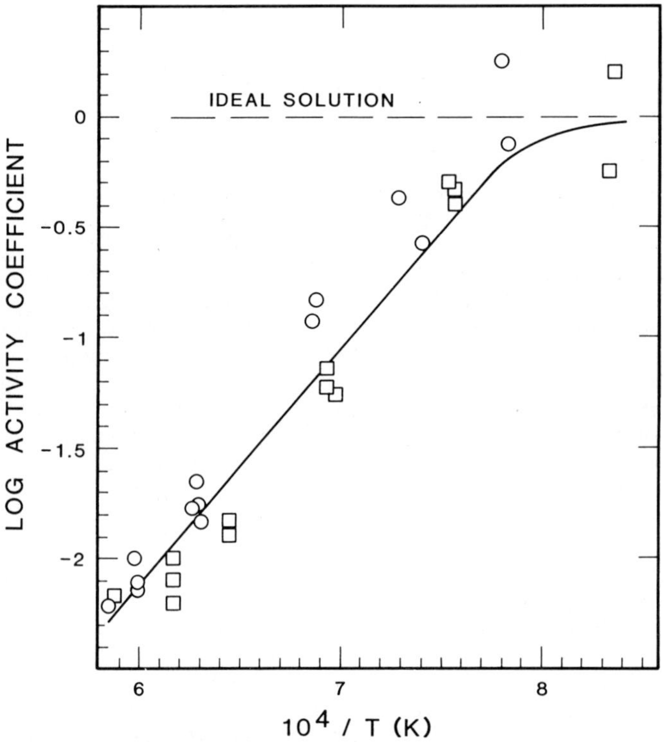

Figure 6. Activity coefficient data (TMS) for Na_2CO_3 (0.45 wt %) in glass (see Fig. 5). ○, run chronology of increasing temperature; and □, run chronology of decreasing temperature.

transpiration data of Preston and Turner (54), indicate several order-of-magnitude differences, with the Charles's results being low. Likewise, the estimates of Spacil and Luthra (35), as based on the phase diagram, yield activities one to two orders of magnitude lower than our experimental values [for details, see (26)].

Table II contains a summary of selected potassium vaporization data for these systems. With a few noted exceptions, the oxygen partial pressures coupled stoichiometrically with potassium, in keeping with reaction [3]. Hence, $K_2O(\ell)$ activities can be derived, to a good to excellent approximation, using the corresponding dissociation-pressure data for pure liquid K_2O. That is,

$$A_{K_2O} = (P_K)^{5/2}/2K_p \qquad (6)$$

where K_p is the dissociation constant (see reaction [3]) for pure liquid K_2O. We have derived a temperature dependent expression for K_p as follows. From JANAF (36), K_p data for $K_2O(s)$ are available. Combining these results with the fusion enthalpy and entropy data of Natola and Touzain (55), and an estimated C_p for liquid K_2O of 25 cal/deg mol, leads to the expression,

$$\log K_p\ [K_2O(\ell)] = \frac{-23751}{T} + 11.7723$$

The potassium vapor pressure data were obtained under neutral conditions using the TMS technique. For most coal-conversion and combustion systems of interest, iron will be present in the slag as Fe_3O_4. Hence the data reported in Table II were obtained, for the most part, at temperatures (and run times) where Fe_2O_3 (slag) had essentially converted to Fe_3O_4 (slag). Evidence of this reduction was conveniently established by monitoring the O_2 pressure. These results can also be applied to systems where O_2 sources other than reaction [3] are present. In this case, the K-pressures are converted to activities using equation (6). The activity data are then combined with the known equilibrium constants for reaction [3], and the assigned O_2-pressures, to yield the new K-pressure data. For relatively low temperature oxidizing conditions, where Fe_2O_3 (slag) may be present, we can reasonably assume that the present alkali activity data will be valid. That is, the various forms of iron oxide do not significantly affect the alkali activity.

Experimental details and an extended discussion of these potassium vapor pressure results have either appeared, or will appear, elsewhere, as indicated in Table II. In general, the potassium vaporization data followed the expected Clausius-Clapeyron behavior, as indicated in Table II. Exceptions to this behavior could be traced to:

(a) residual alkali carbonate impurities resulting either from the sample synthesis method, or from condensation out of MHD plasmas with actual slag samples;
(b) alkali carbonate and sulfate phases in real slag samples;
(c) non-equilibrium effects in the condensed phase;
(d) non-equilibrium between the condensed and vapor phase leading to an unsaturated vapor;
(e) changing phase boundaries due to incongruent vaporization;
(f) rapid loss of alkali pressure with time (T constant), possibly resulting from surface depletion in highly viscous systems; and,
(g) Fe-controlled redox reactions resulting in changing oxygen and, hence, K-partial pressure data with temperature and time.

A systematic study of systems with a progressive increase in the number of components was made in order to isolate and quantify such behavior. Some of these exceptional cases are discussed in the following sections.

K_2O-Al_2O_3-SiO_2 System. The K_2O-Al_2O_3-SiO_2 system has the potential for forming several stable or metastable phase assemblages in which the K_2O activity is fixed according to the phase rule (see Table II). However, in practice, we observe the K pressure (and, hence, K_2O activity) to be dependent on the bulk K_2O concentration. This effect could result from several non-equilibrium factors including, slow condensed phase kinetics attributable to the complex crystal chemistry, changing composition in the β-alumina phase (which extends from K_2O/Al_2O_3 ratios of 1/5 to 1/9), or to dissolution of SiO_2 in the β-alumina phase. The vapor pressure equation in Table II, for the three phase region, is based on the highest alkali-pressures which were effectively independent of composition.

Other evidence for non-equilibrium behavior is shown in Figure 7. In the previous experimental run, alkali-pressure data were obtained as a function of increasing temperature up to 1860 K. On decreasing the temperature, lower pressures were found than for the increased-temperature run, particularly at 1760 K and below. As shown in Figure 7, continuation of this experiment initially produced low pressures over the AB interval. The temperature and K_2O concentration at which this phenomenon was noted is reasonably consistent with the phase diagram (67) which shows eutectic melting at 1829 K and 22 wt.% K_2O. We attribute this loss of alkali-volatility to formation at the sample surface of a frozen eutectic melt which is probably glassy in nature. For an equilibrium system, this melt would recrystallize but for the present experimental conditions there was probably insufficient time. The phase being depleted by vaporization is $KAlSiO_4$ (kalsilite) and the pressure-loss results from the slow alkali transfer rate across the frozen eutectic which results in partial isolation of the remaining $KAlSiO_4$. Note, in Figure 7,

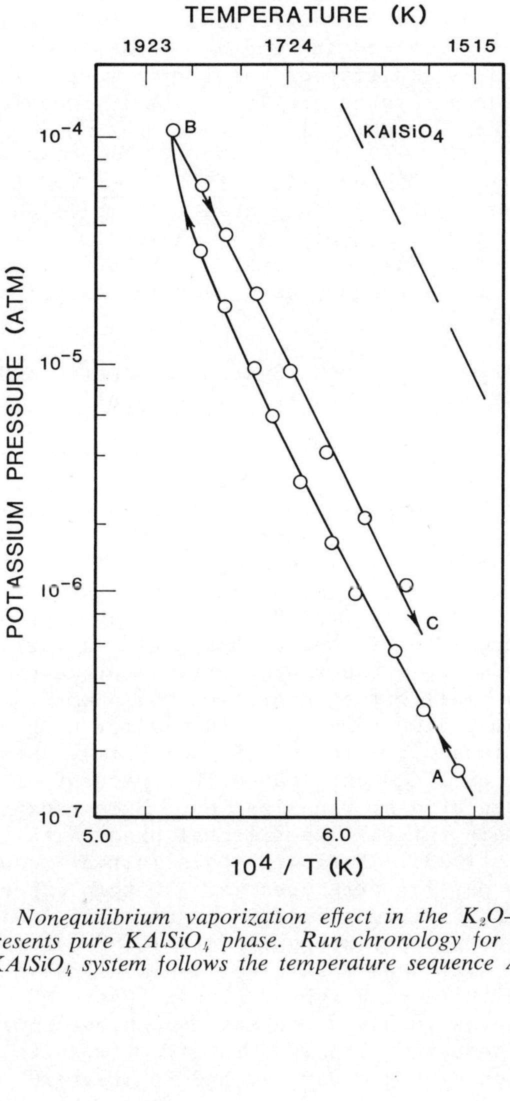

Figure 7. Nonequilibrium vaporization effect in the K_2O–Al_2O_3–SiO_2 system. ––– represents pure $KAlSiO_4$ phase. Run chronology for the partially decomposed $KAlSiO_4$ system follows the temperature sequence ABC (KMS data).

the upward curvature of the AB interval with increasing temperature. This unusual behavior probably results from remelting of the eutectic barrier-phase with an increased rate of alkali transport through this barrier together with incorporation of additional $KAlSiO_4$ in the eutectic melt. The BC interval of Figure 7 represents vaporization from this regenerated phase. This effect was noted only for experiments where a small amount of the $KAlSiO_4$ phase remained and where a potassium-deficient surface glaze acted as a diffusion barrier to vaporization. Representative vapor pressure data for "normal" behavior in this system are summarized in Table II.

K_2O-CaO-Al_2O_3-SiO_2 System (Simplified "Western" Slag). These four component systems, designated as Simplified "Western" slags in Table II, are relatively well behaved in terms of alkali vaporization and are useful model systems for sub-bituminous basic coal slags. The data have been cast in analytical form, as summarized in Table II.

K_2O-CaO-MgO-Al_2O_3-Fe_2O_3-SiO_2 System (Synthetic MHD Channel Slags of "Eastern" and "Western" Coal Types). Literature compositional analyses of several hundred coal ash and a few MHD channel slags, containing potassium seed, have been evaluated for the purpose of selecting representative compositions for modeling. Table II indicates the compositions selected as most representative of MHD channel slags of "Eastern" and "Western" coal types. These slags are non-glassy with $KAlSiO_4$ as the dominant crystalline phase (in both slag types) and small amounts of $KAlSi_2O_6$ and at least one unidentified phase (for "Eastern" slag type only). Thus, the bulk composition is not as meaningful a variable as for homogeneous glassy slags. Vapor pressure measurements were made using a relatively small effusion orifice (0.34 mm diameter) to prevent vapor unsaturation effects found earlier on similar mixtures but with larger orifices (0.5 to 1.0 mm). Steady state pressures were not obtained until about ten percent of the K_2O content had been depleted by vaporization. Excess oxygen was vaporized during this initial experimental phase with reduction of FeO_x (x = 1.5 to 1.33). Following this initial reduction period, the oxygen partial pressures were, within experimental error, what would be expected for K_2O dissociation (reaction [3]).

Figure 8 compares vapor pressure data for "Western" and "Eastern" slags. This comparison clearly demonstrates a basic difference between these slag types in that "Eastern" slags require about twice as much K_2O-content to achieve similar alkali pressures as for "Western" slags. This effect results from the higher concentration of basic CaO and MgO in "Western" slags leading to a less complex silicate structure in the "Western" slag and, hence, a less bound form of K_2O.

The vaporization behavior of the "Eastern" slag, in the composition range 24 to 21 wt.% K_2O, is very different to that

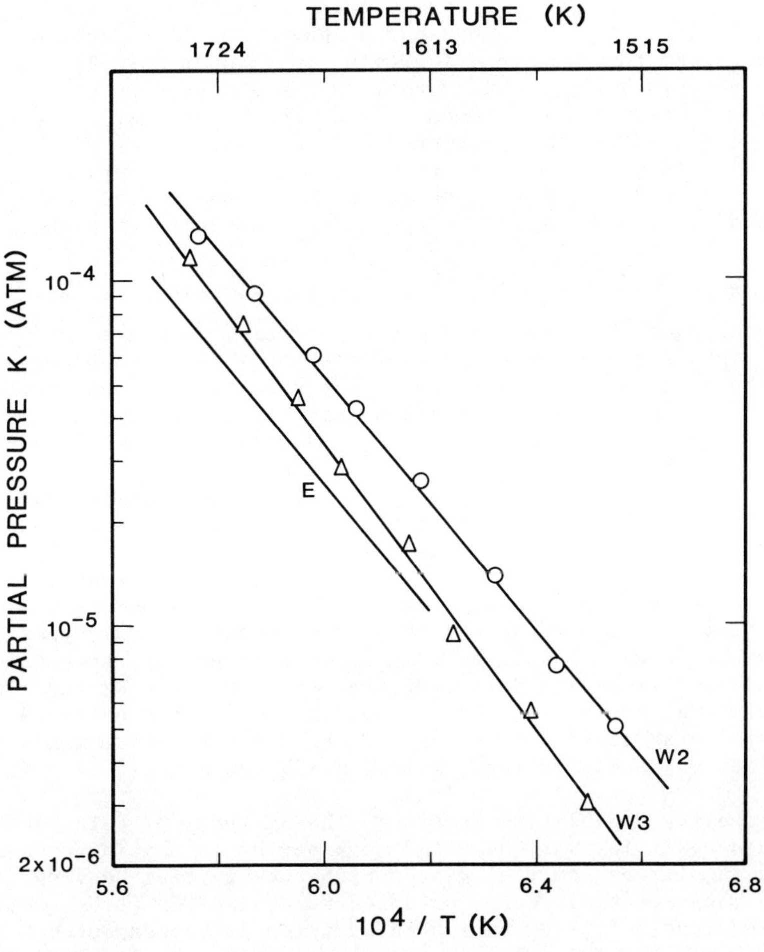

Figure 8. Selected K partial pressure data (KMS data) for synthetic "Western" (W) and "Eastern" (E) MHD-channel slags with K_2O compositions (wt %) of W_2, 19–17.6(○); W_3, 12.9–10.8(△); and E, 23.3–22.1 (——). "Eastern" slag data points omitted for clarity (66).

for the "Western" slag as shown in Figure 9. Here, the vapor pressure curves show a significant break at location X, the position of which depends on the amount of sample vaporized. Note the significant positive deviation of the observed K and O_2-pressures (XC interval) from those obtained by extrapolating the higher temperature linear portion of the respective curves. The curves labeled AB in Figure 9 were obtained prior to the curves labeled BC. At temperatures above 1620 K, these two curve sets are seen to merge into a single linear portion. This non-linear behavior is attributed to a changing oxygen potential in the slag. At the higher temperatures, the oxygen partial pressures were greater than predicted by K_2O dissociation alone (reaction [3]). However, at the lower temperatures (XC region), less oxygen was observed than expected for reaction [3]. Apparently, FeO_x undergoes reduction at temperatures above 1620 K with release of additional O_2 (above the K_2O level). Below this temperature, FeO_x is oxidized by the O_2 resulting from K_2O dissociation. It is significant that thermodynamic equilibrium is maintained during this redox process, as evidenced by a common K_2O activity over the AX and XC intervals (data not shown in Fig. 9). For the lower FeO_x content "Western" slags this effect was not observed.

Real MHD Channel Slag (K_1). Detailed TMS and KMS studies were made of vapor transport over a high liquidus temperature (∿ 1700 K) potassium-enriched coal slag with initial composition as indicated in Table II. This slag sample was obtained by combustion of Illinois No. 6 coal with additional potassium added to the combustor [see (65)]. Note that this slag composition lies between those of the "Eastern" and "Western" coal-types. For identification purposes, this slag is given the designation K_1. X-ray diffraction data indicated that the bulk of the slag potassium was present as the compound $KAlSiO_4$. TMS analysis indicated that about two percent of the slag potassium was present in relatively volatile form, mainly K_2SO_4 and K_2CO_3.

Identity of Volatile Species: The as-received potassium-enriched coal slag was subjected to a series of heating cycles (runs) in nitrogen carrier gas. During the initial heating cycle, mass spectral scans, obtained using the TMS technique, revealed many volatile species in addition to the expected K and Na species. The following species were positively identified: H_2O, CO_2, SO_2, O_2, K, and Na. Some of the other low-intensity ion signals can be very tentatively assigned to the species (some hypothetical): KO or KOH, KS or KSH, SiS, SiSH, H_2S, H_2SO_4, and KSiO. From JANAF (36), we can expect to see KOH under these conditions. Some of these more minor species may result from slag occlusions and metastable phases and most likely do not represent an equilibrium release from the slag. Following this initial heating cycle, the only significant slag vapor species

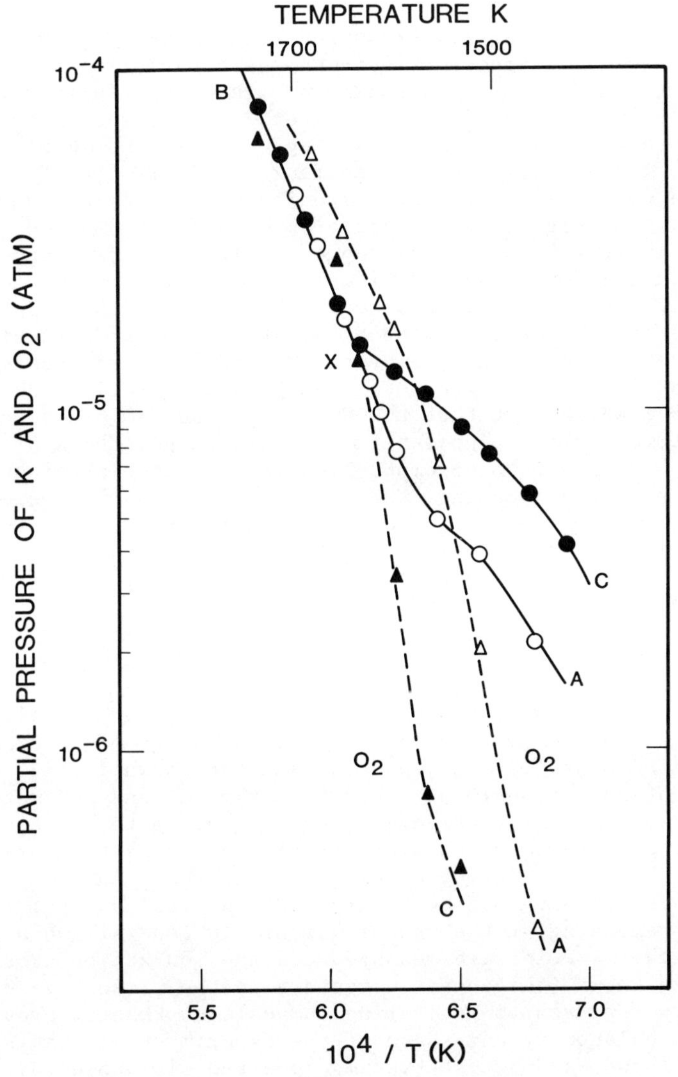

Figure 9. Selected K and O partial pressure data (KMS) for a synthetic "Eastern" MHD-channel slag with composition (K_2O wt %) 23.3–22.8 (AB interval) and 22.8–22.1 (BC interval). Run chronology follows the temperature sequence ABC. Key: –○–, K data with increasing temperature run-chronology and –●–, K data with decreasing temperature run chronology. – – △ – –, O data, with increasing temperature run chronology and – – ▲ – –, O data with decreasing temperature run chronology.

were K and O_2, and these were present in the approximate stoichiometric ratio expected for K_2O decomposition.

Initial Species Partial Pressure--Temperature Dependence: The initial volatiles showed a non-monotonic variation of partial pressure with temperature, as shown in Figure 10. These volatiles constitute only a few percent of the total slag components and are not representative of the bulk slag composition. However, they do provide a sufficiently high flux of alkali (Na, K) and SO_2 to be a potential source of corrosion in downstream MHD components. The high initial partial pressures of SO_2, CO_2, K, and Na are indicative of the presence of alkali sulfate and carbonate in the slag. An additional contribution to low temperature ($T < 1300$ K) alkali release could result from the high H_2O content leading to the formation of volatile hydroxide species (KOH). However, no definitive hydroxide signals were observed. Note that at $T > 1400$ K, the potassium pressures fall below those expected from $KAlO_2$, but that the SO_2, CO_2, and H_2O pressures are still relatively high. Apparently, at this stage, the K produced by sulfate and carbonate decomposition is retained in the bulk slag. After further heating, the sample was virtually depleted of Na, SO_2, and CO_2; H_2O also continued to fall-off in pressure to a negligible level. Following this initial clean-up period, the sample showed a more normal vaporization behavior and representative data are summarized in Table II.

K_2O Activity Coefficients: Most of the bulk-slag composition changes result from K_2O dissociative vaporization. Therefore, a continuous monitoring of the K-partial pressure (and O_2) allows one to calculate the slag composition at any stage of an experiment using relationships (1), (2), and (4) and the known initial sample weight and composition. For both the TMS and KMS methods, the basic experimental requirement for monitoring the bulk composition is the measurement of significant species partial pressures as a function of time during an experimental run. An independent check on this approach can be provided by chemical analysis of the sample remaining at the end of a run. In general, when all the significant species are measured and the ionization cross sections are known (though not necessary in the present case), this <u>in situ</u> approach to monitoring composition changes provides a good mass balance at any stage of the experiment, as was shown for the NaCl and Na_2SO_4 test systems reported elsewhere (<u>8</u>).

Since the mole fraction of K_2O can be defined at any stage of an experiment, it is possible to convert K-partial pressures to K_2O activity coefficients using equation [6]. By varying the amount of K_2O present in the slag <u>during</u> a vaporization run, we were able to follow the dependence of the K_2O "apparent" thermodynamic activity on temperature and composition. The term "apparent" is used to emphasize that the slag system may not always be in a state of complete thermodynamic equilibrium. Typical data,

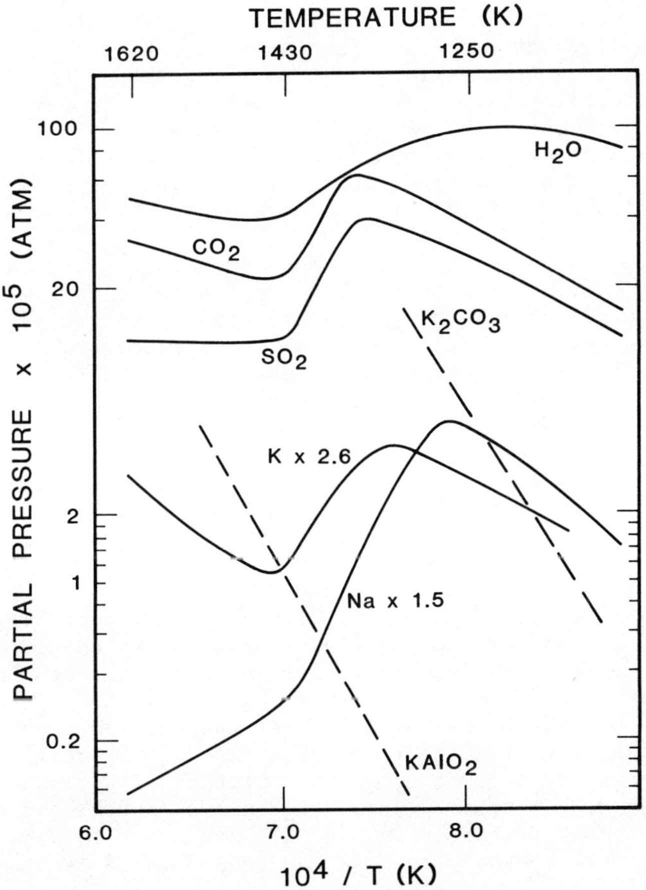

Figure 10. Partial pressure variation of initial volatiles (K_2O, 19.5–19.1 wt %) as a function of temperature and time for the K_1 slag (liquidus temperature ~ 1700 ± 30 K) using TMS approach (run 1). Conditions: 0.5 atm N_2, capillary probe. – – – represents K-pressures over the K_2CO_3 and $KAlO_2$ phases.

expressed in activity coefficient form [$\gamma(K_2O)$], are given in Figure 11. Most of these data were obtained below the liquidus temperature. Good agreement was obtained between the TMS and KMS-based data at relatively high temperatures (∿ 1600 K). From this observation, we can conclude that the high temperature data represent thermodynamic equilibrium because of the greatly different residence time scales involved, i.e., TMS ∿ 10 sec and KMS ∿ 0.04 sec. The largest experimental uncertainty in comparing activity coefficient data from the two different techniques (KMS and TMS) is the accuracy of the K_2O mole fraction, which is probably uncertain by ten percent in each case. Comparison of these data with those for the K_2O-SiO_2 system, for instance, at 10 mole % K_2O in each case and 1700 K, indicates a K_1 slag activity coefficient of about 30 times that for the binary system. This observation is consistent with the more basic character of the K_1 slag.

Note in Figure 11, the non-monotonic nature of the activity coefficient curves. For normal non-ideal solution behavior, we would expect a linear monotonic relationship with a negative slope representing a negative partial molar enthalpy of solution for K_2O in the slag. This type of behavior occurs for segments of each run (see Fig. 11), e.g., for run 1, up to about 1430 K, and for run 2 between 1430 and 1630 K. However, the run 1 data are anomalous as they were obtained during the initial heating period when K_2SO_4 and K_2CO_3 decomposition was a significant source of additional K. The rapid reduction in $\gamma(K_2O)$ as the temperature is increased beyond 1430 K results from the virtually complete depletion of these relatively volatile forms of potassium. For runs 2 and 3, the initial reduction of $\gamma(K_2O)$ with increasing temperature is believed to be due either to diffusion limited (in solid slag) K-transport to the slag surface or to changes in the mode of O_2 release from the slag, for instance, through Fe_3O_4 dissociation. The onset of increasing $\gamma(K_2O)$ with temperature is believed to arise either from an as-yet unspecified physiochemical change in the slag, leading to a less viscous (but still solid) form and increased diffusion, or to depletion of secondary O_2 sources. At higher temperatures, the bulk composition changes rapidly (see mole fractions in Fig. 11) and leads to a peaking in $\gamma(K_2O)$. We believe that the data for runs 2 and 3, at temperatures in excess of 1450 K, represent an equilibrium vaporization condition, particularly as the KMS and TMS data are in agreement for these conditions. Clearly, these unusual trends in the $\gamma(K_2O)$ data indicate the difficulty involved in making *a priori* predictions of real slag vaporization behavior.

Synthetic Low Melting Slag (K_2). A lower liquidus temperature (∿ 1480 K), less viscous (as compared with the K_1 sample) synthetic slag was prepared for studies analogous to those performed for the K_1 system. It was hoped that this slag would not show the same anomalous activity behavior, at lower temperatures, as for the K_1-MHD coal slag sample.

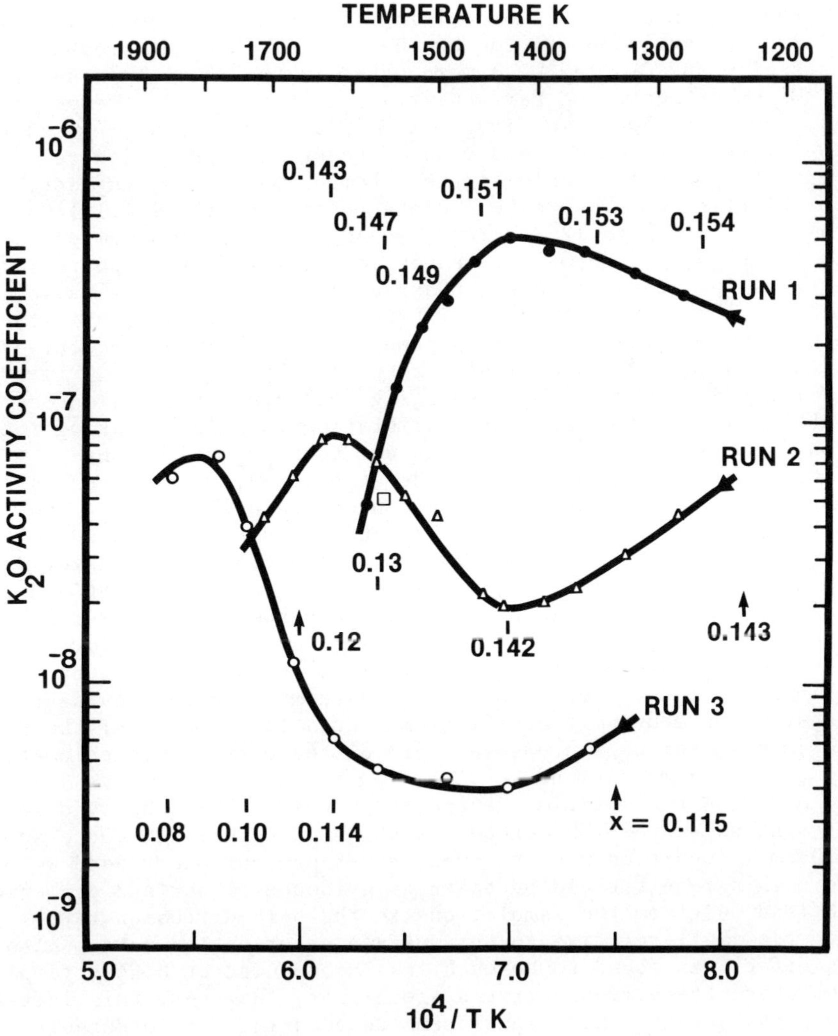

Figure 11. KMS data for variation of K_2O activity coefficient with temperature and composition for the K_1 slag. The numbers, ranging from 0.154 to 0.08, refer to the mole fraction of K_2O (X) remaining in the sample at each measurement point. Runs 1–3 were carried out consecutively on the same sample. The □, data point at 1575 K (run 1) obtained by TMS with additional O_2 present.

Under free vaporization (KMS) or N_2-atmosphere (TMS) conditions, the predominant vapor species in this system are K and O_2, as shown in Figure 12. The initial excess of O_2 present in this slag is believed to result from the following sources. First, the preparation procedure of melting and pouring in room air may have led to oxygen absorption by the sample. Second, pressure bursts of K and O_2 (and CO_2) were noted in the initial phase of the TMS experiments and particularly near the liquidus temperature, e.g., at 1500 K in Figure 12. This effect is attributed to K_2CO_3 impurity. Third, reduction of Fe_2O_3 to Fe_3O_4, with release of excess O_2, is favorable at these temperatures and, in fact, has a similar temperature dependence to that of the initial O_2 data shown in Figure 12. Using the JANAF (36) thermochemical data for Fe_3O_4 and Fe_2O_3, it is possible to predict O_2 partial pressures for given condensed phase activities. On this basis, the initial experimental data of Figure 12 at 1600 K, for instance, are consistent with $\sim$ 50 percent and 25 percent Fe_2O_3 reduction for the KMS and TMS experiments, respectively.

The KMS data were obtained using the integrated ion intensity-weight loss method of pressure calibration [Eq. (2)], taking into account the additional weight loss due to Fe_2O_3 reduction. Calibration of the TMS data, on the other hand, was made using the relative ionization cross section approach [Eq. (3)]. The apparent difference between the KMS and TMS data, indicated in Figure 12, is related to the problem of additional sources of O_2 already mentioned. That is, the TMS data were obtained at an earlier stage of the sample history, where the high O_2 pressure depresses the K-pressure by the mass-action effect. In fact, if the data are converted to K_2O activities, the KMS and TMS data are in satisfactory agreement. Such agreement is good evidence of system thermodynamic equilibrium. At a later phase of the KMS experiments, the O_2/K pressures were of the correct stoichiometry for $K_2O(\ell)$ decomposition.

A second type of time dependent phenomenon was observed for this slag using the TMS method, as shown in Figure 13. Once an isothermal condition was achieved, the K-pressure decreased with time. This result could be taken as evidence of surface depletion of K (and O_2) from the sample, due to the bulk diffusion rate being too small relative to the surface vaporization rate. Also, this effect was found to be much less pronounced at higher temperatures where the diffusion rates are higher. However, this initial interpretation (65) no longer seems reasonable. The order-of-magnitude greater vapor transport rates for KMS vs TMS experiments would indicate lower apparent activities in the former case due to surface depletion effects. However, in practice the KMS activities are somewhat higher. Hence, this time dependent phenomenon is attributed to the combined effects of K_2CO_3 impurity decomposition, as noted previously for the analogous glass system, and, to FeO_x reduction, as noted with the "Eastern" slag data in Figure 9. That this latter effect was more apparent in the TMS, versus the

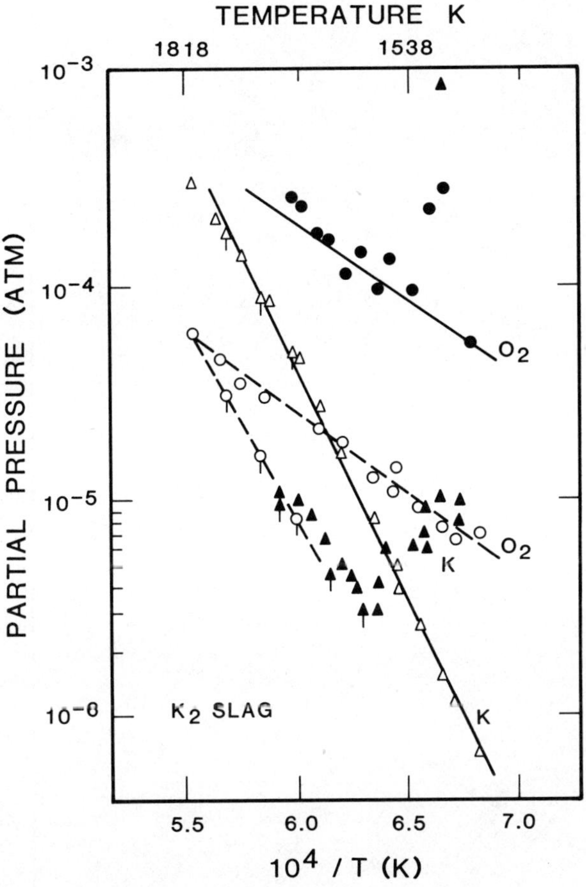

Figure 12. Vaporization of K and O_2 from the K_2 slag. Key: ○, O_2 (KMS); ●, O_2 (TMS); △, K (KMS); and ▲, K (TMS). Chronological order of data taken with increasing temperature except for ticked data points where the temperature was decreasing.

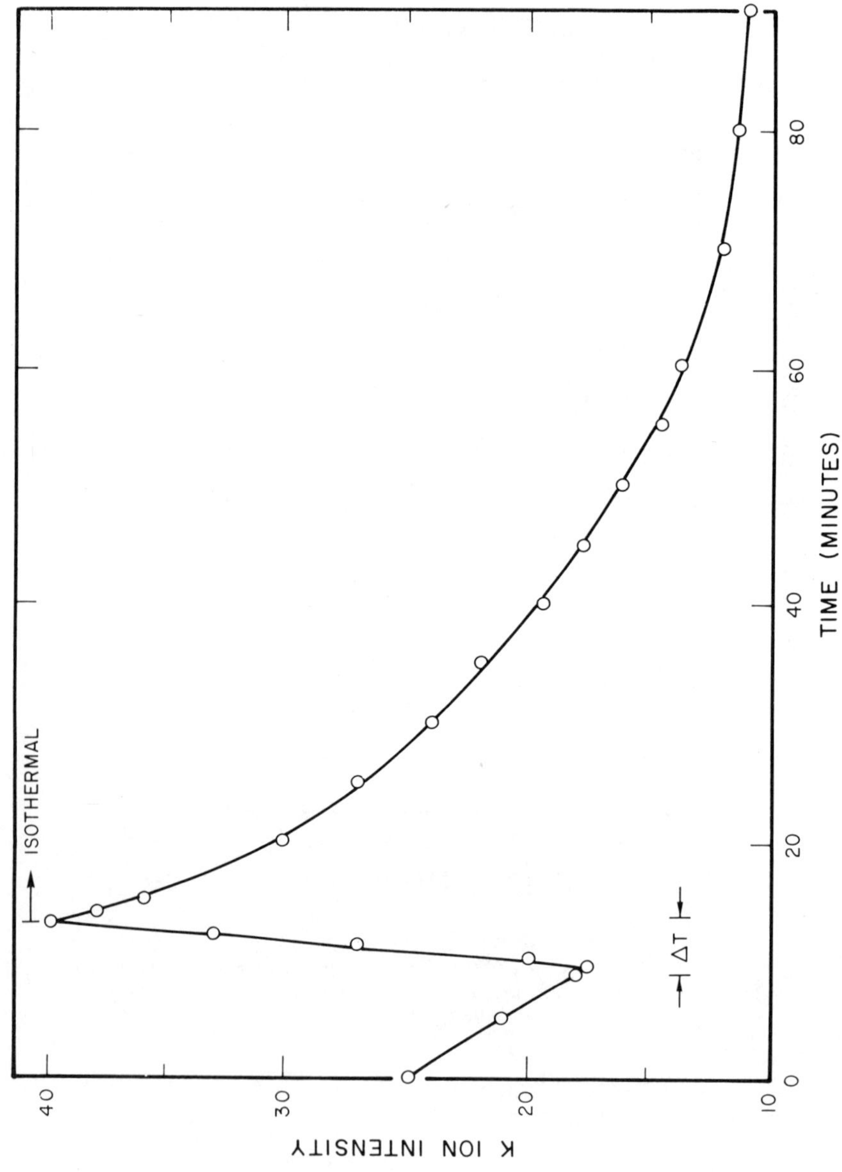

Figure 13. Typical K^+ ion intensity signal-decay with time for K_2 slag. TMS conditions: 1570 K; N_2 carrier gas pressure, 0.21 atm; capillary nozzle.

KMS, experiments can be attributed to the much higher transport rates and shorter FeO_x reduction times for the latter case.

Application of Vaporization Data to Seed-Slag Interaction. A key factor for successful MHD operation is the degree of interaction between plasma potassium seed and the slag medium. Using slag activity data from the present studies, it is possible to predict conditions under which plasma seed will be continuously depleted by slag absorption of alkali. Plante et al. (56) presented similar arguments earlier, based on their data for the binary oxide systems. A more definitive analysis can now be made from the present data on complex synthetic and actual slag systems.

The stability of K_2O in slag solutions can be readily determined by comparison of the K_2O dissociation pressure-product ($P_K{}^2 \cdot P_{O_2}{}^{\frac{1}{2}}$) data in the slag phase with that in the plasma phase, as represented in Figure 14. This dissociation pressure (DP) expression is a convenient representation of slag activity data [activity = DP, soln./DP, $K_2O(\ell)$]. The DP curves for the plasma phase were calculated using a multicomponent equilibrium computer program, assuming stoichiometric combustion of $CH_{0.7}$ with air (4/1 mole ratio N_2/O_2) and KOH, K, KO, K^+, and e as vapor phase species. Pressures of ten and one atmosphere were chosen to be representative of the MHD combustor and the channel-diffuser-downstream seed recovery units, respectively. Corresponding curves for the actual MHD slag (K_1) and the Synthetic "Western" and Simplified "Western" slags, were calculated from the experimental vapor pressure data given in Table II.

Comparison of the DP curves for plasma and slag indicate temperatures for K_2O slag saturation in the range of 2100 to 2300 K at 10 atm, and 1850 to 2050 K at one atm, depending on the slag type and composition. The approximately 200 K equivalence temperature difference between actual and model slags is attributed, primarily, to the low CaO concentration in the former case. Slag absorption of K_2O at the channel and diffuser surfaces depends strongly on the plasma-surface interface temperature. Experimentally, these interface temperatures are difficult to measure though the data reported by Self (68) indicate that they are in the region of 2100 to 2300 K. At these temperatures, and at atmospheric pressure, slags with at least 18 wt.% K_2O are predicted to be stable with respect to the plasma phase, as indicated in Figure 14. Such a prediction is in accord with the observed 19.5 wt.% K_2O content of the actual MHD slag (K_1). Conceivably, calculations of this type, together with actual slag analyses, could be used to infer the actual interface temperatures.

Note that the DP equivalence temperatures in Figure 14 are significantly lower for the atmospheric conditions characteristic of downstream units. Also, as the temperature decreases, the slag dissociation pressure-product decreases much faster than for the plasma, resulting in a super-saturated alkali vapor concentration

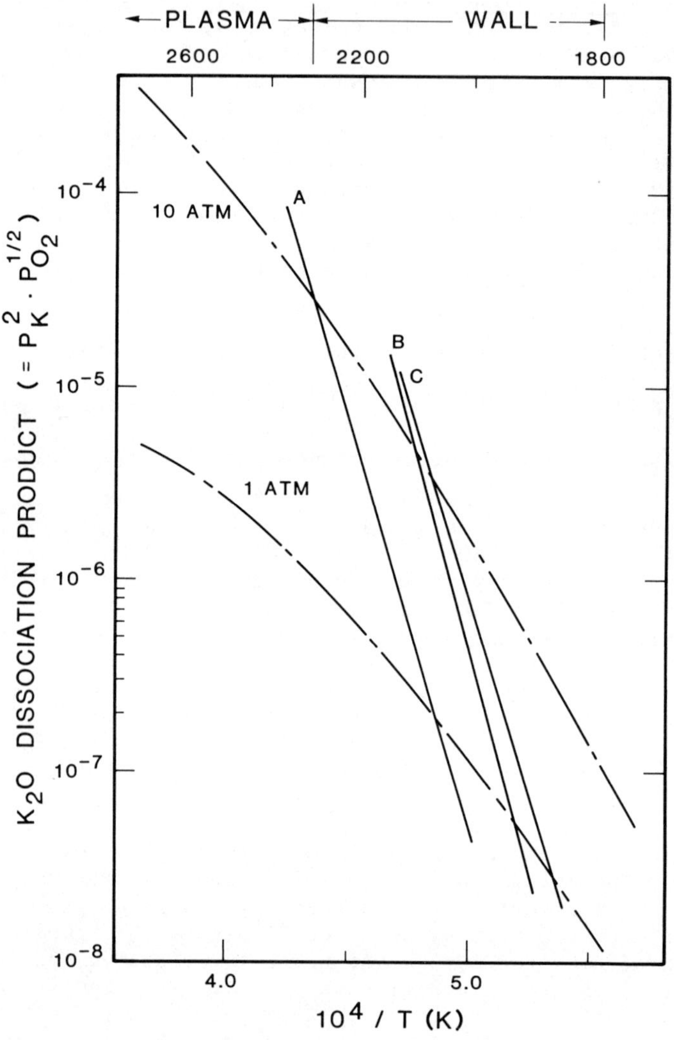

Figure 14. Comparison of K_2O dissociation pressure–product data for the MHD plasma with 1 wt % K_2CO_3 seed (– — –) and slag (——) phases. Curve A, actual slag (K_1) with 14 wt % K_2O content; Curve B, synthetic "Western" slag with 18 wt % K_2O; and Curve C, quaternary simplified "Western" slag with 15 wt % K_2O.

at the plasma-slag interface. Hence, additional K_2O will be absorbed by the slag. Below the dewpoint, this seed will tend to deposit on the slag surface in a relatively non-bound form, such as the alkali sulfate or carbonate, which is desirable from a seed-recovery point of view. As the temperature decreases, the reduced slag species diffusion rates and the increased tendency for phase separation to occur will allow K_2SO_4 and K_2CO_3 deposits to remain at, or near, the slag surface, as is found in practice. Indirect evidence of this type of deposition may be seen in Figure 10.

Salt-Slag Alkali Exchange. The common disposition of alkali in coal minerals is Na as NaCl and K as K_2O--bound in a low-activity silicate phase. Thus, during coal conversion, Na is expected to be released to the vapor phase more readily than K. However, the possibility of NaCl-K_2O (slag) interaction to produce KCl-Na_2O (slag) could greatly enhance K-release to the vapor phase. Also, in MHD slags, where about 20 wt.% K_2O content is possible, the problem of recovering this lost seed could likewise be resolved through replacement by NaCl. The feasibility of such an exchange process was tested by a TMS monitoring of the vapor phase over the system, NaCl + K_1 slag (19.4 wt.% K_2O). Details of this study will be given elsewhere (69), but the main observations are as follows.

When a thin layer of powdered NaCl was present on the surface of the K_1 slag, a rapid exchange reaction occurred near the melting point of NaCl, i.e.,

$$\mathrm{NaCl}(\ell) + \mathrm{K_2O}\ (\mathrm{slag}) = \mathrm{KCl}(\ell) + \mathrm{Na_2O}\ (\mathrm{slag})$$

This result is demonstrated in Figure 15, where the observed partial pressures of NaCl and KCl are expressed in thermodynamic activity form. Note the marked decrease in NaCl activity and concomitant increase in KCl activity just above the melting point of NaCl. After a heating period of about 50 min., the NaCl sample was virtually depleted, as was the KCl product. Insufficient salt was present in the initial mixture to convert all the available K_2O to KCl. However, 90 percent of the initial NaCl was converted to Na_2O (slag) with stoichiometric release of KCl. About six percent of the available K_2O was converted to KCl vapor, and we expect that nearly complete removal of K_2O from the slag would have been possible if sufficient NaCl was present. The remaining ten percent NaCl was lost by vaporization before, and during, the exchange process. During the isothermal, constant activity, phase of the exchange process (20 to 40 min. region of Fig. 15), a potassium vapor transport enhancement factor of,

$$\frac{P_{KCl}}{P_K\ (\mathrm{no\ NaCl})} \sim 10^4$$

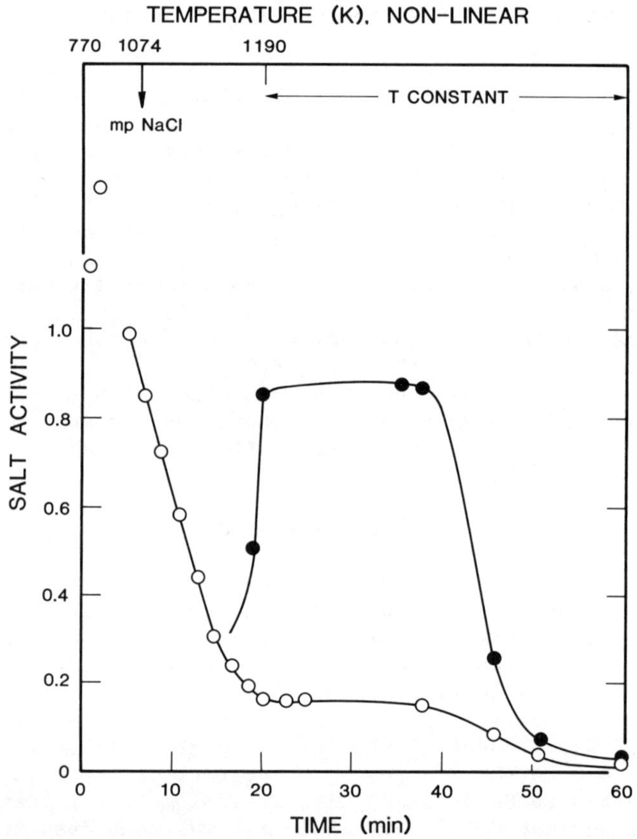

Figure 15. Thermodynamic activities (TMS data) for NaCl (–○–) and KCl (–●–) in the K_1 slag–alkali exchange process. The indicated reference state partial pressures were obtained from Ref. 36. Key: P^0 NaCl = 3 × 10^{-3} atm; P^0 KCl = 5 × 10^{-3} atm.

was observed. Also, during this period, the high KCl activity suggests formation of an essentially ideal solution of KCl-NaCl, as well as the establishment of thermodynamic equilibrium. Note the near unit NaCl activity in the initial phase of the experiment (Fig. 15), which confirms the calibration factors used to convert mass spectral ion intensities to partial pressures and reflects establishment of thermodynamic equilibrium. Additional study of this exchange process is in progress.

Heterogeneous Reactive Gas Systems

Background. In the previous sections, we have considered alkali vapor transport from condensed phase systems in the absence of external influences, such as reactive gases. However, some of the component gases of combustion systems, such as H_2O, HCl, SO_2, O_2, CO, and H_2, can be expected to significantly modify alkali vapor transport through mass action effects or formation of new molecular species. Some representative cases are considered as follows.

Na_2SO_4-NaOH-H_2O System. From a thermodynamic viewpoint, H_2O should react with Na_2SO_4 to form NaOH vapor at high temperatures. However, surprisingly, on the relatively long time scale of TMS measurements, no such reaction was observed. When a small amount (0.25 to 1 percent) of condensed NaOH was mixed with the Na_2SO_4, reaction with H_2O vapor was observed. Also, the alkali vapor pressure enhancement was thermodynamically consistent with the process,

$$H_2O + Na_2SO_4(c) = 2NaOH(g) + SO_2 + 1/2\ O_2$$

Details of this study will appear elsewhere (70).

Illite -H_2O-H_2 System. Vaporization of potassium from the highly acidic illite system, in neutral atmospheres, is expected to provide a relatively insignificant source of alkali in most coal combustion systems. However, in the presence of reactive combustion gases, such as H_2O and H_2, thermodynamic considerations predict a significant KOH partial pressure. In addition, an increase in the K-pressure should result from a reduction in the O_2 pressure, in the presence of H_2. However, KMS experiments did not indicate formation of KOH or additional K in the presence of H_2 gas. Thus, thermodynamic equilibrium does not appear to have been established in this heterogeneous system, even though the temperatures were sufficiently high to have normally ensured a rapid approach to equilibrium.

Further evidence of this lack of thermodynamic equilibrium was provided by monitoring formation of SiO by H_2 reduction of SiO_2, present either as the silicate in illite or as pure silica. Figure 16 shows that SiO production from both forms of SiO_2 is

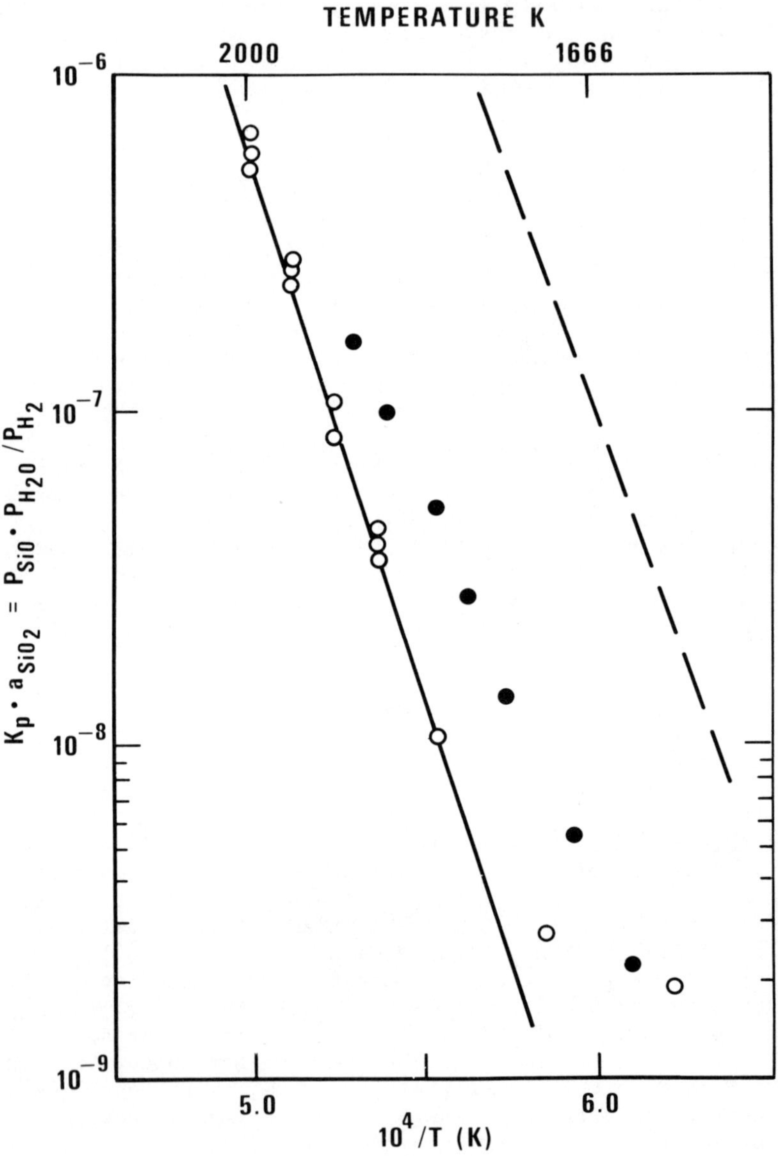

Figure 16. Silica activity function (KMS data). K_p refers to the reaction, $SiO_2(c) + H_2 = SiO + H_2O$. Key: ○, illite; ●, quartz data with much larger sample surface area; and – – –, Ref. 36 data for unit activity silica.

about one to two orders of magnitude less than the equilibrium values represented by the JANAF (36) curve. In the absence of H_2 or H_2O, the SiO and O_2 pressures were reasonably consistent with an equilibrium system.

The kinetic limitation does not appear to result from the heterogeneous reaction process but, rather, from a lack of equilibrium for the homogeneous reaction,

$$H_2O = H_2 + 1/2\ O_2 \qquad [8]$$

as shown in Figure 17. When H_2 was the added reactant, K_p for this reaction was found to be about two orders of magnitude greater than the accepted value. This discrepancy would result if a large fraction of H_2 effused through the Knudsen cell without reaction. However, when H_2O was the added reactant, K_p was also greater than the known value and by an order of magnitude. This case represents closer agreement over the system where H_2 was added. Some of this improvement may result from the lower H_2O pressures used and/or the possibility for H_2O to undergo decomposition on the effusion cell surfaces. To form H_2O, however, H_2 must undergo one or more collisions with the oxide sample surface. The points agreeing most closely with the theoretical H_2O dissociation curve in Figure 17 were obtained by assuming that $P_{H_2} = 2P_{O_2}$, rather than using the measured P_{H_2} to calculate K_p. This assumption would be true for a static system in which H_2 and O_2 are formed only by decomposition of water. For Knudsen effusion, the rate of escape is inversely proportional to the molecular weight and, at steady state conditions, $P_{H_2} = 1/2\ P_{O_2}$. Under these dynamic conditions, K_p would be 1/4 the static value. Values of K_p calculated on the basis of this assumption are not shown in Figure 17. The K_p values for the dissociation of H_2O using either of these two assumptions scatter within a factor of two around the JANAF (36) curve over most of the temperature range. At the highest temperatures, the agreement is less satisfactory because of excess O_2 production from residual illite in the effusion cell. Values of K_p for water addition, where the observed P_{H_2} is used to calculate K_p, indicate that the mass spectrometric method is overestimating the H_2 pressure by almost an order of magnitude. This error could result from H_2O reduction on the furnace element and/or the outer surface of the Knudsen cell. However, we have observed this lack of equilibrium for the H_2O decomposition reaction in other systems and, more particularly, with the TMS method, which has less opportunity for furnace decomposition reactions, as well as a much longer characteristic gas residence time.

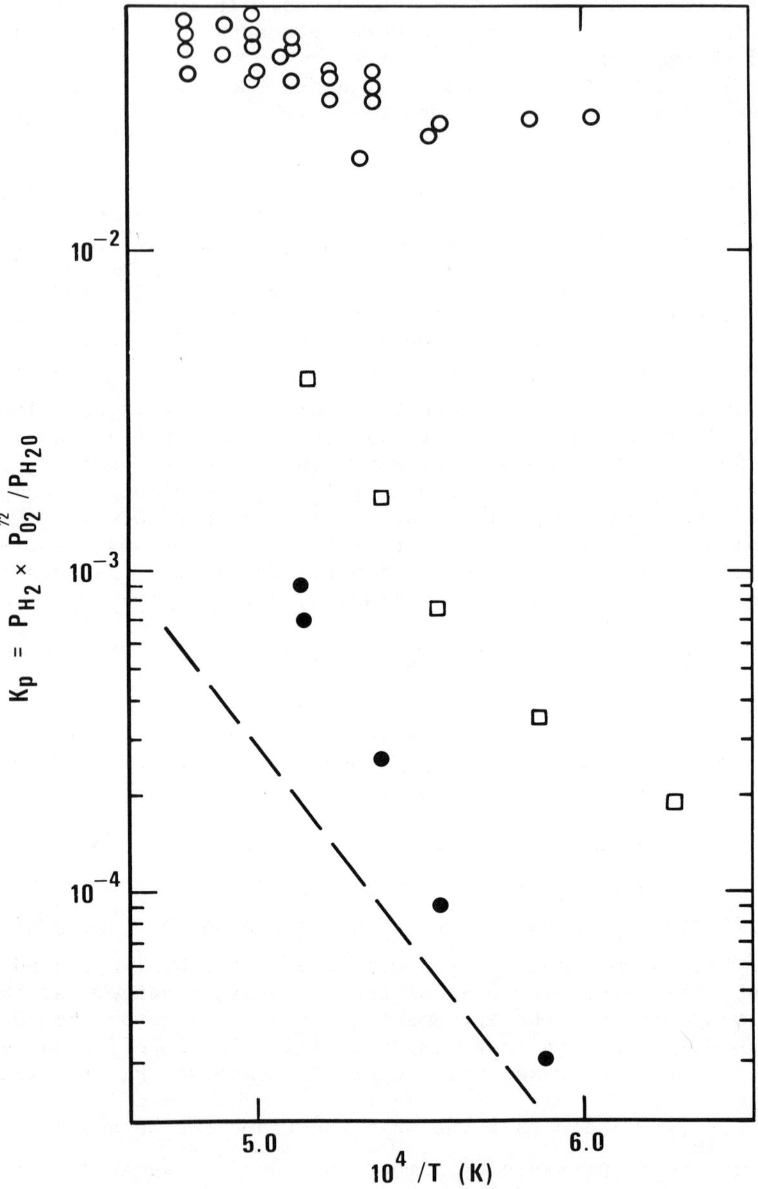

Figure 17. Equilibrium constant for H_2O dissociation (KMS data). Key: ○, data obtained during illite + H_2 experiment; □, H_2O rather than H_2 addition; ●, P_{H_2} assumed equal to $2P_{O_2}$ for K_p calculation; and – – –, Ref. 36 *data.*

If we consider the rate limiting step in the homogeneous H_2O-O_2-H_2 system as,

$$O_2 + H_2 = 2OH$$

then the literature kinetic data (71) indicate reaction times in the approximate range of 10 to 100 sec for our experimental conditions ($\sim$ 1600 K). Thus, the TMS residence time range of 1 to 20 sec is marginal for equilibrium in the H_2O system, and the KMS system is even less favorable (as observed experimentally). However, at temperatures in excess of 2000 K, this limiting rate is several orders of magnitude faster, and equilibrium can be attained, as noted earlier (65).

Evidence of non-equilibrium in H_2O-H_2-condensed phase systems was also noted recently by Sasaki and Belton (72). They observed O_2 partial pressures one to two orders of magnitude less than the equilibrium values at a temperature of 1423 K over liquid Cu. However, in the presence of a Pt wire catalyst, they were able to closely approach equilibrium at residence times not too different from those of the present mass spectrometric studies.

Synthetic Slag (K_2)-H_2O-H_2 System. In order to extend the vapor transport conditions in slag systems to a reducing hydrous environment similar to that present in coal gasification, a series of TMS and KMS measurements were made using H_2 or H_2O as the initial reactant gas. With the TMS system, compositions of H_2-N_2-H_2O up to 10 vol % H_2 were attained prior to hydrogen-induced corrosive loss of the transpiration reactor.

Effect of H_2 on K-Vaporization: As H_2 was introduced to the slag system, the O_2 concentration decreased and K and H_2O increased, as expected for the process,

$$K_2O(slag) + H_2 = 2K + H_2O \qquad [9]$$

Typical TMS data are given in Figure 18 where the H_2 partial pressure was varied over the range 10^{-4} to 10^{-2} atm. Note the pronounced hysterisis effect for increased versus decreased H_2 and H_2O-content. Though not shown here, this effect is also present in the K_2O activity data, as calculated from the observed K and O_2-pressures. Hence the system is not at thermodynamic equilibrium. From the established equilibrium constants for reaction [3] with K_2O (slag) and K_2O (pure liquid), together with the measured K_2O activity data, we calculate,

$$K_p\ [9] = 209 \text{ at } 1650 \text{ K}$$

The corresponding experimental value, obtained from the measured partial pressures of K, H_2, and H_2O, is

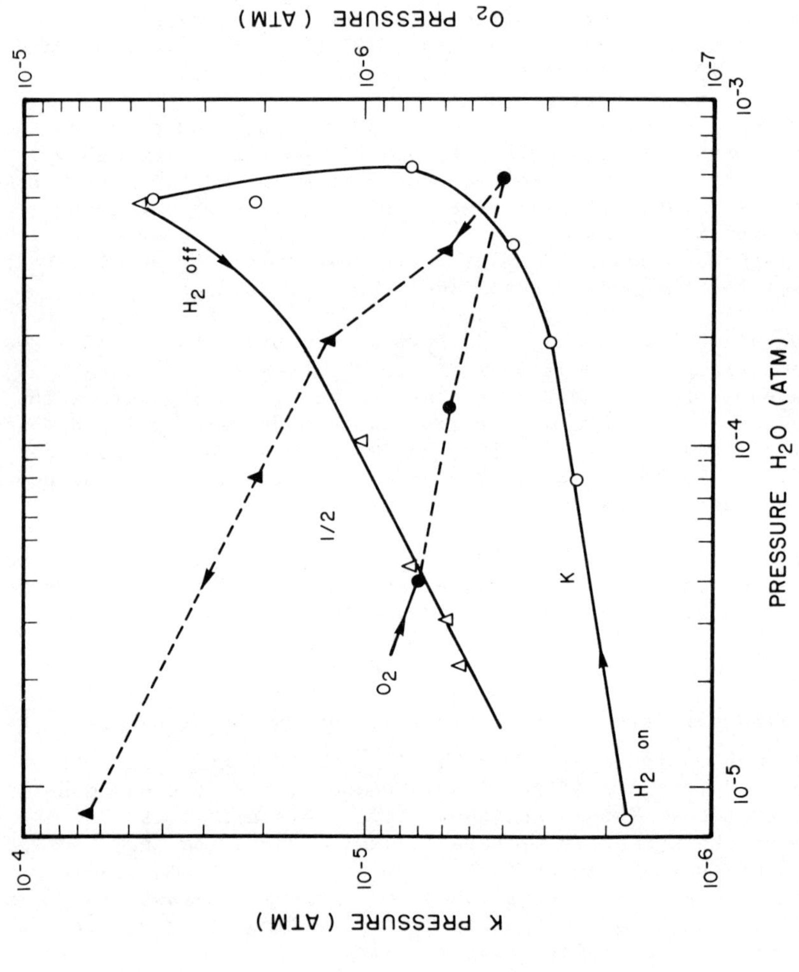

Figure 18. Variation of K pressure (△, ○) and O_2 pressure (▲, ●) with H_2O pressure for K_2 slag in the presence of added H_2. TMS conditions: 1655 K; N_2 carrier gas pressure, 0.18 atm; capillary nozzle; and →, run chronology.

$$K_p \text{ [9], obs.} = 4.2 \text{ at } 1650 \text{ K}$$

Thus the system is far from equilibrium.

Effect of H_2O on K-Vaporization: Similar TMS experiments were performed but with H_2O as the added reactant and a non-reducing atmosphere. An unexpected K-pressure dependence on H_2O was found, as shown in Figure 19. No hysterisis effects were observed in this case. A similar, though less pronounced (factor of four less effect on K-pressure), H_2O-induced K vaporization effect was noted in the more acidic and more viscous MHD (K_1) slag sample (65).

For the H_2O-pressure and temperature conditions used, KOH should have formed according to the process,

$$K_2O \text{ (slag)} + H_2O = 2KOH \qquad [10]$$

However, no KOH was observed in the TMS mass spectra. We also established the K^+ precursor as atomic K, from the pure KOH data and appearance potential measurements. A higher temperature study, using the KMS method, did show the expected formation of KOH in the presence of added H_2O. However, the KOH-pressures were about an order of magnitude below predicted equilibrium values, even though the correct H_2O pressure dependence was found, as shown in Figure 20.

The apparently anomalous H_2O-induced increase in K-pressure can be explained as follows. Literature water solubility data for aluminate and silicate melts [e.g., see (73)] suggest solubilities of at least several hundred ppm for our experimental conditions. Various acid-base reaction mechanisms have been suggested to explain water solubility in silicate melts, as summarized by Turkdogan (74). For basic melts, H_2O acts as an acid and enhances the silicate network structure, and vice-versa for acid (high silica) melts. Though apparently not previously recognized, these structural changes should be reflected in the alkali activity data. Thus, we can reasonably expect an activity increase when water is incorporated into the silicate matrix of the relatively basic K_2 slag. Reaction sets of the type,

$$H_2O + O^{2-} \text{ (slag)} = 2(OH^-)$$

$$K_2O \text{ (slag)} = 2K + 1/2\ O_2, \text{ and}$$

$$O_2 + 2e = 2O^{2-}$$

would be consistent with the observed one-half power dependence of log P_K on log P_{H_2O} (see Fig. 19). These structural changes should also be reflected in viscosity data. The water-solubility viscosity enhancement effects noted by Brower et al. (75) for a similar slag are consistent with the present activity trends.

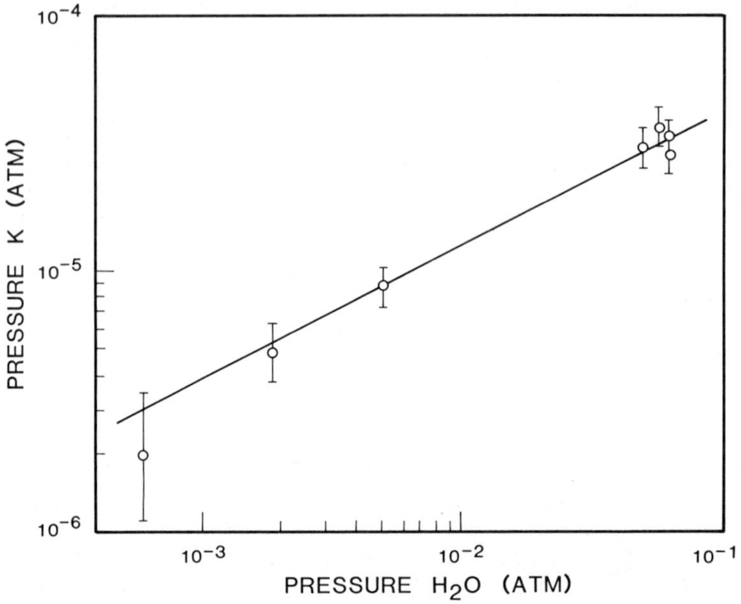

Figure 19. Isothermal (1610 K) dependence of K pressure on H_2O pressure for the K_2 slag with added H_2O. TMS conditions: N_2 carrier gas pressure, 0.21 atm; and capillary nozzle.

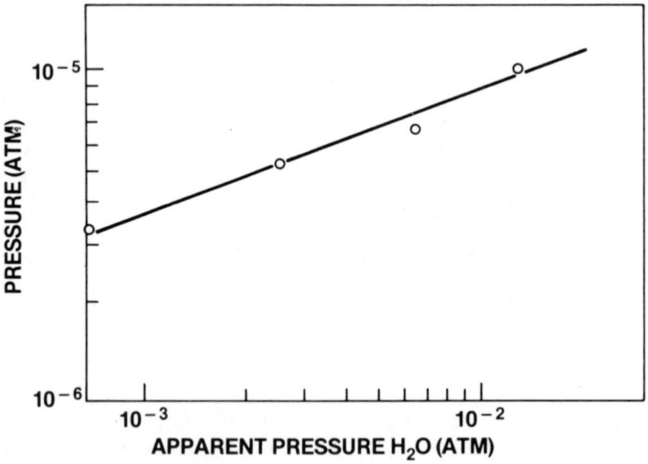

Figure 20. Isothermal dependence of KOH pressure on H_2O apparent pressure for the K_2 slag at 1794 K. KMS data with Pt-cell orifice diameter of 0.34 mm. Curve of slope 0.5 represents the theoretical pressure dependence for reaction 10.

For more basic systems a decreased alkali activity is possible. The recent observations of Gray (76), where water vapor decreased alkali vaporization rates in low silica glasses, could be interpreted in this manner.

We believe that a similar water vapor solubility enhancement of alkali vapor transport is possible in soda-lime-silica glass systems, and work is in progress to verify this. Some of the disparities between various glass vaporization studies may well result from variations in water content and, hence, alkali activities. The common explanation for water vapor enhanced alkali vapor transport over silicates has revolved around formation of volatile NaOH (77) and KOH (53) species. However, no direct test for the presence of these species has been made, and the possibility of water vapor enhancement of atomic Na and K transport exists in these systems.

Mention should also be made of the possible effect of H_2O dissociation (i.e., reaction [8]) to yield vapor phase H_2, which was suggested by Horn et al. (78) as a factor in ceramic degradation. However, in the present system there is an additional source of O_2 and, at thermodynamic equilibrium, K should be H_2O-independent except for the noted H_2O-solubility effect.

Glass-Combustion Gas System. Certain combustion gas components can promote alkali vapor transport in glass systems. Such transport is important in glass melting. Also, glass had been suggested as a medium for trapping particulate material in combustion gas clean-up processes, such as for pressurized fluidized bed combustion (47).

Using our experimental activity data for Na_2O in glass, we have modeled the effect of a typical combustion gas mixture on alkali vaporization (48). For this purpose we have acquired, and adapted to our computers, a code known as SOLGASMIX (79) which is unique in its ability to deal with non-ideal solution multicomponent heterogeneous equilibria. Previous attempts to model this type of problem have been limited to ideal solution assumptions (34). As is demonstrated in Table III, if solution non-ideality is neglected, drastic errors result in the prediction of alkali vapor transport processes. Table III and Figure 21 summarize the predicted alkali species partial pressures. The thermodynamic data base was constructed mainly from the JANAF (36) compilation. Additional details of this study will be presented elsewhere.

The principal results of these calculations can be summarized as follows:

(a) The presence of one percent halogen enhances alkali transport by more than two orders of magnitude at 1200 K, but only by a factor of four at 2000 K.

(b) Only in a N_2 atmosphere is atomic Na a significant vapor species.

(c) If glass is treated as an ideal solution, then halogen has only a slight enhancement effect on alkali transport, i.e., NaOH is the predominant species.

Table III

Effect of Atmosphere and Solution Non-Ideality on Alkali Vapor Species Distribution

Glass: Na_2O (17 wt.%), CaO (12 wt.%), SiO_2 (71 wt.%)

Initial Gas Composition (mole percent): H_2O (4), O_2 (5), CO_2 (12), N_2 (76), SO_2 (2), HCl (1)[d]

Temperature: 1400 K

Total Pressure: one atm

Species	Non-Ideal Soln. N_2 atm. Pressure (atm)	Ideal Soln. Gas atm.[b] Pressure (atm)	Non-Ideal Soln. Gas atm.[b] Pressure (atm)
Na	3.1(-7)[a]	2.2(-4)	1.0(-7)
NaOH	--	8.5(-2)	9.2(-5)
NaCl	--	9.8(-3)	8.8(-3)
Na_2Cl_2	--	7.1(-4)	5.8(-4)
Na_2SO_4	--	4.4(-6)	4.4(-6)
Na_2O Activity[c]	3.4(-8)	1.62(-1)	3.4(-8)
Total Na	3.1(-7)	9.6(-2)	1.1(-2)
No Halogen	3.1(-7)	8.5(-2)	1.0(-2)

[a]Computer notation, e.g., 3.1(-7) = 3.1×10^{-7}.

[b]Refers to mixed gas composition given above.

[c]Infinite supply of glass assumed and activity not affected by vaporization loss.

[d]Halogen content represents an upper limit for combustion of a high-halogen coal in a salty (marine location) atmosphere. For typical low-halogen U. S. coals, the chloride concentration would be ∿ 0.01 mol % and the NaCl pressures correspondingly less.

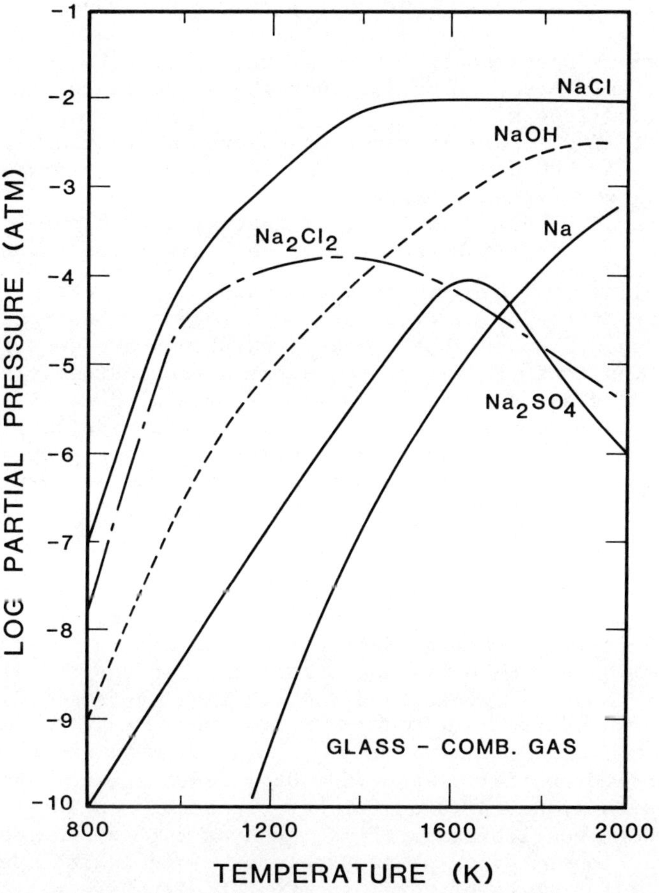

Figure 21. Computer-calculated distribution of alkali-containing vapor species as a function of temperature for the nonideal solution glass–combustion gas system specified in Table III.

(d) Glass non-ideality reduced alkali transport by an order of magnitude in the presence of halogen but by three orders of magnitude in a halogen-free reactive atmosphere.
(e) Sulfur has a much lower effect than halogen, or H_2O, on alkali transport and for the lower temperature range (< 1500 K) most of the sulfur is removed from the gas stream due to $Na_2SO_4(\ell)$ formation. Note, however, (in Fig. 21) the non-monotonic production of Na_2SO_4 vapor species with temperature, leading to significant alkali vapor transport over the temperature interval 1300-2000 K.
(f) At 1200 K, alkali vapor transport covers the range of 7 to 1200 ppm, depending on the absence or presence of halogen, respectively.

From these results, it is clear that alkali release from glass under combustion gas atmospheres will be a significant source of alkali (i.e., > 10^{-6} atm) in the combustion gas stream at temperatures greater than 900 K, even for low-halogen combustion conditions. We should stress that these predictions are sensitive to the assumption that the gas stream attains thermodynamic equilibrium with the glass substrate. Our experimental data on sulfate and complex oxide systems indicates that such an equilibrium is not always readily attained. Incorporation of rate processes into heterogeneous reaction models must await further experiments to develop the necessary data base and mechanistic understanding.

Summary

During the past few years, substantial progress has been made in developing a thermodynamic data base and in providing mechanistic insight into the vaporization processes for alkali metal salt, oxide, silicate, and complex slag systems. Experimental techniques have been improved or newly developed which are suitable for thermodynamic studies of complex, corrosive alkali-containing systems, including coal slags. Computer codes are now available for thermodynamic calculations of heterogeneous non-ideal solution multicomponent equilibrium systems, such as for combustion-coal slag interactions. Vapor phase problems remaining include: (a) possible formation of novel species at very high gas pressures (10 to 100 atm); (b) lack of mechanistic understanding and rate data for non-equilibrium heterogeneous systems; (c) non-availability of computer codes for heterogeneous rate-limited systems; and (d) development of diagnostic methods and data for actual coal-conversion and combustion systems, including MHD channels, coal gasifiers, pressurized fluidized beds, and gas turbines.

Acknowledgments

A significant portion of this work was supported by the Department of Energy. Mr. Art Sessoms provided valuable technical assistance. Valuable discussion with our colleagues, Drs. W. Horton and L. Cook, is also acknowledged.

Literature Cited

1. Stwalley, W. C. and Koch, M. E. Opt. Eng. 1980, 19, 71.
2. Haynes, B. S., Jander, H., and Wagner, H. G. Symp. (Int.) Combust., 17th; The Comb. Inst. Pittsburgh, PA. 1978. p. 1365.
3. Hastie, J. W. "High Temperature Vapors-Science and Technology", Academic Press: New York, NY, 1975.
4. Gangwal, S. K. and Truesdale, R. S. Energy Res. 1980, 4, 113.
5. Rapp, R. Ed. International Conference on High Temperature Corrosion, 1981. San Diego, CA, March 2-6, NACE. Proceedings in press.
6. Hastie, J. W. Combust. Flame 1973, 21, 49.
7. Hastie, J. W. and Bonnell, D. W. "Molecular Chemistry of Inhibited Combustion Systems", 1980. NBSIR 80-2169.
8. Bonnell, D. W. and Hastie, J. W. "Transpiration Mass Spectrometry of High Temperature Vapors", in Characterization of High Temperature Vapors and Gases, Hastie, J. W., Ed. NBS-SP 561, US Government Printing Office: Washington, DC, 1979; p. 357.
9. Plante, E. R. "Vapor Pressure Measurements of Potassium Over K_2O-SiO_2 Solutions by a Knudsen Effusion Mass Spectrometric Method", 1979; p. 265, ibid, (8).
10. Yannopoulos, L. N., Toth, J. L., and Pebler, A. Combust. Flame 1977, 30, 61.
11. Hardesty, D. R. and Pohl, J. H. "The Combustion of Pulverized Coals--An Assessment of Research Needs", 1979, ibid (8).
12. Spacil, H. S. and Luthra, K. L. J. Electrochem. Soc. 1979, 126, C134.
13. Kolodney, M., Yernshalmi, J., Squires, A. M., and Harvey, R. D. Trans and J. Brit. Ceram. Soc. 1976, 75, 85.
14. Crowley, M. S. Am. Ceram. Soc. Bull. 1975, 54, 1072.
15. Vorres, K. S. Energy Res. 1980, 4, 109.
16. Lenzer, C. R. and Laurendeau, M. N. "Gasification of Pulverized Coal Within Swirling Flows: An Interpretive Review", 1976. The Combustion Laboratory School of Mechanical Engineering, Purdue University, West Lafayette, Indiana 47907.
17. Gluskoter, H. J. and Ruch, R. R. Fuel 1971, 50, 65.
18. Shearer, J. A., Johnson, I., and Turner, C. B. Env. Sci. Tech. 1979, 13, 1113.
19. Sarofim, A. F., Howard, J. B., and Padia, A. S. Comb. Sci. Tech. 1977, 16, 187.

20. Boow, J. Fuel 1972, 51, 170.
21. Pressley, H. Trans. Brit. Ceram. Soc. 1970, 69, 205.
22. Raymon, N. S. and Sadler III, L. Y. "Refractory Lining Materials for Coal Gasifiers--A literature Review of Reactions Involving High-Temperature Gas and Alkali Metal Vapors", 1976. Information Circular, 8721. Bureau of Mines.
23. Anthony, A. M. Rev. Int. Hautes Temp. Refract. 1976, 13, 230.
24. Bowen, H. K. "Ceramics for Coal-Fired MHD Power Generation," in Materials Science in Energy Technology, Academic Press: New York, NY, 1979; p. 181.
25. Schneider, S. J., Frederikse, H. P. R., and Negas, T. "Materials for Open Cycle MHD Generators", Current Topics in Materials Science, 1980, 4, p 89.
26. CFCC Development Program, Combustion Chemistry Evaluation 1978. FE-2357-40. General Electric, Schenectady, NY.
27. Clews, J. H., Richardson, H. M., and Green, A. T. Trans. Brit. Ceram. Soc. 1940, 39, 139.
28. Rigby, G. R. and Hutton, R. J. Am. Ceram. Soc. 1962, 45, 68.
29. Venable, C. R., Jr. Am. Ceram. Soc. Bull. 1969, 48, 1114.
30. Luthra, K. L. and Shores, D. A. J. Electrochem. Soc., Solid-State Science and Technology 1980, 127, 2202.
31. Staehle, R. W., Ed. "Materials Problems and Research Opportunities in Coal Conversion", 1974. PB-248 081 (NTIS).
32. Huang, T., Gulbransen, E. A., and Meier, G. H. J. Metals 1979, p. 28.
33. Wilson, J. S. and Redifer, M. W. Trans. ASME, J. Eng. Power 1974, A, 145.
34. Kirkbride, B. J. Glass Tech. 1979, 20, 174.
35. Spacil, H. S. and Luthra, K. L. "Thermochemistry of a Pressurized Fluidized Bed Coal Combustor/Gas Turbine Combined Cycle". 1980. GE Report 80CRD238. See also, CFCC Development Program 1978.
36. JANAF Joint Army, Navy, Air Force Thermochemical Tables, 2nd ed., 1971. NSRDS-NBS 37, US Government Printing Office: Washington, DC. See also later supplements for 1971-1981.
37. Kohl, F. J., Stearns, C. A., and Fryburg, G. C. "Sodium Sulfate: Vaporization Thermodynamics and Role in Corrosive Flames", 1975, NASA TMX-71641.
38. Plante, E. R., Olson, C. D., and Negas, T. "Vapor Pressure Studies of K_2SO_4", 1981, to be published.
39. Lau, K. H., Cubicciotti, D., and Hildenbrand, D. L. J. Electrochem. Soc., Solid State Science and Technology 1979, 126, 490.
40. Efimova, A. G. and Gorokhov, L. N. High Temp. (English Trans.) 1979, 16, 1019.
41. Ficalora, P. J., Uy, O. M., Muenow, D. W., and Margrave, J. L. J. Am. Ceram. Soc. 1968, 51, 574.
42. Hastie, J. W., Bonnell, D. W., and Zmbov, K. "Transpiration Mass Spectrometric Analysis of KOH(ℓ) and KCl(ℓ) Vaporization," 1981, to be published.

43. Gusarov, A. V. and Gorokhov, L. N. Russ. J. Phys. Chem. 1968, 42, 449.
44. Stewart, G. W., Chakrabarti, A., Stinespring, C., and Castleton, K. "Deposition of Alkali in Coal Combustion Streams:," 1980, presented at the Fall Meeting, Western States Section, Comb. Inst., Los Angeles, CA (October).
45. Chakrabarti, A., Bonnell, D. W., Stewart, G., and Hastie, J. W. "Vaporization Studies of Alkali Benzoates in Reactive Atmospheres" 1981, to be published.
46. Hastie, J. W., Plante, E. R., and Bonnell, D. W. "Molecular Basis for Release of Alkali and Other Inorganic Impurities from Coal Minerals and Fly Ash," 1979, Report to DOE, Morgantown, WV, November.
47. Gatti, A., Goldstein, H. W., McCreight, L. R., and Semon, H. W. "Feasibility Study of Coal Slag Based Glasses for Hot Gas Clean-up," 1980, Report, FE-2068-32 to DOE, February.
48. Hastie, J. W., Plante, E. R., Bonnell, D. W., and Horton, W. S. "Molecular Basis for Release of Alkali and Other Inorganic Impurities From Coal Minerals and Fly Ash," 1980, Report to DOE, Morgantown, WV, December.
49. Neudorf, D. A. and Elliott, J. F. Met. Trans. 1980, 11B, 607.
50. Cable, M. and Chaudhry, M. A. Glass Tech. 1975, 16, 125.
51. Sanders, D. M., Blackburn, D. H., and Haller, W. K. J. Am. Ceram. Soc. 1976, 59, 366.
52. Argent, B. B., Jones, K., and Kirkbride, B. J. "Vapors in Equilibrium with Glass Melts", in The Industrial Use of Thermochemical Data, Barry, T. I., Ed. The Chemical Society: London, UK, 1980; p. 379.
53. Charles, R. J. J. Am. Ceram. Soc. 1967, 50, 631.
54. Preston, E. and Turner, W. E. S. J. Soc. Glass Tech. 1933, 17, 122.
55. Natola, F. and Touzain, Ph. Can. J. Chem. 1970, 48, 1955.
56. Plante, E. R., Olson, C. D., and Negas, T. 6th International Conference on Magnetohydrodynamic Electrical Power Generation, II. 1975, p 211, Washington, DC.
57. Plante, E. R. In, Development, Testing and Evaluation of MHD Materials, NBS Quarterly Progress Report to DOE, 1977. H. P. R. Frederikse, T. Negas, and S. J. Schneider, March.
58. Plante, E. R. In, Properties of Electronic Materials, 1979. NBSIR 79-1976, J. R. Manning, Ed.
59. Plante, E. R. In, MHD Materials-Seed/Slag Interactions and Effects, 1979. S. J. Schneider, NBS Report to DOE, December.
60. Plante, E. R. In, MHD Materials-Seed/Slag Interactions and Effects, NBS Quarterly Progress Report to DOE, 1980. S. J. Schneider, March.
61. Plante, E. R. In, MHD Materials-Seed/Slag Interactions and Effects, NBS Quarterly Progress Report to DOE, 1980. S. J. Schneider, June.
62. Plante, E. R. In, MHD Materials-Seed/Slag Interactions and Effects, NBS Quarterly Progress Report to DOE, 1980. S. J. Schneider, September.

63. Plante, E. R. In, MHD Materials-Seed/Slag Interactions and Effects, NBS Quarterly Progress Report to DOE, 1980. S. J. Schneider, December.
64. Plante, E. R. In, MHD Materials-Seed/Slag Interactions and Effects, NBS Quarterly Progress Report to DOE, 1981. S. J. Schneider, March.
65. Hastie, J. W., Bonnell, D. W., Plante, E. R., and Horton, W. S. "Vaporization and Chemical Transport Under Coal Gasification Conditions", 1980. NBSIR 80-2178.
66. Plante, E. R. and Cook, L. P. 17th Symposium on Engineering Aspects of Magnetohydrodynamics, 1978. Stanford, CA. paper C.1.
67. Levin, E. M., Robbins, C. R., and McMurdie, H. F. Phase Diagrams for Ceramists. 1964, p. 407, Am. Ceram. Soc., Columbus, OH.
68. Self, S. A. "Diagnostic Techniques in Combustion MHD Flows", 1979. ibid (8).
69. Hastie, J. W., Plante, E. R., and Bonnell, D. W. "Slag-Alkali Halide Exchange Reactions", 1981, to be published.
70. Bonnell, D. W., Sanders, D. M., and Hastie, J. W. "Alkali Vapor Transport in the Na_2SO_4-H_2O System", 1981, to be published, (High Temperature Science).
71. Baulch, D. L., Drysdale, D. D., Horne, D. G., and Lloyd, A. C. "Evaluated Kinetic Data for High Temperature Reactions", 1972, Vol 1, Butterworths, London and Washington, DC.
72. Sasaki, Y. and Belton, G. R. Met. Trans. 1980, 11B, 221.
73. Schwerdtfeger, K. and Schubert, H. G. Met. Trans. 1978, 9B, 143.
74. Turkdogan, E. T. "Physical Chemistry of High Temperature Technology," Academic Press: New York, NY, 1980.
75. Brower, W. S., Waring, J. L., and Blackburn, D. H. "Slag Characterization: Viscosity of Synthetic Coal Slag in Steam," 1980, NBSIR 80-2124.
76. Gray, W. J. Radioactive Waste Management, 1980, 1, 147.
77. Sanders, D. M. and Haller, W. K. J. Am. Ceram. Soc. 1977, 60, 3.
78. Horn, F. L., Fillo, J. A., and Powell, J. R. J. Nucl. Mater. 1979, 85, 439.
79. Eriksson, G. Chemica Scripta 1975, 8, 100.

RECEIVED September 10, 1981.

35

Studies of Alkali Phosphate and Phosphorus Chemistry Important to Magnetohydrodynamics Combustion

A. FREEDMAN, J. C. WORMHOUDT, and C. E. KOLB

Center for Chemical and Environmental Physics, Aerodyne Research, Inc., Bedford Research Park, Bedford, MA 01730

A preliminary investigation into the thermodynamic properties of phosphorus species found in coal combustion gas streams is described. Species such as HPO_2, HPO_3, KPO_3, and possibly KPO_2 were identified by molecular beam mass spectrometer sampling of gases produced in a laboratory-scale combination burner and insulated channel maintained at 1700 K. Model calculations for an open cycle coal-fired magnetohydrodynamic power generator predict free electron concentrations which are consistent with the limited experimental evidence.

The high temperature gas phase chemistry of phosphorus compounds has attracted attention during the past decade. Investigation into the efficacy of organo-phosphorus compounds as flame inhibitors has indicated that they act as efficient scavengers of free radicals (1). Recent work in this laboratory (2) has also indicated that phosphorus ion species play an important role in coal-fired open cycle magnetohydrodynamic (MHD) generator systems. Modeling efforts (3) indicated that the presence of trace amounts of phosphorus (10-100 ppm) would significantly reduce the operating efficiency of these generators. On the other hand, no attempt was made to include alkali phosphate species in these calculations; there is reason to believe that inclusion of such species in a thermodynamic model would reduce the predicted effects of phosphorus. It is this topic we now address.

The fundamental principles behind an open cycle, coal-fired MHD power generator system are quite simple. Exhaust gases at 2700-2900 K produced by coal combustion are seeded with potassium salts producing an electron-rich plasma with high conductivity. This plasma flows through a channel surrounded by a superconducting magnet; the flow of charged species through a magnetic field produces a direct current which, after conversion

0097-6156/82/0179-0601$05.00/0

to alternating current, can be fed into a power grid network. The work performed by the system lowers the temperature of the exhaust gases to ~2200 K, where the gases can be effectively utilized by conventional steam-driven power generating plants. Thus, by use of an MHD topping cycle, both the first and second law efficiencies of conventional power plants can be improved.

The efficiency of open cycle MHD generator systems depends directly on the plasma conductivity in the generator channel. The core flow conductivity is primarily determined by the temperature, pressure, and alkali (potassium) seed level output by the combustor, with generator inlet temperature being the single most important parameter governing the level of alkali seed ionization and, therefore, free electron production. However, other parameters can also play a significant role in setting core plasma conductivity levels and limits on generator efficiency. In coal-fired systems, one phenomenon which has received considerable study is the effect of electron attachment by gas-phase molecules formed from ash components on plasma conductivities.

Trace Phosphorus Species and MHD Conductivity

The existence of ash-derived negative ions increases the total ionization level of the plasma but reduces the concentration of free electrons. This results in a net decrease in conductivity since free electrons are much more mobile than molecular ions. Furthermore, the increase in both positive and negative ions leads to a shorter electron mean free path, due to Coulombic ion-electron interactions, also reducing conductivity. Mass spectrometric measurements on Montana Rosebud coal combustion plasmas formed in our laboratory (2) along with literature reviews performed by us (3) and other MHD modelers (4) has indicated that PO_2^-, PO_3^-,AlO_2^-, FeO_2^-, and OH^- are major negative ion species which might significantly degrade coal combustion plasma conductivies.

Unfortunately, the molecular parameters for the ash derived ions are currently quite uncertain. The uncertainties in heats of formation are most serious for PO_2^-, PO_3^-, and FeO_2^-. The value of PO_2^- is derived from the flame equilibrium studies of Miller(5) while the PO_3^- value is derived from Miller's value for PO_2^- and measurements of PO_3^-/PO_2^- ratios reported in Reference (2). The stated uncertainties to these measured heats of formation are ±7 kcal/mole and ±12.5 kcal/mole, respectively, but are probably higher. The electron affinty of FeO_2 is known only as a lower bound of 2.5 eV (17). When these ions and phosphorus oxide neutrals were included in MHD generator performance modeling calculations (3), they had the largest negative effect on plasma conductivity and on enthalpy extraction efficiency, even though other ions are formed from more abundant components of coal ash.

A recent experiment in a laboratory MHD channel at Stanford (6) has shown that, under MHD conditions (high potassium loading), the addition of phosphorus, even in amounts much greater than those found in coal, has a much smaller effect on electron concentration and conductivity than would be predicted by the above modeling. That experiment involved simultaneous measurement of electron concentration by submillimeter interferometry, of positive ion concentration by a swept electric probe, and of plasma temperature by the emission-absorption (line-reversal) technique, made in an ethanol-fueled, potassium-seeded combustion plasma.

From spectroscopic analysis of alkali and phosphorus doped $H_2/O_2/N_2$ flames, Hastie and Bonnell deduced the presence of $LiPO_x$ and $NaPO_x$ (x = 2,3). (1) Their observations were consistent with previous vaporization data for $LiPO_2$, $NaPO_2$, and $NaPO_3$ (7-10). Based on the observed stability trends for alkali phosphate molecules, Hastie suggested that inclusion of thermodynamic data for KPO_x molecules in model calculations of electron conductivities would have a significant effect (11). Unfortunately, there has been no direct observation of KPO_2 and measurements of KPO_3 stability (10) are uncertain. Therefore, an experimental investigation of potassium and phosphorus seeded combustion gases was begun using a molecular beam mass spectrometer and atmospheric pressure laboratory scale burner. The intent of these studies was to determine whether phosphorus and potassium/phosphorus chemistry would have a significant effect on MHD channel conductivity. Preliminary efforts are described below.

Experiments and Results

The experimental apparatus is shown in Figures 1 and 2. Complete details are given in References (2) and (12). Briefly, the laboratory-scale burner is designed to burn a methane/air/oxygen mixture with provisions for adding both gaseous and particulate seeds such as phosphine or coal particles. The well-insulated channel is designed to maintain nearly isothermal conditions (temperature gradients are usually less than 100 K over the length of the channel). The gas flow down the 10 cm diameter channel is adjusted to insure ample residence time (~0.5 - 1.0 sec), thereby allowing equilibrium to be reached. Temperatures along the channel and at the exit port are measured with thermocouples. The combustion gases travel down the channel, through the exit port, and out into exhaust pipes which are maintained at slight negative pressure by a blower. The inner core of gases is sampled through a 0.2 mm orifice which leads into the mass spectrometer.

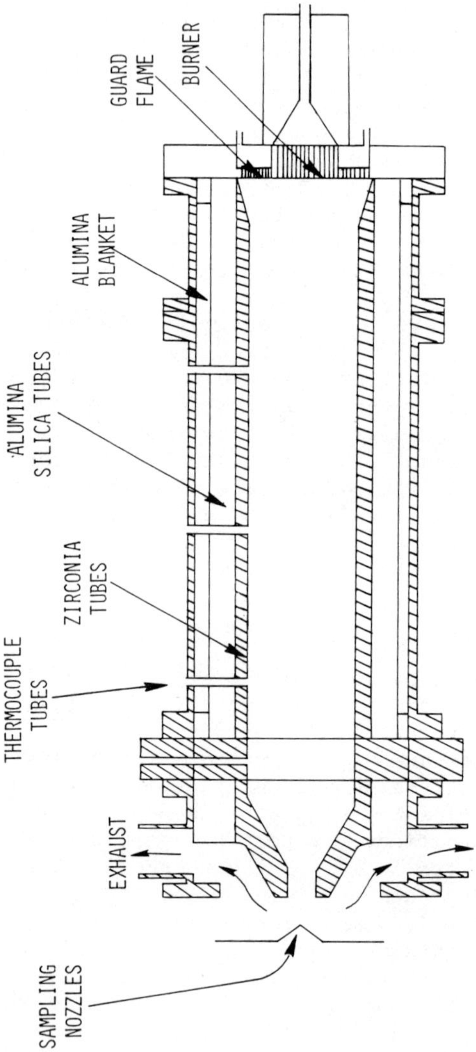

Figure 1. Laboratory scale burner.

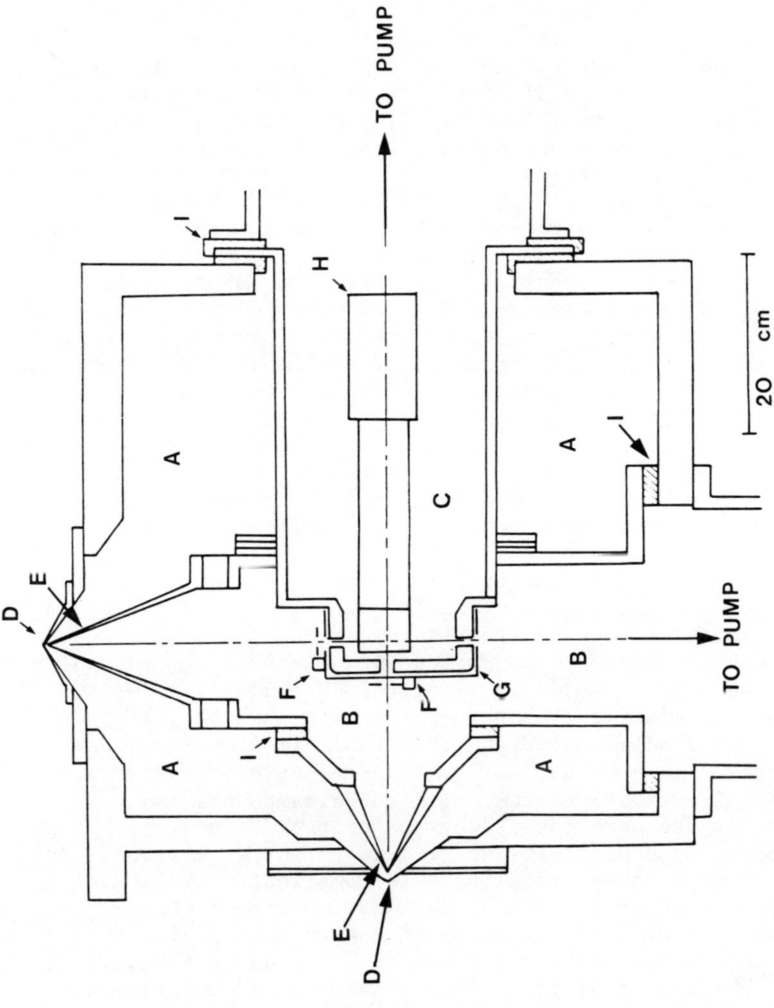

Figure 2. Molecular beam mass spectrometer. Key: A, nozzle-skimmer chamber; B, middle chamber; C, mass spectrometer chamber; D, sampling nozzles; E, skimmers; F, choppers; G, gate valve assembly; H, mass spectrometer; and I, nylon flanges permitting electrical biasing for detection of ions from coal burner.

The sampling molecular beam mass spectrometer is comprised of three chambers which incorporate differential pumping. Under operating conditions, the first vacuum chamber pressure is only 4×10^{-4} Torr, allowing unperturbed free nozzle expansion by the combustion gases, thus freezing the chemical composition of the flow. The molecular beam flow through the skimmer into the second chamber, is modulated by a Bulova tuning fork chopper, and passes into the ionizer and quadrupole filter (Extranuclear Inc.) contained in the innermost chamber. The ions are detected by a Channeltron particle multiplier and the resultant signal is processed by a lock-in amplifier (Ithaco). All surfaces and components are insulated, thus allowing detection of both positive and negative ions as well as neutral species.

Figure 3 is a histogram of mass spectrometer results obtained for a methane/air/phosphine flame (0.09/0.89/0.02 mole fractions) sampled at a temperature of ~1700 K. This spectrum indicates the importance of HPO_2 as a neutral phosphorus species in this system in agreement with similar mass spectrometric observations of Hastie and Bonnell (1). The relative peak height ratio of HPO_2 to PO is consistent with a ΔH_f of -96 kcal/mole for HPO_2 as measured in Reference (1). One should also note the existence of HPO_3; this indicates, based on our calculations, that the chemical bonding is much stronger than in its analog, HNO_3. No peak corresponding to PO_3 (within noise and sensitivity limits) was observed. The relative intensity of PO_2 is much lower than we calculate from data in existing thermochemical literature, but this could be due to dissociative ionization. (This observation is also noted in Reference (1).) Unfortunately, due to the destructive and corrosive effects of the phosphorus compounds on the burner, experimental duration was limited and no absolute calibrations could be obtained.

Further studies were conducted with mixtures of charcoal and KH_2PO_4 which were intended to simulate MHD conditions. A stoichiometric mixture of methane, air, and charcoal was burned with a loading of KH_2PO_4 of 0.005 mole fraction. Mass peaks corresponding to KOH, KPO_2, and KPO_3 were observed at signal-to-noise levels of two. All three compounds had approximately the same mass spectrometer peak height, corresponding to concentrations of 5-10 parts per million. Any interpretation of these magnitudes is complicated by lack of knowledge of the partial cross sections for dissociative ionization. Significant cracking of the parent species is expected due to the ionic nature of the bonding in KPO_2, KPO_3, and KOH. The mass peak identified as KPO_2 could be primarily derived from a KPO_3 parent. However, these observations are sufficient to identify KPO_3 and possibly KPO_2 as key alkali phosphate species for MHD combustion plasmas.

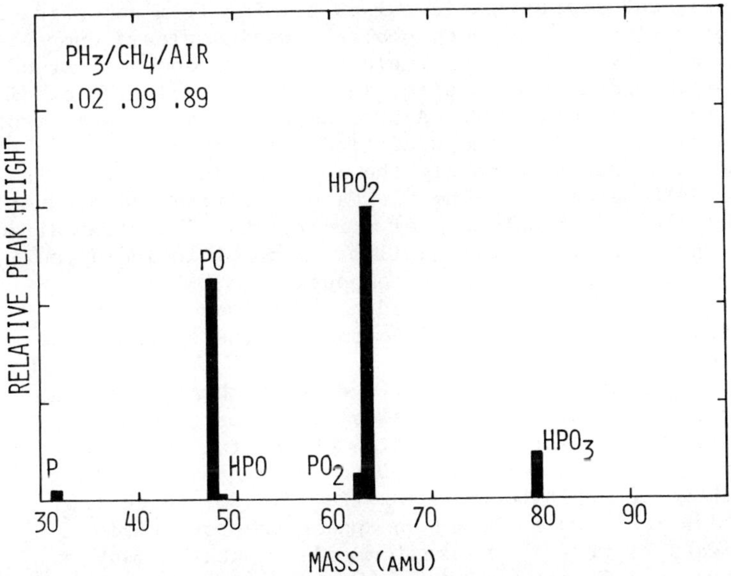

Figure 3. Typical mass spectrum obtained from laboratory-scale burner for phosphine, methane, and air mixture at 1700 K.

Modeling and Discussion

An actual MHD channel contains an extremely varied array of species due to the number of impurities contained in coal (e.g., N, S, Si, Fe, Al, Mg, P, Ca, Na, K, etc.) and the effect of both neutral and ionic phosphorus species on generator efficiency can only be gauged by a complicated calculation which minimizes the free energy of the entire system. Calculations were performed with Aerodyne's PACKAGE Code (18) which has been developed with particular attention to the influence that nonfuel coal components, such as inorganic oxides arising from ash oxidation and vaporization, may have on coal combustion plasma composition and conductivity. This code employs a one-dimensional steady channel flow model, with a specified constant velocity condition used in these calculations. A prespecified channel area profile can be used in place of the specified velocity criterion if desired. The code also models the full MHD interaction with a variable Hall parameter. The plasma composition and conductivity are recalculated at each channel grid point. The chemical model used in this work is a full equilibrium model including an ideal solution model for liquid slag components.

Table I presents the thermodynamic values pertaining to phosphorus species which are important in the systems under discussion.

As can be seen, thermodynamic data (both ΔH_f and structural data) for key species is either uncertain or unavailable. The heat of formation of neutral PO_3 is calculated assuming $D_0(PO_2{-}O) = D_0(PO{-}O) \times D_0(SO_2{-}O)/D_0(SO{-}O)$. An experimental determination of ΔH_f (PO_3) in Reference (10) (ΔH_f = -129 kcal/ mole) based on appearance potential measurements of MPO_3 (M = alkali metals) contains many assumptions and is not used here (fortunately, this species is not of great importance to the calculation). The PO_3^- and PO_3 values imply an electron affinity of 5.7 eV which is comparable to a theoretical prediction of 5.3 eV (15). The only third law determinations of ΔH_f for MPO_2 and MPO_3 species have been made by Gingerich and Miller for the sodium analog (8); assuming D_0 ($K{-}PO_2$) - D_0 ($Na{-}PO_2$) = D_0($K{-}BO_2$) - D_0 ($Na{-}BO_2$), ΔH_f (KPO_2) = -123 ±11 kcal/mole. An upper bound of ΔH_f (KPO_2), -141 kcal/mole, can be obtained by assuming D_0 ($K{-}PO_2$) = D_0($K{-}BO_2$) - EA (BO_2) + EA (PO_2) (EA ≡ electron affinty). The value in Table I is simply the average of these two values. ΔH_f (KPO_3) is assumed to be equal to ΔH_f (KPO_2) + ΔH_f ($NaPO_3$) - ΔH_f ($NaPO_2$) as determined in Reference (8). We have adopted the entropy functions of Hastie (11) for KPO_2 and KPO_3.

Table I represents interim values to be adjusted pending futher work. Reference (16) contains a comparison of MHD channel conductivities calculated from various sets of heats of formation, including those in Table I and values for KPO_x

Table I. Thermochemistry of Phosphorus Species

Species	ΔH_f^o (298)		Comments
P	79.79	±0.05	JANAF
PO	-2.9	±1	JANAF
PO_2	-62.4	±2.7	Ref. (13)
PO_3	-78.4	±25	See discussion
PO^-	-28.0	±1.2	Ref. (14)
PO_2^-	-146	±7	Ref. (5)
PO_3^-	-210	±25	Ref. (2)
HP	+60.6	±8	JANAF
HPO	+0.3	±20	By analogy with HNO
HPO_2	-96	±4	Ref. (1)
HPO_3	-150	±25	See discussion
KPO_2	-132	±25	See discussion
KPO_3	-195	±25	See discussion

species recommended by Hastie (11); the consistency of these data sets with the experimental observations presented above as well as in References (2, 5, and 6) is discussed in detail.

Calculations based on the experimental conditions presented above are consistent with our observations. Modeling of the Stanford experiments (6) represent a great improvement over past efforts although a quantitative match seems to require more stable KPO_x species or lower electron affinities for PO_x species (16). Thus, it appears that our assumed thermodynamic data will provide a qualitatively correct picture although much of it is uncertain by more than ±10 kcal/mole. We now apply our model to realistic MHD generator conditions in order to predict the role of phosphorus chemistry in channel conductivity.

Figures 4 and 5 present model calculations for a Montana Rosebud coal-fired, potassium carbonate seeded combustor operated under slightly fuel-rich conditions (equivalence ratio = 1.09). Note that KPO_2 and KPO_3 are the dominant neutral phosphorus species at all temperatures. Negative ion chemistry is dominated by PO_2^- and PO_3^-; below 2000 K, phosphorus species negative ions outnumber free electrons. The only negative ion which is comparable in concentration to PO_2^- is FeO_2^- and then only at the upper temperature range. The sharp temperature falloff of FeO_2^- is caused by the stability of condensed iron containing species.

These results show that the inclusion of alkali phosphate species in MHD conductivity models causes a large increase in the predicted free electron concentration. Unfortunately, the large uncertainties in the thermodynamic data preclude a quantitative

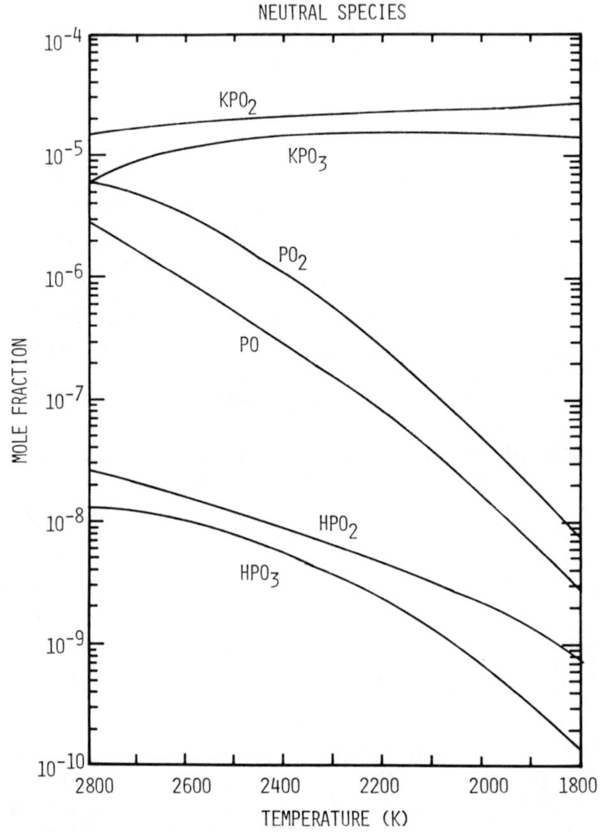

Figure 4. Model calculation of neutral phosphorus species for MHD combustion of Montana rosebud coal under fuel-rich conditions; equivalence ratio = 1.09, O enriched.

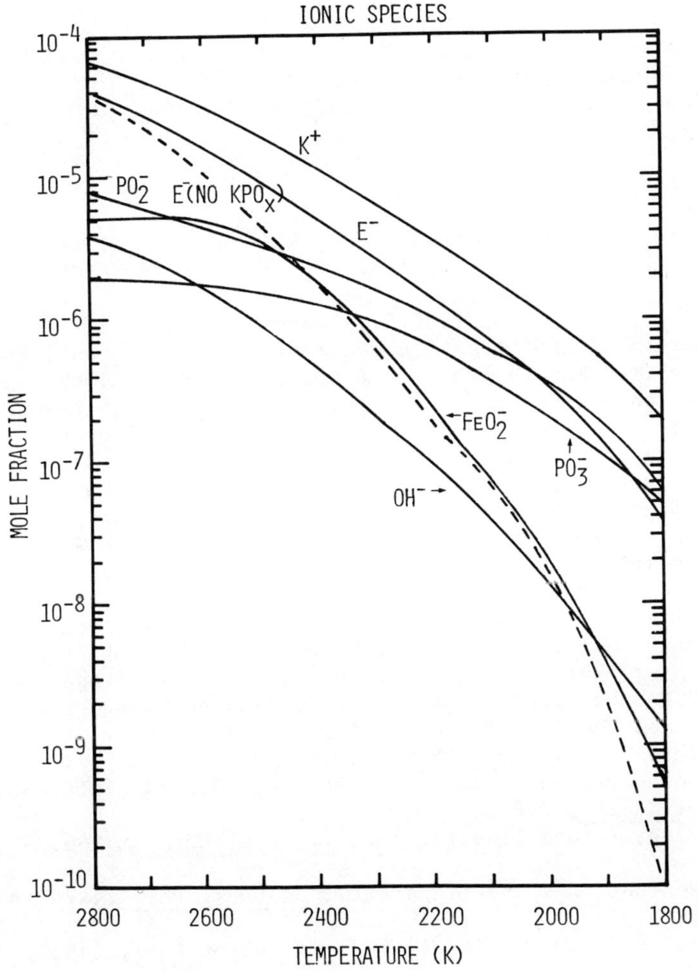

Figure 5. Model calculation of major ionic species for MHD combustion of Montana rosebud coal under fuel-rich conditions, equivalence ratio = 1.09, O enriched.

description. This problem highlights the need for better thermodynamic data for both neutral and ionic species of phosphorus.

Acknowledgments

This work was supported by the Office of Magnetohydrodynamics of the U.S. Department of Energy under Contract No. EX76-C-01-2478. The authors thank J. Hastie, W.J. Miller, and K.D. Annen for their helpful discussions as well as S. Kallelis for his aid in the laboratory.

Literature Cited

1. Hastie, J.W. and Bonnell, D.W., Molecular Chemistry of Inhibited Combustion Systems, National Bureau of Standards 1980 (NBSIR 80-2169).
2. Wormhoudt, J.C., Kolb, C.E., "Mass Spectrometric Determination of Negative and Positive Ion Concentrations in Coal-Fired MHD Plasma," in Proceedings of the 10th Materials Research Symposiums on Characterization of High Temperature Vapors and Gases, NBS Special Publication 561 October 1979, National Bureau of Standards, p 457.
3. Wormhoudt, J., Yousefian, V., Weinberg, M.H., Kolb, C.E., and Sluyter, M.M., "A Review of Plasma Chemical Considerations in MHD Generator Design," at 7th International Conference on MHD Electrical Power Generation, MIT, Cambridge, Mass., June 1980.
4. Sullivan, L.D., Klepeis, J.E., Coderre, W.J., and Fischer, W.H., "Coal-Fired, Open Cycle MHD Combustion Plasmas: Chemical Equilibrium and Transport Properties Workshop Results", Paper 80-0091, AIAA 18th Aerospace Sciences Meeting, Pasadena, California, January 1980.
5. Miller, W., J. Chem. Phys., 1978, 69, 3709.
6. Annen, K.D., Kuzmenko, P.J., Keating, R., and Self, S.A., J. of Energy, 1981, 5, 31.
7. Buchler, A. and Stauffer, J., J. Phys. Chem., 1966, 70, 4092.
8. Gingerich, K.A. and Miller, F., J. Chem. Phys., 1975, 63, 1211.
9. Ogden, J.S. and Williams, S.J., J. Chem. Phys., 1980, 73, 2007.
10. Steblevskii, A.B., Alikhanyan, A.S., Sokolova, I.D., and Gorgoraki, V.I., Russ. J. Inorg. Chem., 1977, 22, 11.
11. Hastie, J., NBS, private communication.
12. Freedman, A., Wormhoudt, J., and Stewart, G., "ARI Modulated Molecular Beam Mass Spectrometer," Aerodyne Research, Inc. Bedford, MA 01730, Report No. TM-13, Feb. 1981.

13. Drowart, J., Myers, C.E., Szwarc, R., Vander Auwera-Mahieu, A., and Uy, O.M., J. Chem. Soc. Faraday Trans. II, 1972, 68, 1749.
14. Zittel, P.F. and Lineberger, W.C., J. Chem. Phys., 1976, 65, 1236.
15. Gutsev, G.L. and Boldyrev, A.I., Chem. Phys., 1981, 56, 277.
16. Wormhoudt, J., Freedman, A., and Kolb, C.E., "Phosphorus Thermochemistry and MHD Plasma Conductivity," 19th Conference on Engineering Aspects of MHD, University of Tennessee, Space Institute, Tullahoma, June 1981.
17. Lineberger, W.C. and Engelking, P.C., "Negative Ion Energetics," Conference on High Temperature Sciences Related to Open Cycle, Coal-Fired MHD Systems; Argonne National Laboratory, April 1977.
18. PACKAGE CODE (Plasma Analysis, Chemical Kinetics, And Generator Efficiency) ARI-RR-177, February 1980, Aerodyne Research, Inc., Bedford, MA 01730 (available from NTIS, U.S. Department of Commerce, Springfield, VA 22161).

RECEIVED August 26, 1981.

INDEX

D

E

H

I

N

T

V

Jacket design by Kathleen Schaner.
Production by Susan Moses and Cynthia E. Hale.

Elements typeset by Service Composition Co., Baltimore, MD.
Printed and bound by Maple Press Co., York, PA.